DIGITAL
ELECTRONICS

FIFTH EDITION

DIGITAL ELECTRONICS

James Bignell
Robert Donovan

THOMSON

DELMAR LEARNING ™

Australia · Brazil · Canada · Mexico · Singapore · Spain · United Kingdom · United States

Digital Electronics

James Bignell and Robert Donovan

Vice President, Technology and Trades Business Unit
David Garza

Director of Learning Solutions:
Sandy Clark

Executive Editor:
Stephen Helba

Senior Product Manager:
Michelle Ruelos Cannistraci

Marketing Director:
Deborah S. Yarnell

Channel Manager:
Dennis Williams

Marketing Coordinator:
Stacey Wiktorek

Editorial Assistant:
Dawn Daugherty

Senior Production Manager:
Larry Main

Production Editor:
Benj Gleeksman

Library of Congress Cataloging-in-Publication Data:
Bignell, James.
 Digital electronics / James Bignell, Robert Donovan.—5th ed.
    p. cm.
 Includes bibliographical references and index.
 ISBN 1-4180-2026-5 (alk. paper)
 1. Digital electronics. I. Donovan, Robert. II. Title.

TK7868.D5B54 2007
621.381—dc22   2006008057

Card Number:
ISBN-13: 9781418020262
ISBN-10: 1418020265

NOTICE TO THE READER

CONTENTS

Digital Electronics, Fifth Edition, is a streamlined, no-nonsense text that is ideal for the community college, Associate of Science degree student who needs a solid, introductory background in digital electronics. No previous knowledge in digital electronics is necessary, although a good working knowledge of dc circuits helps the student feel more comfortable with the concepts of voltage, current, and resistance. Students who complete this course are well prepared for the hardware encountered in a course in microprocessors.

TEXT ORGANIZATION

This book is organized into sixteen chapters, one for each week of a full semester. Each chapter ends with laboratory exercises that closely correlate with the chapter material. It is in these labs that the theory comes alive and practical hands-on skills are learned; a balance is struck between theory and practice. The fifth edition is organized as follows:

Number Systems

Binary, octal, hexadecimal, and binary-coded decimal number systems are covered in Chapter 1, along with binary addition and subtraction.

Basic Gates

Basic gates and exclusive-ORs are covered in Chapter 2 through Chapter 4. Symbols, inverted-logic symbols, Boolean expressions, truth tables, enable/inhibit and gate expansion are stressed. Shift-counter and delayed-clock waveforms are used to introduce waveform analysis. Boolean algebra and Karnaugh map methods are used to implement given truth tables. Exclusive-OR gates are used as parity generators, parity checkers, and magnitude comparators.

Adders

1's and 2's complement method of subtraction is introduced in Chapter 5, along with binary-coded-decimal addition, and signed 2's complement numbers. 1's complement and 2's complement adder/subtractor circuits and binary-coded-decimal adder circuits are created by using the basic gates in conjunction with 4-bit full adders.

Specifications

Totem-pole and open-collector outputs are contrasted in Chapter 6. TTL and CMOS subfamily characteristics and parameters are contrasted. Emitter-coupled logic is introduced. Surface-mount integrated circuit packages and resistors are also discussed.

Flip-Flops

A progression of flip-flops is studied in Chapter 7 and Chapter 8, beginning with crossed NAND and progressing through gated, transparent, data, master-slave, and *JK* flip-flops. *JK* flip-flops and gates are used to create shift-counter and delayed-clock waveforms.

Digital Communications

Integrated circuit serial and parallel shift registers are presented in Chapter 9. The RS-232 standard and ASCII code are studied and a serial receiver is created from flip-flops and gates. Universal serial bus (USB) ports are discussed as well. In a "human-relations" lab exercise, four students work as a team to create a serial receiver from flip-flops, gates, and shift register integrated circuits. The system includes shift-counter and delayed-clock circuits studied in Chapter 8. The lab is complete when each member of the group is able to receive and decode the RS-232 ASCII signals from a computer.

TIMING CIRCUITS

Decode-and-clear and synchronous counters are presented in Chapter 10. Both integrated circuit counters and counters created from flip-flops and gates are studied. The student learns to design and create synchronous counters that count in any sequence.

Schmitt-trigger gates are introduced in Chapter 11. Schmitt-trigger gates, 555 timers, CMOS gates, and crystals are used to create a variety of clock circuits.

Triggerable and non-retriggerable one-shot circuits are covered in Chapter 12. Both integrated circuit one-shots and one-shots created from Schmitt-trigger gates and 555 timers are studied.

Interface Circuits

Chapter 13 begins a sequence of topics concerned with interfacing digital control circuits with the external world.

Digital-to-analog and analog-to-digital converters are covered in Chapter 13. Count-up and compare, flash converters, and successive approximation converters are created with flip-flops, gates, and voltage comparators. The successive approximation circuit begins with the shift-counter and delayed-clock circuits developed in Chapter 8. Integrated circuit converters are presented.

In Chapter 14, the concept of decoding is expanded into multiplexers and demultiplexers. Integrated circuit multiplexers and demultiplexers are presented. LED and liquid crystal seven-segment displays are introduced.

Chapter 15 introduces tri-state gates and bus drivers. Examples are given of interfacing control circuits with high-current, high-voltage devices.

Introduction to Microcomputers

Chapter 16 is a bridge from digital electronics into microcomputers. The basic parts of a microcomputer are discussed. Memory-integrated circuits are presented. This chapter also discusses flash memory ICs—and their use in SmartMedia cards.

In this text a continuity is established in which the skills developed in one chapter are used and augmented in subsequent chapters. For example, in Chapter 3, delayed-clock and shift-counter waveforms are used as inputs to study waveform analysis of basic gates.

This topic is continued in Chapter 8, where delayed-clock and shift-counter waveforms are constructed as an application of flip-flops. In Chapter 9, delayed-clock and shift-counter waveforms are incorporated into a serial receiver system; and in Chapter 13, a delayed-clock circuit is used in a successive approximation analog-to digital circuit. Likewise, basic gates are used to introduce exclusive-ORs and flip-flops. Exclusive-ORs are used to introduce parity generators, comparators, and adders. Flip-flops are used to introduce serial receivers.

Appendixes

Appendix A contains plans and schematics for construction of a lab trainer. Appendix B contains a list of materials needed to construct the lab circuits. Pinouts for the integrated circuits used in the lab exercises are shown in Appendix C. Although handy, these pinouts are no substitute for good TTL and CMOS specification manuals (data books). It is recommended that data books be obtained from one or more of the major integrated circuit manufacturers.

HOW TO USE THE BOOK

1. An *Outline* and a list of *Key Terms* begins each chapter, highlighting main topics and the new terms the students will learn.

2. The *Objectives* for each chapter identifies the skills that the student will acquire after reading the material.

3. Each chapter contains *Self-Check* questions to keep the students focused on the material and to provide immediate feedback on their progress. The answers to the Self-Checks are included at the back of the book.

4. *Examples* are given to enhance the presentation of new material and to guide the student in solving problems.

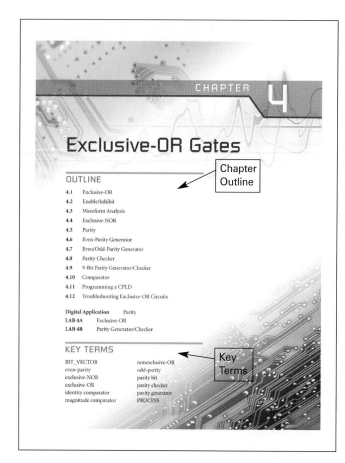

CHAPTER 4

Exclusive-OR Gates

Chapter Outline

OUTLINE

KEY TERMS

Key Terms

BIT_VECTOR   nonexclusive-OR
even-parity   odd-parity
exclusive-NOR   parity bit
exclusive-OR   parity checker
identity comparator   parity generator
magnitude comparator PROCESS

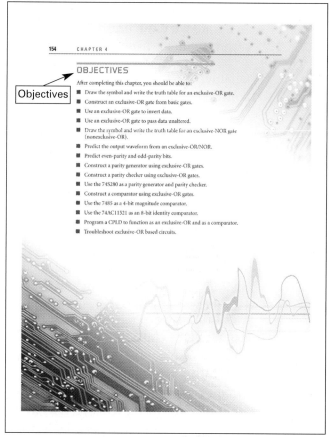

OBJECTIVES

Objectives

After completing this chapter, you should be able to:

- Draw the symbol and write the truth table for an exclusive-OR gate.
- Construct an exclusive-OR gate from basic gates.
- Use an exclusive-OR gate to invert data.
- Use an exclusive-OR gate to pass data unaltered.
- Draw the symbol and write the truth table for an exclusive-NOR gate (nonexclusive-OR).
- Predict the output waveform from an exclusive-OR/NOR.
- Predict even-parity and odd-parity bits.
- Construct a parity generator using exclusive-OR gates.
- Construct a parity checker using exclusive-OR gates.
- Use the 74S280 as a parity generator and parity checker.
- Construct a comparator using exclusive-OR gates.
- Use the 7485 as a 4-bit magnitude comparator.
- Use the 74AC11521 as an 8-bit identity comparator.
- Program a CPLD to function as an exclusive-OR and as a comparator.
- Troubleshoot exclusive-OR based circuits.

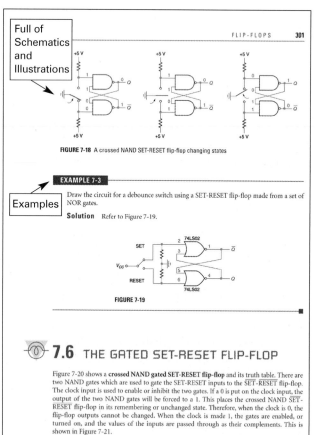

Full of Schematics and Illustrations

FIGURE 7-18 A crossed NAND SET-RESET flip-flop changing states

EXAMPLE 7-3

Examples

Draw the circuit for a debounce switch using a SET-RESET flip-flop made from a set of NOR gates.

Solution   Refer to Figure 7-19.

FIGURE 7-19

7.6   THE GATED SET-RESET FLIP-FLOP

Figure 7-20 shows a **crossed NAND gated SET-RESET flip-flop** and its truth table. There are two NAND gates which are used to gate the SET-RESET inputs to the $\overline{\text{SET}}$-$\overline{\text{RESET}}$ flip-flop. The clock input is used to enable or inhibit the two gates. If a 0 is put on the clock input, the output of the two NAND gates will be forced to a 1. This places the crossed NAND $\overline{\text{SET}}$-$\overline{\text{RESET}}$ flip-flop in its remembering or unchanged state. Therefore, when the clock is 0, the flip-flop outputs cannot be changed. When the clock is made 1, the gates are enabled, or turned on, and the values of the inputs are passed through as their complements. This is shown in Figure 7-21.

Figure 5-41A shows the simulation waveforms for this project with **a** and **b** and **sum** in composite form; Figure 5-41B shows those waveforms expanded so that waveforms for each bit are shown.

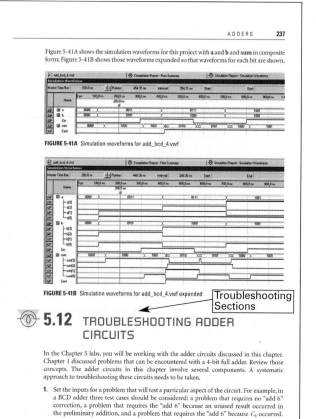

FIGURE 5-41A Simulation waveforms for add_bcd_4.vwf

FIGURE 5-41B Simulation waveforms for add_bcd_4.vwf expanded

Troubleshooting Sections

5.12   TROUBLESHOOTING ADDER CIRCUITS

In the Chapter 5 labs, you will be working with the adder circuits discussed in this chapter. Chapter 1 discussed problems that can be encountered with a 4-bit full adder. Review those concepts. The adder circuits in this chapter involve several components. A systematic approach to troubleshooting these circuits needs to be taken.

1. Set the inputs for a problem that will test a particular aspect of the circuit. For example, in a BCD adder three test cases should be considered: a problem that requires no "add 6" correction, a problem that requires the "add 6" because an unused result occurred in the preliminary addition, and a problem that requires the "add 6" because C_4 occurred.

ADDERS **231**

Solution $S = 0001$ selects $\overline{A + B}$. Corresponding bits are NORed together.

$A = 0110$, $B = 1100$, and $F = 0001$.

Here are some other arithmetic logic unit ICs:

74381	4-bit ALU
74382	4-bit ALU with overflow output for 2's complement
74881	4-bit ALU
74582	4-bit BCD ALU
74583	4-bit BCD adder
74882	32-bit look ahead carry generator

SELF-CHECK 6

Given these inputs, predict the outputs from a 74181.

1. $A = 0111$, $B = 1001$, $M =$ LOW. $S = 0110$, $C_n =$ LOW (1 has been added to A.)

2. $A = 1100$, $B = 0100$, $M =$ HIGH, $S = 0110$

Self Check

 5.11 PROGRAMMING A CPLD

In this section we will program a CPLD to function as a 4-bit adder, a 4-bit full adder, and a 4-bit BCD adder. Follow the procedure presented in Chapters 2 and 3:

Optional Sections on CPLDs.

Step 1. Create a new folder within the Altera folder and name it **bin_add_4**.

Step 2. Run Quartus II.

Step 3. Run "New Project Wizard."

Step 4. Create a new .vhd file and name it **bin_add_4.vhd**.

Step 5. Write a VHDL program that will add two four-binary numbers (a_3 a_2 a_1 a_0 and b_3 b_2 b_1 b_0) and output the sum (sum$_3$ sum$_2$ sum$_1$ sum$_0$).

In the ARCHITECTURE statement, all we have to do is define sum as $a + b$ (sum $<= a + b$). For VHDL to perform that four-bit addition, it needs access to the ieee procedures for arithmetic operations for unsigned numbers. In order to invoke the ieee library, we need to begin this program with these three lines:

LIBRARY ieee;
USE ieee.std_logic_1164.all;
USE ieee.std_logic_unsigned.all;

It is the third line that defines the arithmetic operations. The second line defines standard logic. Whereas the BIT and BIT_VECTOR signals we have used so far can have the values of

---

318 CHAPTER 7

 DIGITAL APPLICATION Mechanical Flip-Flop

Digital Applications

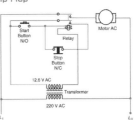

Have you ever seen a control box for a large machine with a start button and a stop button to start and stop the machine? The electrical circuit that controls the machine is an electromechanical SET-RESET flip-flop. The schematic shown is a typical start/stop circuit for a large machine, such as a large compressor for an air conditioner.

The circuit uses a relay which can be latched on by pushing the start button (SET input) and delatched by pushing the stop button (RESET). The relay contacts can handle the large ac currents needed to turn the large machine on and off, but the current needed to run the relay coil is small. do you think you could design a digital circuit using NAND gates to run the relay?

SUMMARY Summary

- Flip-flops are logic circuits designed to store one bit of a binary number.

  The outputs of a flip-flop are called Q and $\overline{Q}$ and should always have different logic states from one another. There are several types of flip-flops which are used for different things.

- The crossed NAND has active LOW inputs called the $\overline{\text{SET}}$ and $\overline{\text{RESET}}$, while the crossed NOR has active HIGH inputs called SET-RESET.

  These flip-flops are called SET-RESET flip-flops and contain an unused state in their truth tables. Flip-flops of this type are often used for debouncing switches and storing logic states in more complex circuits.

- Gated flip-flops are SET-RESET flip-flops that have a pair of gates such as NAND or NOR gates to gate the SET and RESET inputs.

  This gives the gated flip-flop a new input called a clock. The clock will enable the flip-flop or inhibit it. When enabled, the flip-flop can change states, but when inhibited the Q and $\overline{Q}$ can not change states.

DIGITAL-TO-ANALOG AND ANALOG-TO-DIGITAL CONVERSIONS **485**

SUMMARY

- There are two main resistor networks used for digital-to-analog conversion (binary ladder and the $2R$ ladder).

  The binary ladder can be used for small and simple D-to-A converters. Because the binary ladder uses precise resistor values that are multiples of each other, it is impractical to make very large or accurate binary ladder networks. The $2R$ ladder needs only two values of resistors and can be as large as needed.

- The conversion from TTL voltage levels to the voltages needed by the resistor network can be done reasonably well using an open-collector output and a pull-up resistor.

  This method does introduce some error in the final output voltage due to the added resistance of the pull-up resistor and the fact that the open-collector will not pull to exact ground.

- Analog-to-digital conversion can be accomplished by three different methods—flash conversion, count-up and compare, and successive approximation.

  Flash conversion is by far the fastest of the three and is often used for such devices as frame grabbers, which capture TV pictures into a computer. The count-up and compare method is simple to design but requires the most time to do the conversion. The successive approximation method is faster than the count-up and compare but takes much more hardware to implement.

- Today, the two main A-to-D conversion methods used in ICs are the flash and successive approximation methods.

  Most D-to-A and A-to-D ICs are designed for use with computers and have extra interfacing logic added to them to interface with the computer bus.

QUESTIONS AND PROBLEMS Questions and Problems

1. Draw the logic diagram for a TTL $2R$ D-to-A converter with 256 increments and 10 volts of maximum output.

2. What is the output voltage if the binary number 0110 is put on the TTL inputs of the D-to-A converter in Figure 13-8?

3. What is the voltage increment of the D-to-A converter in Figure 13-8 if the supply voltage to the $2R$ network is changed to 5 volts, 10 volts, and 32 volts?

4. What is the binary number output of the A-to-D converter in Figure 13-10 if 6.3 volts are put on the analog input?

5. Draw the waveform for the analog output of the successive approximation A-to-D converter in Figure 13-14 with 5.5 volts applied.

6. Why is the 7406 open-collector inverter used in Figure 13-8?

7. What is the purpose of the op amp-buffer in Figure 13-9?

8. Draw the logic diagram for an A-to-D converter like the one in Figure 13-10 but make the output four bits wide.

LAB 5A ADDERS

Hands-on Labs and Multisim Labs

OBJECTIVES

After completing this lab, you should be able to:

- Draw the logic diagram for a BCD adder.
- Construct and use a BCD adder.
- Draw the logic diagram for a 1's complement adder/subtractor.
- Construct and use a 1's complement adder/subtractor.

COMPONENTS NEEDED

1	7408 IC
1	7432 IC
2	7483 ICs
2	7486 ICs
5	LEDs
5	330-Ω resistors

PREPARATION

Test a BCD adder as follows:

1. Put 0000 and 0011 into the BCD adder input. Check to see if the output of the first 7438 IC is 0011. If it is, check the second 7483 IC for the sum of 0011.

2. If the adder works with sums of 9 or less but does not work with sums of 10 or more, then the ADD 6 part of the adder is not functioning.

3. If the sum of the two inputs 0000 and 0011 is 1001, then the ADD 6 part of the BCD adder is turned on when it should be off.

Troubleshoot the rest of Lab 5A if it is needed. It is a good idea to write the steps you use to troubleshoot a circuit in a notebook for reference during the troubleshooting procedure and for use at a later time.

Review the lab safety rules under SAFETY ADVICE in the PREPARATION section of Lab 1A, Chapter 1.

5. A *Chapter Summary* lists pertinent facts for quick review and reinforcement.

6. End-of-chapter *Questions and Problems* offer review of material and practice at putting the material to work. The answers to odd-numbered problems are included at the back of the book.

7. Two *Labs* are included at the end of each chapter. The first is a *hands-on* construction project and the second is a Multisim analysis and troubleshooting project. Troubleshooting competence is developed in the first lab by wiring and troubleshooting the lab circuits, and in the second lab by analyzing and troubleshooting those Multisim circuits. By using both types of labs, a balance is struck between physically constructing a circuit and analyzing/troubleshooting circuits using computer simulation tools, such as Multisim. Students need to experience both approaches to fully understand circuit problems.

8. Many of the words and phrases used in this text are briefly defined in the *Glossary*. Refer to it often to familiarize yourself with the text's terminology.

9. A *concise, easy-to-read style* presents fundamental digital concepts in clear, understandable terms.

10. This comprehensive text is accompanied by *schematics and illustrations* that help clarify material.

NEW TO THIS EDITION

1. In Chapter 1, calculator sequences are presented for the TI-86 calculators and for the scientific calculator included with Windows operating systems. These calculators are used to perform operations in binary, octal, hexadecimal, and decimal conversions between number systems.

2. Optional sections covering Complex Programmable Logic Devices (CPLDs) have been integrated into this Fifth Edition. In Chapter 2, Section 2.19 (Programming a PLD), a thirteen-step procedure is presented that guides the student through the complete process of developing a project in which the gates within a CPLD are configured to implement a particular function. Steps 7 through 10 involve creating a vector waveform file that simulates the operation of the circuit. Later, in Chapter 3, these steps are presented in detail—after waveform analysis has been studied. The thirteen-step procedure is reinforced in Chapters 4, 5, 7, 8, 9, 10, and 14.

   Steps 4 and 5 of the thirteen-step procedure involve writing a program that describes the hardware configuration desired.

   The programs are developed using Altera Corporation's Quartus II® development system. Quartus II, version 5.1, is included on a CD-ROM inside the back cover of this book. The programming language used in Quartus II is Very High-speed Hardware Description Language (VHDL). In each of the chapters mentioned above, VHDL programs are created, compiled, simulated, and downloaded to the CPLD.

   The examples in this text refer to the RSR PLDT-2® trainer that was purchased from the Electronix Express catalog for approximately $80.00. The PLDT-2 contains a CPLD of the MAX 7000S family, an EPM7128SLC84-15. However, the programming procedure is easily adapted to other trainers such as the University® boards available from Altera.

In each chapter containing a section on programming a CPLD (Chapters 2, 3, 4, 5, 7, 8, 9, 10, and 14), one or more projects are developed. In the Questions and Problems section at the end of each chapter, problems have been added that ask the student to write a program that uses VHDL to implement a particular circuit. Any of these assignments can be expanded into a project that results in programming and testing the actual CPLD. All of these projects (over 70) have been included on a CD at the back of the book.

As in the Fourth Edition, a simple programmable logic device (SPLD) is used (Chapter 3, Section 3.9) to explain how AND-OR combinations are programmed to perform specific logic functions. An example of programming an SPLD using CUPL is included. In each example thereafter, VHDL is used to program CPLDs.

If the reader chooses to skip these sections concerning programmable logic devices, they can do so without losing the continuity of the remaining material.

3. In Chapter 6, specifications have been added for three additional CMOS subfamilies: Advanced High-speed CMOS (AHC), Advanced High-speed CMOS-TTL Compatible (AHCT), and Advanced Low-voltage TTL Compatible (ALVT). Problems 46 and 47 ask the student to research the 7aACTQ subfamily and the configuration and availability of single-gate ICs.

Also in Chapter 6, the coverage of surface-mount packages has been expanded to include surface-mount resistors as well as SOIC-, SOP-, SSOP-, and TSSOP-style packages and their pin pitch.

4. Chapter 9 includes a section on Universal serial bus (USB) ports. Coverage includes the ports' connectors and cabling, method of transmission, data encoding and waveforms, and process for extracting the clock from the data stream.

5. Chapter 16 contains a new section on the flash memory used in digital cameras and music recording devices. Waveforms are presented that explain how blocks of data are programmed, read, and erased. Flash memory ICs are explained and their use in SmartMedia memory cards is discussed.

6. Each chapter concludes with two lab exercises. The first lab is constructed on a protoboard, and the second is performed using Multisim. The Multisim labs have been upgraded to version 9.0. Those Multisim circuits are included on a CD in the back of the book. Also, as mentioned earlier, any of the VHDL programming exercises in the Questions and Problems section at the end of Chapters 2, 3, 4, 5, 7, 8, 9, 10, and 14—and the CPLD projects developed in those chapters—can be assigned as a lab exercise that results in programming and testing the actual CPLD. All of these projects (over 70) have been included on a CD at the back of the book.

7. The accompanying CD includes the following components:
   - MultiSIM circuit files allow troubleshooting practice. MultiSIM circuit files are created using versions 5, 8, and 9, and can be used for the textbook and the lab manual.
   - Quartus II Web Edition Software allows users who would like to teach the optional "Programming a CPLD" sections to create, compile, simulate and download VHDL programs onto a CPLD.
   - Quartus files provide examples of how to program using VHDL. These files can be also be used as lab projects.

COURSE OPTIONS

This text has been used three different ways at Manatee Community College.

1. *Digital Electronics* was originally written to be used as a second semester text for students who had completed DC/AC circuits during their first semester. This worked well except, with digital being introduced in the second semester, only semesters three and four remained for microprocessor-based course work.

2. In order to introduce digital electronics earlier in the curriculum, it was moved to the first semester and taught concurrently with DC/AC circuits. Both formats worked well, but digital electronics was offered only to Electronics Engineering Technology majors.

3. A new format has now evolved, offering a survey of digital and microprocessors to networking and programming students. This Fifth Edition text is used as one of two books in a two-semester sequence, Digital/Microprocessors I and II. In the first course, Digital/Micro I, number systems, gates, and flip-flops (Chapters 1 through 5, 7, and 8) are covered during the first half of the semester. Assembly language programming of a microprocessor, associated hardware, and interfacing are taught during the second half of that first semester. Students majoring in electronics return for Digital/Micro II and study the rest of *Digital Electronics* (Chapter 6 and Chapters 9 through 14) and additional microprocessor work, including interfacing processors to serial ports, parallel ports, programmable counters, and stepper motors. This new format is working well at Manatee Community College.

SUPPLEMENTARY PACKAGE

The majority of our students find this text refreshing and challenging—and they are excited about continuing their studies in electronics. To augment the learning process, we now offer these useful supplements to accompany *Digital Electronics.*

1. The *Laboratory Manual* contains additional activities to supplement the lab exercises contained in the text. For each chapter of *Digital Electronics,* the *Laboratory Manual* contains three or four lab exercises. The first two contain circuits that are constructed on a protoboard and analyzed. The third lab is based on Multisim Student Edition, version 9.0. Chapter 2, 3, 4, 5, 7, 8, 9, 10, and 14 contain a fourth lab that lists projects that students can develop using the QuartusII development system from Altera Corporation and VHDL programming. These lab chapters correspond to the text chapters that present VHDL and CPLD programming. The labs are well correlated with the text material. (ISBN: 1418020273)

2. The *Instructor's Guide* offers teaching hints for the text and lab manual and the answers to all text and lab manual questions and problems. (ISBN: 1418020281)

3. An e.resource CD for instructors includes helpful teaching tools, such as the ExamView testbank with over 300 questions, as well as an Image Library with all images from the textbook to create transparency masters or to customize presentations. (ISBN: 1418064998)

ACKNOWLEDGMENTS

We gratefully acknowledge the contributions of the following reviewers, whose valuable comments helped shape this revision:

Seddik Benhamida, DeVry University, Arlington, VA

Billy Graham, Northwest Technical Institute, Springdale, AR

Stanley Krause, St. Philip's College, San Antonio, TX

Byron Paul, Bismarck State College, Bismarck, ND

Michael Pelletier, Northern Essex Community College, Haverville, MA

We would like to thank the following companies for use of schematics: National Semiconductor Corporation, Lattice Semiconductor Corporation, and Precise Power Corporation.

MARKET FEEDBACK

The authors can be reached for questions, comments, or suggestions at the following e-mail addresses. Robert Donovan can be contacted at donovar@mccfl.edu, and James Bignell can be contacted at biggy01@aol.com. We hope you enjoy using this text as much as we do.

James Bignell
Robert Donovan

Number Systems

OUTLINE

KEY TERMS

anode

binary

Binary-Coded Decimal
  (BCD)

bit

carry-in

carry-out

cathode

hexadecimal

light-emitting diode
  (LED)

octal

OBJECTIVES

After completing this chapter, you should be able to:

- ■ Count in binary, octal, hexadecimal, and binary-coded decimal (BCD).
- ■ Convert from decimal to binary and binary to decimal.
- ■ Convert from binary to octal and octal to binary.
- ■ Convert from binary to hexadecimal and hexadecimal to binary.
- ■ Convert from decimal to BCD and BCD to decimal.
- ■ Add and subtract binary numbers.

1.1 BINARY NUMBER SYSTEM

Digital electronics makes extensive use of the **binary** number system. Binary is useful in electronics because it uses only two digits, 1 and 0. Binary digits are used to represent the two voltage levels used in digital electronics, HIGH or LOW. In most digital systems, a high voltage level is represented by 1; a low voltage level or zero volts is represented by 0. A switch, a light, or a transistor can be ON and represented by 1 or OFF and represented by 0. A decimal number like 32 must be converted into binary and represented by ones and zeros before it can be manipulated by a digital computer.

Since we use the decimal number system daily, we are most familiar with it. First, we will examine a characteristic of the decimal number system and then compare the binary system to it. In the decimal system, we work with ten different digits, zero through nine. These ten digits make the decimal system a base-ten system. In the binary system, we work with two different digits, 0 and 1. These two digits make the binary system a base-two system.

To count in the decimal system, start in the first column or decimal place with 0 and count up to 9. Once the first place is "full," reset the first column to 0 and add 1 to the next column to the left. After 9 comes 10. Now the first column can be "filled" again. After 10 comes 11, 12, 13, etc. When the first column is full again, reset to 0 and add 1 to the next column to the left. After 19 comes 20. When both columns are full, reset both to zeros and add 1 to the next column on the left. After 99 comes 100.

To count in binary, start in the first column or binary place with 0 and count up to 1. The first column is full. Reset and add 1 to the next binary place on the left. After 0 comes 1, 10. Now the first column can be filled again. After 10 comes 11. Both columns are full. Reset both and add one to the next binary place to the left. After 11 comes 100. Now the first column can be filled again. After 100 comes 101, 110, 111, 1000, 1001, 1010, 1011, 1100, 1101, and so on. Counting in binary we have

0	
1	First column is full.
10	Reset and add 1 to second column.
11	First two columns are full.
100	Reset and add 1 third column.
101	
110	
111	First three columns are full.
1000	Reset and add 1 to fourth column.
1001	
1010	
1011	
1100	
1101	
1110	
1111	First four columns are full.

10000   Reset and add 1 to fifth column.
10001
10010
10011
10100
10101

Try writing the binary numbers from 11111 to 1000000.

The word **bit** is a contraction of the words *binary digit*. Each place in a binary number is called a bit. The binary number 10110 is a 5-bit binary number. The first place on the right is called the least significant bit (LSB) and the left-most place is called the most significant bit (MSB).

─────Most Significant Bit (MSB)

10110 is a 5-bit binary number

Least Significant Bit (LSB)─────

Using three bits we can count in binary to 111 or 7. Including 000, we have eight different combinations. In general, with N bits we can count up to $2^N - 1$ for a total of 2^N different numbers.

maximum count $= 2^N - 1$
where N is the number of bits

number of combinations $= 2^N$
where N is the number of bits

EXAMPLE 1-1

How high can you count using a 4-bit number?

Solution   With $N = 4$, we can count up to $2^4 - 1 = 15$.

EXAMPLE 1-2

How many different numbers can be represented with six bits?

Solution   With $N = 6$, there are 2^N combinations, $2^6 = 64$.

1.2 BINARY TO DECIMAL CONVERSION

In the decimal system, the first decimal place to the left of the decimal point is called the one's or unit's place. Each column to the left increases by a factor of ten (base-ten system). Moving to the left from the decimal the values can be expressed in terms of the base-ten as

$10^0, 10^1, 10^2, 10^3$, and so on. The decimal number 3954 means

$$
\begin{array}{cccc}
3 & 9 & 5 & 4 \\
(3 \times 10^3) & + (9 \times 10^2) + (5 \times 10^1) + (4 \times 10^0) \\
(3 \times 1000) & + (9 \times 100) + (5 \times 10) + (4 \times 1) \\
3000 \ + & 900 \ + & 50 \ + & 4 \ = 3954
\end{array}
$$

In the binary system, the first binary place to the left of the binary point is still the one's or unit's place. Each column to the left increases by a factor of 2 (base-two system). Moving to the left from the binary point the column values are 1, 2, 4, 8, 16, 32, 64, 128, 256, 512, 1024, and so on. These values can be represented in terms of the base two as $2^0, 2^1, 2^2, 2^3, 2^4, 2^5, 2^6, 2^7, 2^8$, $2^9, 2^{10}$, and so on. The binary number 10110 means

$$
\begin{array}{ccccc}
1 & 0 & 1 & 1 & 0 \\
(1 \times 2^4) & + (0 \times 2^3) + (1 \times 2^2) + (1 \times 2^1) + (0 \times 2^0) \\
(1 \times 16) & + (0 \times 8) + (1 \times 4) + (1 \times 2) + (0 \times 1) \\
16 \ + & 0 \ + & 4 \ + & 2 \ + & 0 \ = 22
\end{array}
$$

The binary number 10110 is the same as the decimal number 22. A binary number is often distinguished from a decimal number by writing the base as a subscript. Thus

$$10110_2 = 22_{10}$$

To convert a binary number to a decimal number, list the value of each place; then total the values that are represented by ones.

EXAMPLE 1-3

Convert 1000111_2 to a decimal number.

Solution List the value of each place.

1	0	0	0	1	1	1
2^6	2^5	2^4	2^3	2^2	2^1	2^0
64	32	16	8	4	2	1

Total the values that are represented by ones.

$64 + 4 + 2 + 1 = 71$
$1000111_2 = 71_{10}$

EXAMPLE 1-4

Convert 101011_2 to a decimal number.

Solution

1	0	1	0	1	1
2^5	2^4	2^3	2^2	2^1	2^0
32	16	8	4	2	1

$32 + 8 + 2 + 1 = 43$
$101011_2 = 43_{10}$

EXAMPLE 1-5

Convert 11001100_2 to a decimal number.

Solution

1	1	0	0	1	1	0	0
128	64	32	16	8	4	2	1

$128 + 64 + 8 + 4 = 204$

$11001100_2 = 204_{10}$

 # 1.3 DECIMAL TO BINARY CONVERSION

Two methods are presented for converting decimal numbers to binary numbers.

Method 1

Label the binary places until you reach the place with a value which exceeds the decimal number to be converted. For example, to convert 23_{10} to a binary number:

32	16	8	4	2	1

There are no 32s in 23, but there is a 16. Place a 1 in the 16's column, and subtract 16 from 23 to see how much is left to convert.

	1				
32	16	8	4	2	1

$23 - 16 = 7$

There are no 8s in 7, but there is a 4. Place a 0 in the 8's column and a 1 in the 4's place and subtract 4 from 7 to see what remains.

	1	0	1		
32	16	8	4	2	1

$7 - 4 = 3$

There is a 2 in 3. Place a 1 in the 2's column and subtract 2 from 3 to see what remains.

	1	0	1	1	
32	16	8	4	2	1

$3 - 2 = 1$

Place a 1 in the 1's column and subtract 1 from 1 to see what remains.

	1	0	1	1	1
32	16	8	4	2	1

$1 - 1 = 0$ The process is finished.

$23_{10} = 10111_2$

EXAMPLE 1-6

Convert 45_{10} to a binary number.

Solution

1	0	1	1	0	1
64	32	16	8	4	2	1

$45 - 32 = 13$
$13 - 8 = 5$
$5 - 4 = 1$
$1 - 1 = 0$
$45_{10} = 101101_2$

EXAMPLE 1-7

Convert 132_{10} to a binary number.

Solution

1	0	0	0	0	1	0	0
256	128	64	32	16	8	4	2	1

$132 - 128 = 4$
$4 - 4 = 0$
$132_{10} = 10000100_2$

Method 2

Divide the decimal number to be converted successively by 2, ignoring the remainders, until you have a quotient of 0. The remainders will be used later to determine the answer. For example, to convert 101_{10} to a binary number:

$101 \div 2 = 50$ remainder 1 LSB
$50 \div 2 = 25$ remainder 0
$25 \div 2 = 12$ remainder 1
$12 \div 2 = 6$ remainder 0
$6 \div 2 = 3$ remainder 0
$3 \div 2 = 1$ remainder 1
$1 \div 2 = 0$ remainder 1 MSB

Read the remainders from bottom to top to form the answer.

    1100101

Therefore,

    $101_{10} = 1100101_2$

EXAMPLE 1-8

Convert 291_{10} to a binary number.

Solution

$$
\begin{array}{lll}
291 \div 2 = 145 \text{ remainder } 1 & \quad \text{LSB} \\
145 \div 2 = 72 \text{ remainder } 1 \\
72 \div 2 = 36 \text{ remainder } 0 \\
36 \div 2 = 18 \text{ remainder } 0 \\
18 \div 2 = 9 \text{ remainder } 0 \\
9 \div 2 = 4 \text{ remainder } 1 \\
4 \div 2 = 2 \text{ remainder } 0 \\
2 \div 2 = 1 \text{ remainder } 0 \\
1 \div 2 = 0 \text{ remainder } 1 & \quad \text{MSB}
\end{array}
$$

$291_{10} = 100100011_2$

EXAMPLE 1-9

Convert 1024_{10} to a binary number.

Solution

$$
\begin{array}{lll}
1024 \div 2 = 512 \text{ remainder } 0 & \quad \text{LSB} \\
512 \div 2 = 256 \text{ remainder } 0 \\
256 \div 2 = 128 \text{ remainder } 0 \\
128 \div 2 = 64 \text{ remainder } 0 \\
64 \div 2 = 32 \text{ remainder } 0 \\
32 \div 2 = 16 \text{ remainder } 0 \\
16 \div 2 = 8 \text{ remainder } 0 \\
8 \div 2 = 4 \text{ remainder } 0 \\
4 \div 2 = 2 \text{ remainder } 0 \\
2 \div 2 = 1 \text{ remainder } 0 \\
1 \div 2 = 0 \text{ remainder } 1 & \quad \text{MSB}
\end{array}
$$

$1024_{10} = 10000000000_2$

✔ SELF-CHECK 1

1. Write the binary numbers from 11111 to 1000000.

2. How high can you count using a 6-bit number?

3. How many different numbers can be represented using six bits?

4. Convert. $\qquad 10110_2 = \underline{\hspace{2cm}}_{10}$

5. Convert. $\qquad 110001_2 = \underline{\hspace{2cm}}_{10}$

6. Convert using Method 1. 402_{10} = _____$_2$

7. Convert using Method 1. 79_{10} = _____$_2$

8. Convert using Method 2. 598_{10} = _____$_2$

9. Convert using Method 2. 126_{10} = _____$_2$

Although binary numbers are ideal for digital machines, they are cumbersome for humans to manipulate. It is very difficult to copy a string of 8-bit binary numbers without losing or transposing a 1 or 0. Octal and hexadecimal number systems are used as an aid in handling binary numbers. First, we will examine the characteristics of octal numbers and use them to represent binary numbers. Then we will examine hexadecimal numbers and use them to represent binary numbers.

1.4 OCTAL NUMBER SYSTEM

Octal is a base-eight number system. There are eight different digits to work with, zero through seven. To count in octal, start in the first column to the left of the octal point and count from zero to seven. The first column is full, so reset to zero and add one to the next column. After 7 comes 10. Now fill the first column again. After 10 comes 11, 12, 13, 14, 15, 16, 17. The first column is again full, so reset and add one to the next column to the left. After 17 comes 20, 21, 22, and so on. When the first two columns are full, reset both and add one to the next column. After 77 comes 100, 101, 102, and so on. After 757 comes 760, 761, 762, and so on.

EXAMPLE 1-10

Count in octal from 666_8 to 710_8.

Solution

666	
667	First column is full.
670	Reset and add 1 to second column.
671	
672	
673	
674	
675	
676	
677	First two columns are full.
700	Reset and add 1 to third column.
701	
702	
703	
704	

705	
706	
707	First column is full again.
710	Reset and add 1 to second column.

In the octal system, the first place to the left of the octal point is the one's or unit's place. Each column to the left increases by the factor of 8 (base-eight system). Moving to the left from the octal point the values of the columns are 1, 8, 64, 512, 4096, and so on. These values can be expressed in terms of the base number 8 as $8^0, 8^1, 8^2, 8^3, 8^4$, and so on. The octal number 6405_8 means

$$
\begin{array}{cccc}
6 & 4 & 0 & 5 \\
(6 \times 8^3) & + (4 \times 8^2) + & (0 \times 8^1) + & (5 \times 8^0) \\
(6 \times 512) & + (4 \times 64) + & (0 \times 8) \ + & (5 \times 1) \\
3072 & + \quad 256 + & 0 \ + & 5 \ = 3333
\end{array}
$$

An octal number is distinguished from a decimal number by writing the base as a subscript.

$$6405_8 = 3333_{10}$$

Comparing decimal, binary, and octal we have

Decimal	Binary	Octal
0	000	0
1	001	1
2	010	2
3	011	3
4	100	4
5	101	5
6	110	6
7	111	7
8	1000	10
9	1001	11
10	1010	12
11	1011	13
12	1100	14

Notice that three binary bits correspond perfectly with one octal digit. That is, it takes exactly three bits to count from zero to seven.

1.5 BINARY TO OCTAL CONVERSION

The fact that three binary bits represent eight different octal digits yields an easy method for converting from binary to octal. Starting at the binary point and moving to the left, mark off groups of three. Add leading zeros in the most significant group as needed to complete a group of three bits. Then, using weights of 4, 2, and 1, convert each group into the corresponding octal digit.

EXAMPLE 1-11

Convert 10111101_2 to an octal number.

Solution

010	111	101
2	7	5

Note that the most significant group only had two bits. A leading zero was added to complete a group of three bits.

$$10111101_2 = 275_8$$

An 8-bit binary number can be represented by three octal digits, which are much easier to handle.

EXAMPLE 1-12

Convert 10101010_2 to an octal number.

Solution

010	101	010
2	5	2

$$10101010_2 = 252_8$$

EXAMPLE 1-13

Convert $11010100110111101001000_2$ to an octal number.

Solution

011	010	100	110	111	101	001	000
3	2	4	6	7	5	1	0

$$11010100110111101001000_2 = 32467510_8$$

 # 1.6 OCTAL TO BINARY CONVERSION

To convert from octal back to binary is just as easy. For each octal digit write the three corresponding binary bits. For example, to convert 3062_8 to a binary number:

3	0	6	2
011	000	110	010

The number $3062_8 = 011000110010_2$. Note that the 2 is written 010 with a leading 0 added to complete the three bits and that 0 is written 000 to hold three places. On the most significant digit, leading zeros can be suppressed. The 3 can be written as 11 or 011.

EXAMPLE 1-14

Convert 377_8 to a binary number.

Solution

3	7	7
011	111	111

$377_8 = 11111111_2$

EXAMPLE 1-15

Convert 647015_8 to a binary number.

Solution

6	4	7	0	1	5
110	100	111	000	001	101

$647015_8 = 110100111000001101_2$

1.7 HEXADECIMAL NUMBER SYSTEM

An alternate method for handling binary numbers is to use the **hexadecimal** number system. Hexadecimal is a base-sixteen number system, which means that a choice of 16 digits is available for each column. Those 16 are 0, 1, 2, 3, 4, 5, 6, 7, 8, 9, A, B, C, D, E, and F. To count in the hexadecimal system, start in the first column to the left of the hexadecimal point and count from 0 to F. Once the first column is full, reset and add one to the second column. After 18, 19, 1A, 1B, 1C, 1D, 1E, 1F comes 20, 21, and so on. After 9FFF comes A000, and so on.

EXAMPLE 1-16

Count in the hexadecimal number system from AE9 to B00.

Solution

AE9	First column is not full.
AEA	Continue with A through F.
AEB	
AEC	
AED	
AEE	
AEF	First column is now full.
AF0	Reset and add 1 to second column.
AF1	Fill first column again.
AF2	

AF3

.

.

.

AF9
AFA
AFB
AFC
AFD
AFE
AFF   First two columns are full.
B00   Reset and add 1 to third column.

The first column to the left of the hexadecimal point is the one's or unit's place. Moving to the left, each column increases by a factor of 16, giving values of 1, 16, 256, 4096, 65536, 1048576, and so on. The hexadecimal number $A6F0_{16}$ means

$$
\begin{array}{cccc}
A & 6 & F & 0 \\
(10 \times 16^3) & + (6 \times 16^2) & + (15 \times 16^1) & + (0 \times 16^0) \\
(10 \times 4096) & + (6 \times 256) & + (15 \times 16) & + (0 \times 1) \\
40960 + & 1536 + & 240 & + \; 0 = 42736
\end{array}
$$

A hexadecimal number is distinguished from a decimal number by writing the base number as a subscript.

$$A6F0_{16} = 42736_{10}$$

Comparing decimal, binary, and hexadecimal numbers, we have

Decimal	Binary	Hexadecimal
0	0000	0
1	0001	1
2	0010	2
3	0011	3
4	0100	4
5	0101	5
6	0110	6
7	0111	7
8	1000	8
9	1001	9
10	1010	A
11	1011	B
12	1100	C
13	1101	D
14	1110	E
15	1111	F
16	10000	10

1.8 BINARY TO HEXADECIMAL CONVERSION

Notice that four binary bits correspond perfectly with one hexadecimal digit. That is, it takes exactly four binary bits to count from 0 to F. To represent binary numbers as hexadecimal numbers, mark off groups of four, starting at the binary point and moving to the left. Then convert each group into the corresponding hexadecimal digit. Until you learn the binary to hexadecimal conversions, refer to the chart in Section 1.7 or, better yet, make your own chart in the margin of your paper. With use, the conversions become automatic.

EXAMPLE 1-17

Convert 10111001_2 to a hexadecimal number.

Solution

$$
\begin{array}{cc}
1011 & 1001 \\
B & 9
\end{array}
$$
$$10111001_2 = B9_{16}$$

An 8-bit binary number can be represented quite nicely with two hexadecimal digits.

EXAMPLE 1-18

Convert 01011110_2 to a hexadecimal number.

Solution

$$
\begin{array}{cc}
0101 & 1110 \\
5 & E
\end{array}
$$
$$01011110_2 = 5E_{16}$$

EXAMPLE 1-19

Convert 11110000001110_2 to a hexadecimal number.

Solution

$$
\begin{array}{cccc}
0011 & 1100 & 0000 & 1110 \\
3 & C & 0 & E
\end{array}
$$
$$11110000001110_2 = 3C0E_{16}$$

1.9 HEXADECIMAL TO BINARY CONVERSION

Conversion from hexadecimal back to binary is just as easy. For each hexadecimal digit write the corresponding four binary bits. Refer to the chart in Section 1.7 until you learn the conversions.

EXAMPLE 1-20

Convert $C3A6_{16}$ to a binary number.

Solution

C	3	A	6
1100	0011	1010	0110

$C3A6_{16} = 1100001110100110_2$

Note that 3 is written 0011 to complete the four bits required and 6 is written 0110. The leading zeros must be added to work with groups of four bits.

EXAMPLE 1-21

Convert $48BA_{16}$ to a binary number.

Solution

4	8	B	A
0100	1000	1011	1010

$48BA_{16} = 100100010111010_2$

EXAMPLE 1-22

Convert $1FC02_{16}$ to a binary number.

Solution

1	F	C	0	2
0001	1111	1100	0000	0010

$1FC02_{16} = 11111110000000010_2$

1.10 BINARY-CODED DECIMAL (BCD)

Some binary machines represent decimal numbers in codes other than straight binary. One such code is **Binary-Coded Decimal (BCD)**. In BCD, each decimal digit is represented with four binary bits, according to the weighted 8, 4, 2, 1 system that you have already learned.

EXAMPLE 1-23

Convert 3906_{10} to BCD.

Solution

$$
\begin{array}{cccc}
3 & 9 & 0 & 6 \\
0011 & 1001 & 0000 & 0110
\end{array}
$$

$3906_{10} = 0011100100000110_{BCD}$

Note that leading zeros are added to ensure that each digit is represented by four bits.

EXAMPLE 1-24

Convert 5437_{10} to BCD.

Solution

$$
\begin{array}{cccc}
5 & 4 & 3 & 7 \\
0101 & 0100 & 0011 & 0111
\end{array}
$$

$5437_{10} = 0101010000110111_{BCD}$

Converting BCD back to decimal is just as easy. Starting at the BCD point and moving to the left, mark off groups of four. Convert each group of four bits to the corresponding decimal digit.

EXAMPLE 1-25

Convert 11010010011_{BCD} to a decimal number.

Solution

$$
\begin{array}{ccc}
0110 & 1001 & 0011 \\
6 & 9 & 3
\end{array}
$$

$11010010011_{BCD} = 693_{10}$

Using four bits, we can count from 1 to 15. The six numbers over 9 are not valid BCD numbers because these numbers do not convert to a single decimal digit. Care must be taken that these numbers are not used in the BCD system. For example, 1010 is not a legitimate BCD number because 1010 does not convert to a single decimal digit. Figure 1-1 lists the ten valid BCD numbers and the six that are invalid and must be avoided.

```
0000 ⎫
0001 ⎪
0010 ⎪
0011 ⎪
0100 ⎬   Valid — These ten 4-bit numbers are
0101 ⎪              used in BCD numbers.
0110 ⎪
0111 ⎪
1000 ⎪
1001 ⎭

1010 ⎫
1011 ⎪   Invalid — These six 4-bit numbers are
1100 ⎬            not used in BCD numbers
1101 ⎪            because their decimal conversions
1110 ⎪            exceed the value of decimal digit "9".
1111 ⎭
```

FIGURE 1-1 Valid and invalid BCD numbers

The flowchart in Figure 1-2 summarizes the conversions that have been discussed so far. From any of the five number systems you can convert to any or all of the others. There is no line directly from octal to BCD because that direct conversion was not presented. To convert from octal to BCD, convert to binary, then decimal, then BCD.

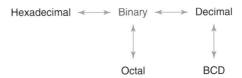

FIGURE 1-2 Conversion flowchart

EXAMPLE 1-26

Convert 157_8 to BCD.

Solution

Convert octal to binary.

$157_8 = 1101111_2$

Convert binary to decimal.

$1101111_2 = 111_{10}$

Convert decimal to BCD.

$111_{10} = 100010001_{BCD}$

EXAMPLE 1-27

Convert 362_8 to hexadecimal, binary, decimal, and BCD.

Solution

Convert octal to binary.

$$362_8 = 11110010_2$$

Convert binary to hexadecimal and binary to decimal.

$$11110010_2 = F2_{16}$$
$$11110010_2 = 242_{10}$$

Convert decimal to BCD.

$$242_{10} = 1001000010_{BCD}$$

Some scientific calculators have binary, octal, and hexadecimal modes. Converting from one base to another is easy on these calculators.

 SELF-CHECK 2

1. Count in octal from 760 to 1000.

2. Convert. 111110010_2 = _____ $_8$

3. Convert. 76540_8 = _____ $_2$

4. Count in hexadecimal from F0F to F20.

5. Convert. 111000011000_2 = _____ $_{16}$

6. Convert. $4CB0_{16}$ = _____ $_2$

7. Convert. 258_{10} = _____ $_{BCD}$

8. Convert. 100100000100_{BCD} = _____ $_{10}$

9. Convert. 370_8 = _____ $_{16}$

10. Convert. $AE0_{16}$ = _____ $_{10}$

11. Convert. 10010110_{BCD} = _____ $_{16}$

12. Convert. 254_8 = _____ $_{BCD}$

The TI-86 calculator can be set to interpret and display numbers in four number systems: decimal, binary, octal, and hexadecimal.

To set the number base, press **2nd** and then **MODE**. Use the arrow key to move the cursor down to the number base row and then across to highlight the number system of choice. Press **ENTER** and then **EXIT** to leave the menu function. If binary is chosen, a "b" appears as a suffix with each answer; an "o" appears for octal; "h" appears for hexadecimal; and no suffix is added for decimal numbers.

When the TI-86 is set to binary, binary numbers can be entered using the 1 and 0 keys on the keypad, and operations can be performed. The result appears with a "b" suffix. For example, 1000 + 1000 yields 10000b. If a key other than 1 or 0 is entered, then when the ENTER key is pressed, the message "ERROR 08 NUMBER BASE" appears. Press **F5** (QUIT) to continue.

In the hex mode, the A–F keys on the key pad **do not** function as hex digits. Instead, press **2ⁿᵈ BASE** (written above the 1 key). Select **F1** (A–F). A and B appear in a column above the F1 key. C through F appear above the function keys F2 through F5. Hex digits A through F are entered using these function keys. To enter hex digit A, enter M1 by pressing **2ⁿᵈ F1**. For example, C0 + C0 yields 180h.

To enter a number from a number system other than the default system set in the MODE menu, enter the digits and then identify the number system by pressing **2ⁿᵈ BASE**. Press **F2** (TYPE) and then function keys F1–F4 for b, h, o, d, respectively. For example, suppose the TI-81 is set to decimal mode. To add binary numbers 1000 and 111, enter this sequence: **1000 2ⁿᵈ BASE TYPE F1 + 111 F1**. 1000b + 111b should be displayed. Press **ENTER**, and the result will appear in decimal—15—since decimal base was set in the MODE menu.

EXAMPLE 1-28

Add A9h + 10010b and display the result in decimal.

Solution   Set the TI-86 to decimal mode.

> **2ⁿᵈ MODE Dec ENTER EXIT**

Enter the problem, providing the proper suffixes.

> **2ⁿᵈ BASE F1** (to enter hex digits A–F)
> **2ⁿᵈ F1** (to enter A)
> **9**
> **EXIT** (to back out one level)
> **F2** (TYPE) **F2** (to designate hexadecimal)
> **+10010 F1** (to designate binary)
> **ENTER**

187 will appear in the display. No suffix indicates that the answer is displayed in decimal.

To convert from one number system to another, enter the number to be converted, using a suffix if needed. Press **2ⁿᵈ BASE**. Press **F3** to choose CONV (conversion). Press **F1** to **F4** to convert to binary, hexadecimal, octal, or decimal, respectively. Press **ENTER**.

EXAMPLE 1-29

Convert 1111111b to decimal and to hexadecimal.

Solution   Press **1111111 2ⁿᵈ BASE F2** (TYPE) **F1** (binary)

Press **EXIT** (to back out one menu)
Press **F3** (CONV)
Press **F4** (to convert into decimal)

1111111b > Dec will appear.

Press **ENTER.**

127 will appear.

Press **F2** (to convert the answer to hex)

The display will show Ans > Hex. Press **ENTER.**
The display will show 7Fh.
So, 1111111b = 127 = 7Fh.

SELF-CHECK 3

1. Add 76o + 1BEh + 101101b and display the result in decimal.

2. Set the TI-86 to binary mode. Add these binary numbers:

   100101 + 111011 + 1101.

3. Set the TI-86 to hexadecimal mode. Combine these hexadecimal numbers:

   A6 + 2F + 47 − C2.

4. Set the TI-86 to decimal mode. Add these numbers:

   13o + 4Bh + 101101b.

In problems 5–10, make the indicated conversions.

5. 76 = _____b

6. 4Dh = _____b

7. 1011101b = _____decimal

8. 101decimal = _____h

9. ABCDEFh = _____decimal

10. 10101010b = _____h = _____decimal

In a Windows operating system, the scientific calculator available under Accessories can be used to convert from one number system to another. To access the calculator, click **Start** and then choose **Programs Accessories Calculator**. Display the scientific version by clicking **View Scientific**.

Choose **Hex, Dec, Oct,** or **Bin** as the default number system. In Bin mode, notice that only keypad digits 0 and 1 are highlighted, as they are the only digits allowed in binary. In Oct mode, 0 through 7 are highlighted; in Dec mode, 0 through 9 are highlighted; and in Hex mode, 0 through 9 and A through F are highlighted.

When octal, binary, or hexadecimal are chosen, a display size menu appears, prompting you to choose the size of calculations to be performed. For example, in binary mode with "Byte" chosen as the word size, only 8 bits can be entered for each number. If the result of an operation contains more than 8 bits, the excess bits are not displayed. This causes the calculator to mimic the operation of an 8-bit processor, where only 8 bits are available to store the result. With "Word" as the display size, 16 bits can be entered on displayed. With "Dword" (double-word), 32 bits can be entered and displayed; and with "Qword" (quadruple-word), 64 bits can be entered and displayed.

In the hexadecimal mode, "Byte" display size limits the entries and results to two hex digits, since each hex digit represents 4 bits. "Word" limits to four hex digits, "Dword" to eight, and "Qword" to 16 hex digits. To convert from one number system to another, select the base of the number to be converted, enter the number, and select the base of the result.

EXAMPLE 1-30

Convert $A5C7_{16}$ to decimal.

Solution   Set the calculator to hexadecimal.

Click **Hex** and **Dword** or **Qword**. Enter **A5C7** and click **Dec**.
42439 should appear.
$A5C7_{16} = 42439_{10}$

■

 SELF-CHECK 4

Perform these conversions:

1. $8B_{16} =$ _____$_{10}$

2. $8B_{16} =$ _____$_{8}$

3. $8B_{16} =$ _____$_{2}$

4. $557_{10} =$ _____$_{16}$

5. $557_{10} =$ _____$_{2}$

6. $557_{10} =$ _____$_{8}$

7. $1001001001001_{2} =$ _____$_{10}$

8. $1001001001001_{2} =$ _____$_{8}$

9. $1001001001001_{2} =$ _____$_{16}$

1.11 BINARY ADDITION

The table in Figure 1-3 summarizes the results that can occur when adding two binary bits, A and B. The outputs are listed as a sum and a carry. The carry indicates whether a 1 has to be added to the next column to the left. The first three lines are exactly what you would expect. In the last line, $1 + 1 = 2$, and 2 in binary is 10_2. Therefore, the sum is 0 and the carry is 1.

$$\begin{array}{r} A \\ + B \\ \hline \text{Carry} \quad \text{Sum} \end{array}$$

Inputs		Outputs	
A	**B**	**Carry**	**Sum**
0	0	0	0
0	1	0	1
1	0	0	1
1	1	1	0

FIGURE 1-3 Binary addition

The table in Figure 1-4 covers the situation in which a carry from a previous column, the **carry-in**, is added to A and B. The outputs are sum and **carry-out**. In the top four lines of the table in Figure 1-4, the carry-in is zero and the results are the same as those shown in Figure 1-3. In the last four lines, the carry-in is 1. On line 8, $1 + 1 + 1 = 3$, and 3 in binary is 11_2. Therefore, the sum is 1 and the carry-out is 1.

$$\begin{array}{r} \text{Carry-in} \\ A \\ + B \\ \hline \text{Carry-out} \quad \text{Sum} \end{array}$$

Inputs			Outputs	
Carry-in	**A**	**B**	**Carry-out**	**Sum**
0	0	0	0	0
0	0	1	0	1
0	1	0	0	1
0	1	1	1	0
1	0	0	0	1
1	0	1	1	0
1	1	0	1	0
1	1	1	1	1

FIGURE 1-4 Binary addition with carry-in

EXAMPLE 1-31

Add 11110_2 and 1100_2.

Solution

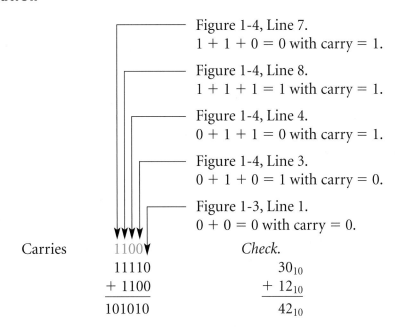

Figure 1-4, Line 7.
$1 + 1 + 0 = 0$ with carry $= 1$.

Figure 1-4, Line 8.
$1 + 1 + 1 = 1$ with carry $= 1$.

Figure 1-4, Line 4.
$0 + 1 + 1 = 0$ with carry $= 1$.

Figure 1-4, Line 3.
$0 + 1 + 0 = 1$ with carry $= 0$.

Figure 1-3, Line 1.
$0 + 0 = 0$ with carry $= 0$.

$$
\begin{array}{rr}
\text{Carries} \quad 1100 & \textit{Check.} \\
11110 & 30_{10} \\
+\ 1100 & +\ 12_{10} \\
\hline
101010 & 42_{10}
\end{array}
$$

This approach is tedious, but it shows how to apply the information in Figure 1-3 and Figure 1-4. The next example shows three numbers plus a carry being added. These addition facts are not covered in Figure 1-3 and Figure 1-4, so a different approach will be used.

EXAMPLE 1-32

Add 1011_2, 101_2, and 1001_2.

Solution

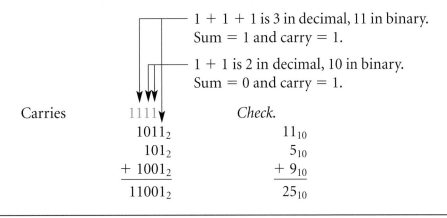

$1 + 1 + 1$ is 3 in decimal, 11 in binary.
Sum $= 1$ and carry $= 1$.

$1 + 1$ is 2 in decimal, 10 in binary.
Sum $= 0$ and carry $= 1$.

$$
\begin{array}{rr}
\text{Carries} \quad 1111 & \textit{Check.} \\
1011_2 & 11_{10} \\
101_2 & 5_{10} \\
+\ 1001_2 & +\ 9_{10} \\
\hline
11001_2 & 25_{10}
\end{array}
$$

1.12 BINARY SUBTRACTION

The table in Figure 1-5 summarizes the results that can occur when subtracting two binary bits, A and B. The outputs are listed as a difference and a borrow. The borrow indicates whether a 2 has to be borrowed from the column on the left to complete the subtraction. The second line is the most difficult to understand. To subtract 1 from 0, we must borrow from the next column on the left, which makes the problem:

$$10_2 - 1_2 \text{ or } 2 - 1$$

which is equal to 1. We borrowed 1 and had a difference of 1.

	Inputs		Outputs	
Borrow	**A**	**B**	**Borrow**	**Difference**
	0	0	0	0
A	0	1	1	1
− B	1	0	0	1
Difference	1	1	0	0

FIGURE 1-5 Binary subtraction

EXAMPLE 1-33

Subtract 1001_2 from 10011_2.

Solution

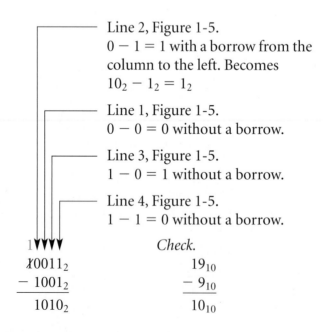

Line 2, Figure 1-5.
$0 - 1 = 1$ with a borrow from the column to the left. Becomes $10_2 - 1_2 = 1_2$

Line 1, Figure 1-5.
$0 - 0 = 0$ without a borrow.

Line 3, Figure 1-5.
$1 - 0 = 1$ without a borrow.

Line 4, Figure 1-5.
$1 - 1 = 0$ without a borrow.

$$
\begin{array}{r}
1\;\blacktriangledown\blacktriangledown\blacktriangledown\blacktriangledown \\
\cancel{1}0011_2 \\
-\ \ 1001_2 \\
\hline
1010_2
\end{array}
\qquad
\begin{array}{r}
Check. \\
19_{10} \\
-\ 9_{10} \\
\hline
10_{10}
\end{array}
$$

EXAMPLE 1-34

Subtract 1010_2 from 10001_2.

Solution

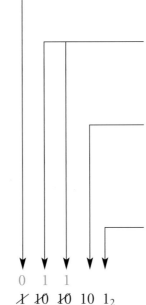

1 is borrowed from this column, leaving 0.

When 1 is borrowed from column on left, this column becomes 10_2 or 2_{10}. When 1 is borrowed from this column, it becomes 1.

Line 2, Figure 1-5.
$0 - 1 = 1$ with borrow. The borrow must be made from the 16s place, causing the chain reaction shown, $10_2 - 1_2 = 1_2$.

Line 3, Figure 1-5.
$1 - 0 = 1$ without borrow.

$$\begin{array}{r} 0 \quad 1 \quad 1 \\ \cancel{1}\ \cancel{10}\ \cancel{10}\ 10\ 1_2 \\ -\quad\ 1\ \ 0\ \ 1\ \ 0_2 \\ \hline 0\ \ 0\ \ 1\ \ 1\ \ 1_2 \end{array}$$

$$\textit{Check.} \quad \begin{array}{r} 17_{10} \\ -10_{10} \\ \hline 7_{10} \end{array}$$

EXAMPLE 1-35

Subtract 1001_2 from 1110_2.

Solution

$$\begin{array}{r} 1110_2 \\ -1001_2 \\ \hline 0101_2 \end{array}$$

$$\textit{Check.} \quad \begin{array}{r} 14_{10} \\ -9_{10} \\ \hline 5_{10} \end{array}$$

Some scientific calculators can be used to perform arithmetic operations in the different number systems. For example, to subtract two binary numbers, set the calculator to BIN mode and follow the same sequence used for subtraction in the decimal mode.

EXAMPLE 1-36

Subtract. $1011010_2 - 10001_2$

Solution   Sequence entered on TI-86 calculator: 2nd MODE Bin Enter EXIT.

$1011010 - 10001 =$

The result is 1001001.

Work several of the binary addition and subtraction problems that have appeared in this chapter on a calculator.

Using the truth table to subtract parallels the longhand method used in decimal arithmetic. It is possible to program a machine to subtract this way, but most computers utilize a complement method of subtraction that converts the problem into addition. 1's and 2's complement subtraction will be studied in Chapter 5.

SELF-CHECK 5

1. Add these binary numbers.

11000	1010
1101	1101
1011	

2. Subtract these binary numbers.

10110	100111
− 1001	− 11100

1.13 TROUBLESHOOTING A 4-BIT ADDER

In the Chapter 1 labs, you will be working with 4-bit adder integrated circuits (ICs): the 7483 in Lab 1A and the 4008 in Lab 1B. These ICs provide the sum of two 4-bit numbers, $A_4 A_3 A_2 A_1$ and $B_4 B_3 B_2 B_1$, and C_0.

$$
\begin{array}{ccccc}
& & C_0 & & \\
& A_4 & A_3 & A_2 & A_1 \\
+ & B_4 & B_3 & B_2 & B_1 \\
\hline
C_4 & \Sigma_4 & \Sigma_3 & \Sigma_2 & \Sigma_1
\end{array}
$$

C_0 (pronounced "C sub zero") is a carry from a previous column, column zero. Some manufacturers call C_0, C_{IN}. It is added to the column one bits A_1 and B_1. The column one sum is called Σ_1 (Greek letter upper case sigma, pronounced "summation"). The carry from column 1 is called C_1. It is added to column two internally and does not appear on an output pin. C_2 and C_3 are also handled internally. In column four, the sum of A_4, B_4, and C_3 produces Σ_4 and C_4, both outputs. So, the 7483 adds the two 4-bit numbers together with C_0 and produces outputs $\Sigma_4 \Sigma_3 \Sigma_2 \Sigma_1$ and a carry-out, C_4. You will provide the nine inputs by connecting the pins of the IC to +5 volts for a 1 and to ground for a 0, and observe the outputs by connecting them to **light-emitting diodes** (**LEDs**). Having C_0 available makes the 7483 and the 4008 "full adders." Full adders will be studied in Chapter 5. The preparation section of Lab 1A describes how to connect ICs and light-emitting diodes.

What kind of difficulties can you expect? For an integrated circuit to function, the connections to the dc power supply need to be made. TTL ICs have a V_{CC} input pin that has to be connected to the +5 volt terminal of the dc power supply, and a ground pin that has to be connected to the negative terminal of the dc power supply. Likewise, CMOS ICs have a V_{DD} input pin that must be connected to the positive dc supply and a V_{SS} pin that must be connected to ground. If you are constructing a circuit, make these connections first. If your IC is not behaving properly, check these power supply connections before worrying about anything else. Measure the voltages right on the pin of the IC itself using an oscilloscope or voltmeter. On all TTL integrated circuits, V_{CC} should measure close to +5 volts, and the ground pin of the IC should measure 0 volts. CMOS supply voltage varies according to the subfamily. In the 4000 series used in Lab 1B, V_{DD} can range from +3 volts to +15 volts. If the power connections are not made properly, your circuit cannot work. Trace the wiring or printed circuits back to the power supply if they do not measure correctly.

If the power pins measure correctly, but the outputs are not correct, take a look at the voltage levels on the input pins. Don't forget C_0. In Lab 1A you can use switches to supply 1s and 0s to be added. If you don't understand the operation of the switches, you might be supplying the wrong voltages to the input pins.

If the input levels are correct on the pins of the IC itself, and the outputs are not correct, either the IC is bad or something connected to the output pin is keeping it from operating correctly. For example, if the outputs are connected to indicators like light-emitting diodes, a fault in an LED circuit could hold the output at an improper voltage level.

ICs can have internal faults. An input pin can be shorted to (connected to) another input or the V_{CC} pin or the ground pin. Or an input can be open; the small internal conductor connecting the pin to the internal chip is broken or not connected. Likewise, an output pin can be shorted to another pin or open (not connected to anything). Your job is to find these faults efficiently and cheerfully. With practice, your troubleshooting skills will improve.

EXAMPLE 1-37

Troubleshoot the 7483 4-bit full adder circuit shown on page 28.

Solution   This circuit is adding $8 + 3 + 0$ ($A + B + C_0$). The LEDs show a result of 9 instead of 11. The Σ_2 output is not turning on its light-emitting diode. Several possibilities exist: the LED is bad; the LED is connected into the circuit backward; the output Σ_2 is bad internally; the input voltage levels are not correct; the power supply is not connected to the IC.

Step 1.   Confirm proper supply voltage.

V_{CC} pin measures +5 V. Ground pin measures 0 V.

Step 2.   Confirm input levels. A_4, B_2, and B_1 measure +5 V. A_3, A_2, A_1, B_4, B_3, and C_0 all measure 0 V on the pins of the IC. The inputs are correct.

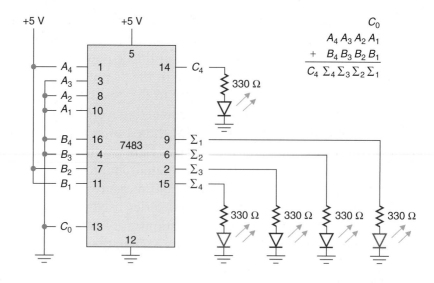

Step 3.   Check the voltage at Σ_2 output. The voltage is 0 V. Either Σ_2 is not functioning or the LED is loading the output down. Disconnect the LED. Σ_2 remains low. Output Σ_2 is shorted to ground. The IC must be replaced. If Σ_2 goes HIGH with the lamp removed, the LED circuit is faulty, not the IC.

In Lab 1B, circuit files 1B-2.ms9 through 1B-5.ms9 contain faults for you to find.

DIGITAL
APPLICATION Binary and Hexadecimal Number Systems

Computer programs consist of instructions and data stored in memory. The computer uses binary for both the memory location (address) and the contents of that address (instruction or data). Figure 1-6 shows a listing of a portion of an assembly-language program written for a Motorola 68HC11 microcontroller unit (MCU). This program produces pulse width modulated signals to control external devices. The left-hand column is the address written as a 4-digit hexadecimal number. The next columns list the op-code—the instruction to be run. Some instructions are a single byte and take two hex digits; some are two-byte instructions and take four hex digits; and some are three-byte instructions and take six hex digits to list.

Find memory location c227. The byte stored there is cb. It is the first byte in a two-byte instruction, cb40. The second byte, 40, is stored at location c228. [2] indicates that this instruction takes two clock cycles to run. ADDB #%01000000 is the same instruction, cb40, written in assembly language. It means add the following number to accumulator B. % indicates the number is written in binary. Note that this binary number converted to hex is 40, the second byte in the op-code. The remaining text on this line is a comment written by the programmer to explain the operation of the program.

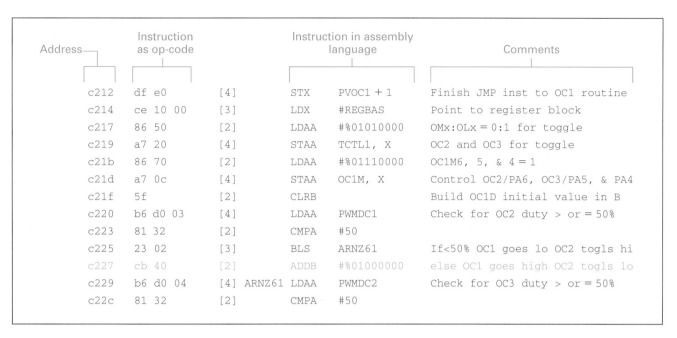

Address	Instruction as op-code		Instruction in assembly language		Comments
c212	df e0	[4]	STX	PVOC1 + 1	Finish JMP inst to OC1 routine
c214	ce 10 00	[3]	LDX	#REGBAS	Point to register block
c217	86 50	[2]	LDAA	#%01010000	OMx:OLx = 0:1 for toggle
c219	a7 20	[4]	STAA	TCTL1, X	OC2 and OC3 for toggle
c21b	86 70	[2]	LDAA	#%01110000	OC1M6, 5, & 4 = 1
c21d	a7 0c	[4]	STAA	OC1M, X	Control OC2/PA6, OC3/PA5, & PA4
c21f	5f	[2]	CLRB		Build OC1D initial value in B
c220	b6 d0 03	[4]	LDAA	PWMDC1	Check for OC2 duty > or = 50%
c223	81 32	[2]	CMPA	#50	
c225	23 02	[3]	BLS	ARNZ61	If<50% OC1 goes lo OC2 togls hi
c227	cb 40	[2]	ADDB	#%01000000	else OC1 goes high OC2 togls lo
c229	b6 d0 04	[4] ARNZ61	LDAA	PWMDC2	Check for OC3 duty > or = 50%
c22c	81 32	[2]	CMPA	#50	

FIGURE 1-6 Portion of an assembly-language program

SUMMARY

- Binary is a base-two number system with digits 0 and 1.

- Octal is a base-eight number system with digits 0–7.

- Hexadecimal is a base-sixteen number system with digits 0–9 and A–F.

- To convert binary to decimal, list the value of each place, then total the values that are represented by 1s.

- There are two methods for converting from decimal to binary.

  1. Find the highest binary place value that does not exceed the decimal number being converted. Put a 1 in that place. Subtract that value from the decimal number being converted. Repeat the process for the remainder until nothing is left. Fill in the other places with 0s.

  2. Divide successively by 2, ignoring the remainders, until a quotient of 0 is reached. The remainders, last to first, form the answer.

- To convert binary to octal, mark off groups of three. Use weights of 4, 2, and 1 to convert each group into the corresponding octal digits.

- To convert octal to binary, write the three corresponding binary digits for each octal digit.

- To convert binary to hexadecimal, mark off groups of four. Use weights of 8, 4, 2, 1 to convert each group into corresponding hexadecimal digits.

- To convert hexadecimal back to binary, write the four corresponding binary digits for each hex digit.

- In binary coded decimal (BCD), each decimal digit is represented with four bits according to the weighted 8, 4, 2, 1 system.

- When constructing a circuit, connect the V_{CC} and ground connections first.

- When troubleshooting a circuit, if an IC is not behaving properly, measure V_{CC} and ground potentials right on the pins of the IC.

- If V_{CC} and ground are correct, measure the input voltage levels right on the pins of the IC.

- If the power and inputs are correct and the output is not, either the IC is bad or something wired to the output is loading it down. Both possibilities must be considered.

QUESTIONS AND PROBLEMS

1. Write the binary numbers from 100_2 to 1000_2.

2. Write the binary numbers from 1011_2 to 10101_2.

3. Write the octal numbers from 66_8 to 110_8.

4. Write the octal numbers from 767_8 to 1010_8.

5. Write the hexadecimal numbers from DD_{16} to 101_{16}.

6. Write the hexadecimal numbers from EFD_{16} to $F10_{16}$.

7. Write the BCD numbers from 10001001_{BCD} to 100000001_{BCD}.

8. Write the BCD numbers from 1101000_{BCD} to 10010000_{BCD}.

9. a. How high can you count with a 4-bit binary number?

   b. How many different numbers are represented?

A. a. How high can you count with an 8-bit binary number?

   b. How many different numbers are represented?

B. a. How high can you count with a 16-bit binary number?

   b. How many different numbers are represented?

C. a. How many different digits are used in the octal number system?

   b. How many different digits are used in the hexadecimal number system?

   c. How many different digits are used in the BCD number system?

D. Complete the chart by converting the given number to each of the other number systems.

Octal	Hexadecimal	Binary	Decimal	BCD
36				
	A9			
		10010		
			99	
				1100111

E. Complete the chart by converting the given number to each of the other number systems.

Octal	Hexadecimal	Binary	Decimal	BCD
54				
	3C			
		1011100		
			100	
				10000001

F. Add in binary.

a. 1001_2
$+1101_2$

b. 1_2
1011_2
$+1001_2$

c. 10010_2
1100_2
$+11101_2$

d. 10001_2
1101_2
$+10101_2$

10. Add in binary.

a. 110_2
$+101_2$

b. 1110_2
$+110_2$

c. 1_2
1101_2
$+1101_2$

d. 1010_2
110_2
$+1011_2$

11. Subtract in binary.

a. 1001_2
-110_2

b. 10101_2
-1110_2

c. 1101_2
-100100_2

d. 10010100_2
-1010010_2

12. Subtract in binary.

a. 1100_2
-101_2

b. 11010_2
-1011_2

c. 1101_2
-100111_2

d. 101_2
-10010_2

13. Why do computers make extensive use of the binary number system?

14. Discuss the use of binary and hexadecimal number systems in digital work.

7483 4-BIT FULL ADDER

OBJECTIVES

After completing this lab, you should be able to:

- Connect a 7483 IC for proper operation.
- Use a 7483 as an adder.
- Cascade two 7483s to operate as an 8-bit adder.

COMPONENTS NEEDED

2	7483 ICs
9	LEDs
9	330-Ω resistors

PREPARATION

A 7483 4-bit adder will be used in this lab to add binary numbers. The 7483 integrated circuit will be studied in detail in Chapter 5, but you can use it here to confirm what you have learned in Chapter 1. The 7483 is a TTL integrated circuit. Detailed specifications of the TTL family of integrated circuits will be presented in Chapter 6. The facts you need to know about TTL to perform this lab are listed below.

1. The power supply connections are called V_{CC} and ground. V_{CC} is always connected to +5 V and ground is connected to 0 V.

2. A legitimate 0 on an output pin can range from 0 V to 0.4 V. A legitimate 1 on an output pin can range from 2.4 V to 5 V.

3. TTL input pins that are left unconnected are usually taken by the IC to be at a 1 level. In this lab, input signals will be connected to +5 V for a 1 and to ground for a 0.

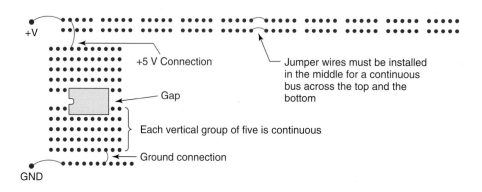

You will wire this lab on the protoboard of your TTL/CMOS trainer. The protoboard has buses across the top that you can connect to +5 V and ground. Each bus is in two halves. You must install jumper wires across with hookup wire to have a continuous bus. ICs are plugged into the protoboard so that they span the gap. You then have access to each pin via the vertical groups of 5 continuous connections. Short wires are connected to the power supply connection of each IC to be used. Switches can be used to supply 1s (+5 V) and 0s (ground) to the inputs, or inputs can be wired directly to +5 V or ground with hook-up wire. Outputs can be connected to light-emitting diodes (LEDs) through current-limiting resistors of approximately 330 Ω, as shown.

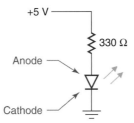

Four ways to tell the **anode** lead from the **cathode** lead are listed.

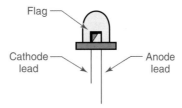

1. The cathode lead has the flag (thick portion).

2. The anode lead is usually longer than the cathode lead.

3. The flat edge of the package is on the cathode side of the LED.

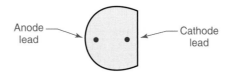

4. Connect the LED in a test circuit. If it does not light, turn the LED around. When lit, the lead connected to ground is the cathode lead. Do not forget the current-limiting resistor or the LED will become a dark-emitting diode.

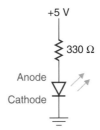

The power supply has a fixed 5-V output for work in TTL or a variable supply for work in CMOS. Be sure that the supply is set to TTL. If your trainer or dc power supply does not have a fixed 5-V output, connect a voltmeter across the output of a variable supply and adjust it to 5 V. A higher supply voltage can destroy a TTL integrated circuit. Switch off supply voltages before inserting or removing integrated circuits.

Integrated circuit pins are arranged in a definite pattern. One end of the top of the IC has a notch or indentation. Starting from the notch the pins are numbered counterclockwise. The IC shown is a 16-pin Dual In-line Package or DIP. The 14-, 20-, 24-, and 40-pin ICs are also common.

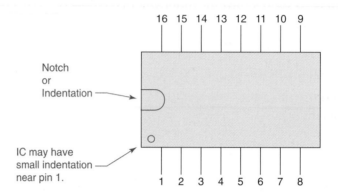

ICs are identified by a number code stamped on the top. The prefix is a manufacturer's code. The next two numbers denote the family of ICs such as TTL or CMOS. If letters follow, they indicate the subfamily of the IC. The next numbers indicate the function of the IC, and the last letters indicate the package style. For example:

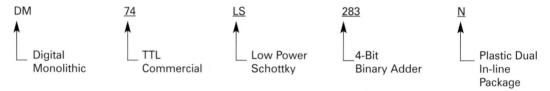

DM	74	LS	283	N
Digital Monolithic	TTL Commercial	Low Power Schottky	4-Bit Binary Adder	Plastic Dual In-line Package

Do not confuse the IC number with the date code that is also often stamped on the IC. The number 7436 indicates that the IC was manufactured in 1974 during the 36th week.

Integrated circuits are divided into categories according to their complexity. Small-scale integration (SSI) circuits such as gates and flip-flops have circuitry that is equivalent to less than 12 gates. Medium-scale integration (MSI) circuits, such as decoders, counters, multiplexers, and the adder IC that you will use in this lab, have circuitry that is equivalent to 12 or more gates but less than 100. Large-scale integration (LSI) circuits contain the equivalent of 100 or more gates.

In a TTL spec book, find the IC to obtain the pinout (where the power supply, input signals, and output signals are connected). Plug your IC into the protoboard and make the power connections V_{CC} (+5 V) and ground (0 V). Get in the habit of making those connections first. Many hours are lost troubleshooting the circuit when the only problem is faulty power supply. This IC adds two 4-bit numbers, $A_4\ A_3\ A_2\ A_1$ and $B_4\ B_3\ B_2\ B_1$, plus a carry C_0 from some previous addition.

$$\begin{array}{r} C_0 \\ A_4\ A_3\ A_2\ A_1 \\ +\ B_4\ B_3\ B_2\ B_1 \end{array}$$

You are going to supply those inputs, +5 V for a 1, 0 V for a zero.

Don't forget C_0. TTL inputs left floating (unconnected) are interpreted by the IC as 1s. If you do not want a carry into the first column ($C_0 = 0$), then ground it. The outputs are the answers to the addition. They will be connected to LEDs to monitor the outputs. A 1 output should light the LED; a zero should not. Note that a carry from a previous addition is called C_0. The carries, C_1, C_2, and C_3, are handled inside the IC and C_4 represents the overflow or carry to the next column. The outputs other than C_4 are labeled Σ (sigma) for summation. Σ_4, Σ_3, Σ_2, Σ_1 represent the sums of columns 4, 3, 2, and 1.

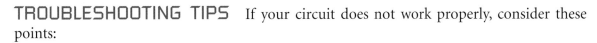

$$\begin{array}{ccccc} & & & & C_0 \\ & A_4 & A_3 & A_2 & A_1 \\ & B_4 & B_3 & B_2 & B_1 \\ \hline C_4 & \Sigma_4 & \Sigma_3 & \Sigma_2 & \Sigma_1 \end{array}$$

Labs are hard work. Do not get frustrated. Avoid the trap of rushing through a lab to get home early and missing the point. You must pay close attention to details, but don't miss the overall concepts involved.

TROUBLESHOOTING TIPS

If your circuit does not work properly, consider these points:

1. V_{CC} and ground. Use a voltmeter to check for $+5$ V on V_{CC} and 0 V on ground directly on the pins of the IC. If it measures otherwise, trace the wiring back to find the fault.

2. Inputs. Use a voltmeter to check that each input is at the level you expected. Check directly on the pins of the IC itself. Correct any discrepancies. Since these inputs are supplied directly from the switches or power supply buses, a 1 should be close to $+5$ V and 0 close to ground.

3. Outputs. Use a voltmeter to check the outputs directly on the pins of the IC (2.4 V to 5 V for a 1, 0 V to 0.4 V for a 0). If steps 1 and 2 check and step 3 does not, then either the IC is bad or something connected to the output is loading it down. A common beginning mistake is to tie the outputs, especially C_4, to ground or V_{CC}. Disconnect the wires from an output pin that does not check and see if the proper value is restored.

4. Pinout. Are you using the right pinout for your IC? Consult your spec book.

5. Think and act. You cannot correct a circuit by staring at it. Use your voltmeter. Get involved. Discuss it with your lab partner.

6. Hookup wires are sometimes shoved too far into the protoboards so that the insulation prevents electrical connections. You should be able to track down such a situation with your voltmeter.

7. If you don't understand after really trying, ask!

SAFETY ADVICE

LABORATORY SAFETY RULES

Here are some general rules to keep in mind that will make your lab sessions safe for you and those working near you. Your instructor may have some additional, more specific rules that apply to your laboratory.

1. Be aware of the fire extinguishers available in and near your lab. Type C fire extinguishers are suitable for electrical fires. Type ABC extinguishers are suitable for use against all fires. Know where the extinguishers are and how to use them.

2. Be aware of the main power disconnect switches that can be used to kill the electricity to the outlets in the lab. If someone is in trouble (electrically speaking) hit the "kill" switch and then try to help.

3. Most of the digital circuits in this book operate on +5 volts. We all get careless working with 5-volt circuits because of the low voltage level. Do not become sloppy in your work habits so that you endanger yourself or others when +5 volts is used to control 120 volts ac.

    a. Jewelry can be a good conductor of current. Severe burns can result if your jewelry becomes part of the current path.

    b. If your signal generator and dc power supply are not isolated from ground, your oscilloscope ground can only be connected to their common ground. Otherwise the circuit will be modified and heavy currents can flow.

    c. If electrolytic capacitors are installed backwards (polarity reversed), they can become hot and explode. Concentrate on your work.

4. Wear safety glasses to solder or unsolder components. Eyes often are very close to the work being soldered. A solder splash in the eye is not only extremely painful, it can also endanger your vision.

5. Solder in a well ventilated area to avoid breathing flux fumes.

6. Protect your eyes and the eyes of those around you by holding on to the end of a lead to be cut with diagonal cutters.

7. Avoid the temptation to play around in lab. A little common sense goes a long way. Playing tricks on your colleagues has no place in the lab.

PROCEDURE

1. Connect the IC as shown below.

2. Let $A = 10_{10}$ and $B = 6_{10}$ and $C_0 = 0$.

3. Verify the output.

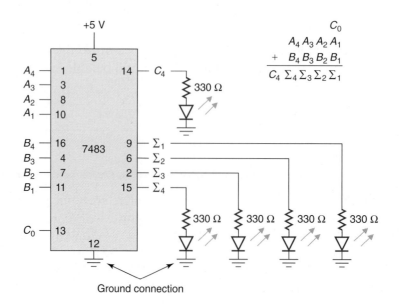

Ground connection

4. Let $A = 7_{10}$ and $B = 4_{10}$ and $C_0 = 1$.

5. Verify the output.

6. Try several other combinations.

7. Interconnect two 7483 ICs as shown to form an 8-bit full adder.

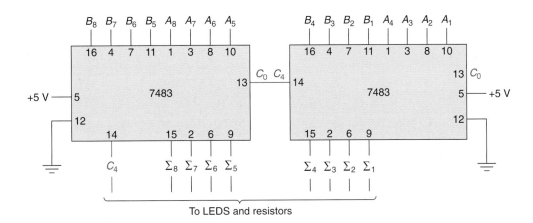

8. Work the following problems. Verify your answers using the 7483s.
   a. $150_{10} + 201_{10} =$ _____ $(C_0 = 0)$.
   b. $255_{10} + 1_{10} =$ _____ $(C_0 = 1)$.

LAB 1B

4008 4-BIT FULL ADDER

OBJECTIVES

After completing this lab, you should be able to use Multisim to:

- Analyze an adder circuit.

- Connect an adder circuit.

- Troubleshoot an adder circuit.

PREPARATION

Multisim provides a means by which to connect, troubleshoot, and analyze a circuit. In Part 1 of this lab, we will analyze a 4-bit binary full adder circuit. The 4008 adds two 4-bit binary numbers, $A_3\,A_2\,A_1\,A_0$ and $B_3\,B_2\,B_1\,B_0$, and a carry-in, C_{IN}, and produces the 4-bit sum $S_3\,S_2\,S_1\,S_0$ and a carry-out, C_{OUT}. V_{DD} and V_{SS} are the dc voltage supply connections for this IC.

V_{DD} is $+15$ V and V_{SS} is 0 V. The input pins are connected to V_{DD} to represent a 1 and to V_{SS} to represent a 0.

$$
\begin{array}{ccccc}
 & & & C_{IN} & \\
 & A_3 & A_2 & A_1 & A_0 \\
+ & B_3 & B_2 & B_1 & B_0 \\
\hline
C_{OUT} & S_3 & S_2 & S_1 & S_0
\end{array}
$$

In Part 2 you will connect a 2nd 4008 to produce an 8-bit full adder.

$$
\begin{array}{ccccccccc}
 & & & & & & & C_{IN} & \\
 & A_7 & A_6 & A_5 & A_4 & A_3 & A_2 & A_1 & A_0 \\
+ & B_7 & B_6 & B_5 & B_4 & B_3 & B_2 & B_1 & B_0 \\
\hline
C_{OUT} & S_7 & S_6 & S_5 & S_4 & S_3 & S_2 & S_1 & S_0
\end{array}
$$

In Part 3 you will troubleshoot the 4-bit full-adder circuit.

PROCEDURE

PART 1

Using the website information in the Preface, open the circuit named File 1B-1.ms9. Turn on the power switch, and answer these questions:

1. What two 4-bit numbers are being added?

2. What is the state of C_{IN}?

3. Is the result correct?

4. What is the maximum number this IC can output?

Now use the 4-bit adder to work these problems.

5. Let $A = 0101_2$ and $B = 0111_2$ and $C_{IN} = 1$.

◀ Result _____

6. Let $A = 1101_2$ and $B = 0111_2$ and $C_{IN} = 1$.

◀ Result _____

7. Let $A = E_{16}$ and $B = B_{16}$ and $C_{IN} = 0$.

◀ Result _____

8. Let $A = 9_{16}$ and $B = 7_{16}$ and $C_{IN} = 1$.

◀ Result _____

PART 2

Expand the circuit 1B-1.ms9 by adding a second 4008 in cascade with the first to produce an 8-bit full adder. What happens to C_{OUT} and C_{IN}? Get organized and remember which 4008 is handling the most significant 4 bits (nibble), and which IC is handling the least significant nibble. Refer to the following procedure as needed while connecting the second 4008.

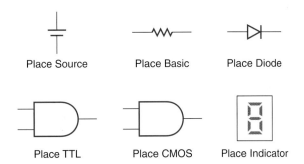

Place Source   Place Basic   Place Diode

Place TTL   Place CMOS   Place Indicator

Step 1. Select and place the IC.

On the component tool bar, click **Place CMOS**. The "Select a Component" window opens. Under "Family:" choose **CMOS_15V**. From the component list choose 4008BP_15 and click **OK**. Click in the Circuit Window to place the 4008 IC.

Step 2. Connect the power supply.

The power pins V_{DD} (15 V) and V_{SS} (ground, 0 V) are not shown. They are automatically connected to the power supply.

Step 3. Connect C_{OUT} of the least significant 4008 to C_{IN} of the most significant 4008.

Step 4. Connect the outputs to indicators.

On the component tool bar, click **Place Indicator**.

Under "Family:" click **PROBE**.

Under "Component:" choose a probe and place it in the circuit.

Connect probes to C_{OUT}, S_7, S_6, S_5, and S_4.

Step 5. Wire the inputs of the IC.

Connect each input A_7 through A_4 and B_7 through B_4 to the V_{DD} bus to input a 1—or to V_{SS} to input a 0.

Step 6. Turn on the power switch and observe the outputs.

Now use the 8-bit adder to work these problems.

1. Let $A = 10011101_2$ and $B = 10000111_2$ and $C_{IN} = 1$.

◀ Result _____

2. Let $A = 3F_{16}$ and $B = 7B_{16}$ and $C_{IN} = 1$.

◀ Result _____

3. Let $A = AA_{16}$ and $B = BB_{16}$ and $C_{IN} = 0$.

◀ Result _____

PART 3

Each of the circuits in files 1B-2.ms9 through 1B-5.ms9 contains one or no faults. First, determine whether the circuit is performing correctly. Try test problems in order to exercise C_{OUT} as well as $S_3\ S_2\ S_1\ S_0$. If the circuit has a problem, use the multimeter to isolate the fault. The multimeter is the top instrument in the column of instruments along the right-hand column of the screen. Click and release the multimeter icon. Click in the Circuit Window to place the multimeter. Wire the $+$ and $-$ inputs in order to test points as needed. Double-click on the multimeter to expand its control panel. Select **DC** (——) and **Volts** (V). Follow the guidelines in the troubleshooting section of Chapter 1. Keep a log of the steps performed. Looking back at the log can help you analyze your efficiency in isolating the fault.

Logic Gates

OUTLINE

Digital Application   Gate Symbols
LAB 2A   Gates
LAB 2B   Gates

KEY TERMS

ARCHITECTURE
active HIGH
active LOW
AND
Boolean algebra
bubble
enable
ENTITY
gate
inhibit
inverter

NAND
NOR
OR
Programmable Logic Device
    (PLD)
truth table
unique state
Very High-speed Hardware
    Description Language
    (VHDL)

OBJECTIVES

After completing this chapter, you should be able to:

- Draw a logic symbol for each gate.
- Write the Boolean expression for the output of each gate.
- Write the truth table for each gate.
- Draw the inverted logic symbol for each gate.
- Write the Boolean expression for the output of each inverted logic gate.
- Predict the output of each gate, given inputs.
- Draw the IEC symbol for each gate.
- Describe how to enable and inhibit each of the two-input gates.
- Use a NAND and a NOR as inverters.
- Expand each of the two input gates.
- Write a program in VHDL.
- Program a PLD.

2.1 GATES

Gates are circuits that are used to combine digital logic levels (1s and 0s) in specific ways. A system called **Boolean algebra** is used to express the output in terms of the inputs. The basic gates are the **inverter, OR, AND, NAND**, and **NOR**.

2.2 INVERTERS

The *inverter* is a single-input gate whose output is the complement of the input. It inverts the signal on the input. The symbol for the inverter is shown in Figure 2-1. If A is 0 then Y is 1 and if A is 1 then Y is 0. The operation of the inverter can be summarized in a **truth table** by listing all possible inputs and the corresponding outputs, as in Figure 2-2.

FIGURE 2-1 Inverter

Input	Output
A	Y
0	1
1	0

FIGURE 2-2 Inverter truth table

The small circle on the output of the symbol in Figure 2-1 is called a **bubble**. The input is not bubbled. The symbol is read, "1 in, 0 out." The bubble on the output indicates that the output is **active LOW**, and the absence of a bubble on the input indicates that the input is **active HIGH**. The input is "looking for" a 1 level to produce a 0, active low, output. The Boolean expression for the output is $\overline{A}$, which is read "A complement" or "A not" or "not A."

An alternate symbol, called an inverted logic symbol or equivalent logic symbol, Figure 2-3, has a bubble on the input but none on the output. This symbol is read "0 in, 1 out." The result is the same either way. Both symbols are used on schematics and both should be learned. Equivalent inverter symbols are shown in Figure 2-4.

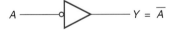

FIGURE 2-3 Inverted logic symbol for inverter

FIGURE 2-4 Equivalent inverter symbols

Inverters are available in 14-pin DIP packages in both TTL and CMOS. In the TTL family, the 7404 is a hex inverter. *Hex* signifies that six inverters are contained on the same IC. Each is independent from the others and each can be used in a different part of the circuit. The supply voltage, V_{CC}, is +5 V and is applied to pin 14 with pin 7 connected to ground, as shown in Figure 2-5.

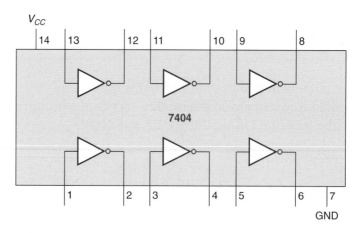

FIGURE 2-5 Hex inverter pinout

The 7404 is an example of a standard TTL IC. Over the years, subfamilies of TTL have been developed that have superior characteristics. Letters following the 74 denote the subfamily. For example, 74LS04 indicates a TTL hex inverter in **L**ow-**P**ower **S**chottky technology. 74ALS04 indicates **A**dvanced **L**ow-**P**ower **S**chottky TTL technology. These subfamilies will be covered in Chapter 6.

If the letters following 74 contain a C, then the IC is in the CMOS family. The CMOS family also contains several subfamilies. For example, 74AC04 is a hex inverter of **A**dvanced **C**MOS technology. 74HC04 is a hex inverter of the **H**igh-speed **C**MOS subfamily. The 74HCT04 is a **H**igh-speed **C**MOS-**T**TL compatible hex inverter. The pinouts of all 74xxx04 ICs, regardless of family or subfamily, are the same. See Figure 2-5.

The original CMOS family was numbered 4xxx. For example, the 4069 is a hex inverter. Most 4xxx ICs have a different pinout from their 74xxx counterparts. The pinout of the 4069 happens to be the same as the 7404. The power pin on the 4xxx series is labeled V_{DD} instead of V_{CC}; and the ground pin is labeled V_{SS}. V_{DD} can range from +3 volts to +15 volts. Some of the available inverter ICs are listed in Table 2-1.

TABLE 2-1   Inverter ICs

Number	Family	Description
7404	TTL	Hex Inverter
74LS04	Low-Power Schottky TTL	Hex Inverter
74HC04	High-Speed CMOS	Hex Inverter
4069	CMOS	Hex Inverter

In addition to the conventional logic symbol shown in Figure 2-5, the IEC (International Electrotechnical Commission) and the IEEE (Institute of Electrical and Electronics Engineers) have developed a system of logic symbols that shows the relationship of each input to each output, without showing the internal circuitry.

The IEC symbol for a 7406 hex inverter is shown in Figure 2-6. Since each inverter functions independently of the others, each is drawn in its own rectangle. The "1" in the top rectangle indicates that one input must be active to produce the output. The triangle on the right is equivalent to the bubble in the conventional symbol. A HIGH input produces a LOW output.

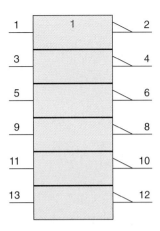

FIGURE 2-6 IEC symbol—
7406 hex inverter

2.3 OR GATES

The *OR gate* is a circuit that produces a 1 on its output when any of its inputs are 1. Figure 2-7 shows the symbol for a two-input OR gate with inputs *A* and *B* and output *Y*.

The Boolean expression for the output is $A + B$, which is read "*A* OR *B*." The output *Y* is 1 when *A* is 1 or *B* is 1 or both. The truth table in Figure 2-8 summarizes the operation of the OR gate. All possible input combinations are listed by counting in binary from 00 to 11.

FIGURE 2-7 The two-input OR gate

Inputs		Output
B	*A*	*Y*
0	0	0
0	1	1
1	0	1
1	1	1

FIGURE 2-8 Truth table for two-input OR gate

The symbol shown in Figure 2-7 represents an OR function. Since there are no bubbles shown on the inputs or outputs, the symbol is read "1 OR 1 in, 1 out." This statement summarizes the last three lines of the truth table in Figure 2-8. The first line of the truth table shows the only condition in which the output is 0. This is called the **unique state** of the gate.

EXAMPLE 2-1

Predict the output of each gate.

Solution   In the first gate one input is 1 (line 2 or 3 of the truth table), and the output is 1. In the second gate, both inputs are 0 (line 1 of the truth table), and the output must be 0.

Figure 2-9 shows an alternate symbol for the two-input OR gate called the inverted logic symbol. The shape of the symbol represents the AND function. Both the inputs and outputs are bubbled. This represents the first line of the truth table, whereas the symbol in Figure 2-7 represents the last three lines. The alternate symbol can be read "0 AND 0 in, 0 out."

FIGURE 2-9 Two-input OR inverted logic symbol

The Boolean expression for the output of the gate in Figure 2-9 is developed as follows:

1. Since input A is bubbled, write A complement, $\overline{A}$.

2. Since input B is bubbled, write B complement, $\overline{B}$.

3. Since the shape of the gate is AND, which is written as a multiplication sign (or omitted), write $\overline{A} \cdot \overline{B}$ or $\overline{A}\,\overline{B}$.

4. Since output Y is bubbled, complement the whole expression, $\overline{\overline{A} \cdot \overline{B}}$.

Since the symbols shown in Figure 2-7 and Figure 2-9 are equivalent, the outputs are equivalent and $A + B = \overline{\overline{A} \cdot \overline{B}}$. Both symbols are used in schematics and both should be learned. Equivalent OR gate symbols are shown in Figure 2-10.

FIGURE 2-10 Equivalent OR gate symbols

Note that the alternate gate can be drawn by changing everything:

1. Change the OR shape to AND.

2. Change the inputs from no bubbles to bubbles.

3. Change the output from no bubble to bubble.

EXAMPLE 2-2

Predict the output of each gate.

Solution   This alternate symbol states that a 0 AND a 0 in will produce a 0 out. In the first gate the inputs are not both 0s, and the output is a 1. In the second gate the inputs are both 0s, and the output is 0.

The inverted logic symbols appear on schematics because of the nature of the signals that the gates combine.

Some signals are normally HIGH and go to LOW when they are active. These are called active LOW signals. Others are normally LOW and go HIGH when they are active. They are called active HIGH signals. A gate that is combining active HIGH signals is usually drawn without bubbles on the inputs. A gate that is combining active LOW inputs is sometimes drawn in the inverted logic form with bubbles on its inputs.

A good example of the use of an inverted logic symbol occurs when a Z-80 microprocessor needs to store a word in memory; see Figure 2-11. It issues two control signals, MEMORY REQUEST and WRITE, both active LOW. The complement over their names indicates that they are active LOW. These two signals need to be combined to produce a new active LOW signal called MEMORY WRITE. MEMORY WRITE should go LOW when both inputs are LOW. The inverted logic OR symbol fits this situation perfectly. MEMORY WRITE will be LOW when MEMORY REQUEST and WRITE are LOW.

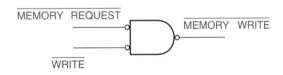

FIGURE 2-11

A variety of forms of OR gates are available in TTL and CMOS. The 7432 is a quad (meaning four gates) two-input TTL OR gate IC. The four are independent. Each can be used in a different part of a circuit without feedback. Power is supplied to the IC through a V_{CC} (+5 V) and ground connection. The 4072 is a dual (meaning two gates) four-input CMOS OR gate IC. The symbol and truth table for a four-input OR gate are shown in Figure 2-12.

The pinout for a 7432, including all the TTL subfamilies and the 74 CMOS series subfamilies, is shown in Figure 2-13A. The pinout for the 4072 is shown in Figure 2-13B. Some of the available OR gate ICs are listed in Table 2-2.

Inputs				Output
D	C	B	A	Y
0	0	0	0	0
0	0	0	1	1
0	0	1	0	1
0	0	1	1	1
0	1	0	0	1
0	1	0	1	1
0	1	1	0	1
0	1	1	1	1
1	0	0	0	1
1	0	0	1	1
1	0	1	0	1
1	0	1	1	1
1	1	0	0	1
1	1	0	1	1
1	1	1	0	1
1	1	1	1	1

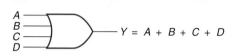

$$Y = A + B + C + D$$

FIGURE 2-12 Symbol and truth table for four-input OR gate

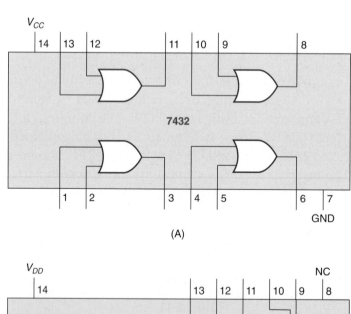

(A)

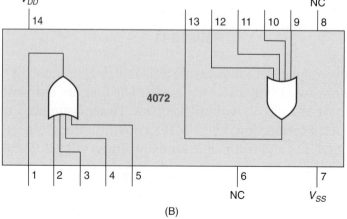

(B)

FIGURE 2-13 OR gate pinouts

TABLE 2-2 OR Gates

Number	Family	Description
7432	TTL	Quad 2-input OR
74LS32	Low-Power Schottky TTL	Quad 2-input OR
74HC32	High-Speed CMOS	Quad 2-input OR
4071	CMOS	Quad 2-input OR
4072	CMOS	Dual 4-input OR

The IEC symbol for a 7432 quad two-input OR gate is shown in Figure 2-14. The ≥ 1 sign indicates that one or more inputs must be active (HIGH in this case) to produce an active output (also HIGH). Since there are no triangles on the inputs or outputs, they are all active HIGH. One or more 1s into each gate produces a 1 out. Since each gate functions independently of the others, each is drawn in its own rectangle.

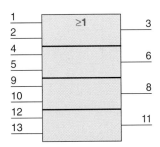

FIGURE 2-14 IEC symbol—7432 quad two-input OR gate

2.4 AND GATES

An *AND gate* is a circuit that produces a 1 on its output only when all of its inputs are 1. A two-input AND gate, with inputs A and B and output Y, is shown in Figure 2-15.

The Boolean expression for the output is $A \cdot B$, or just AB, which is read "A AND B." The output Y is 1 only when both A and B are ones. The truth table, Figure 2-16, summarizes the operation of the gate. All possible input combinations are listed by counting in binary from 00 to 11.

Inputs		Output
B	*A*	*Y*
0	0	0
0	1	0
1	0	0
1	1	1

FIGURE 2-16 Two-input AND truth table

FIGURE 2-15 Two-input AND gate

The AND symbol describes the operation of the gate. Since there are no bubbles on the input or the output, the gate is read "1 AND 1 in, 1 out." This statement describes the last line of the truth table and the only situation in which the output is 1. This is the unique state of the AND gate.

EXAMPLE 2-3

Predict the output of each gate.

Solution In the first gate the inputs are different (line 2 or 3 of the truth table), and the output is 0. In the second gate, both inputs are 1 (last line of the truth table), and the output is 1.

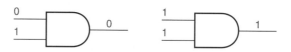

The top three lines of the AND gate truth table show that if a 0 is present on A or B (or both) then the output will be 0. This is summarized as "0 OR 0 in, 0 out." The symbol that represents this statement is shown in Figure 2-17. Both inputs and outputs are bubbled in this inverted logic symbol.

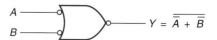

FIGURE 2-17 Inverted logic symbol for two-input AND gate

The Boolean expression for the inverted logic symbol is developed as follows:

1. Since the A input is bubbled, write $\overline{A}$.

2. Since the B input is bubbled, write $\overline{B}$.

3. The shape of the gate is OR, which is written $+$.

4. Write $\overline{A} + \overline{B}$. Since the output is bubbled, $\overline{A} + \overline{B}$ is the Boolean expression for $\overline{Y}$. $\overline{Y} = \overline{A} + \overline{B}$.

5. To find Y, complement the whole expression $\overline{\overline{A} + \overline{B}}$.

Since the symbols shown in Figures 2-15 and 2-17 are equivalent, their outputs are equivalent and $A \cdot B = \overline{\overline{A} + \overline{B}}$. Both symbols are used on schematic diagrams and both should be learned. Equivalent AND gate symbols are shown in Figure 2-18.

FIGURE 2-18 Equivalent AND gate symbols

EXAMPLE 2-4

Predict the output of each gate.

Solution   This alternate symbol states that any 0 in yields a 0 out. In gate 1, there is a 0 input, and the output must be a 0. In the second gate, there are no zeros in, and the output must be a 1.

A variety of forms of AND gates are available in TTL and CMOS. The 7408 IC is a quad (meaning four gates) two-input AND gate IC. There are four independent two-input gates. The 7411 is a triple (meaning three gates) three-input TTL AND gate IC, and the 4082 is a dual (meaning two gates on one IC) four-input CMOS AND gate IC.

The symbol and truth table for the three-input AND gate are shown in Figure 2-19. The symbol and truth table for a four-input AND gate are shown in Figure 2-20.

$$Y = A \cdot B \cdot C$$

Inputs			Output
C	B	A	Y
0	0	0	0
0	0	1	0
0	1	0	0
0	1	1	0
1	0	0	0
1	0	1	0
1	1	0	0
1	1	1	1

FIGURE 2-19 Symbol and truth table for three-input AND gate

Inputs				Output
D	*C*	*B*	*A*	*Y*
0	0	0	0	0
0	0	0	1	0
0	0	1	0	0
0	0	1	1	0
0	1	0	0	0
0	1	0	1	0
0	1	1	0	0
0	1	1	1	0
1	0	0	0	0
1	0	0	1	0
1	0	1	0	0
1	0	1	1	0
1	1	0	0	0
1	1	0	1	0
1	1	1	0	0
1	1	1	1	1

FIGURE 2-20 Symbol and truth table for four-input AND gate

The pinouts for the 7408, 7411, and 4082 are shown in Figure 2-21. Some of the available AND gates are listed in Table 2-3.

TABLE 2-3 AND Gates

Number	Family	Description
74ALS08	Advanced Low-Power Schottky TTL	Quad 2-input AND
74ACT08	Advanced CMOS-TTL Compatible	Quad 2-input AND
74HCT11	High-Speed CMOS-TTL Compatible	Triple 3-input AND
4081	CMOS	Quad 2-input AND
4082	CMOS	Dual 4-input AND

The IEC chose the ampersand (&) sign to represent the AND function. Figure 2-22A shows the IEC symbol for a 7408 quad two-input AND gate integrated circuit. The symbol for a 4082 dual four-input AND gate integrated circuit is shown in Figure 2-22B.

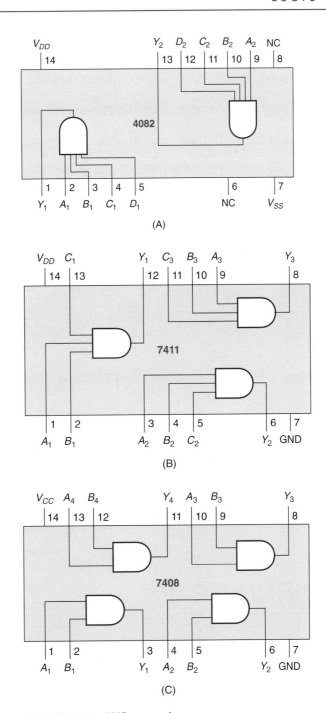

FIGURE 2-21 AND gate pinouts

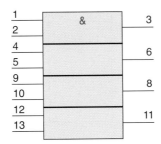

FIGURE 2-22A IEC symbol—7408 quad two-input AND gate

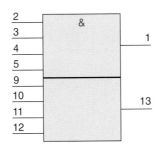

FIGURE 2-22B IEC symbol—4082 dual four-input AND gate

SELF-CHECK 1

1. Write the symbol, truth table, and Boolean expression for an inverter.

2. Write the symbol, truth table, and Boolean expression for a two-input OR gate.

3. Write the symbol, truth table, and Boolean expression for a two-input AND gate.

4. Draw the inverted logic symbol and Boolean expression for an inverter.

5. Draw the inverted logic symbol and Boolean expression for an OR gate.

6. Draw the inverted logic symbol and Boolean expression for an AND gate.

7. Predict the outputs.

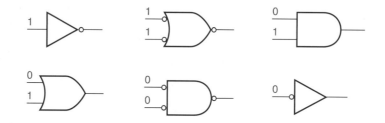

8. Draw the IEC symbols for an inverter, AND, and OR gate.

9. What is the unique state of an AND gate?

10. What is the unique state of an OR gate?

 # 2.5 NAND GATES

A *NAND gate* is a circuit that produces a 0 on its output only when all of its inputs are 1s. NAND is the contraction of the words "not" and "and." Its symbol is the AND symbol with an inverted (bubbled) output, as shown in Figure 2-23.

The truth table for a NAND gate is shown in Figure 2-24. Notice that its output is the complement of the AND gate output.

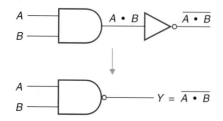

FIGURE 2-23 Two-input NAND gate

Inputs		Output
B	A	Y
0	0	1
0	1	1
1	0	1
1	1	0

FIGURE 2-24 Truth table for two-input NAND gate

The symbol describes the operation of the gate. Since the inputs are not bubbled and the output is, the symbol is read "1 AND 1 in, 0 out." This describes line four of the truth table, which is the unique state of this gate (the only situation that produces a 0).

EXAMPLE 2-5

Predict the output of each gate.

Solution   The truth table shows that the output of a NAND gate is 0 only when all inputs are 1s. This is the situation for the last gate. For the first two gates a 0 is present on the inputs, and the output must be a 1.

The top three lines of the truth table are described by the inverted logic symbol, Figure 2-25, and state that a 0 on *A* or *B* (or both) produces a 1 on the output. This is read "0 OR 0 in, 1 out," or "any 0 in, 1 out."

FIGURE 2-25 Inverted logic symbol
for two-input NAND gate

The Boolean expression for the inverted logic symbol is developed as follows:

1. Since *A* is bubbled, write *A* complement, $\overline{A}$.

2. Since *B* is bubbled, write *B* complement, $\overline{B}$.

3. Since the shape of the gate is OR, write $\overline{A} + \overline{B}$.

The Boolean expression for the output is $\overline{A} + \overline{B}$, which is read "*A* complement OR *B* complement."

Both symbols represent the NAND gate, both are used in schematics, and both should be learned. Since the symbols shown in Figure 2-23 and Figure 2-25 are equivalent, their outputs are equivalent and $\overline{A \cdot B} = \overline{A} + \overline{B}$. Equivalent NAND gate symbols are shown in Figure 2-26.

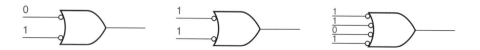

FIGURE 2-26 Equivalent NAND gate symbols

EXAMPLE 2-6

Predict the output of each gate.

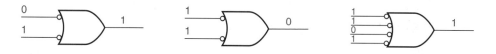

Solution   The alternate symbol for the NAND gate states any 0 in, 1 out. The first and last gates of this example have 0s in and their outputs are 1s. The middle gate has no 0s in, and its output must be 0.

The pinouts for some common NAND gates are shown in Figure 2-27.

The 7400 is a quad two-input TTL NAND gate IC, and the 7410 is a triple three-input NAND gate IC. Their pinouts are shown in Figure 2-27. The 74C30 is an eight-input CMOS NAND gate. Figure 2-28 shows the symbol and the truth table for a three-input NAND gate.

The NAND gate is available in many forms in TTL and CMOS, as shown in Table 2-4.

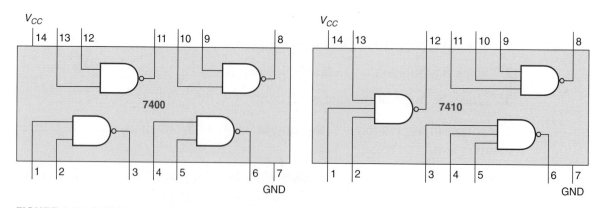

FIGURE 2-27 NAND gate pinouts

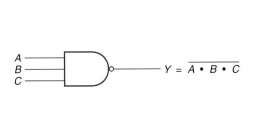

Inputs			Output
C	*B*	*A*	*Y*
0	0	0	1
0	0	1	1
0	1	0	1
0	1	1	1
1	0	0	1
1	0	1	1
1	1	0	1
1	1	1	0

FIGURE 2-28 Symbol and truth table for three-input NAND gate

TABLE 2-4   NAND Gates

Number	Family	Description
74HCT00	High-Speed CMOS-TTL Compatible	Quad 2-input NAND
74ALS10	Advanced Low-Power Schottky TTL	Triple 3-input NAND
74LS20	Low-Power Schottky TTL	Dual 4-input NAND
7430	TTL	8-input NAND
74ALS133	Advanced Low-Power Schottky TTL	13-input NAND
4011	CMOS	Quad 2-input NAND
4012	CMOS	Dual 4-input NAND
4023	CMOS	Triple 3-input NAND

The IEC symbol for a 7400 quad two-input NAND gate is shown in Figure 2-29. The triangle on the output of each gate represents the bubble on the standard symbol. The symbol shows that inputs 1 and 2 are ANDed together, and if inputs 1 and 2 are HIGH, output pin 3 will be LOW.

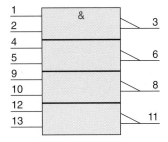

FIGURE 2-29 IEC symbol—
7400 quad two-input
NAND gate

2.6 NOR GATES

A *NOR gate* is a circuit that produces a 0 on its output when one or more of its inputs are 1. NOR is the contraction of the words "not" and "or." Its symbol is the OR symbol with an inverted or bubbled output, shown in Figure 2-30.

The truth table for a NOR gate is shown in Figure 2-31. Notice that its output is the complement of the OR gate output.

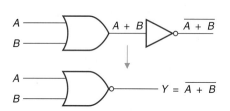

FIGURE 2-30 Two-input NOR gate

Inputs		Output
B	A	Y
0	0	1
0	1	0
1	0	0
1	1	0

FIGURE 2-31 Two-input NOR truth table

The symbol describes the operation of the gate. Since the inputs are not bubbled and the output is, the symbol is read "1 OR 1 in, 0 out." This describes the last three lines of the truth table in Figure 2-31.

EXAMPLE 2-7

Predict the output of each gate

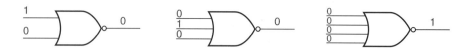

Solution The symbol states that 1 OR 1 in yields a 0 out. The first two gates have 1s on their inputs, and their outputs must be 0s. The last gate has no 1s in, and its output must be a 1.

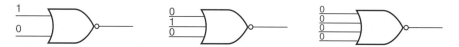

The top line of the truth table is described by the inverted logic symbol, Figure 2-32, which states that 0 on *A* and *B* gives a 1 out on *Y*. The symbol is read "0 AND 0 in, 1 out." This is the unique state of the NOR gate.

FIGURE 2-32 Inverted logic symbol for two-input NOR gate

The Boolean expression for the output is developed as follows:

1. Since A is bubbled, write A complement, $\overline{A}$.

2. Since B is bubbled, write B complement, $\overline{B}$.

3. Since the shape of the gate is AND, write $\overline{A} \cdot \overline{B}$.

Both symbols represent the NOR gate, both are used in schematics, and both should be learned. Since the symbols shown in Figure 2-30 and Figure 2-32 are equivalent, their outputs are equivalent, and $\overline{A + B} = \overline{A} \cdot \overline{B}$. Equivalent NOR gate symbols are shown in Figure 2-33. The pinouts for some common NOR gates are shown in Figure 2-34.

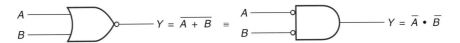

FIGURE 2-33 Equivalent NOR gate symbols

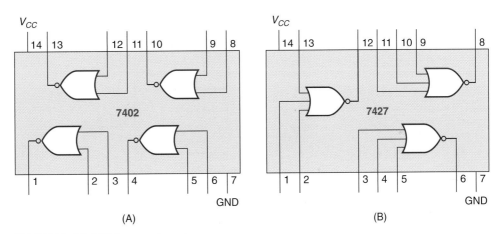

FIGURE 2-34 NOR gate pinouts

Some of the available NOR gate ICs are listed in Table 2-5.

TABLE 2-5 NOR Gates

Number	Family	Description
74LVQ02	Low-Voltage Quiet TTL	Quad 2-input NOR
7425	TTL	Dual 4-input NOR
74ALS27	Advanced Low-Power Schottky TTL	Triple 3-input NOR
4001	CMOS	Quad 2-input NOR
4002	CMOS	Dual 4-input NOR
4025	CMOS	Triple 3-input NOR

EXAMPLE 2-8

Predict the outputs of each gate.

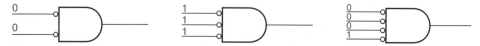

Solution   The alternate symbol for the NOR states that all 0s in yields a 1 out. The first gate has all 0s in; the others do not.

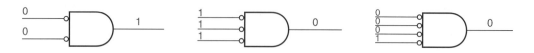

The IEC symbol for a 7427 triple three-input NOR gate integrated circuit is shown in Figure 2-35. The ≥1 sign indicates that one or more inputs must be active to cause an active output. Since the outputs are active LOW (triangles), if one or more inputs are HIGH the corresponding output goes LOW.

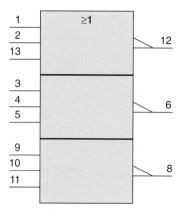

FIGURE 2-35 IEC symbol—
7427 triple three-input NOR gate

Figure 2-36 summarizes the symbols discussed in this chapter.

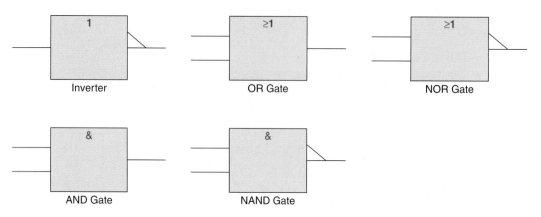

FIGURE 2-36 IEC symbol—basic gates

SELF-CHECK 2

1. Write the symbol, truth table, and Boolean expression for a NAND gate.

2. Write the symbol, truth table, and Boolean expression for a NOR gate.

3. Draw the inverted logic symbol and Boolean expression for a NAND gate.

4. Draw the inverted logic symbol and Boolean expression for a NOR gate.

5. Predict the outputs.

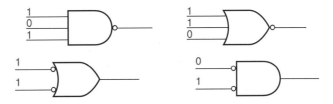

6. Draw the IEC symbols for a NAND and a NOR gate.

7. What is the unique state of a NAND gate?

8. What is the unique state of a NOR gate?

 # 2.7 DATA CONTROL ENABLE/INHIBIT

One of the common uses of gates is to control or gate the flow of data from the input to the output. In that mode of operation, one input is used as the control and the other presents the data to be passed to the output. If the data is allowed to pass through, the gate is said to be **enabled**. If the data is not allowed to pass through, the gate is said to be **inhibited**.

 # 2.8 AND GATE ENABLE/INHIBIT

If the signal on the control input of an AND gate is 0 (top two lines of the truth table in Figure 2-37), the output of the gate is 0 regardless of the data present on the data input. The data does not pass through the gate, and the gate is said to be inhibited. The output is "locked up" in the 0 state.

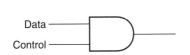

	Inputs		Output	
	Control	*Data*	*Y*	
Inhibit	0	0	0	Output locked at 0
	0	1	0	
Enable	1	0	0	Data passes through unaltered
	1	1	1	

FIGURE 2-37 AND enable/inhibit

If the signal on the control input is 1 (bottom two lines of the truth table in Figure 2-37), then whatever is present on the data input appears on the output and the gate is said to be enabled. The data "passes through" the gate.

EXAMPLE 2-9

Predict the output of each AND gate.

Solution In each case use the waveform as the data and the static (unchanging) signal as the control input. In the first case, the 1 enables the gate and data passes through unaltered. In the second case, the 0 inhibits the gate and the output is locked at 0. The data input is ignored.

2.9 NAND GATE ENABLE/INHIBIT

If the signal on the control input of a NAND gate is 0 (top two lines of the truth table in Figure 2-38), the signal on the data input is ignored and the output is "locked up" in the 1 state. The gate is said to be inhibited even though the output is 1.

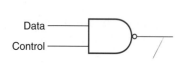

	Inputs		Output	
	Control	*Data*	*Y*	
Inhibit	0	0	1	Output locked at 1
	0	1	1	
Enable	1	0	1	Data passes through inverted
	1	1	0	

FIGURE 2-38 Nand enable/inhibit

If the signal on the control input is 1 (bottom two lines of the truth table in Figure 2-38), the signal on the data input is passed through the gate but is inverted in the process. The gate is said to be enabled.

EXAMPLE 2-10

Predict the output of each NAND gate.

Solution   In each case use the waveform as the data and the static (unchanging) signal as the control. In the first case the 1 enables the gate, and the data passes through inverted. In the second case the 0 inhibits the gate, and the output is locked at 1. The data input is ignored.

 2.10 OR GATE ENABLE/INHIBIT

If the signal on the control input of an OR gate is 0 (top two lines of the truth table in Figure 2-39), the signal on the data input passes through to the output and the gate is enabled.

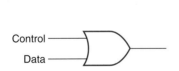

	Inputs		Output	
	Control	*Data*	*Y*	
Enable	0	0	0	Data passes through unaltered
	0	1	1	
Inhibit	1	0	1	Output locked at 1
	1	1	1	

FIGURE 2-39 OR enable/inhibit

If the signal on the control input is 1 (bottom two lines of the truth table in Figure 2-39), the signal on the data input is ignored and the output is "locked up" in the 1 state. The gate is said to be inhibited.

EXAMPLE 2-11

Predict the output of each OR gate.

Solution In each case use the waveform as the data and the static (unchanging) signal as the control. In the first case the 0 enables the gate and data passes through unaltered. In the second case the 1 inhibits the gate and the output is locked at 1. The data input is ignored.

2.11 NOR GATE ENABLE/INHIBIT

If the signal on the control input of a NOR gate is 0 (top two lines of the truth table in Figure 2-40), whatever is present on the data input appears at the output inverted. The gate is enabled.

	Inputs		Output	
	Control	*Data*	*Y*	
Enable	0	0	1	Data passes through inverted
	0	1	0	
Inhibit	1	0	0	Output locked at 0
	1	1	0	

FIGURE 2-40 NOR enable/inhibit

If the signal on the control input is 1 (bottom two lines of the truth table in Figure 2-40), the output of the gate is 0 regardless of the data present at the data input. The gate is said to be inhibited.

EXAMPLE 2-12

Predict the output of each gate.

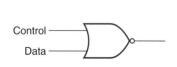

Solution   In each case use the waveform as the data and the static (unchanging) signal as the control. In the first case the 0 enables the gate and the data passes through inverted. In the second case the 1 inhibits the gate and the output is locked at 0. The data input is ignored.

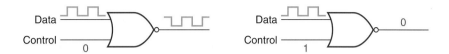

 # 2.12 SUMMARY ENABLE/INHIBIT

Each gate is enabled or inhibited in its own fashion. There is no need to memorize the function of each gate. By examining the truth table of each gate, you can determine its method of operation. The operation of each is summarized in Table 2-6.

TABLE 2-6   Enable/Inhibit

Gate	Control Input	Gate Condition	Output
AND	0	Inhibit	0
	1	Enable	Data
NAND	0	Inhibit	1
	1	Enable	$\overline{Data}$
OR	0	Enable	Data
	1	Inhibit	1
NOR	0	Enable	$\overline{Data}$
	1	Inhibit	0

Note: Data—Data passes through unaltered.
$\overline{Data}$—Data passes through inverted.

 SELF-CHECK 3

1. A 1 on the control input of a NAND (enables, inhibits) the gate.

2. When a NOR gate is enabled data passes through (unaltered, inverted).

3. When an OR gate is inhibited with a (0, 1) on the control input, the output is locked (high, low).

4. To enable an AND gate, place a (0, 1) on the control input. Data will pass through (inverted, unaltered).

2.13 NAND AS AN INVERTER

Suppose we apply the same signal to both inputs of a two-input NAND. Then either both inputs are 0 or both inputs are 1. If A is 0 then the output is 1. If A is 1 then the output is 0. The output is always the complement of the input. Figure 2-41 shows a NAND as an inverter.

FIGURE 2-41 NAND—as an inverter

2.14 NOR AS AN INVERTER

If we apply the same signal to both inputs of a two-input NOR, then either both inputs are 0 or both inputs are 1. In either case the output is always the complement of the input. Figure 2-42 shows a NOR as an inverter.

FIGURE 2-42 NOR—as an inverter

Why would we use a two-input NAND or NOR for an inverter? Sometimes there is an extra NAND or NOR gate on an IC that has been used in a circuit. It is better to use the extra gate as an inverter than to add an inverter IC. The extra IC would take up space on the circuit board (real estate), consume extra power, generate extra heat, and add extra expense.

2.15 EXPANDING AN AND GATE

A three-input AND gate can be created from two two-input gates as shown in Figure 2-43. The output is the same as if we had fed A, B, and C into a three-input AND, $Y = A \cdot B \cdot C$. You can expand an AND with another AND.

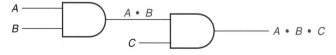

FIGURE 2-43 Expanding an AND

2.16 EXPANDING A NAND GATE

The output from a three-input NAND would be $\overline{A \cdot B \cdot C}$. At first you might think that you could expand a NAND gate with another NAND. But watch what happens in Figure 2-44. $\overline{\overline{A \cdot B} \cdot C}$ does not give the desired output, $\overline{A \cdot B \cdot C}$.

FIGURE 2-44 Attempt to expand a NAND with a NAND

Consider the circuit in Figure 2-45. This yields the desired output. You can expand a NAND with an AND, but you cannot expand a NAND with another NAND.

A ——[AND]—— A • B ——[NAND]——o —— $\overline{A \cdot B \cdot C}$
B ——   C ——

FIGURE 2-45 Expanding a NAND

2.17 EXPANDING AN OR GATE

A three-input OR gate can be created from the two-input gates as shown in Figure 2-46. You can expand an OR with another OR.

A ——[OR]—— A + B ——[OR]—— A + B + C
B ——   C ——

FIGURE 2-46 Expanding an OR

2.18 EXPANDING A NOR GATE

The output from a three-input NOR gate is $\overline{A + B + C}$. As with the NAND, you cannot expand a NOR with another NOR. A NOR can be expanded as shown in Figure 2-47. This yields the desired output. You can expand a NOR with an OR, but you cannot expand a NOR with another NOR.

FIGURE 2-47 Expanding a NOR

SELF-CHECK 4

1. How can a NAND be used as an inverter?

2. How can a NOR be used as an inverter?

3. How can a NOR be expanded?

4. How can an AND be expanded?

5. How can an OR be expanded?

6. How can a NAND be expanded?

 2.19 PROGRAMMING A PLD (Optional)

Suppose you had all the AND, NAND, OR, NOR gates, and inverters that you needed on one IC, and suppose you could connect them however you needed to for a specific application and could change the configuration as needed. This is the idea behind a **programmable logic device (PLD)**. The internal connections are established via a program that is downloaded to the PLD. The PLD retains its program and performs those functions until it is reprogrammed. So, the PLD can replace numerous ICs containing the basic gates.

Here we will use a programming language called Very High-speed Hardware Description Language (VHDL) to configure the gates within a PLD in a particular manner. We will program the PLD to implement these functions:

Two-input AND	$v = a\,b$
Three-input OR	$w = a + b + c$
Four-input NAND	$x = \overline{c\,d\,e\,f}$
Five-input NOR	$y = \overline{a + b + d + e + f}$
Inverter	$z = \overline{a}$

A VHDL program contains an **ENTITY** statement in which all the input and output signals (ports) are identified. We will name this program and entity "gates".

```
ENTITY gates IS
  PORT(a,b,c,d,e,f:IN BIT;
       v,w,x,y,z:OUT BIT);
END gates;
```

Notice the structure of the ENTITY statement:

- The entity declaration has a BEGIN and an END.

- The entity name "gates" appears in the END statement also.

- After PORT is an open parenthesis; the closing parenthesis is after OUT BIT and before the semicolon.

- a,b,c,d,e,f are identified as input signals of the BIT type. This means that those signals can have the bit values of "0" or "1."

- v,w,x,y,z are identified as the output signals of the BIT type.

- The input- and output-signal names are entered in lowercase letters, as is the name of the entity "gates." However, VHDL is not case-sensitive.

- There is a semicolon after the inputs or outputs are identified, and after the END statement.

Following the ENTITY statement is the **ARCHITECTURE** body. It declares the relationships between the inputs and outputs. The architecture statement uses the same name as the VHDL file, the ENTITY statement, and the folder containing the project—in this case **gates**.

```
ARCHITECTURE a OF gates IS
BEGIN
    v <= a AND b;
    w <= a OR b OR c;
    x <= NOT(c AND d AND e AND f);
    y <= NOT(a OR b OR d OR e OR f);
    z <= NOT a ;
END a;
```

Notice the structure of the ARCHITECTURE statement:

- ARCHITECTURE has a BEGIN and an END (ENTITY did not have a BEGIN).

- The name of the ARCHITECTURE appears in the END statement also. In this case "a." "ARCHITECTURE a" ends with "END a".

- Signals are defined with <=.

- Each equation ends with a semicolon.

- The END statement ends with a semicolon.

- Parentheses are needed in the definition of x and y to complement the whole expression.

The VHDL program looks like this:

```
ENTITY gates IS
    PORT(a,b,c,d,e,f:IN BIT;
        v,w,x,y,z:OUT BIT);
END gates;
ARCHITECTURE a OF gates IS
BEGIN
    v <= a AND b;
    w <= a OR b OR c;
    x <= NOT (c AND d AND e AND f);
    y <= NOT (a OR b OR d OR e OR f);
    z <= NOT a ;
END a;
```

The following procedure demonstrates how to use QuartusII® development system from Altera in order to compile, simulate, and finally program a PLD to implement this program in hardware. It will take some time to work through all the steps, but it is time well invested. In the following chapters we will build on this framework to implement a wide variety of circuits.

Each step begins with a statement describing **what** to do. The following indented text explains **how** to do it.

Step 1. Create a new folder within your Altera folder. Name it **gates**. This folder will contain all the files generated in the development of this project.

- Run **My Computer**.

- Navigate to your Altera folder.

- Right-click and select **New/Folder**.

- In the "New Folder" box, type **gates** and then press **Enter** on the keypad.

Step 2. Run Quartus II.exe.

Quartus II.exe can be found in the Altera /Quartus42sp1/bin folder. Double-click on quartus.exe.

Step 3. Run the **New Project Wizard**.

- Click on **File/New Project Wizard**.

- On the introduction page click **Next**.

- On page 1 of 5, in the "working directory" window, enter **C:/altera/gates**. Press the "Tab" key to enter the middle window. For the name of the project, type **gates**, and then click **Next**.

- On page 2, click **Next**.

- On page 3, choose the device to be programmed. For example, on the RSR PLDT-2 trainer, choose the Max 7000S Family. Scroll down to EPM7128SLC84-15 and click **Next**.

- On page 4, click **Next**.

- On page 5, click **Finish**.

Step 4. Create a new VHDL file.

- Click **File/New**. Scroll down to "VHDL file" and click **OK**.

- A new file will open named Vhdl1.vhd. Change its name to **gates.vhd**.

  ○ Click **File/Save As**. Type **gates** if necessary, and click **save**.

  ○ The name of your new file should appear in the upper left hand corner as gates.vhd.

Step 5. Type in your VHDL text.

- Declare your input and output signals (PORTS) in the ENTITY statement.

- Describe your circuit with an ARCHITECTURE statement. Save often.

Step 6. Compile your file.

- Save your VHDL file.

- Click the ▶ button on the toolbar or click **Processing/Start Compilation**.

- As your file compiles, messages will be printed out at the bottom of the screen in the status window. When the compiler finishes, a report will appear stating the number of errors and number of warnings.

- Click **OK**. If errors have occurred, click in the message window. Use the arrow keys to scroll through the messages. Double-click on a message that includes a line number; your VHDL program will open with that line highlighted.

- Correct your code and recompile until there are no errors. A missing semicolon at the end of a statement or a missing parenthesis can cause multiple errors to occur. You might have to correct and recompile your file many times before all errors are eliminated. Be patient and keep at it.

- The output of the compiler is a programmer object file (.pof); in this case, gates.pof.

Steps 7, 8, 9, and 10 involve creating a vector waveform file (.vwf) to simulate your circuit before downloading it to the PLD. Creating and using these waveforms will make more sense after studying waveform analysis in Chapter 3. Steps 7–10 will be described in detail at the end of Chapter 3. Now we will skip to Step 11.

Step 11. Assign input and output signals to pins on the PLD.

- Click **Assignments/Timing Closure Floorplan**. Then click **View/Package Top**. A top view of the PLC IC is shown.

- To zoom in or out on the top view, click the "zoom" button (looks like a magnifying glass) on the vertical toolbar on the left. Place the cursor over the area to be magnified. A left-click on the mouse will zoom in; a right-click will zoom out.

- Pins which have been assigned by the fitter function of Quartus appear as dark green circles. As we change pin assignments to fit our external hardware, our pin placements will appear in dark purple. (Click **View/Pin Legend** to see the code used in the top view.) Above the top view of the PLC is a chart listing all the unassigned pins. Below the top view is a chart listing all the assigned pins. Each signal in the top chart must be assigned to a pin on the IC. On the RSR PLDT-2 trainer, the 8 switches of S1 (8-bit DIP switch) are wired to jumper switch HD1. When jumpers are installed on HD1, these switches are connected to pins 34, 33, 36, 35, 37, 40, 39, and 41 of the PLD. We will use six of these switches as input signals **a** through **f**. Pins 44, 45, 46, 48, 49, 50, 51, and 52 are wired to jumper switch HD2. When jumpers are installed on HD2, these outputs are connected to red LEDs 1 through 8. We will use five of these LEDs to monitor output signals **v, w, x, y**, and **z**.

- To change from "zoom" to "select," click the select arrow on the vertical toolbar on the left.

- Highlight (left-click) signal **a** in the unassigned chart. Click on signal **a** again and a "drag" message appears with the pointer. Drag signal **a** to the desired pin on the IC outline. The "drag" message changes to "+." The signal is placed. The pin turns dark purple and signal **a** appears in the "Assigned Pin" table. Assign input signals **a** through **f** to the pins associated with S1 and HD1 (listed above).

- Place the select arrow over one of the pins and the name of the signal assigned and the pin number will appear. The pins you have assigned will be labeled "user placed."

- Assign output signals **v** through **z** to five of the output pins listed above.

- If you change your mind during the placement process, click and drag the signal from the "Assigned Pin" chart to the new pin. The "Assigned Pin" chart will update to reflect the change.

- Recompile the program.

- View the resulting pin placements. Notice that the pins we selected have changed from "user placed" to "fitter placed."

- Record the pins assigned to each input signal and each output signal for later use.

Step 12.   Program the PLD.

- Click **Tools/Programmer**. The programming window opens. The currently selected hardware is listed along the top of the window. For the RSR trainer we need the Byte-BlasterMV [LPT1] hardware. If the hardware listed is not correct for your situation, click **Hardware Setup**. From that window select the hardware to be used. If it does not appear, try the **Add Hardware** button. If your hardware is not listed, a driver may have to be installed. Driver software is found in the **Altera\Drivers\i386** folder.

- The name of the compiled program to be downloaded to the PLD should be listed in the File column; in this case, **gates.pof**.

- Verify that the IC listed in the Device column is the device to be used.

- Click **Program/Configure**. This tells the programmer function to ship the gates.pof file down to the PLD.

- Click **Verify**. After programming is complete, the program residing in the PLD will be checked against the .pof file.

- Click **Start**.

- A message appears when programming is complete.

- **Caution:** If you are using the PLDT-2 board from RSR® Electronics, remove all eight jumpers from HDI before programming the PLD.

Step 13.   Test the hardware.

- Use six switches to represent inputs **a, b, c, d, e**, and **f**. Connect the switches to the input pins shown in the "Package Top" view in step 11.

- Use five LEDs to display the output values of **v, w, x, y**, and **z**. Connect the LEDs to the output pins shown in the "Package Top" view in step 11.

- Vary the switches to thoroughly test the PLD.

Congratulations! You're done!

2.20 TROUBLESHOOTING GATES

Many of the ideas presented in Chapter 1 about troubleshooting adder ICs apply to the basic gate ICs as well and to ICs in general. If positive voltage supply connections (V_{CC} or V_{DD}) and ground connections (ground or V_{EE}) are not made correctly, the IC will not function. If your circuit is faulty, check the power connections first. Use a voltmeter or oscilloscope to check for +5 volts (or other supply voltage for CMOS) right on the pin of the IC itself. Measure the ground pin voltage also to make sure it is a solid zero and not floating. It is a waste of time to check other voltages in the circuit if the supply pins are not correct.

This chapter presented integrated circuits that contain multiple gates. Each gate is independent and can be used in a different part of the circuit. When a gate IC fails, it can affect all of the gates, some of the gates, or just an individual gate.

An IC can fail when the wire that connects the external pin of the IC to the internal circuitry on the silicon chip breaks or is missing. This causes an "open." If the open is between an input or output pin of one of the gates and the internal chip, that gate will be affected but the other gates can function properly. If the open is on the power supply input, all gates will be affected.

When there is an open between the input pin and the internal chip, that input to the internal circuitry is not connected to anything; it is floating. A TTL gate treats the floating input as a HIGH. On a two-input AND gate, the open (floating) input enables the gate. The data on the other input can pass through to the output unaltered. On an OR gate the floating input causes the gate to be inhibited with the output locked HIGH. An open on an input to a CMOS gate causes that input to drift. It can act as a LOW and then drift to a HIGH state. Its effect on the gate is unpredictable.

An open on the input of a gate is fairly easy to detect by using the enable/inhibit ideas discussed in this chapter. For example, to test a NAND gate, enable the gate by placing a HIGH on one input. As the other input is pulsed, the output should respond by inverting the pulse. (When a NAND is enabled with a HIGH, the data on the other input passes through the gate inverted.)

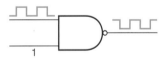

The pulse can be applied with a pulser, a signal generator, or simply by connecting the data input to V_{CC} and then to ground. If the output is not connected to a load (another circuit, an LED circuit, something) and does not respond to the pulse, the gate is bad and must be replaced. If there is a load connected to the output, the load circuit may be loading down the output and keeping it from responding to changes in the input. That possibility must be considered.

An open can also occur externally to the IC due to a broken conductor or a cold solder joint. If an input signal does not make it to the IC due to an open, the pin of the IC is floating. On a TTL gate the open input measures about 1.6 V and the floating input is taken as a HIGH. On a CMOS gate the floating input can assume any value and can drift. Unused inputs on CMOS gates being used in the circuit must be tied to V_{DD} (positive supply voltage) or V_{SS} (ground). Note that Multisim treats open inputs as zeroes, unlike TTL or CMOS ICs.

Those same connecting wires can connect two or more pins together inside the IC. This causes a "short." An input or output pin can be shorted to the supply voltage V_{CC} or to ground. Shorts are often more catastrophic than opens since heavy currents can flow. Components get hot, circuit boards get scorched, fuses blow. Shorts can sometimes be located by using your sense of smell or touch to detect the hot area. External shorts can be caused by solder blobs connecting traces together or by small pieces of wire lying across the circuit. A visual inspection can often identify these problems before damage is done.

In Part 3 of Lab 2B you will be asked to troubleshoot several gates. Use the enable/inhibit ideas discussed in this section.

DIGITAL APPLICATION

Gate Symbols

The circuit given below was designed by Zilog to interface the Z180 microprocessor unit (MPU) to a Z8530 serial communications controller (SCC) operating at 10 MHz. Note that the schematic uses both the standard and inverted logic symbols. Find both symbols for the inverter, the two-input NOR, and the three-input NOR. One input of a two-input NOR is grounded. Is that gate enabled or inhibited? Does the data on the other input pass inverted or unaltered?

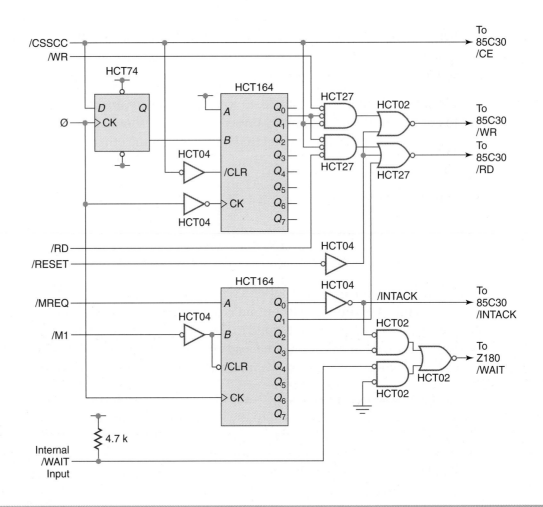

SUMMARY

- Gates are used to combine signals in specific ways.

- Gates are also used to control the flow of data from input to output.

- When a gate is inhibited, data cannot pass through.

- When a gate is enabled, data can pass through.

- Each gate responds differently to enable/inhibit signals, as summarized in the chart on page 76.

- NANDs and NORs can be wired as inverters.

- An AND can be expanded with an AND.

- A NAND can be expanded with an AND.

- An OR can be expanded with an OR.

- A NOR can be expanded with an OR.

- A break in the path of a signal is called an open.

- When an open occurs between an input pin and the internal chip, the input signal is ignored. A TTL gate will process the open condition as a HIGH. A CMOS gate can function as if the open is drifting or oscillating between a HIGH and a LOW.

- If a gate is being used in a circuit, any of its unused inputs should be tied HIGH, LOW, or to another input on the same gate.

- When an input of a functioning TTL gate is left unconnected, the floating input measures about 1.3 V and is processed as a HIGH.

- When an input of a functioning CMOS gate is left unconnected, the floating input can drift or oscillate between a HIGH and a LOW.

- A PLD is an integrated circuit that contains many logic gates. A program is downloaded to the PLD to configure its internal connections and cause it to function in a particular manner. It can be reprogrammed as needed.

- VHDL is a powerful language that is used to program PLDs.

- A VHDL program contains an ENTITY statement in which the input and output signals are declared.

- A VHDL program contains an ARCHITECTURE statement that defines the relationships between the input and output signals.

- The PLD programming process includes specifying the circuit needed, writing the VHDL (.vhd) program, compiling the program, simulating the circuit (not covered in this chapter), and programming the PLD.

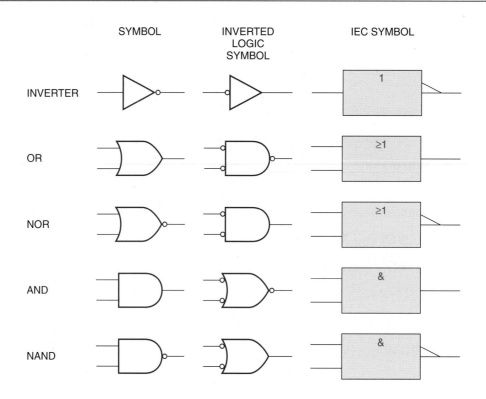

	SYMBOL	INVERTED LOGIC SYMBOL	IEC SYMBOL
INVERTER			1
OR			≥1
NOR			≥1
AND			&
NAND			&

ENABLE/INHIBIT SUMMARY

Gate	Control Input	Gate Condition	Output
AND	0	Inhibit	0
	1	Enable	Data
NAND	0	Inhibit	1
	1	Enable	$\overline{\text{Data}}$
OR	0	Enable	Data
	1	Inhibit	1
NOR	0	Enable	$\overline{\text{Data}}$
	1	Inhibit	0

Note: Data—Data passes through unaltered.
$\overline{\text{Data}}$—Data passes through inverted.

QUESTIONS AND PROBLEMS

1. Draw the symbol for each of the following gates, label the inputs, and write the Boolean expression for the output.

    a. Inverter   **d.** AND
    b. OR   **e.** NAND
    c. NOR

2. For each gate, draw the equivalent logic symbol and write the Boolean expression for the output.

3. Write the truth table for each two-input gate.

4. Write the truth table for three-input AND, NAND, OR, and NOR gates.

5. For a two-input AND gate

    a. Draw the symbol and write the Boolean expression for the output.
    b. Draw the inverted logic symbol and write the Boolean expression for the output.
    c. Write the truth table and indicate the line that represents the unique state.

6. Repeat problem 5 for a two-input NAND.

7. Repeat problem 5 for a two-input OR.

8. Repeat problem 5 for a two-input NOR.

9. Predict the output of each gate.

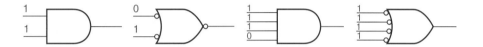

10. Predict the output of each gate.

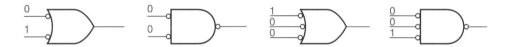

11. Predict the output of each gate.

12. Predict the output of each gate.

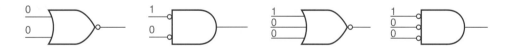

13. Predict the output of each gate.

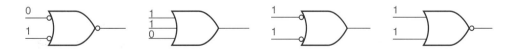

14. Predict the output of each gate.

15. A 0 on the control input of a NOR gate (enables, inhibits) the gate.

16. When a NAND gate is enabled, the data passes through (unaltered, inverted).

17. How do you inhibit an OR gate?

18. How do you enable an AND gate?

19. When an AND gate is inhibited, the output is a (0, 1).

20. When a NOR gate is inhibited, what is the state of the output?

21. When a NOR gate is enabled, the data passes through (unaltered, inverted).

22. How do you inhibit a NOR gate?

23. How do you enable a NAND gate?

24. When a NAND gate is inhibited, what is the state of the output?

25. When an OR gate is inhibited, what is the state of the output?

26. When an OR gate is enabled, the data passes through (unaltered, inverted).

27. When an AND gate is enabled, data passes through (inverted, unaltered).

28. Predict the output of each gate.

29. Predict the output of each gate.

30. Predict the output of each gate.

31. Predict the output of each gate.

32. Wire two gates from a two-input NAND IC to make an AND gate and show the pin numbers.

33. Wire a two-input NOR gate to act as an inverter and show the pin numbers.

34. Wire two gates from a two-input NOR IC to make an OR gate and show the pin numbers.

35. Wire gates from a two-input NAND IC to make a three-input NAND gate (expand a NAND) and show the pin numbers.

36. Wire gates from a two-input AND IC to make a three-input AND and show the pin numbers.

37. What is the function of each of the following ICs?

   a. 7427
   b. 4025
   c. 74C20
   d. 7410
   e. 4081
   f. 4069

38. How does the IEC specify the following:

   a. Active LOW input
   b. Active HIGH input
   c. Inverter function
   d. OR function
   e. AND function
   f. NOR function
   g. NAND function

39. Draw the IEC symbol for each of the following integrated circuits:

   a. Eight-input NAND gate—7430
   b. Dual four-input AND gate—4082
   c. Triple three-input NAND gate—7410
   d. Dual four-input OR gate—4072
   e. Triple three-input NOR gate—4025

40. If two active low signals $\overline{\text{IO REQUEST}}$ and $\overline{\text{WRITE}}$ from a processor are LOW at the same time, an active HIGH signal called IO WRITE is produced. What gate should be used? Draw the gate and label the inputs and output. Write a Boolean expression for the output in terms of the input.

41. If three signals, C1, D1, and E4, are all HIGH at the same time, an active LOW output called $\overline{\text{F6}}$ is produced. What gate should be used? Draw the gate and label the inputs and output. Write the Boolean expression for the output in terms of the inputs.

42. If one or more of three inputs, M2, N1, or 06, goes LOW, an alarm signal, STOP, goes HIGH. What gate should be used? Draw the gate and label the inputs and output. Write the Boolean expression for the output in terms of the inputs.

43. If pump motor is M1 is running, signal M1 is HIGH. Likewise, M2 indicates that pump motor M2 is running. If the water in the reservoir rises to within four feet from the top, signal TOOHIGH goes HIGH. If the water in the reservoir falls to a depth of one foot or less, signal TOOLOW goes HIGH. If either of the following conditions occurs, produce an active low signal called ALARM that will turn on a red lamp.

   a. The water level is too high and both motors are on.
   b. The water level is too low and either pump is still running. (You don't want to burn up the pump motors.) Draw the required logic circuit. Label the inputs and output. Write the Boolean expression for the output in terms of the inputs.

44. Discuss the use of the enable/inhibit concept in testing a two-input NAND gate.

45. Discuss the use of the enable/inhibit concept in testing a two-input AND gate.

46. Discuss the use of the enable/inhibit concept in testing a two-input NOR gate.

47. Discuss the use of the enable/inhibit concept in testing a two-input OR gate.

48. If an input to a TTL OR gate is left floating, what will be the state of the output?

49. If an input to a TTL NOR gate is left floating, what will be the state of the output?

50. If an input to a TTL AND gate is left floating, what will be the state of the output?

51. If an input to a TTL NAND gate is left floating, what will be the state of the output?

52. Draw the circuits described by this VHDL program.

```
ENTITY logic IS
      PORT(a,b,c:IN BIT;
           x,y,z:OUT BIT);
END logic;
ARCHITECTURE a OF logic IS
BEGIN
      x <=  a AND b;
      y <=  NOT (a OR b OR c);
      z <=  NOT c ;
END a;
```

53. Draw the circuits described by this VHDL program.

```
ENTITY chap2 IS
      PORT(rd,wr,memreq : IN BIT;
           memrd,memwr : OUT BIT);
END chap2;
ARCHITECTURE a OF chap2 IS
BEGIN
      memrd <=  memreq AND rd;
      memwr <=  memreq AND wr;
END a;
```

54. Draw the circuits described by this VHDL program.

```
ENTITY logic IS
      PORT(a,b,c,d,e,f:IN BIT;
           w,x,y,z:OUT BIT);
END logic;
ARCHITECTURE a OF logic IS
BEGIN
      w <= NOT (a AND c AND d AND f);
      x <= a AND b AND f;
      y <= NOT (a OR b OR c OR e OR f);
      z <= NOT d;
END a;
```

55. This VHDL program contains four errors. Find and correct them.

```
ENTITY nand IS
BEGIN
     PORT(a,b,c:IN BIT;
          x,y:OUT BIT);
END nand;
ARCHITECTURE nand OF chap2 IS
BEGIN
     x <=  a AND b;
     y <=  NOT (a OR b OR c);
     z <=  NOT c ;
END nand;
```

56. This VHDL program contains four errors. Find and correct them.

```
ENTITY and IS;
     PORT(a,b,c:IN BIT;
          x,y,z:OUT BIT;
END and;
ARCHITECTURE logic OF and IS
BEGIN
     x = a AND b;
     y = NOT (a OR b OR c);
     z =  NOT d;
END logic;
```

57. Write a VHDL program that will implement these functions:

X is A and D and E NORed together.

Y is A and E ANDed together.

Z is the complement of A.

58. Write a VHDL program that will implement these functions:

$x = \overline{a \cdot b}$

$y = c + d$

$z = \overline{a + d}$

$w = a \cdot b \cdot c \cdot d \cdot e \cdot f \cdot g \cdot h$

$v = \overline{f}$

59. Write a VHDL program that will implement these functions:

x is a, b, c, d, e, and f NANDed together.

y is a, c, d, and f ORed together.

Z is the complement of c.

OBJECTIVES

After completing this lab, you should be able to:

- Determine the truth table of a gate.
- Use each gate in the enable/inhibit mode.
- Use a NAND as an inverter.
- Use a NOR as an inverter.
- Expand a NAND.
- Expand a NOR.

COMPONENTS NEEDED

1	7400 IC
1	7402 IC
1	7404 IC
1	7408 IC
1	7411 IC
1	7432 IC
1	4001 IC
1	4011 IC
1	4069 IC
1	4071 IC
1	4081 IC
1	LED
1	330-Ω resistor

PREPARATION

In Lab 2A both CMOS and TTL gates will be used. On TTL ICs, an input left floating (disconnected) is usually taken as a 1. This is not true in CMOS. Due to the extremely high input impedance of a CMOS gate, unused inputs can drift between a 1 level and a 0 level. Tie all unused inputs on the gates to be used to the power supply voltage or ground. Supply $+5$ V to V_{DD} and connect V_{SS} to ground.

For CMOS Operating at +5 V:

1. A legitimate 1 input can range from 3.5 V to 5 V. A legitimate 0 input can range from 0 V to 1.5 V.

2. With inputs maintained within these ranges, the outputs should remain within 0.05 V of the supply levels. A 0 output should not rise above 0.05 V and a 1 output should not fall below 4.95 V.

For TTL ICs:

1. A legitimate 1 input can range from 2.0 V to 5 V, and a legitimate 0 input can range from 0 V to 0.8 V.

2. A legitimate 1 output can range from 2.4 V to 5 V, and a legitimate 0 output can range from 0 V to 0.4 V.

SAFETY ADVICE

PRECAUTIONS FOR HANDLING CMOS ICs   Care must be taken in the handling of CMOS ICs since they can be destroyed from excessive static build-up between pins. These guidelines should be followed:

1. Store CMOS ICs in antistatic tubes or in black conductive foam. Never push CMOS ICs into Styrofoam. They can be wrapped in aluminum foil.

2. In low humidity environments where static build-up is a problem, avoid touching the pins of a CMOS IC when they are removed from storage unless precautions have been taken to bleed off the static charge. Conductive wrist straps connected through a resistor to ground is one method used.

3. Apply dc voltage to the CMOS circuit before signals are applied.

4. Remove signal sources before the dc supply is switched off.

5. Switch off supply voltages before inserting or removing CMOS devices from a circuit.

In Part 1 of this lab, static voltage levels (unchanging 1s or 0s) will be used to verify the truth table of a variety of gates.

In Part 2 of this lab, the square wave generator on the trainer will be used to generate the signal for the data input of the gate. An oscilloscope will be used to compare the data input signal and the output signal. Use both channels of your scope, synchronize on the data input signal, and use dc coupling so that the 1 and 0 levels can be detected.

Part 3 of this lab calls for logic diagrams to be drawn for various configurations. A logic diagram shows the symbol for the gates used, the pin numbers, and the IC numbers.

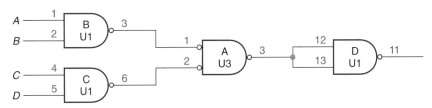

If an accurate logic diagram is drawn before connecting the circuit, it can be used as a guide in troubleshooting. Since the gates are independent and can be used in different parts of a circuit, they are often identified with a letter, A, B, C, etc., and IC number, U1, U2, etc. The three NAND gates are from the same IC U1 and the OR gate is from IC U3.

Review the lab safety rules under SAFETY ADVICE in the PREPARATION section of Lab 1A, Chapter 1.

PROCEDURE

PART 1

To determine the truth table of a gate:

1. Connect the inputs of the gate to the switches provided on a trainer or hook them directly to the $+5$ V supply or ground as required.

2. Connect the outputs to the LED monitors on a trainer or hook them to an LED and current-limiting resistor of approximately 330 Ω.

3. Write the input portion of the truth table by counting in binary with one bit for each input. For example,

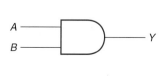

Inputs		Output
B	A	Y
0	0	
0	1	
1	0	
1	1	

4. Determine the output of the truth table by supplying the inputs called for in each line of the truth table. Use this procedure to write the truth table for a gate in a 7408, 7411, 7432, 7404, 7400, 7402, 4001, 4069, 4071, 4081, and 4011.

PART 2

For one of the gates on each of the ICs 7400, 4001, 4071, and 7408, verify the enable/inhibit operation as follows:

1. Use the square wave generator on your trainer as a source for the data input on your gate. Use the 10 kHz setting.

2. Monitor the data input and output with channel 1 and 2 of your scope. Use dc coupling so that 1 and 0 levels can be detected. Sync on the data input signal.

3. Supply a 1 or 0 to the control input to enable or inhibit the gate. Observe the relationships between input and output on the scope.

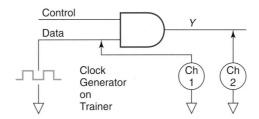

4. Summarize the operation with a truth table and written description of the operation.

PART 3

In each of the following, draw a logic diagram, including pin numbers and IC numbers, of the circuit. Then connect the circuit and verify the operations.

1. Use a NAND as an inverter.

2. Use a NOR as an inverter.

3. Expand a two-input NAND into a three-input NAND.

4. Expand a two-input NOR into a three-input NOR.

5. Using one two-input NAND gate chip, wire a three-input NAND gate.

LAB 2B   GATES

OBJECTIVES

After completing this lab, you should be able to use Multisim to:

• Select the proper gate for a given situation.

• Design and construct a 3-bit binary-to-decimal decoder.

• Troubleshoot basic gate ICs.

PREPARATION

In Part 1 of this lab you are given several statements. Study each statement and determine which gate is needed to build the circuit. Confirm your design using Multisim. In Part 2 you are asked to construct a 3-bit binary to decimal decoder in Multisim. Your circuit will have three inputs, C, B, and A, and eight outputs, 0 through 7. Your circuit will cause one of the outputs to go HIGH depending on the binary input. For example, if 100 is input, output 4 should go HIGH, and the other outputs should go LOW. In Part 3 you are asked to troubleshoot three ICs: a 7400 quad two-input NAND, a 7432 quad two-input OR, and a 4001 quad two-input NOR.

PROCEDURE

PART 1

Study each statement and predict which gate is needed to perform the function described. Draw the gate on paper and label the inputs and output. Then draw the gate in Multisim. Confirm its operation by connecting its inputs to +5 V for a 1 and to ground for a 0. Log your results.

1. A LOW-going signal called Memory Write ($\overline{\text{MWR}}$) is produced when two signals, Memory Request ($\overline{\text{MREQ}}$) and Write ($\overline{\text{WR}}$), both go LOW.

2. When Memory Read ($\overline{\text{MRD}}$) or Memory Write ($\overline{\text{MWR}}$) goes LOW, produce a LOW signal called Enable ($\overline{\text{EN}}$).

3. If IO Read ($\overline{\text{IORD}}$) and Device Select Pulse 1 ($\overline{\text{DS1}}$) go LOW, produce a HIGH-going signal to latch the data at input port 1 (INP1).

4. If the water level in a reservoir gets too close to the top, a pressure sensor produces an active high signal called High. When pump motor 1 is on, an active high signal is produced called M1; likewise, M2 indicates that pump motor 2 is running. If the water level in the reservoir gets too low, a pressure sensor produces an active high signal called Low. As the level gets low the pump motors should turn off automatically.

   a. If the water level is close to the top and both pumps are already on, produce an active low output signal called Alarm-Hi that will be used to turn on a red lamp.
   b. If the water level is too shallow and one or both of the pumps have not turned off, produce an active low signal called Alarm-Lo that will be used to turn on a red lamp (two gates required).

PART 2

Design and construct a 3-bit binary-to-decimal decoder. Use the red lamp indicators or LEDs and resistors for outputs. The Word Generator instrument can be used to cycle through the eight input combinations. Here is a procedure for using the Word Generator.

In the instrument toolbar to the right of the circuit window, click the **Word Generator** icon (eighth from top). Click in the circuit window to place the Word Generator. Connect the three least significant outputs (the ones on the right) to the inputs of our circuit. To initialize the Word Generator, double-click on the icon. Under "Display" click **Hex**. Click in the top box and enter **1**. Use the down arrow to move to the second box and enter **2**—and so on—to **7**. Click in the small square to the left of line **7** and check **set final position**. Click in the small square to the left of line **1** and check **set initial position**. Set the frequency to 20 Hertz. Under "Controls" click **Cycle**. The output should sequence from 0 to 7 and repeat. In the "Step" mode the word generator advances to the next line each time **Step** is clicked.

Describe your circuit and results.

PART 3

Download the three circuits named Files 2B-1.ms9 through 2B-3.ms9. Each circuit contains one or more faults. Keep a log of the steps performed to solve each circuit.

Circuit 2B-1.   Open circuit File 2B-1.ms9. If the circuit Description Box is not shown, click **View**, then click **Circuit Description Box**. Read the description of the circuit and procedure to follow. Record your results.

Circuit 2B-2.   Open circuit File 2B-2.ms9. If the circuit Description Box is not shown, click **View**, then click **Circuit Description Box**. Read the description of the circuit and procedure to follow. Sketch a logic diagram of the circuit. What is the function of this circuit? Record your results.

Circuit 2B-3.   Open circuit File 2B-3.ms9. If the circuit Description Box is not shown, click **View**, then click **Circuit Description Box**. Read the description of the circuit and procedure to follow. Sketch a logic diagram of the circuit. What is the function of this circuit? Record your results.

Waveforms and Boolean Algebra

OUTLINE

KEY TERMS

AND-OR-INVERT

Boolean algebra

cells

combinational
    (combinatorial) logic

complex programmable logic
    device (CPLD)

delayed clock

DeMorgan's theorems

field programmable gate
    array (FPGA)

generic array logic (GAL)

Karnaugh map

nonoverlapping clock

programmable array
    logic (PAL)

programmable logic
    device (PLD)

shift counter

simple programmable logic
    device (SPLD)

subcube

trailing edge

OBJECTIVES

After completing this chapter, you should be able to:

■ Predict the output waveforms for each of the gates, given input waveforms.

■ Combine signals from a shift counter and predict the outputs.

■ Select shift counter signals and gates to produce required outputs.

■ Develop the Boolean expression for the output of a combinational logic circuit.

■ Use Boolean algebra to reduce expressions to minimal terms.

■ Use DeMorgan's theorems to change the form of a Boolean expression.

■ Design and construct a logic circuit to implement a given truth table using Boolean algebra.

■ Design and construct a logic circuit to implement a given truth table using a Karnaugh map.

■ Design and construct a logic circuit to implement a given truth table using the logic converter instrument in Multisim.

■ Reduce Boolean expressions using a Karnaugh map.

■ Contrast SPLDs, CPLDs, and FPGAs.

■ Program a CPLD to implement combinational logic.

■ Troubleshoot logic circuits.

3.1 WAVEFORM ANALYSIS

In Chapter 2 you learned the truth tables for the basic gates. Once you know the truth table of a gate it becomes an easy chore to predict the output waveforms from given inputs. First determine the unique state of the gate. Then find all times at which those inputs occur. Graph the unique output for those times and its complement at all other times.

AND Gate

The unique state of the AND is "all 1s in, 1 out." Find those times at which all inputs are HIGH. The output is HIGH during those times and LOW at all other times.

EXAMPLE 3-1

Refer to Figure 3-1. If *A* and *B* are as shown, predict the output *Y*.

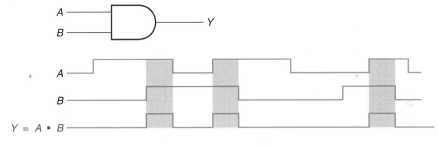

FIGURE 3-1

Solution Watch for areas where both *A* and *B* are HIGH. The output is HIGH during those times and LOW at all other times. The waveform for *Y* is shown in Figure 3-1.

NAND Gate

The unique state of the NAND is "all 1s in, 0 out." Find those times at which all inputs are HIGH. The output is LOW only during those times.

EXAMPLE 3-2

Refer to Figure 3-2. If *A*, *B*, and *C* are as shown, predict the output *Y*.

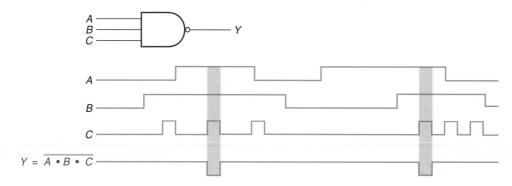

FIGURE 3-2

Solution   The areas shaded are those in which all three inputs are HIGH. At those times the output goes LOW. At all other times the output is HIGH. The waveform for *Y* is shown in Figure 3-2.

OR Gate

The unique state of an OR gate is "all 0s in, 0 out." Find those times at which all inputs are LOW. The output is LOW only during those times.

EXAMPLE 3-3

Refer to Figure 3-3. If *A*, *B*, and *C* are as shown, predict the output *Y*.

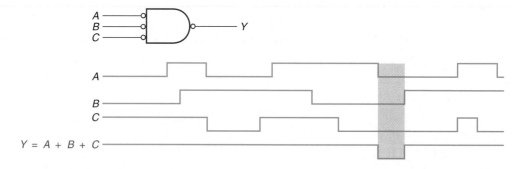

FIGURE 3-3

Solution   There is only one period of time at which all three inputs are LOW. At that time, the output goes LOW. At all other times the output is HIGH. The waveform for *Y* is shown in Figure 3-3.

NOR Gate

The unique state of the NOR is "all 0s in, 1 out." Find those times at which all inputs are LOW. The output is HIGH only during those times.

EXAMPLE 3-4

Refer to Figure 3-4. If *A*, *B*, and *C* vary as shown, predict the output *Y*.

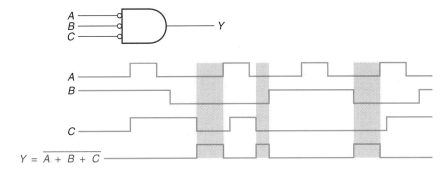

FIGURE 3-4

Solution   The areas shaded are those in which all three inputs are LOW. At those times the output goes HIGH. At all other times the output is LOW. The waveform for *Y* is shown in Figure 3-4.

3.2 DELAYED-CLOCK AND SHIFT-COUNTER WAVEFORMS

The input signals used in the last section come from a wide variety of sources. In later chapters, we will build circuits that produce a **delayed-** or **nonoverlapping-clock** system. The output waveforms are shown in Figure 3-5 as *CP* for clock pulse and *CP'* for the delayed or nonoverlapping clock. Also shown in Figure 3-5 are the output waveforms, *A*, $\overline{A}$, *B*, $\overline{B}$, *C*, $\overline{C}$, from a **shift counter**. The outputs from the shift counter change on the **trailing edge** (HIGH to LOW transition) of *CP*. Time 1 is at the trailing edge of clock pulse 1. Time 2 is at the trailing edge of pulse 2, and so on. These waveforms are continuous. After *CP* reaches 6 it starts again at 1. This circuit will also be constructed in a later chapter. A great variety of control signals can be generated from these waveforms. Refer to the graphs in Figure 3-5 often as you master the following eight examples. In the first four examples the input waveforms and gates are given, and the output waveforms are predicted. In the last four examples the desired output is given, and the task is to predict which gate and which input signals must be used to produce that output. Both types of problems should be mastered.

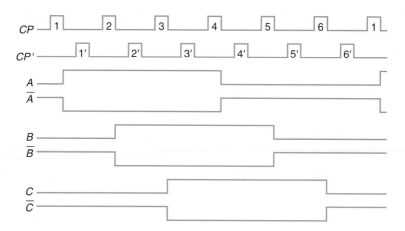

FIGURE 3-5 Delayed-clock and shift-counter waveforms

EXAMPLE 3-5

Suppose we AND *A* and *C* together. What does the output look like?

Solution   The unique state of an AND gate is "all 1s in, 1 out." *A* goes HIGH at 1 and enables the AND gate. When *C* goes HIGH at 3, the output goes HIGH. The output stays HIGH until 4, when *A* goes LOW and inhibits the output. In other words, *A* and *C* are both HIGH from 3 to 4, and the output can be represented as shown in Figure 3-6.

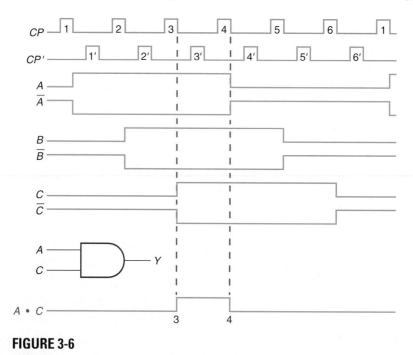

FIGURE 3-6

EXAMPLE 3-6

Suppose we AND $\overline{A}$ and *CP′* together. What does the output look like?

Solution $\overline{A}$ goes HIGH at time 4, which is after the 3′ pulse and before the 4′ pulse, and goes back LOW at time 1, which is after the 6′ pulse. $\overline{A}$ enables the gate during the 4′, 5′, and 6′ pulses, and they appear at the output Y. See Figure 3-7.

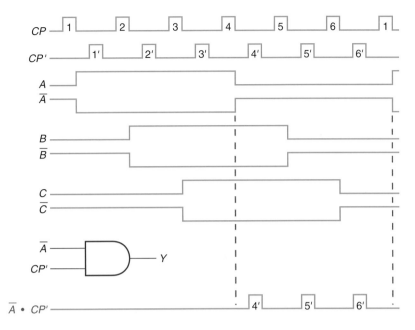

FIGURE 3-7

EXAMPLE 3-7

Now let's NOR B and $\overline{C}$ together. What does the output look like?

Solution The unique state of a NOR gate is "all 0s in, 1 out." $\overline{C}$ goes LOW at 3 and enables the gate. When B goes LOW at 5, both inputs are LOW and the output goes HIGH. The output stays HIGH until $\overline{C}$ goes HIGH at 6 and inhibits the output. In other words, B and $\overline{C}$ are both LOW from 5 to 6. The output can be represented as shown in Figure 3-8.

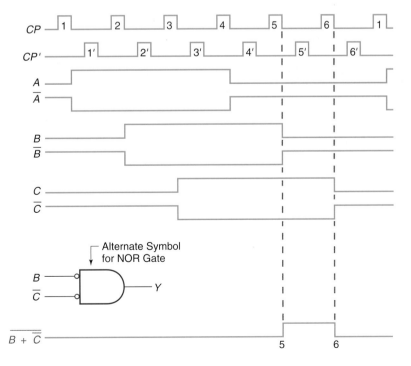

FIGURE 3-8

EXAMPLE 3-8

Suppose that we NAND three signals together: CP', $\overline{A}$, and $\overline{B}$.

Solution   The unique state of a NAND is "all 1s in, 0 out." First, let's find when $\overline{A}$ and $\overline{B}$ are both 1s. $\overline{A}$ goes HIGH at 4, $\overline{B}$ follows at 5. Both are HIGH from 5 until 1 when $\overline{A}$ goes LOW again. Between times 5 and 1, CP' goes HIGH in pulses 5' and 6'. During those times all three inputs are HIGH and the output is LOW. The output can be represented as shown in Figure 3-9.

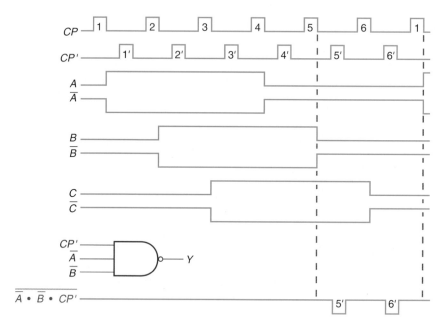

FIGURE 3-9

EXAMPLE 3-9

Find a combination that yields only the 6' pulse.

Solution   An AND gate yields a 1 out only when all inputs are 1. To get the 6' pulse out, one input must be CP'. The other two inputs must enable and inhibit the three-input gate at the proper times. For the 6' pulse to be enabled, one of the inputs must go HIGH at 6 to enable the gate. The other must already be HIGH at 6 to go back LOW at 1 to keep any other pulses from appearing at the output. $\overline{C}$ goes HIGH at 6, $\overline{A}$ goes back LOW at 1. The inputs are $\overline{A}$, $\overline{C}$, and CP'. See Figure 3-10.

EXAMPLE 3-10

Find a combination that will produce an output of 2' and 3'.

Solution   Using a 3-input AND gate, B goes HIGH at 2 to enable the gate and allow 2' to appear at the output. A goes LOW at 4 after 3' has appeared at the output and inhibits the gate so that 4' cannot pass. The other input must be CP'. See Figure 3-11.

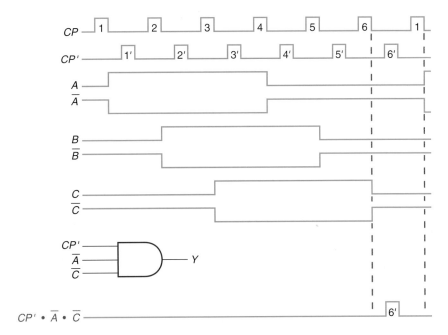

FIGURE 3-10

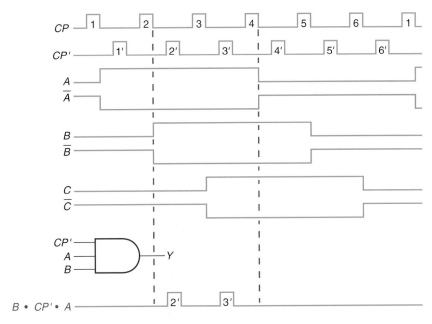

FIGURE 3-11

EXAMPLE 3-11

Find two combinations that will produce an output that is LOW from 3 to 5.

Solution 1 The unique state of an OR gate is "all 0s in, 0 out." Find one signal that goes LOW at time 3 to enable the OR and another signal that is already LOW and goes HIGH at time 5 to inhibit the gate. $\overline{C}$ goes LOW at 3, and $\overline{B}$ goes back HIGH at 5. $\overline{B} + \overline{C}$ solves the problem. See Figure 3-12.

Solution 2 The unique state of a NAND gate is "all 1s in, 0 out." Find one signal that goes HIGH at time 3 to enable the NAND and another that is already HIGH and goes LOW at time 5 to inhibit the gate. C goes HIGH at 3, and B goes LOW at 5. $\overline{B \cdot C}$ will solve the problem. See Figure 3-12.

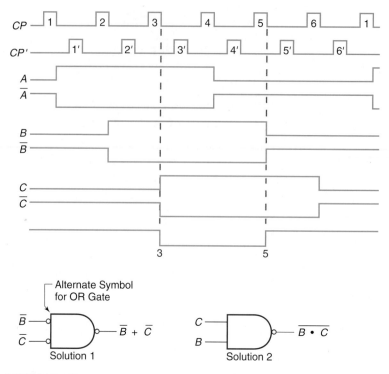

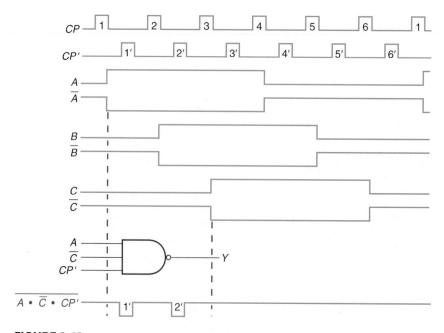

FIGURE 3-12

EXAMPLE 3-12

Find a combination that will yield LOW-going pulses at 1′ and 2′.

Solution   All 1s into a NAND produce a 0 out. Find a signal that goes HIGH before 1′ and one that is already HIGH and goes LOW after 2′. A and $\overline{C}$ satisfy these conditions. Put A, $\overline{C}$, and CP' into a NAND gate. See Figure 3-13.

FIGURE 3-13

3.3 COMBINATIONAL LOGIC

To produce a required output waveform, it is often necessary to use a combination of gates. For example, suppose we needed a control signal from the delayed-clock and shift-counter waveforms that consisted of the 2′ pulse and the 5′ pulse. It takes one 3-input AND gate to isolate the 2′ pulse and another 3-input AND to isolate the 5′ pulse. These two individual outputs must be ORed together to produce the desired output waveform. The output will be HIGH either during the 2′ pulse OR during the 5′ pulse. Combining gates to produce the required output is called **combinational logic** or **combinatorial logic**.

EXAMPLE 3-13

Combine the delayed-clock and shift-counter signals to produce a pulse on the output when either the 2′ pulse or the 5′ pulse occurs.

Solution

AND gate 1 produces a pulse when 2′ occurs.
AND gate 2 produces a pulse when 5′ occurs.

See Figure 3-14.

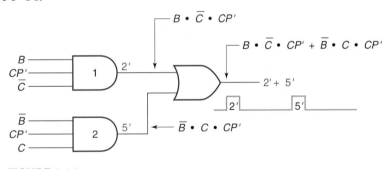

FIGURE 3-14

A Boolean expression can be written for the output of the combination of gates used in the last example. First write the Boolean expression for the output of each AND gate. Use those expressions as inputs into the following gate. The output of AND gate 1 is $B \cdot \overline{C} \cdot CP'$. The output of AND gate 2 is $\overline{B} \cdot C \cdot CP'$. The overall output of the circuit is $B \cdot \overline{C} \cdot CP' + \overline{B} \cdot C \cdot CP'$.

EXAMPLE 3-14

Write the Boolean expression for the output of the circuit in Figure 3-15.

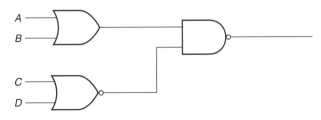

FIGURE 3-15

Solution   Write the Boolean expression for the output of the OR gate, $A + B$. Write the expression for the output of the NOR gate, $\overline{C + D}$. Use these two expressions as inputs into the NAND. The final expression is $\overline{(A + B) \cdot \overline{C + D}}$. See Figure 3-16.

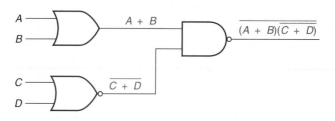

FIGURE 3-16

SELF-CHECK 1

1. Predict the output of the NAND gate in Figure 3-17.

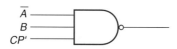

FIGURE 3-17

2. Choose a gate and inputs from the shift counter to produce an output that is LOW from time 2 to time 4.

3. Predict the output of the NOR gate in Figure 3-18.

FIGURE 3-18

3.4 BOOLEAN THEOREMS

We will examine two methods of developing the required logic diagram from a given truth table. The first method requires **Boolean algebra** and **DeMorgan's theorems** to reduce the expressions produced to lowest terms (minimal expressions). The second method is a variation of the first and uses a tool called the **Karnaugh map**.

In Boolean algebra ORed terms are commutative; the order in which they are written is not critical.

$$A + B = B + A$$

For example, $XY + Y$ is the same as $Y + XY$.

Likewise, ANDed terms are commutative; the order in which they are written is not critical.

$$AB = BA$$

For example, ZYX is the same as XZY or XYZ.

Boolean algebra is based on the following set of eleven fundamental theorems and the two DeMorgan theorems. Each is discussed or proved.

Theorem 1. $\overline{\overline{A}} = A$ (Refer to Figure 3-19)

A is either 0 or 1.
Case I: If $A = 0$, then $\overline{A} = 1$ and $\overline{\overline{A}} = 0$.
Case II: If $A = 1$, then $\overline{A} = 0$ and $\overline{\overline{A}} = 1$.
In either case, $A = \overline{\overline{A}}$.

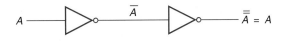

FIGURE 3-19

Theorem 2. $A \cdot 0 = 0$ (Refer to Figure 3-20)

The 0 inhibits the AND gate and the output will always be 0.

FIGURE 3-20

EXAMPLE 3-15

Using Boolean algebra, write an equivalent expression for $Y \cdot Z \cdot 0$.

Solution

$Y \cdot Z \cdot 0 = 0$ by Theorem 2

Theorem 3. $A + 0 = A$ (Refer to Figure 3-21)

The 0 input enables the gate.
Case I: If $A = 1$, the output is 1.
Case II: If $A = 0$, the output is 0.
The output is always the same as A.

FIGURE 3-21

EXAMPLE 3-16

Using Boolean algebra, write an equivalent expression for $Y + \overline{Z} + 0$.

Solution

$$Y + \overline{Z} + 0 = Y + \overline{Z} \quad \text{by Theorem 3}$$

Theorem 4. $A \cdot 1 = A$ (Refer to Figure 3-22)

The 1 input enables the gate.
Case I: If $A = 1$, the output is 1.
Case II: If $A = 0$, the output is 0.
The output is always the same as A.

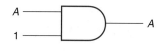

FIGURE 3-22

EXAMPLE 3-17

Using Boolean algebra, write an equivalent expression for $\overline{D} \cdot E \cdot 1$.

Solution

$$\overline{D} \cdot E \cdot 1 = \overline{D} \cdot E \quad \text{by Theorem 4}$$

Theorem 5. $A + 1 = 1$ (Refer to Figure 3-23)

The 1 input inhibits the gate and "locks up" the output at 1. The output does not respond to changes in A.

FIGURE 3-23

EXAMPLE 3-18

Using Boolean algebra, write an equivalent expression for $\overline{E} + H + N + 1$.

Solution

$$\overline{E} + H + N + 1 = 1 \quad \text{by Theorem 5}$$

Theorem 6. $A + A = A$ (Refer to Figure 3-24)

        Case I: If $A = 0$, then $0 + 0 = 0$.
        Case II: If $A = 1$, then $1 + 1 = 1$.
        In either case, the output follows A.

FIGURE 3-24

EXAMPLE 3-19

Using Boolean algebra, write an equivalent expression for $M \cdot \overline{N} + M \cdot \overline{N}$.

Solution

    $M\overline{N} + M\overline{N} = M\overline{N}$ by Theorem 6

Theorem 7. $A \cdot A = A$ (Refer to Figure 3-25)

        Case I: If $A = 0$, then two 0s into an AND yields 0 out.
        Case II: If $A = 1$, then two 1s in yields 1 out.
        In either case, the output is the same as the inputs.

FIGURE 3-25

EXAMPLE 3-20

Using Boolean algebra, write an equivalent expression for $C \cdot C \cdot \overline{D} \cdot E \cdot \overline{D}$.

Solution

    $C \cdot C \cdot \overline{D} \cdot E \cdot \overline{D} = C \cdot C \cdot \overline{D} \cdot \overline{D} \cdot E$ ANDed terms are commutative
    $C \cdot C \cdot \overline{D} \cdot \overline{D} \cdot E = C \cdot \overline{D} \cdot E$ by Theorem 7 (twice)

Theorem 8. $A + \overline{A} = 1$ (Refer to Figure 3-26)

        Case I: If $A = 1$, then the output is 1.
        Case II: If $A = 0$, then $\overline{A} = 1$ and the output is 1.
        The output is always 1.

FIGURE 3-26

EXAMPLE 3-21

Using Boolean algebra, write an equivalent expression for $\overline{AB} + AB$.

Solution

$\overline{AB} + AB = 1$ by Theorem 8

EXAMPLE 3-22

Using Boolean algebra, write an equivalent expression for $A\overline{B} + \overline{A}B$.

Solution

$A\overline{B}$ and $\overline{A}B$ are not the complements of each other, as shown by this truth table. Avoid the trap of reducing this expression to 1. It does not reduce.

A	B	$\overline{A}$	$\overline{B}$	$A\overline{B}$	$\overline{A}B$
0	0	1	1	0	0
0	1	1	0	0	1
1	0	0	1	1	0
1	1	0	0	0	0

Not complements of each other

EXAMPLE 3-23

Using Boolean algebra, write an equivalent expression for $X + Y + \overline{X}$.

Solution

$$X + Y + \overline{X} = X + \overline{X} + Y = 1 + Y \qquad \text{by Theorem 8}$$
$$= 1 \qquad \text{by Theorem 5}$$

Theorem 9. $A \cdot \overline{A} = 0$ (Refer to Figure 3-27)

Case I: If $A = 1$, then $\overline{A} = 0$ and the output is 0.
Case II: If $A = 0$, then the output is 0.
In either case, the output is 0.

FIGURE 3-27

EXAMPLE 3-24

Using Boolean algebra, write an equivalent expression for $A \cdot B \cdot \overline{D} \cdot \overline{A}$.

Solution

$$A \cdot B \cdot \overline{D} \cdot \overline{A} = A \cdot \overline{A} \cdot B \cdot \overline{D} = 0 \cdot B \cdot \overline{D} \qquad \text{by Theorem 9}$$
$$= 0 \qquad \text{by Theorem 2}$$

Theorem 10. $A \cdot B + A \cdot C = A(B + C)$ (Refer to Figure 3-28)

One way to prove this theorem is to examine its validity for all possible combinations of A, B, and C. An organized approach is to develop a truth table for each side of the equation and see if they are identical. To develop the truth table for a complex expression, start with each term. Write a column for $A \cdot B$, then $A \cdot C$, and finally "OR" those two columns together to produce the left side of the equation. To develop the right side, first write a column for $B + C$, then "AND" with A. The truth tables are identical, so the equation holds true. Notice the way that we "built up" the final expressions for each side.

A	B	C	$A \cdot B$	$A \cdot C$	$A \cdot B + A \cdot C$	$B + C$	$A(B+C)$ AND Function
0	0	0	0	0	0	0	0
0	0	1	0	0	0	1	0
0	1	0	0	0	0	1	0
0	1	1	0	0	0	1	0
1	0	0	0	0	0	0	0
1	0	1	0	1	1	1	1
1	1	0	1	0	1	1	1
1	1	1	1	1	1	1	1

$$A \cdot B + A \cdot C \qquad = \qquad A(B+C)$$

FIGURE 3-28

EXAMPLE 3-25

Using Boolean algebra, write an equivalent expression for $\overline{A}B + \overline{A}C$.

Solution

$$\overline{A}B + \overline{A}C = \overline{A}(B + C) \quad \text{by Theorem 10}$$

EXAMPLE 3-26

Using Boolean algebra, write an equivalent expression for $XYZ + X\overline{Y}Z$.

Solution

$$XYZ + X\overline{Y}Z = XZ(Y + \overline{Y}) \qquad \text{by Theorem 10}$$
$$= XZ(1) \qquad \text{by Theorem 8}$$
$$= XZ \qquad \text{by Theorem 4}$$

EXAMPLE 3-27

Using Boolean algebra, write an equivalent expression for $AB + AC + AD$.

Solution

$$AB + AC + AD = A(B + C + D) \quad \text{by Theorem 10}$$

Theorem 11. $A + \overline{A} \cdot B = A + B$ (Refer to Figure 3-29)

Once again using truth tables, the left side of the equation is shown to be identical to the right side.

A	B	$\overline{A}$	$\overline{A} \cdot B$	$A + \overline{A} \cdot B$	$A + B$
0	0	1	0	0	0
0	1	1	1	1	1
1	0	0	0	1	1
1	1	0	0	1	1

$$A + \overline{A} \cdot B = A + B$$

FIGURE 3-29

EXAMPLE 3-28

Using Boolean algebra, write an equivalent expression for $\overline{X} + XZ$.

Solution

$$\overline{X} + XZ = \overline{X} + Z \quad \text{by Theorem 11}$$

EXAMPLE 3-29

Using Boolean algebra, write an equivalent expression for $D + \overline{D}\,\overline{E}\,\overline{F}$.

Solution

$$D + \overline{D}\,\overline{E}\,\overline{F} = D + \overline{E}\,\overline{F} \quad \text{by Theorem 11}$$

EXAMPLE 3-30

Using Boolean algebra, reduce the expression $\overline{A}\,\overline{B}\,\overline{C} + \overline{A}\,B\,\overline{C} + A\,\overline{B}\,\overline{C}$.

Solution

$$
\begin{aligned}
\overline{A}\,\overline{B}\,\overline{C} + \overline{A}\,B\,\overline{C} + A\,\overline{B}\,\overline{C} &= \overline{A}\,\overline{C}\,(\overline{B} + B) + A\,\overline{B}\,\overline{C} &&\text{by Theorem 10} \\
&= \overline{A}\,\overline{C}\,(1) + A\,\overline{B}\,\overline{C} &&\text{by Theorem 8} \\
&= \overline{A}\,\overline{C} + A\,\overline{B}\,\overline{C} &&\text{by Theorem 4} \\
&= \overline{C}\,(\overline{A} + A\,\overline{B}) &&\text{by Theorem 10} \\
&= \overline{C}\,(\overline{A} + \overline{B}) &&\text{by Theorem 11}
\end{aligned}
$$

$$\overline{A}\,\overline{B}\,\overline{C} + \overline{A}\,B\,\overline{C} + A\,\overline{B}\,\overline{C} = \overline{C}\,(\overline{A} + \overline{B})$$

3.5 DEMORGAN'S THEOREMS

DeMorgan's theorems are as follows:

Theorem 1. $\overline{A \cdot B} = \overline{A} + \overline{B}$ (Refer to Figure 3-30)

The truth table shows that the left side of the equation is equal to the right side of the equation.

A	B	A · B	$\overline{A \cdot B}$	$\overline{A}$	$\overline{B}$	$\overline{A} + \overline{B}$
0	0	0	1	1	1	1
0	1	0	1	1	0	1
1	0	0	1	0	1	1
1	1	1	0	0	0	0

$$\overline{A \cdot B} \quad = \quad \overline{A} + \overline{B}$$

FIGURE 3-30

This theorem reinforces the fact that a NAND is the same as inverting the inputs into an OR. See Figure 3-31.

FIGURE 3-31

Theorem 2. $\overline{A + B} = \overline{A} \cdot \overline{B}$ (Refer to Figure 3-32)

The truth table shows that the left side of the equation is equal to the right side of the equation.

A	B	A + B	$\overline{A + B}$	$\overline{A}$	$\overline{B}$	$\overline{A} \cdot \overline{B}$
0	0	0	1	1	1	1
0	1	1	0	1	0	0
1	0	1	0	0	1	0
1	1	1	0	0	0	0

$$\overline{A + B} \quad = \quad \overline{A} \cdot \overline{B}$$

FIGURE 3-32

This theorem supports the fact that a NOR is the same as inverting the inputs into an AND. See Figure 3-33.

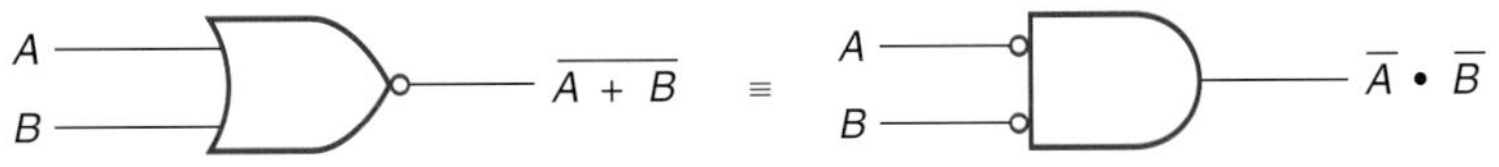

FIGURE 3-33

To apply DeMorgan's theorems successfully, you must be able to identify the terms of an expression. In the expression $AB + CDE + FG$, the three terms AB, CDE, and FG are ORed together. In the expression $(A + B)(C + D)$, the two terms $A + B$ and $C + D$ are ANDed together. In the expression $(A + B)(C + D) + EFG$, the two terms $(A + B)(C + D)$ and EFG are ORed together. In the expression ABC the three terms A, B, and C are ANDed together.

To implement DeMorgan's theorems, follow these three steps:

Step 1. Complement the entire expression.

Step 2. Change the function between each term.

Step 3. Complement each term.

EXAMPLE 3-31

Apply the three steps to $\overline{A \cdot B}$.

Solution

Step 1. $\overline{\overline{A \cdot B}} = A \cdot B$ Complement the entire expression.

Step 2. $A + B$ Change AND to OR.

Step 3. $\overline{A} + \overline{B}$ Complement each term.

$$\overline{A \cdot B} = \overline{A} + \overline{B}$$

EXAMPLE 3-32

Apply the three steps to $\overline{A} \cdot \overline{B}$.

Solution

Step 1. $\overline{\overline{A} \cdot \overline{B}}$ Complement the entire expression.

Step 2. $\overline{\overline{A} + \overline{B}}$ Change AND to OR.

Step 3. $\overline{\overline{\overline{A}} + \overline{\overline{B}}} = \overline{A + B}$ Complement each term.

$$\overline{A} \cdot \overline{B} = \overline{A + B}$$

EXAMPLE 3-33

Change the form of $A \cdot B \cdot C$ by using DeMorgan's theorems.

Solution

Step 1. $\overline{A \cdot B \cdot C}$ Complement the entire expression.

Step 2. $\overline{A + B + C}$ Change AND to OR.

Step 3. $\overline{\overline{A} + \overline{B} + \overline{C}}$ Complement each term.

$$A \cdot B \cdot C = \overline{\overline{A} + \overline{B} + \overline{C}}$$

EXAMPLE 3-34

Change the form of $ABC + \overline{A}\,\overline{B}\,\overline{C}$ by using DeMorgan's theorems and draw the logic diagram of each.

Solution

Step 1. $\overline{\overline{ABC + \overline{A}\overline{B}\overline{C}}}$ Complement the entire expression.

Step 2. $\overline{\overline{ABC} \cdot \overline{\overline{A}\overline{B}\overline{C}}}$ Change OR to AND.

Step 3. $\overline{\overline{ABC} \cdot \overline{\overline{A}\overline{B}\overline{C}}}$ Complement each term.

Now apply DeMorgan's theorems to each term, $\overline{ABC}$ and $\overline{\overline{A}\,\overline{B}\,\overline{C}}$, and substitute the equivalent expressions.

Step 1. $\overline{\overline{ABC}} = ABC$ Complement the entire expression.

Step 2. $A + B + C$ Change the ANDs to ORs.

Step 3. $\overline{A} + \overline{B} + \overline{C}$ Complement each term.

Step 1. $\overline{\overline{\overline{A}\,\overline{B}\,\overline{C}}} = \overline{A}\,\overline{B}\,\overline{C}$ Complement the entire expression.

Step 2. $\overline{A} + \overline{B} + \overline{C}$ Change the ANDs to ORs.

Step 3. $\overline{\overline{A}} + \overline{\overline{B}} + \overline{\overline{C}} = A + B + C$ Complement each term.

$\overline{\overline{ABC} \cdot \overline{\overline{A}\,\overline{B}\,\overline{C}}}$ becomes $\overline{(\overline{A} + \overline{B} + \overline{C}) \cdot (A + B + C)}$.

Figure 3-34A and Figure 3-34B show the original and modified circuits.

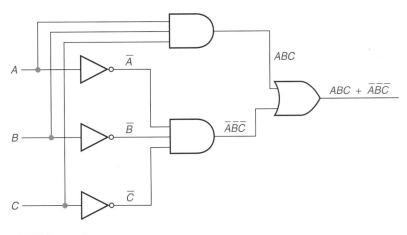

FIGURE 3-34A

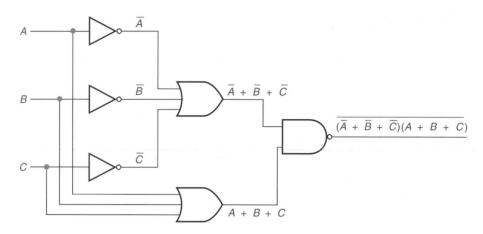

FIGURE 3-34B

To confirm that these two circuits are equivalent, develop a truth table for each circuit. The two truth tables should be identical.

C B A	ABC	$\overline{A}\,\overline{B}\,\overline{C}$	$ABC + \overline{A}\,\overline{B}\,\overline{C}$	$\overline{A} + \overline{B} + \overline{C}$	$A + B + C$	$\overline{(\overline{A} + \overline{B} + \overline{C})\,(A + B + C)}$
0 0 0	0	1	1	1	0	1
0 0 1	0	0	0	1	1	0
0 1 0	0	0	0	1	1	0
0 1 1	0	0	0	1	1	0
1 0 0	0	0	0	1	1	0
1 0 1	0	0	0	1	1	0
1 1 0	0	0	0	1	1	0
1 1 1	1	0	1	0	1	1

original circuit   equivalent circuit

The truth tables are identical: The circuits are equivalent. $ABC + \overline{A}\,\overline{B}\,\overline{C}$ is in "sum of products" form and is constructed from two three-input AND gates and a two-input OR gate. $(\overline{A} + \overline{B} + \overline{C})\,(A + B + C)$ is in "product of sums" form and is constructed from two three-input OR gates and a two-input NAND gate.

EXAMPLE 3-35

Use DeMorgan's theorems to reduce the circuit in Figure 3-35.

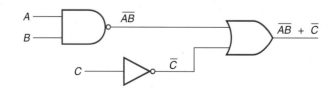

FIGURE 3-35

Solution Write the Boolean expression for the output $\overline{AB} + \overline{C}$.
Apply DeMorgan's theorem.

Step 1. $\overline{\overline{AB} + \overline{C}}$ Complement the entire expression.

Step 2. $\overline{\overline{AB} \cdot \overline{\overline{C}}}$ Change OR to AND.

Step 3. $\overline{\overline{\overline{AB}}\ \overline{\overline{C}}}$ Complement each term.

$$\overline{AB} + \overline{C} = \overline{ABC}$$

See Figure 3-36.

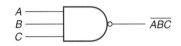

FIGURE 3-36

To confirm that these two circuits are equivalent, develop a truth table for each circuit. The two truth tables should be identical.

C B A	AB	$\overline{AB}$	$\overline{C}$	$\overline{AB} + \overline{C}$	A B C	$\overline{ABC}$
0 0 0	0	1	1	1	0	1
0 0 1	0	1	1	1	0	1
0 1 0	0	1	1	1	0	1
0 1 1	1	0	1	1	0	1
1 0 0	0	1	0	1	0	1
1 0 1	0	1	0	1	0	1
1 1 0	0	1	0	1	0	1
1 1 1	1	0	0	0	1	0

original circuit   equivalent circuit

The truth tables are identical: the circuits are equivalent.

 SELF-CHECK 2

Reduce using Boolean algebra.

1. $1 + \overline{B} + C$

2. $D \cdot \overline{C} \cdot 0$

3. $\overline{A} + B + A$

4. $A + \overline{A} B C$

5. $\overline{A} B C + \overline{A} B \overline{C}$

6. $\overline{C} B A + \overline{C} \overline{B} A + \overline{C} B \overline{A}$

Apply DeMorgan's theorems.

7. $A + \overline{C}$

8. $\overline{A} + \overline{B\,C}$

9. $\overline{A}\,\overline{(B + C)}$

 # **3.6** DESIGNING LOGIC CIRCUITS

Method 1—Boolean Algebra

The following examples demonstrate a method used to develop the Boolean expression for a logic diagram that will behave according to a given truth table. An equation is written from the truth table and then reduced using the Boolean algebra theorems.

EXAMPLE 3-36

Design a logic circuit that will behave according to the truth table shown in Figure 3-37.

Solution We want the output Y to be 1 when $A = 0$ and $B = 0$ (1st line) OR when $A = 0$ and $B = 1$ (3rd line). Working with the first line, if $A = 0$ then $\overline{A} = 1$ (if we don't have A then we have $\overline{A}$) and if $B = 0$ then $\overline{B} = 1$. We want Y to be 1 when we have $\overline{A}$ AND $\overline{B}$. The first line is represented by $\overline{A} \cdot \overline{B}$. Working with the third line, if $A = 0$ then $\overline{A} = 1$. We want Y to be 1 when we have $\overline{A}$ AND B. The third line is represented by $\overline{A} \cdot B$. The output Y is 1 when the first line is 1 or when the third line is 1. This results in the equation $Y = \overline{A} \cdot \overline{B} + \overline{A} \cdot B$. This equation can be reduced using the theorems.

Inputs		Output
B	A	Y
0	0	1
0	1	0
1	0	1
1	1	0

FIGURE 3-37

$$Y = \overline{A} \cdot \overline{B} + \overline{A} \cdot B$$
$$Y = \overline{A}\,(\overline{B} + B) \qquad \text{by Theorem 10}$$
$$Y = \overline{A}\,(1) \qquad \text{by Theorem 8}$$
$$Y = \overline{A} \qquad \text{by Theorem 4}$$

The desired output is the complement of A and can be produced by inverting the A input. See Figure 3-38.

$$A \longrightarrow \!\!\!\triangleright\!\!\circ\longrightarrow Y = \overline{A}$$

FIGURE 3-38

EXAMPLE 3-37

Given the truth table shown in Figure 3-39, design the logic circuit.

Solution We want the output Y to be a 1 when $C = 0$, $B = 1$, and $A = 1$ (4th line) OR when $C = 1$, $B = 1$, and $A = 0$ (7th line) OR when $C = 1$, $B = 1$, and $A = 1$ (8th line). Working with line 4, if $C = 0$ then $\overline{C} = 1$; also $B = 1$ and $A = 1$. We want Y to be 1 when we have $\overline{C}$ AND B AND A. Line 7 is represented by $C \cdot B \cdot \overline{A}$, and line 8 is represented by $C \cdot B \cdot A$. This results in the equation $Y = \overline{C}BA + CB\overline{A} + CBA$. This equation can be reduced using the theorems.

Inputs			Output
C	B	A	Y
0	0	0	0
0	0	1	0
0	1	0	0
0	1	1	1
1	0	0	0
1	0	1	0
1	1	0	1
1	1	1	1

FIGURE 3-39

$Y = \overline{C}BA + CB\overline{A} + CBA$

$Y = \overline{C}BA + CB\,(\overline{A} + A)$ by Theorem 10

$Y = \overline{C}BA + CB\,(1)$ by Theorem 8

$Y = \overline{C}BA + CB$ by Theorem 4

$Y = B\,(\overline{C}A + C)$ by Theorem 10

$Y = B\,(C + \overline{C}A)$ ORed terms are commutative

$Y = B\,(C + A)$ by Theorem 11

$Y = B\,(A + C)$ ORed terms are commutative

Build the logic circuit so that $Y = B\,(A + C)$. See Figure 3-40.

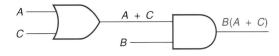

FIGURE 3-40

The truth table can be implemented using one two-input OR gate and one two-input AND gate.

This solution can be checked by hooking up the circuit and verifying that the output follows the truth table (a good exercise) or by developing a truth table for the expression $B\,(A + C)$. See Figure 3-41.

C	B	A	$A + C$	$B(A+C)$
0	0	0	0	0
0	0	1	1	0
0	1	0	0	0
0	1	1	1	1
1	0	0	1	0
1	0	1	1	0
1	1	0	1	1
1	1	1	1	1

FIGURE 3-41

This is identical to the original truth table.

Try creating your own three-input truth table and designing the logic circuit to implement it.

EXAMPLE 3-38

Design a circuit that will implement the truth table shown in Figure 3-42A.

Inputs			Output
C	B	A	Y
0	0	0	1
0	0	1	1
0	1	0	0
0	1	1	1
1	0	0	1
1	0	1	0
1	1	0	1
1	1	1	1

FIGURE 3-42A

Solution We want Y to be 1 when we have:

$\overline{A}\,\overline{B}\,\overline{C}$ (first line) or $A\,\overline{B}\,\overline{C}$ (second line) or
$A\,B\,\overline{C}$ (fourth line) or $\overline{A}\,\overline{B}\,C$ (fifth line) or
$\overline{A}\,B\,C$ (seventh line) or $A\,B\,C$ (last line)

$$Y = \overline{A}\,\overline{B}\,\overline{C} + A\,\overline{B}\,\overline{C} + A\,B\,\overline{C} + \overline{A}\,\overline{B}\,C + \overline{A}\,B\,C + A\,B\,C$$

$$Y = \overline{B}\,\overline{C}\,(\overline{A} + A) + A\,B\,\overline{C} + \overline{A}\,\overline{B}\,C + \overline{A}\,B\,C + A\,B\,C \qquad \text{by Theorem 10}$$

$$Y = \overline{B}\,\overline{C} + A\,B\,\overline{C} + \overline{A}\,\overline{B}\,C + \overline{A}\,B\,C + A\,B\,C \qquad \text{by Theorem 8}$$

$$Y = \overline{B}\,\overline{C} + A\,B\,(\overline{C} + C) + \overline{A}\,\overline{B}\,C + \overline{A}\,B\,C \qquad \text{by Theorem 10}$$

$$Y = \overline{B}\,\overline{C} + A\,B + \overline{A}\,\overline{B}\,C + \overline{A}\,B\,C \qquad \text{by Theorem 8}$$
$$Y = \overline{B}\,\overline{C} + A\,B + \overline{A}\,C\,(\overline{B} + B) \qquad \text{by Theorem 10}$$
$$Y = \overline{B}\,\overline{C} + A\,B + \overline{A}\,C \qquad \text{by Theorem 8}$$

The desired output can be produced as shown in Figure 3-42B.

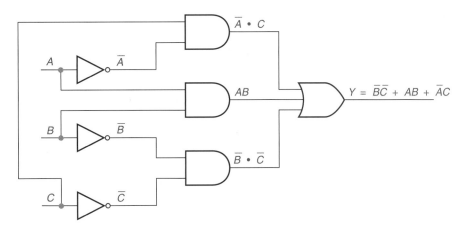

FIGURE 3-42B

Alternative Solution The first solution was a lot of work because there were six 1s in the output column of the truth table. An alternative approach is to work with the two 0s in the output column by writing an output expression for $\overline{Y}$, simplifying it, and then taking the complement to obtain Y. See Figure 3-42A.

When Y is 0, $\overline{Y}$ is 1. We want $\overline{Y}$ to be 1 when we have $\overline{C}\overline{B}\overline{A}$ (3rd line) OR $C\overline{B}A$ (6th line).

$\overline{Y} = \overline{C}\overline{B}\overline{A} + C\overline{B}A$. Both sides of this expression must be complemented to obtain an expression for Y.

$$\overline{Y} = \overline{C}\overline{B}\overline{A} + C\overline{B}A$$
$$Y = \overline{\overline{C}\overline{B}\overline{A} + C\overline{B}A}$$

See Figure 3-42C.

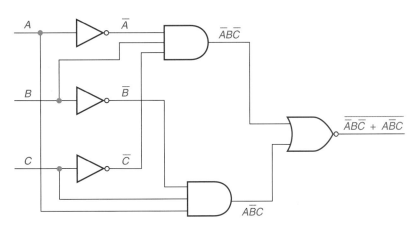

FIGURE 3-42C

EXAMPLE 3-39

Design a circuit that will implement the truth table shown in Figure 3-43A.

Solution

$$Y = A\,\overline{B}\,\overline{C} + \overline{A}\,\overline{B}\,C + A\,\overline{B}\,C + \overline{A}\,B\,C$$
$$Y = A\,\overline{B}\,(\overline{C} + C) + \overline{A}\,\overline{B}\,C + \overline{A}\,B\,C$$
$$Y = A\,\overline{B} + \overline{A}\,\overline{B}\,C + \overline{A}\,B\,C$$
$$Y = A\,\overline{B} + \overline{A}\,C\,(\overline{B} + B)$$
$$Y = A\,\overline{B} + \overline{A}\,C$$

The required circuitry is shown in Figure 3-43B.

Inputs			Output
C	B	A	Y
0	0	0	0
0	0	1	1
0	1	0	0
0	1	1	0
1	0	0	1
1	0	1	1
1	1	0	1
1	1	1	0

FIGURE 3-43A

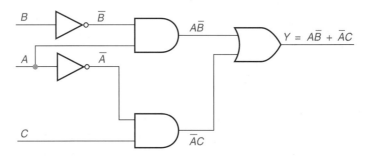

FIGURE 3-43B

EXAMPLE 3-40

Design a circuit that will behave according to the truth table shown in Figure 3-44A.

Solution

$$Y = \overline{A}\,B\,\overline{C} + A\,B\,\overline{C} + \overline{A}\,\overline{B}\,C + A\,\overline{B}\,C$$
$$Y = \overline{A}\,B\,\overline{C} + A\,B\,\overline{C} + \overline{B}\,C\,(\overline{A} + A)$$
$$Y = \overline{A}\,B\,\overline{C} + A\,B\,\overline{C} + \overline{B}\,C$$
$$Y = B\,\overline{C}\,(\overline{A} + A) + \overline{B}\,C$$
$$Y = B\,\overline{C} + \overline{B}\,C$$

The circuit can be constructed as shown in Figure 3-44B.

Inputs			Output
C	B	A	Y
0	0	0	0
0	0	1	0
0	1	0	1
0	1	1	1
1	0	0	1
1	0	1	1
1	1	0	0
1	1	1	0

FIGURE 3-44A

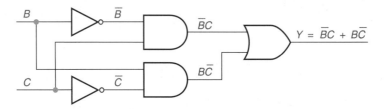

FIGURE 3-44B

Notice that the *A* input did not appear in the final circuit. It was not needed to produce the required output.

EXAMPLE 3-41

Design a circuit that will implement the truth table shown in Figure 3-45.

Inputs			Output
C	*B*	*A*	*Y*
0	0	0	1
0	0	1	1
0	1	0	0
0	1	1	1
1	0	0	1
1	0	1	0
1	1	0	1
1	1	1	1

FIGURE 3-45

Solution

$$Y = \overline{A}\,\overline{B}\,\overline{C} + A\,\overline{B}\,\overline{C} + A\,B\,\overline{C} + \overline{A}\,\overline{B}\,C + \overline{A}\,B\,C + A\,B\,C$$
$$Y = \overline{B}\,\overline{C}(\overline{A} + A) + A\,B\,\overline{C} + \overline{A}\,\overline{B}\,C + \overline{A}\,B\,C + A\,B\,C$$
$$Y = \overline{B}\,\overline{C} + A\,B\,\overline{C} + \overline{A}\,\overline{B}\,C + B\,C(\overline{A} + A)$$
$$Y = \overline{B}\,\overline{C} + A\,B\,\overline{C} + \overline{A}\,\overline{B}\,C + B\,C$$
$$Y = \overline{C}\,(\overline{B} + BA) + \overline{A}\,\overline{B}\,C + B\,C$$
$$Y = \overline{C}\,(\overline{B} + A) + \overline{A}\,\overline{B}\,C + B\,C$$
$$Y = \overline{B}\,\overline{C} + A\,\overline{C} + \overline{A}\,\overline{B}\,C + B\,C$$
$$Y = \overline{B}\,\overline{C} + A\,\overline{C} + C(\overline{A}\,\overline{B} + B)$$
$$Y = \overline{B}\,\overline{C} + A\,\overline{C} + C(B + \overline{B}\,\overline{A})$$
$$Y = \overline{B}\,\overline{C} + A\,\overline{C} + C(B + \overline{A})$$
$$Y = \overline{B}\,\overline{C} + A\,\overline{C} + B\,C + \overline{A}\,C$$

This same truth table appeared in a previous example. Although no errors have occurred in this solution and no further reductions can be made, the resulting expression contains more terms and requires more circuitry than the previous solution. Following the rules of Boolean algebra does not guarantee a minimal solution. The following method of designing logic circuits, Karnaugh (Car' no) maps, offers better control of the result and guarantees a minimal output expression.

Method 2—Karnaugh Map

The **Karnaugh** method uses a table or "map" to reduce its expressions. Each position in the tables is called a cell. **Cells** are filled with 1s and 0s according to the expression to be reduced. Adjacent 1s are grouped together in clusters, called **subcubes**, following definite rules.

A subcube must be of size 1, 2, 4, 8, 16, etc. All 1s must be included in a subcube of maximum size. These rules will be expounded upon and explained by working several examples.

EXAMPLE 3-42

Design a circuit that will behave according to the truth table shown in Figure 3-46.

Inputs			Output
C	B	A	Y
0	0	0	0
0	0	1	0
0	1	0	1
0	1	1	1
1	0	0	0
1	0	1	1
1	1	0	1
1	1	1	1

FIGURE 3-46

Solution

Step 1. Draw the table. Choose two of the variables to use as column headings across the top. We will use C and B. Form all combinations of C and $\overline{C}$ with B and $\overline{B}$. Each column heading should differ from the adjacent column by one variable only.

Part 1

$\overline{C}\,\overline{B}$	$\overline{C}B$	CB	$C\overline{B}$

Start with $\overline{C}\,\overline{B}$ and change $\overline{B}$ to B to form the heading for column 2, $\overline{C}B$. Then change $\overline{C}$ to C for the third column CB, and finally $C\overline{B}$. The fourth column wraps around to the first column and should differ by one variable only, which it does.

Part 2

	$\overline{C}\,\overline{B}$	$\overline{C}B$	CB	$C\overline{B}$
$\overline{A}$				
A				

Use the third variable, A, for row headings $\overline{A}$ and A.

Step 2. Fill the table with 1s and 0s from the truth table. The output Y is 1 in line 3 when we have $\overline{C}$ and B and $\overline{A}$. Place a 1 in the table in cell $\overline{C}B\overline{A}$. The output Y is also 1 on line 4, which is represented by $\overline{C}BA$; on line 6, which is $C\overline{B}A$; on line 7, which is $CB\overline{A}$; and on line 8, which is CBA. Fill those cells with ones and the remaining cells with zeros.

	$\overline{C}\,\overline{B}$	$\overline{C}\,B$	$C\,B$	$C\,\overline{B}$
$\overline{A}$	0	1	1	0
A	0	1	1	1

Step 3.   Combine adjacent cells that contain ones in subcubes of maximum size. The four ones in the center of the table compose a subcube of size 4.

	$\overline{C}\,\overline{B}$	$\overline{C}\,B$	$C\,B$	$C\,\overline{B}$
$\overline{A}$	0	1	1	0
A	0	1	1	1

The 1 in cell $CB\overline{A}$ has not been included in a subcube so it is used with its adjacent 1 in a subcube of size 2.

Step 4.   Write the expression that each subcube represents. In the subcube of size 4, find the variables that occur in all four cells. In this case, B is the only variable that appears in all four cells. The subcube of size 4 represents B. In the subcube of size two, A and C appear in each cell, so the subcube represents AC.

Step 5.   Form the output expression. The output Y is the expression from each subcube ORed together. In this case $Y = B + AC$.

The truth table can be implemented by the logic diagram.

Check:

How do we know this circuit is correct?

In the circuit, when A and C are both 1s, the output of the AND will also be 1. A 1 into an OR gives a 1 out. In the truth table, A and C are both 1s on lines 6 and 8, and the required output is 1. In the circuit, any time B is 1 the output is 1. In the truth table, B is 1 on lines 3, 4, 7, and 8, and the required output is 1. The rest of the time both inputs into the OR gate will be 0, and the result will be 0. This occurs on lines 1, 2, and 5 of the truth table, where the output is 0. In all cases, the circuit produces the results required by the truth table.

EXAMPLE 3-43

Use a Karnaugh map to design a logic diagram to implement the following truth table.

	Inputs			Output
	C	B	A	Y
	0	0	0	0
	0	0	1	1
	0	1	0	0
	0	1	1	0
	1	0	0	0
	1	0	1	1
	1	1	0	1
	1	1	1	1

Solution

Step 1. Draw the table.

	$\overline{C}\,\overline{B}$	$\overline{C}\,B$	$C\,B$	$C\,\overline{B}$
$\overline{A}$	0	0	1	0
A	1	0	1	1

Step 2. Fill the table with 1s and 0s from the truth table.

Step 3. Combine adjacent cells that contain 1s into subcubes (size 1, 2, 4, or 8).

	$\overline{C}\,\overline{B}$	$\overline{C}\,B$	$C\,B$	$C\,\overline{B}$
$\overline{A}$	0	0	1	0
A	1	0	1	1

The right side of the table "wraps around" to the other side so that the table is continuous. The 1s in the lower corners form a subcube of size 2. The two subcubes "cover" the map in that all 1s are contained in subcubes. Any additional subcubes drawn would add unneeded terms to the final expression.

Step 4. Write the expression that each subcube represents. In the vertical subcube, C and B remain constant. In the horizontal subcube, $\overline{B}$ and A are constant.

Step 5. Form the output expression.

$$Y = BC + A\overline{B}$$

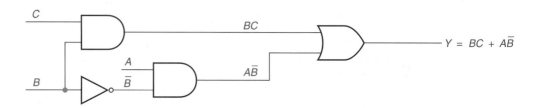

EXAMPLE 3-44

Use a Karnaugh map to design a logic diagram to implement the following truth table.

Inputs				Output
D	C	B	A	Y
0	0	0	0	1
0	0	0	1	0
0	0	1	0	1
0	0	1	1	0
0	1	0	0	0
0	1	0	1	1
0	1	1	0	0
0	1	1	1	0
1	0	0	0	1
1	0	0	1	0
1	0	1	0	1
1	0	1	1	0
1	1	0	0	1
1	1	0	1	1
1	1	1	0	1
1	1	1	1	1

Solution

Step 1. Draw the table. Since four variables are needed, use two across the top and two down the side.

	$\overline{D}\,\overline{C}$	$\overline{D}C$	DC	$D\overline{C}$
$\overline{B}\,\overline{A}$	1	0	1	1
$\overline{B}A$	0	1	1	0
BA	0	0	1	0
$B\overline{A}$	1	0	1	1

Step 2. Fill the table with 1s and 0s from the truth table.

Step 3. Combine adjacent cells that contain 1s into subcubes of size 1, 2, 4, 8, or 16.

	$\overline{D}\,\overline{C}$	$\overline{D}C$	DC	$D\overline{C}$	
$\overline{B}\,\overline{A}$	1	0	1	1	$\leftarrow \overline{A}C$
$\overline{B}A$	0	1	1	0	
BA	0	0	1	0	
$B\overline{A}$	1	0	1	1	

$A\overline{B}\overline{C}$ ——   ——CD

Since the map is continuous top to bottom and side to side, the four corners are adjacent and form a subcube of size 4. The DC column forms another subcube of size 4. One cell remains uncovered. $\overline{DC}\overline{B}A$ forms a subcube of size 2 with the cell on its right.

Step 4. Write the expression that each subcube represents. The subcube formed by the four corners represents the term $\overline{A}\,\overline{C}$. The vertical subcube represents the expression CD, and the subcube of size 2 represents the expression $A\overline{B}C$.

Step 5. Form the output expression.

$$Y = CD + \overline{A}\,\overline{C} + A\overline{B}C$$

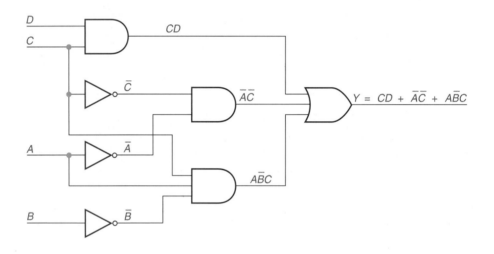

 SELF-CHECK 3

C	B	A	Y
0	0	0	1
0	0	1	0
0	1	0	1
0	1	1	1
1	0	0	1
1	0	1	0
1	1	0	1

1. Use the Boolean algebra method to develop a circuit to implement this truth table.

2. Use the Karnaugh map method to develop a circuit to implement this truth table.

3.7 AND-OR-INVERT GATES

In most of the design exercises in the last section, the resulting expressions have been in a "sum of products" form. For example, the output expression in Figure 3-42B is $Y = \overline{B}\,\overline{C} + AB + \overline{A}\,C$. Groups of variables are ANDed and those results are ORed together. ICs called **AND-OR-INVERT** are available to make implementing this circuitry easy. The 74F51 Dual 2-wide 2-input, 2-wide 3-input AND-OR-INVERT gate is shown in Figure 3-47. The 74F64 4-2-3-2 Input AND-OR-INVERT gate is shown in Figure 3-48. Note how the two IEC symbols denote the difference in these two ICs.

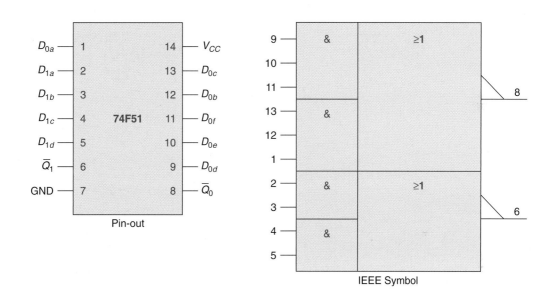

Pin-out

IEEE Symbol

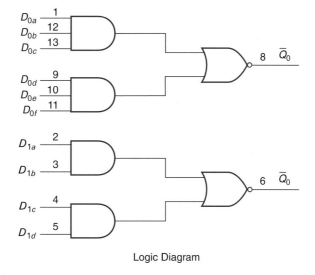

Logic Diagram

FIGURE 3-47 74F51 dual 2-wide 2-input, 2-wide 3-input AND-OR-INVERT gate

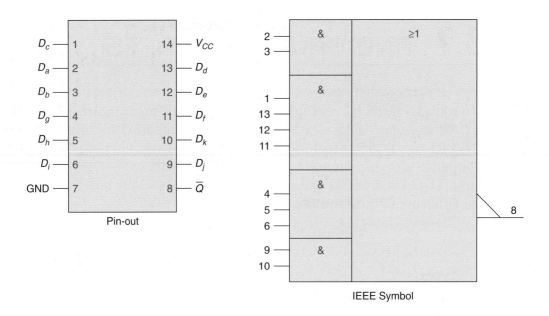

Pin-out

IEEE Symbol

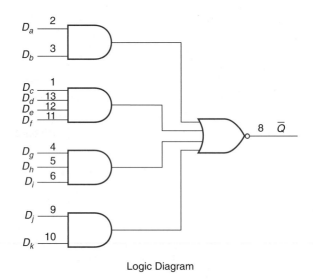

Logic Diagram

FIGURE 3-48 74F64 four-two-three-two AND-OR-INVERT gate

EXAMPLE 3-45

Use an AND-OR-INVERT gate IC to implement the resulting expression in the last example, $Y = CD + \overline{A}\,\overline{C} + A\overline{B}C$.

Solution   Use the 74F64. The output must be inverted. The unused inputs should be grounded. See Figure 3-49.

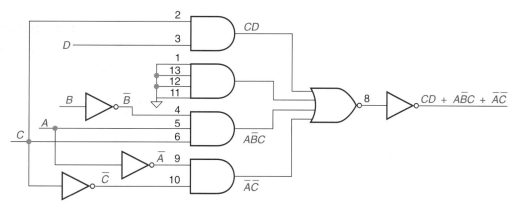

FIGURE 3-49

EXAMPLE 3-46

Use an AND-OR-INVERT gate IC to implement the logic circuit shown in Figure 3-42C, $Y = \overline{\overline{A}\,B\overline{C} + A\overline{B}C}$.

Solution   The circuit requires a 2-wide, 3-input AND-OR-INVERT gate IC. Use the 74F51. See Figure 3-50.

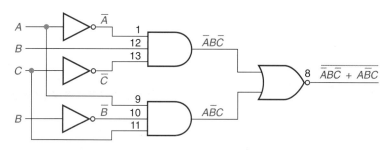

FIGURE 3-50

 3.8 REDUCING BOOLEAN EXPRESSIONS USING KARNAUGH MAPS

We have used Karnaugh maps to implement truth tables. They can also be used to reduce Boolean expressions that are written in sum-of-product form. Represent each product (term) in the map with one or more 1s. Group the 1s into subcubes of size 1, 2, 4, or 8. Write the resulting minimal expression. For example, in a map with three variables A, B, and C, shown below, $\overline{A}\,B\overline{C}$ is represented by a 1 in the single cell where $\overline{A}$, B, and $\overline{C}$ intersect. AC is represented by 1s in the two cells that contain A and C. B is represented by 1s in each of the four cells that contain B.

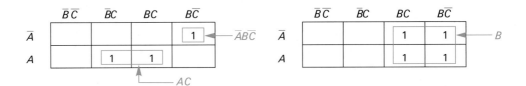

	$\bar{B}\bar{C}$	$\bar{B}C$	BC	$B\bar{C}$
$\bar{A}$				1
A		1	1	

$\leftarrow \overline{ABC}$

AC

	$\bar{B}\bar{C}$	$\bar{B}C$	BC	$B\bar{C}$
$\bar{A}$			1	1
A			1	1

$\leftarrow B$

EXAMPLE 3-47

Using a Karnaugh map, reduce the expression $Y = A + A\bar{B}C + AB$.

Solution $A\bar{B}C$ is represented by a 1 at the bottom of the fourth column. AB is represented by 1s at the bottom of the middle two columns. A is represented by four 1s across the bottom.

	$\bar{C}\bar{B}$	$\bar{C}B$	CB	$C\bar{B}$
$\bar{A}$				0
A	1	1	1	1

$\leftarrow A$

All 1s are included in a subcube of size 4, and A is common to each of the cells in that subcube. $A + A\bar{B}C + AB = A$

If the expression to be reduced is not in sum-of-product form, convert it by clearing the parentheses (Theorem 10) or by applying DeMorgan's Theorem.

EXAMPLE 3-48

Reduce by Karnaugh map $Y = (\bar{A}BC + D)(\bar{A}D + \bar{B}\bar{C})$.

Solution   First convert to sum-of-product form by clearing the parentheses.

$$Y = \bar{A}BCD + \bar{A}BC\bar{B}\bar{C} + \bar{A}DD + \bar{B}\bar{C}D$$

$$Y = \bar{A}BCD + 0 + \bar{A}D + \bar{B}\bar{C}D$$

$$= \bar{A}BCD + \bar{A}D + \bar{B}\bar{C}D$$

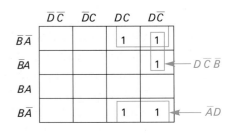

	$\bar{D}\bar{C}$	$\bar{D}C$	DC	$D\bar{C}$
$\bar{B}\bar{A}$			1	1
$\bar{B}A$				1
BA				
$B\bar{A}$			1	1

$\leftarrow D\bar{C}\bar{B}$

$\leftarrow \bar{A}D$

$$Y = \bar{A}D + D\bar{C}\bar{B}$$

SELF-CHECK 4

Reduce these expressions and use an AND-OR-INVERT gate IC to implement the reduced expressions.

1. $Y = A\overline{B}C + AB\overline{C} + \overline{A}BC + ABC$

2. $Y = (\overline{A} + B + C)(A + \overline{B} + C)(A + B + \overline{C})$

 3.9 PROGRAMMABLE LOGIC DEVICES

PLDs are divided into three broad categories: simple programmable logic devices (SPLDs), complex programmable logic devices (CPLDs), and field programmable gate array devices (FPGAs). In this section we will look at the internal configuraton of an SPLD. One type of SPLD is called **programmable array logic (PAL)**. Another is called **generic array logic (GAL)**. A GAL contains AND-OR combinations hard wired inside, but the user can select which inputs are to be connected to which AND gate. Figure 3-51 shows a small segment of a typical logic diagram of a GAL or PAL.

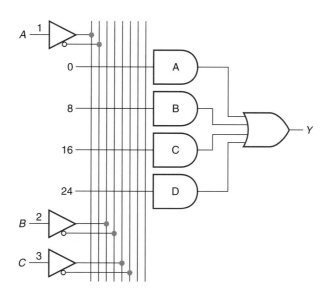

FIGURE 3-51 AND/OR portion of a PLD

Three inputs, *A*, *B*, and *C*, and one output, *Y*, are shown. Each input enters a buffer that produces two outputs, the variable, and its complement. Each input and its complement are connected internally to a column of the array. Find the columns for inputs *A*, *B*, and *C*, and their complements $\overline{A}$, $\overline{B}$, and $\overline{C}$. Each AND gate is connected to its own horizontal row of the array. The vertical lines in the array provide a means to connect the input signals to the AND gates. For simplicity and clarity, the conventional method of depicting the internal logic of PLDs is to represent the multiple-input AND gates with a single input line. In Figure 3-52A the AND gate is connected to inputs *A*, $\overline{B}$, and *C*. Figure 3-52B shows the actual configuration.

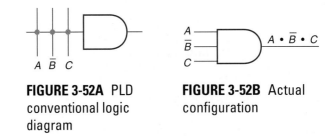

FIGURE 3-52A PLD conventional logic diagram

FIGURE 3-52B Actual configuration

Each intersection of a vertical column and horizontal row is a potential connection. These intersections are called fuses. In Figure 3-51 the fuses are numbered in decimal, beginning with 0 and ending with 31. The number of the first fuse in each row is shown.

EXAMPLE 3-49

Figure 3-53 is the same as Figure 3-51 with connections made at some of the junctions. Write the Boolean expression implemented by the PLD.

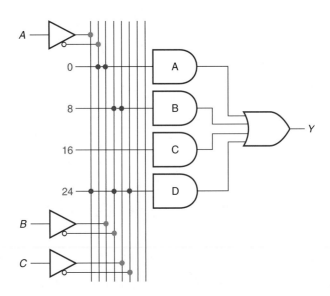

FIGURE 3-53 AND/OR portion of a PLD

Solution   Fuses 1 and 2 connect inputs $\overline{A}$ and B to AND gate A. Fuses 11 and 12 connect inputs $\overline{B}$ and C to AND gate B. Fuses 24, 27, and 29 connect inputs A, $\overline{B}$, and $\overline{C}$ to AND gate D. The outputs of the AND gates are ORed together.

$$Y = \overline{A}B + \overline{B}C + A\overline{B}\,\overline{C}$$

The connections inside an electrically erasable PLD can be erased and the IC can be programmed in a different configuration. When erased, all junctions between a horizontal row and vertical columns are connections. In the programming process, the connections not needed are eliminated.

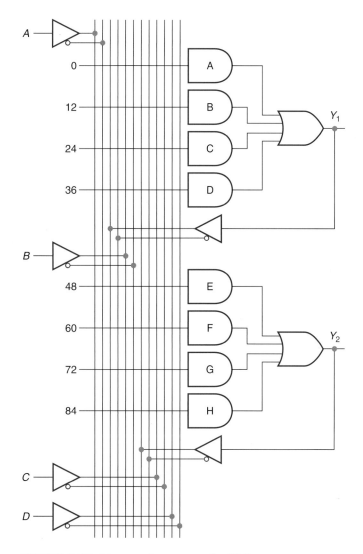

FIGURE 3-54 4 inputs, 2 outputs of a PLD

Figure 3-54 shows a larger section of a sample PAL, with four inputs and two outputs. Note that the output of each AND/OR combination and its complement are fed back into the array. Y_1, Y_2, and their complements each has its own column. The output of one combination can be used as an input into the next AND/OR combination. In this manner, the output of the second AND/OR combination can have more terms than there are AND gates in the combination.

EXAMPLE 3-50

For the PAL shown in Figure 3-54, show the connections needed to implement this Boolean expression.

$$Y = A\overline{B}D + \overline{A}BC + A\overline{B}\,\overline{C} + \overline{A}B\overline{D} + B\overline{C}\,\overline{D} + \overline{A}\,\overline{C}\,\overline{D}$$

Use the top AND/OR combination to produce the first four terms of the expression. Use the Y_1 output as an input to the second AND/OR section.

Solution   See Figure 3-55. Fuses 0, 5, and 10 connect inputs A, $\overline{B}$, and D to AND gate A. Fuses 13, 16, and 20 connect inputs $\overline{A}$, B, and C to AND gate B. Fuses 24, 29, and 33 connect inputs A, $\overline{B}$, and $\overline{C}$ to AND gate C. Fuses 37, 40, and 47 connect inputs $\overline{A}$, B, and $\overline{D}$ to AND gate D. AND gates A, B, C, and D are ORed together to produce output Y_1. Y_1 is fed back into the array and is used as an input to AND gate G. Fuses 52, 57, and 59 connect inputs B, $\overline{C}$, and $\overline{D}$ to AND gate E. Fuses 61, 69, and 71 connect inputs $\overline{A}$, $\overline{C}$, and $\overline{D}$ to AND gate F. AND gate H is not used. AND gates E, F and G are ORed together to produce Y_2.

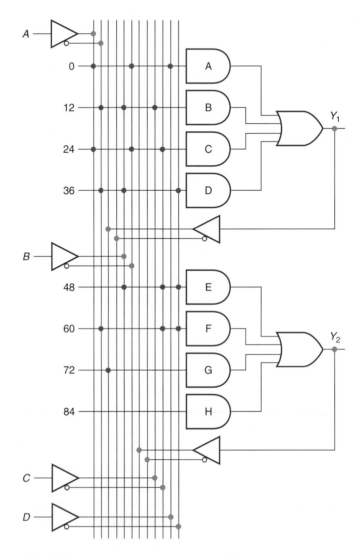

FIGURE 3-55 4 inputs, 2 outputs of a PLD

A CPLD consists of many SPLD circuits within a single IC. Each of the SPLD circuits is called a macrocell. Macrocells are connected together internally as needed to produce the required circuits. In this text we are programming a CPLD manufactured by Altera Corporation—the EPM7128SLC84. It is an 84-pin IC that contains 128 macrocells with 2500 usable gates, arranged into 8 logic array blocks (16 macrocells per block). An FPGA is an integrated circuit that produces the required output by storing truth tables instead of logic circuitry. Each truth table is called a look-up table (lut).

 SELF-CHECK 5

1. In Figure 3-51, each AND gate has how many inputs?

2. In Figure 3-51, the OR gate has how many inputs?

3. Which fuses must be intact in the GAL shown in Figure 3-51 to implement this Boolean expression?

$$Y = A\overline{B}\,\overline{C} + \overline{A}B\overline{C} + ABC + \overline{A}\,\overline{B}C$$

4. In Figure 3-54, each AND gate has how many inputs?

5. In Figure 3-54, each OR gate has how many inputs?

6. Which fuses must be intact in the GAL shown in Figure 3-54 to implement this Boolean expression?

$$Y_2 = \overline{A}\,\overline{B}\,\overline{C}\,\overline{D} + \overline{A}\,\overline{B}C D + \overline{A}B\overline{C}\,D + \overline{A}BC\overline{D} + AB\overline{C}\,\overline{D} + A\overline{B}\,\overline{C}\,D + A\overline{B}C\overline{D}$$

 # 3.10 PROGRAMMING A PLD (Optional)

Refer to Section 2.19, "Programming a PLD," as needed. The steps involved in simulating the circuits 7 through 10 were not covered in Chapter 2. They are discussed in detail here.

Step 1. Create a new folder within the Altera folder and name it **and_or**.

Step 2. Run QuartusII.

Step 3. Run "New Project Wizard."

Step 4. Create a new .vhd file and name it **and_or.vhd**.

Step 5. Write a VHDL program that will implement these circuits:

x = ab + cd

y = abc + d

z = $\overline{\text{abcd}}$

Note: To produce the output x, inputs a and b should be ANDed, inputs c and d should be ANDed, and the results should be ORed. In order for the QuartusII compiler to group the signals as intended, parentheses are needed around a AND b and around c AND d. Likewise, to produce y, inputs a, b, and c should be ANDed and the result should be ORed with d.

Parentheses are needed around a AND b AND c. To produce the output z, all four inputs must be ANDed and the result complemented. z = NOT (a AND b AND c AND d). Here is a listing of **and_or.vhd**. Try writing it by yourself before looking at the answer.

```
ENTITY and_or IS
     PORT (a,b,c,d:IN BIT;
          x,y,z:OUT BIT);
END and_or;
ARCHITECTURE a OF and_or IS
BEGIN
     x <= (a AND b) OR (c AND d);
     y <= (a AND b AND c) OR (d);
     z <= NOT (a AND b AND c AND d);
END a;
```

Step 6. Compile your program.

Step 7. Create a vector waveform file (.vwf) to use in simulation.

- Click **File/New**.

- In the "New File" window, click the tab "Other Files."

- At the bottom of the list highlight (click) **Vector Waveform File** and **OK**.

- Name your file **and_or.vwf**.

  Click **File/Save As**. Make sure the file is saved in the **and_or** folder under the name **and_or.vwf**.

Step 8. Configure the and_or.vwf file for use in this project. Specify the input and output signals to be used.

- Click **Edit/Insert Node or Bus**.

- In the "Insert Node or Bus" dialog box, click **Node Finder**.

- In the "Filter" dialog box, click the drop-down arrow. Highlight **Pins-all** and click the "List" button. All signals are listed in the window on the left as "Nodes Found."

- Select (highlight) the input signals a, b, c, and d, and click the ">" arrow in the middle of the windows to move those inputs to the "Select Nodes" window on the right.

- Now repeat that process for the output signals x, y, and z. If you select the inputs first, they will appear at the top of the resulting chart and the outputs will be at the bottom.

- Click **OK** and **OK**.

- Set the horizontal timing axis of the chart.

  Click **Edit/Grid Size**.

  In the "Grid Size" dialog box, set the period to 50 nanoseconds and leave the phase at 0 ns. Click OK.

Click **Edit/End Time**.

In the "End Time" dialog box, set the End Time to 1000 nanoseconds (1 microsecond). Click **OK**.

Click **View/Fit in Window**

Your chart should have the signals listed vertically on the left—and a timing grid marked across the top—that ends at 1 microsecond.

- Save your file as **and_or.vwf**.

Step 9.   Use the Waveform Editing Tool to draw input waveforms for signals a, b, c, and d.

- On the vertical toolbar on the left of the chart, click ⇥⇤ "Waveform Editing Tool."

- To modify a signal, click on the start of the interval to be changed and drag across to the end of the interval. Draw waveform **a** to toggle each 50 ns, **b** to toggle each 100 ns, **c** to toggle each 200 ns, and **d** to toggle each 400 ns.

- An alternate procedure for entering a waveform is to click on the selection tool instead of the Waveform Editing Tool, and highlight the interval to be changed. Force the highlighted interval LOW by clicking ⊥ "Forcing Low (0)." Likewise, to force the highlighted area HIGH, click ⌐ "Forcing High (1)."

- Your input waveforms **a**, **b**, **c**, and **d** should look like those shown in Figure 3-56. Output waveforms **x**, **y**, and **z** will be generated by the simulation tool in Step 10.

- Save your waveform as **and_or.vwf**.

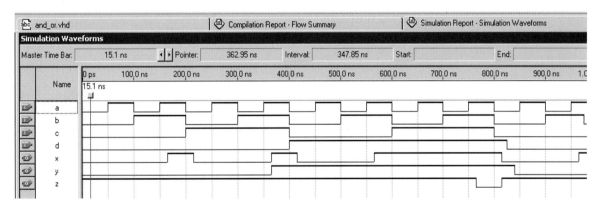

FIGURE 3-56 Simulation waveforms

Step 10.   Simulate the circuit.

- Click **Processing/Start Simulation** or click the ▶ "Start Simulation" button on the tool bar.

- When the simulation is complete, a report will show the number of errors encountered during simulation. If no errors were found, click **OK** and study the simulated output in order to see whether the circuit behaves correctly.

- If all is well, save the project. Click **File/Save Project**.

Step 11. Assign pins to the input and output signals. Recompile.

Step 12. Program your PLD. **Caution:** If you are using the PLDT-2 board from RSR Electronics, remove all 8 jumpers from HDI before programming the PLD.

Step 13. Test the operation of the PLD using input switches and LEDs.

3.11 TROUBLESHOOTING COMBINATIONAL LOGIC CIRCUITS

In section 3.6, the circuit in Figure 3-57A was designed to implement the truth table in Figure 3-57B.

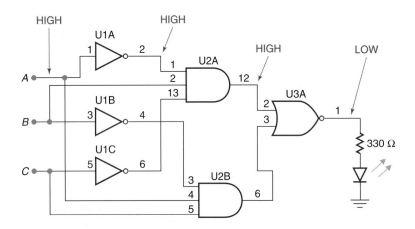

FIGURE 3-57A Logic diagram

FIGURE 3-57B Truth table

Inputs	Output
C B A	Y
0 0 0	1
0 0 1	1
0 1 0	0
0 1 1	1
1 0 0	1
1 0 1	0
1 1 0	1
1 1 1	1

To construct this circuit three ICs are required, a 7404 inverter labeled U1, a 7411 three-input AND labeled U2, and a 7402 two-input NOR labeled U3. See Figure 3-57C. Pin numbers have been transferred from the schematic to the logic diagram.

The circuit is tested by supplying input values for *C*, *B*, and *A* corresponding to each line of the truth table. The first three lines of the truth table test good, but with *C* = 0, *B* = 1, and *A* = 1 (line four of the truth table), the output of the circuit measures LOW instead of HIGH. The LED is OFF. What steps should be taken?

Step 1. V_{CC} and ground measure good at the power pins of each IC.

Step 2. The inputs to the circuit *C*, *B*, and *A* measure good (LOW, HIGH, and HIGH, respectively).

Step 3. Line four of the truth table calls for the output of the NOR gate U3A to be HIGH. For this to happen, both of its inputs should be LOW. Measure the inputs on pins 2 and 3 of the NOR gate. Pin 2 measures HIGH, causing the output to go LOW. The NOR gate is

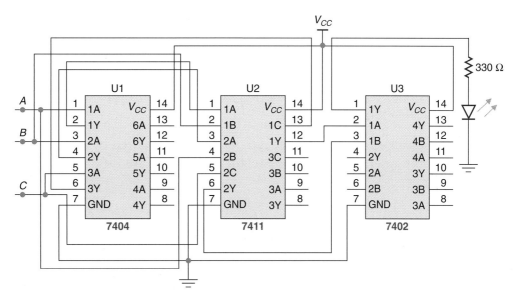

FIGURE 3-57C Schematic diagram

functioning correctly. Why is there a HIGH on input pin 2? The inputs into AND gate U2A need to be investigated. Note that the faulty input to U3A pin 2 has caused us to investigate U2A instead of U2B.

Step 4.   Measure the inputs to U2A. Pins 1, 2, and 13 are HIGH. The output on pin 12 of U2A is HIGH because each of the inputs measures HIGH. The input on pin 1 should be the complement of A and should be LOW. Either gate U1A output is faulty with its output clamped HIGH or U2A is faulty with input pin 1 clamped HIGH. Either fault would produce this result.

Step 5.   If the circuit is on a protoboard or on a printed circuit board with ICs in sockets, kill the power supply, remove U2, and restore power. If the output of U1A returns LOW as it should, U2 is the faulty IC. Otherwise, U1 is at fault. If your circuit is on a protoboard, switch to one of the unused gates on U1 or U2. On a printed circuit board, replace the faulty IC.

Part 3 of Lab 3B has circuits for you to troubleshoot—circuit files 3B-1 through 3B-5.ms9. Remember that Multisim treats an open input (floating input) as a 0. In an actual TTL circuit, a floating input measures about 1.6 volts and the IC often takes it as a HIGH level. In CMOS ICs, a floating input can drift between a HIGH and a LOW level.

DIGITAL APPLICATION Boolean Algebra

PROGRAM 3.1 is written in a language called CUPL. This program is used to configure the gates inside an SUPL into a motor control circuit. The circuit is used by Precise Power Corporation to control large single-phase motors (20-, 40-, and 60-horsepower). The input section defines which variables will be input to which input pins. For instance, the signal RPM_3500 will be connected to pin 7. These inputs are active LOW. RPM_3500 remains HIGH until the motor reaches 3500 rpm, then it snaps LOW. The output section defines which output variables will appear on which pins. For instance, the output signal RUN will appear at pin 15. The logic section defines the relationships between the inputs and outputs in Boolean algebra.

The Boolean symbols used here are the symbols required by the compiler program that will design the internal connections to the PLD. They are different than those learned in this chapter.

$$! = \text{NOT} \qquad \# = \text{OR} \qquad \& = \text{AND}$$

$$\text{MEX} = !\text{RPM\_3150} \& !\text{SYNC} \text{ translates into MEX} = \overline{\text{RPM\_3150}} \cdot \overline{\text{SYNC}}$$

Translate the logic statements and draw the circuits required to implement PROGRAM 3-1.

```
        /*
         * Inputs: define inputs
         */

        Pin 5 = RPM_2200;
        Pin 6 = RPM_3150;
        Pin 7 = RPM_3500;
        Pin 9 = TRANS;
        Pin 11 = ILINE;

        /*
         * Outputs: define outputs as active HI levels
         */

        Pin 12 = C2;
        Pin 13 = C1;
        Pin 14 = RESTART;
        Pin 15 = RUN;
        Pin 16 = LATCH;
        Pin 18 = SYNC;
        Pin 19 = MEX;

        /*
         * Logic: ! = NOT # = or & = AND $ = EXOR
         *
         */

        C1 = !SYNC;
        C2 = RPM_3500;
        MEX = !RPM_3150 & !SYNC;
        RUN = RPM_3150 & !SYNC;
        RESTART = !RPM_2200 & !SYNC;
        SYNC = !RPM_3500 & ILINE;
        LATCH = !TRANS;
```

PROGRAM 3-1

SUMMARY

- Use the unique state of each gate to predict output waveforms.

  Unique states:
  AND—all 1s in, 1 out
  NAND—all 1s in, 0 out
  OR—all 0s in, 0 out
  NOR—all 0s in, 1 out

- Delayed-clock and shift-counter waveforms can be used to create a wide variety of timing or control signals.

- Combinations of gates are often needed to produce a required output.

- These Boolean theorems can be used to reduce an expression to minimal terms:

  1. $\overline{\overline{A}} = A$

  2. $A \cdot 0 = 0$

  3. $A + 0 = A$

  4. $A \cdot 1 = A$

  5. $A + 1 = 1$

  6. $A + A = A$

  7. $A \cdot A = A$

  8. $A + \overline{A} = 1$

  9. $A \cdot \overline{A} = 0$

  10. $A \cdot B + A \cdot C = A(B + C)$

  11. $A + \overline{A}B = A + B$

- DeMorgan's theorems can be used to write an expression in equivalent form.

  1. $\overline{A \cdot B} = \overline{A} + \overline{B}$

  2. $\overline{A + B} = \overline{A} \cdot \overline{B}$

- A circuit can be designed to implement a truth table by these two methods:

  1. Write a Boolean expression for the output of the truth table to be implemented. Reduce the expression to minimal terms using Boolean algebra. Construct a circuit to represent the reduced expression.

  2. Draw a Karnaugh map to represent the truth table to be implemented. Combine the 1s into the largest subcubes possible of size 1, 2, or 4. Write the expression represented by the subcubes and construct the circuit.

- An AND-OR-INVERT gate can be used to implement a Boolean expression in sum-of-products form.

- PALs and GALs are examples of SPLDs.

- PALs and GALs contain programmable AND/OR arrays.

- In a PLD, each input and its complement are attached to unique columns in the array.

- In a PLD, each AND gate is attached to its own row in the array.

- An intact fuse at the intersection of a row and a column makes a connection.

- Some outputs of a PLD are fed back into the array to be used as inputs to the AND gates.

- SPLDs, CPLDs, and FPGAs are types of PLDs.

QUESTIONS AND PROBLEMS

1. In Figure 3-58, use the waveforms A, B, and C to determine the waveforms for each expression shown.

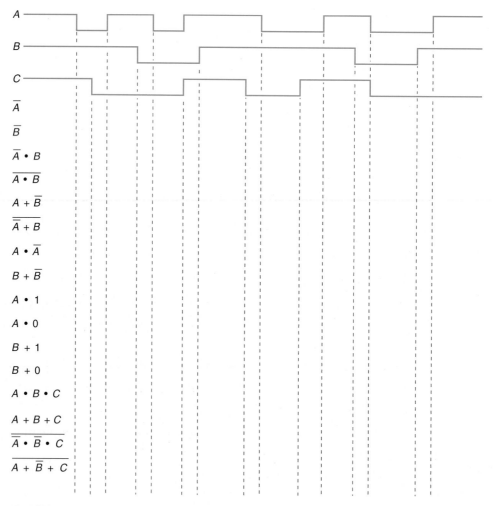

FIGURE 3-58

2. In Figure 3-59, use the waveforms A, B, and C to determine the waveforms for each expression shown.

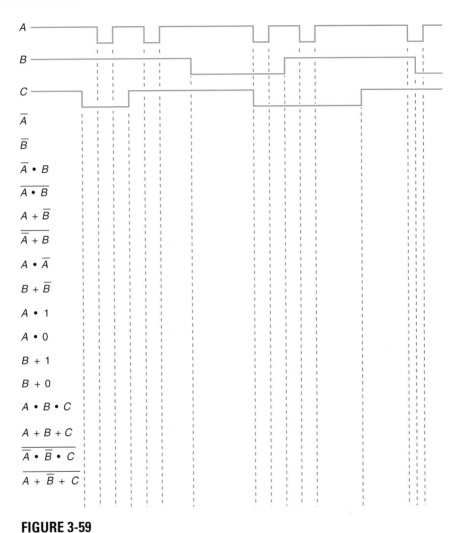

FIGURE 3-59

3. Use the delayed-clock, shift-counter waveforms in Figure 3-5 (page 94) to determine the missing inputs and outputs for the figures listed.

a. Figure 3-60   **d.** Figure 3-63   **f.** Figure 3-65
b. Figure 3-61   **e.** Figure 3-64   **g.** Figure 3-66
c. Figure 3-62

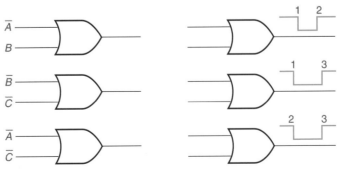

FIGURE 3-60   **FIGURE 3-61**

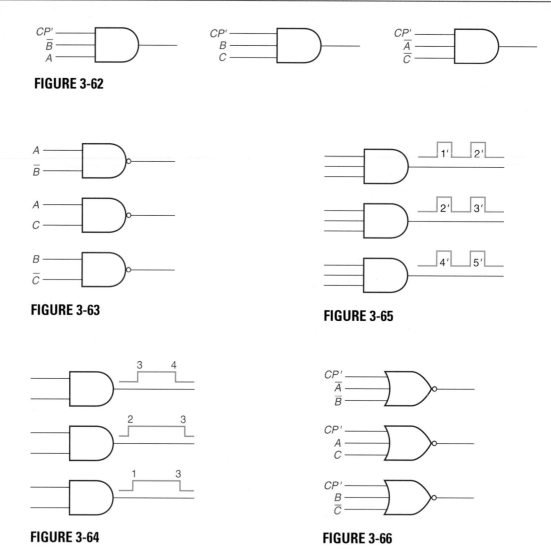

FIGURE 3-62

FIGURE 3-63

FIGURE 3-65

FIGURE 3-64

FIGURE 3-66

4. Use the waveforms in Figure 3-5 (page 94) to determine the missing inputs and outputs for the figures listed.

   a. Figure 3-67   **d.** Figure 3-70   **f.** Figure 3-72
   b. Figure 3-68   **e.** Figure 3-71   **g.** Figure 3-73
   c. Figure 3-69

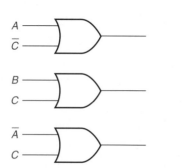

FIGURE 3-67

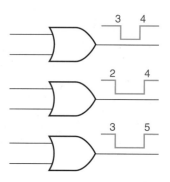

FIGURE 3-68

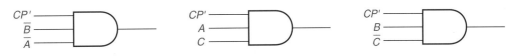

FIGURE 3-69

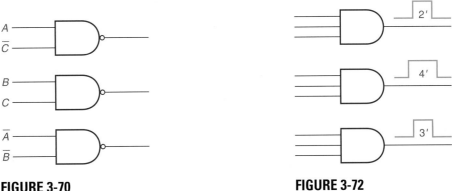

FIGURE 3-70

FIGURE 3-72

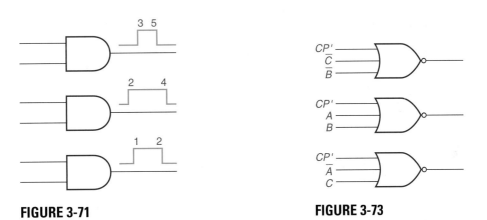

FIGURE 3-71

FIGURE 3-73

5. Design a circuit to implement the truth table shown in Figure 3-74.

Inputs			Output
C	B	A	Y
0	0	0	0
0	0	1	0
0	1	0	1
0	1	1	1
1	0	0	0
1	0	1	0
1	1	0	1
1	1	1	0

FIGURE 3-74

6. Design a circuit to implement the truth table shown in Figure 3-75.

Inputs			Output
C	B	A	Y
0	0	0	1
0	0	1	1
0	1	0	0
0	1	1	1
1	0	0	0
1	0	1	0
1	1	0	0
1	1	1	1

FIGURE 3-75

7. Design a circuit to implement the truth table shown in Figure 3-76.

   a. Use Boolean algebra.

   b. Use the Karnaugh map method.

C	B	A	Y
0	0	0	1
0	0	1	0
0	1	0	1
0	1	1	0
1	0	0	1
1	0	1	1
1	1	0	1
1	1	1	0

FIGURE 3-76

8. Design a circuit to implement the truth table shown in Figure 3-77.

   a. Use Boolean algebra.

   b. Use the Karnaugh map method.

C	B	A	Y
0	0	0	1
0	0	1	0
0	1	0	0
0	1	1	0
1	0	0	1
1	0	1	1
1	1	0	1
1	1	1	0

FIGURE 3-77

9. Reduce these expressions.

   a. $\overline{A} B + \overline{A} C$

   b. $A B + A \overline{B}$

   c. $A + \overline{A} D$

   d. $A \cdot \overline{A} \cdot B \cdot C$

   e. $\overline{A} + A B$

10. Reduce these expressions.

   a. $X Y Z + X \overline{Y}$

   b. $B C D + B \overline{C}$

   c. $B + \overline{B} E$

   d. $A \cdot C \cdot \overline{A}$

   e. $A B + C + 1$

11. Reduce these expressions.

   a. $A B C + A B \overline{C} + \overline{B}$

   b. $A \overline{B} \overline{C} + A \overline{C} + C$

   c. $A B \overline{C} + A \overline{B} \overline{C} + A B C$

   d. $A \overline{B} C + A B \overline{C} + A \overline{B} \overline{C}$

12. Reduce these expressions.

   a. $\overline{A} \overline{B} C + A \overline{B} C + \overline{C}$

   b. $A B C + \overline{A} \overline{B} C + \overline{A} B C$

   c. $A + A \overline{B} C + A B$

   d. $A \overline{C} + A \overline{B} + A B$

13. Group the 1s into subcubes and write the output expression.

	$\overline{C} \overline{B}$	$\overline{C} B$	$C B$	$C \overline{B}$
$\overline{A}$	1	0	1	1
A	1	0	0	1

	$\overline{C} \overline{B}$	$\overline{C} B$	$C B$	$C \overline{B}$
$\overline{A}$	1	1	0	0
A	0	1	1	0

14. Group the 1s into subcubes and write the output expressions.

	$\overline{C} \overline{B}$	$\overline{C} B$	$C B$	$C \overline{B}$
$\overline{A}$	0	1	1	1
A	0	0	1	1

	$\overline{C} \overline{B}$	$\overline{C} B$	$C B$	$C \overline{B}$
$\overline{A}$	0	1	1	1
A	1	0	0	0

15. Group the 1s into subcubes and write the output expression.

	$\overline{D} \overline{C}$	$\overline{D} C$	$D C$	$D \overline{C}$
$\overline{B} \overline{A}$	1	0	0	1
$\overline{B} A$	0	1	1	0
$B A$	0	1	1	0
$B \overline{A}$	1	0	0	1

16. Group the 1s into subcubes and write the output expression.

	$\overline{D} \overline{C}$	$\overline{D} C$	$D C$	$D \overline{C}$
$\overline{B} \overline{A}$	0	1	1	1
$\overline{B} A$	0	0	0	0
$B A$	0	1	0	0
$B \overline{A}$	0	1	1	1

17. Rework Problem 11 using Karnaugh map.

18. Rework Problem 12 using Karnaugh map.

19. Show the fuses needed in Figure 3-51 to implement this truth table.

C	B	A	Y
0	0	0	1
0	0	1	0
0	1	0	1
0	1	1	1
1	0	0	0
1	0	1	1
1	1	0	0
1	1	1	0

20. Write the Boolean expression for output Y in Figure 3-78.

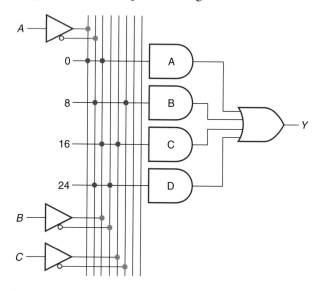

FIGURE 3-78

21. Show the fuses needed in Figure 3-54 to implement this truth table.

D	C	B	A	Y
0	0	0	0	1
0	0	0	1	0
0	0	1	0	1
0	0	1	1	0
0	1	0	0	0
0	1	0	1	1
0	1	1	0	0
0	1	1	1	1
1	0	0	0	0
1	0	0	1	1
1	0	1	0	0
1	0	1	1	1
1	1	0	0	0
1	1	0	1	0
1	1	1	0	0
1	1	1	1	1

22. Write the Boolean expression for outputs Y_1 and Y_2 in Figure 3-79.

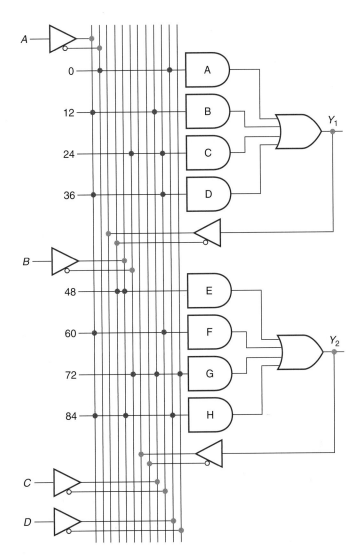

FIGURE 3-79

23. Draw the circuits described by this VHDL program.

```
ENTITY logic IS
     PORT(a,b,c:IN BIT;
          x,y,z:OUT BIT);
END logic;
ARCHITECTURE a OF logic IS
BEGIN
        x <= (NOT a) AND (NOT b);
        y <= (NOT a) OR b OR (NOT c);
        z <= NOT ((NOT a) OR b OR c);
END a;
```

24. Draw the circuits described by this VHDL program.

```
ENTITY chap3 IS
      PORT(a,b,c,d,e,f,g:IN BIT;
            x,y,z:OUT BIT);
END chap3;
ARCHITECTURE a OF chap3 IS
BEGIN
      x <= NOT ((a AND b) OR (c AND d) OR (e AND f) OR g);
      y <= (c OR (NOT d)) AND (g OR (NOT b));
      z <= NOT (a OR b OR c OR d OR e OR f OR g);
END a;
```

25. This VHDL program contains four errors. Find and correct them.

```
ENTITY combo IS
BEGIN
            PORT(a,b,c:IN;
               x,y,z:OUT);
END combo;
ARCHITECTURE b OF combo IS;
BEGIN
      x <=  a AND b;
      y <=  NOT (a OR b OR c);
      z <=  NOT d OR a AND b OR c;
END b;
```

26. This VHDL program contains four errors. Find and correct them.

```
ENTITY alarm IS
      (a,b,c:IN BIT;
        x,y,z:OUT BIT);
END alarm;
ARCHITECTURE c OF alarm IS
BEGIN c
      x <= (NOT a) AND (NOT b);
      y <= NOT (a OR b OR c);
      z <= NOT b;
END;
```

27. Write a VHDL program that will implement these Boolean expressions:

$x = a \cdot \overline{c} \cdot e + \overline{b} \cdot c \cdot d + c \cdot \overline{d} \cdot e.$

$y = \overline{a} \cdot \overline{c} \cdot e + \overline{b} \cdot c \cdot \overline{d} + c \cdot \overline{d} \cdot e.$

$z = \overline{a} \cdot \overline{c} \cdot \overline{e} + \overline{b} \cdot c \cdot \overline{d} + c \cdot d \cdot e.$

28. Write a VHDL program that will implement the circuit shown in Figure 3-42B.

29. Write a VHDL program that will implement the circuit shown in Figure 3-42C.

30. Write a VHDL program that will implement the truth table shown in Figure 3-74.

31. Write a VHDL program that will implement the truth table shown in Figure 3-75.

32. Write a VHDL program that will implement the truth table shown in Figure 3-76.

Problems 33 through 36 refer to Figures 3-57A, B, and C.

33. When U2 was plugged into its socket, pin 5 was bent under the IC and did not make contact with the socket. What effect will this have on the circuit?

34. Input C splits at a node and goes to pin 5 of U2 and pin 5 of U1. A cold solder joint leaves pin 5 of U1 floating or open, but the signal is present on pin 5 of U2. What effect will the floating input have on the circuit?

35. Pin 2 of U3 is bent under the socket. What effect will this have on the circuit?

36. A solder blob shorts pin 14 and pin 13 of U2 together. What effect will this have on the circuit?

LAB 3A BOOLEAN ALGEBRA

OBJECTIVES

After completing this lab, you should be able to:

- Write and verify truth tables for combinational logic circuits.
- Design and construct a decimal-to-BCD encoder.
- Design and construct a circuit to implement a truth table.

COMPONENTS NEEDED

1	7400 IC
1	7402 IC
1	7404 IC
1	7408 IC
1	7411 IC
3	7432 IC
4	LEDs
4	330-Ω resistors

PREPARATION

In Part 1 and Part 2 of Lab 3A, you will be asked to write the Boolean expression and truth table for two circuits. To develop the Boolean expression, write the output expression for the first gates encountered. Use those expressions as inputs to the following gates. For example,

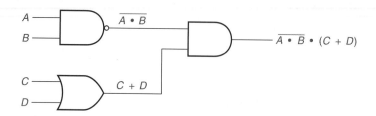

Assign pin numbers to your logic diagram before you begin wiring. To develop the truth table, list all the inputs and all the possible combinations (count in binary). Write a column for each term in the expression. Combine the terms into a final expression.

D	C	B	A	A • B	$\overline{A \cdot B}$	C + D	$\overline{A \cdot B} \cdot (C + D)$
0	0	0	0	0	1	0	0
0	0	0	1	0	1	1	1
	etc.						

To verify the truth table, connect the inputs to switches to supply the required 1s and 0s.

Part 4 asks you to design and build an encoder. An encoder changes a decimal number into another number system or code. A decoder converts a number or code back into decimal. On your encoder, your 10 inputs represent the ten decimal digits. Only one should go HIGH at a time. The outputs are active HIGH also and should produce the corresponding BCD number. For example, if 5 goes HIGH on the input, then outputs 1 and 4 should go HIGH in response.

Draw a logic diagram with pin numbers and IC numbers before beginning construction. Insert your hookup wires away from the IC pins so that you have room to check voltages on the pins of the IC.

Review the lab safety rules presented in the PREPARATION section of Lab 1A, Chapter 1.

PROCEDURE

PART 1

1. Write the Boolean expression and the truth table for the output.
2. Connect the circuit and verify the output.

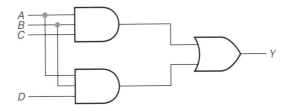

PART 2

1. Write the Boolean expression and truth table for the output.

2. Connect the circuit and verify the output.

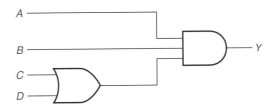

PART 3

1. Write the truth table for the Boolean expression, $Y = \overline{(A + B)C}$.

2. Connect the logic circuit and verify the output.

PART 4

Construct a decimal-to-BCD encoder using three two-input OR gate integrated circuits.

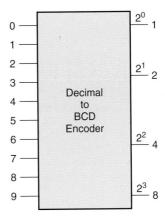

HINT: The 1 output should go HIGH for each odd decimal number input, 1 OR 3 OR 7 OR 9, so OR those inputs together to produce the 1 output (expand an OR with another OR). The 2 output should go HIGH for inputs 2, 3, 6, 7, so OR these inputs together to produce the 2 output, etc.

PART 5

1. Write a Boolean expression and write a truth table for this circuit.

2. Reduce the circuit. Construct and verify.

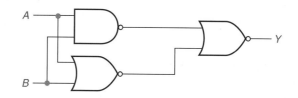

PART 6

1. Design a circuit to implement this truth table.

2. Construct and verify.

Inputs			Output
C	B	A	Y
0	0	0	0
0	0	1	1
0	1	0	0
0	1	1	0
1	0	0	1
1	0	1	1
1	1	0	1
1	1	1	1

LAB 3B   LOGIC CONVERTER

OBJECTIVES

After completing this lab, you should be able to use Multisim to:

- Create a truth table from a given circuit.
- Create a Boolean expression for the output of a given circuit.
- Simplify a Boolean expression to minimal terms.
- Create the NAND gate version of a given circuit.
- Design a circuit to implement a 4-variable truth table.
- Troubleshoot combinational logic circuits.

PREPARATION

Parts 1 and 2 will use the logic converter instrument. It performs much of the work studied in this chapter. For example: from a given truth table, the logic converter can write the Boolean expression, reduce it to minimal terms, draw the logic circuit, and draw the logic circuit using only NAND gates.

PROCEDURE

PART 1

1. Construct this circuit in Multisim.

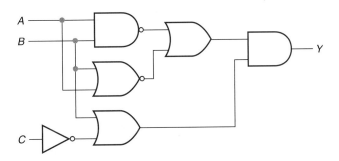

2. Connect the inputs of the logic converter to the inputs of the circuit, and the output of the logic converter to the output of the circuit.

   In the instrument toolbar to the right of the circuit window, click the "Logic Converter" icon (tenth from the top). Click in the circuit window to place the logic converter. From left to right are input terminals A through H. Wire A, B, and C to the inputs of your circuit. The right-most terminal is the output. Wire it to the output of your circuit.

3. Expand the logic converter instrument by double-clicking its icon. Click on the "circuit to truth table" button (the top button in the "conversion" column). Record the truth table.

4. Click on the "truth table to Boolean" button. Record the Boolean expression.

5. Click on the "truth table to simplified Boolean" button. Record the minimized Boolean expression.

6. Click on the "Boolean to circuit" button. Draw the reduced circuit. How does it compare to the original?

7. Finally, click on the "Boolean to NAND circuit" button. Record the NAND gate version of the circuit.

8. Write the Boolean expression for the output of the NAND gate version. Reduce to minimal terms. How does it compare with the logic converter's simplified Boolean expression?

PART 2

1. Use Boolean algebra or a Karnaugh map to design a circuit to implement this 4-variable truth table.

A	B	C	D	Y
0	0	0	0	1
0	0	0	1	1
0	0	1	0	0
0	0	1	1	0
0	1	0	0	1
0	1	0	1	1
0	1	1	0	0
0	1	1	1	0
1	0	0	0	1
1	0	0	1	1
1	0	1	0	0
1	0	1	1	0
1	1	0	0	0
1	1	0	1	1
1	1	1	0	0
1	1	1	1	0

2. Now enter the truth table into the logic converter. (Click on the circles above the variables A, B, C, and D and the logic converter will count in binary for you.) Click on a number in the output column to change the number's value.

3. Have the logic converter write the Boolean expression.

4. Have the logic converter simplify the Boolean expression.

5. Have the logic converter draw the circuit.

6. How do your results compare to the logic converter's?

PART 3

1. There are 5 circuits for you to download and troubleshoot, Files 3B-1.ms9 through 3B-5.ms9. Sketch a logic diagram of the circuit under study to understand how it is connected, and transfer the pin numbers from the Multisim schematic to your logic diagram. Use the word generator to supply inputs to the circuit. Choose Window Description to see the truth table that is being implemented. Step through the truth table until a fault is encountered (the output does not agree with the truth table). Leave the word generator in that state while you investigate. Log the steps taken.

Exclusive-OR Gates

OUTLINE

KEY TERMS

BIT_VECTOR

even-parity

exclusive-NOR

exclusive-OR

identity comparator

magnitude comparator

nonexclusive-OR

odd-parity

parity bit

parity checker

parity generator

PROCESS

OBJECTIVES

After completing this chapter, you should be able to:

- Draw the symbol and write the truth table for an exclusive-OR gate.

- Construct an exclusive-OR gate from basic gates.

- Use an exclusive-OR gate to invert data.

- Use an exclusive-OR gate to pass data unaltered.

- Draw the symbol and write the truth table for an exclusive-NOR gate (nonexclusive-OR).

- Predict the output waveform from an exclusive-OR/NOR.

- Predict even-parity and odd-parity bits.

- Construct a parity generator using exclusive-OR gates.

- Construct a parity checker using exclusive-OR gates.

- Use the 74S280 as a parity generator and parity checker.

- Construct a comparator using exclusive-OR gates.

- Use the 7485 as a 4-bit magnitude comparator.

- Use the 74AC11521 as an 8-bit identity comparator.

- Program a CPLD to function as an exclusive-OR and as a comparator.

- Troubleshoot exclusive-OR based circuits.

4.1 EXCLUSIVE-OR

The **exclusive-OR** is a two-input gate that produces a 1 on its output when its inputs are different and a 0 if they are the same. The symbol and the truth table for an exclusive-OR are shown in Figure 4-1. Notice that the output Y is 1 if A is 1 or if B is 1, but not if both A and B are 1. If A and B are both 1s, the 1s are excluded from the output; hence the name of the gate is the exclusive-OR. The output Y is sometimes written $A \oplus B$ which is read "A exclusive-OR B."

Inputs		Output
B	A	Y
0	0	0
0	1	1
1	0	1
1	1	0

FIGURE 4-1 Exclusive-OR symbol and truth table

An exclusive-OR gate is not one of the basic gates, but is constructed from a combination of the basic gates. To design an exclusive-OR, first write a Boolean expression for the truth table in Figure 4-1. The output is 1 for the conditions on line 2 or line 3. On line 2 we have $A \cdot \overline{B}$ and on line 3 we have $\overline{A} \cdot B$. The output Y is 1 for $\overline{A} \cdot B + A \cdot \overline{B}$. The logic diagram for this expression is shown in Figure 4-2. This solution requires two AND gates, one OR gate, and, if the complemented inputs are not available, two inverters.

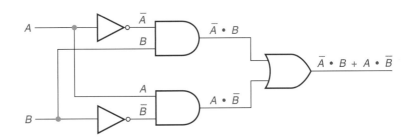

FIGURE 4-2 Exclusive-OR logic diagram

An equivalent logic diagram can be developed by using Boolean algebra and DeMorgan's Theorems.

First apply DeMorgan's theorem to the overall expression.

$$Y = \overline{A}\,B + A\,\overline{B}$$

Step 1. $\overline{\overline{A}\,B + A\,\overline{B}}$ Complement the entire expression.

Step 2. $\overline{\overline{A}\,B \cdot A\,\overline{B}}$ Change the OR to AND.

Step 3. $\overline{\overline{\overline{A}\,B} \cdot \overline{A\,\overline{B}}}$ Complement each term.

Now apply DeMorgan's theorems to each term, $\overline{\overline{A}\,B}$ and $\overline{A\,\overline{B}}$, and substitute the equivalent expressions.

$$\overline{\overline{A}\,B}$$

Step 1. $\overline{\overline{\overline{A}B}} = \overline{A}\,B$ Complement the entire expression.

Step 2. $\overline{A} + B$ Change the AND to OR.

Step 3. $\overline{\overline{A} + B} = A + \overline{B}$ Complement each term.

$$\overline{A\,\overline{B}}$$

Step 1. $\overline{\overline{A\,\overline{B}}} = A\,\overline{B}$ Complement the entire expression.

Step 2. $A + \overline{B}$ Change the AND to OR.

Step 3. $\overline{A} + \overline{\overline{B}} = \overline{A} + B$ Complement each term.

So, $\overline{\overline{A}\,B} \cdot \overline{A\,\overline{B}}$ becomes $\overline{(A + \overline{B})} \cdot (\overline{A} \cdot B)$

By Theorem 10,

$$Y = (A + \overline{B}) \cdot (\overline{A} + B) = A\,\overline{A} + A\,B + \overline{A}\,\overline{B} + \overline{B}\,B$$

$$= 0 + A\,B + \overline{A}\,\overline{B} + 0$$

$$= A\,B + \overline{A}\,\overline{B}$$

The logic diagram for this expression is shown in Figure 4-3. This solution requires two NOR gates and an AND gate.

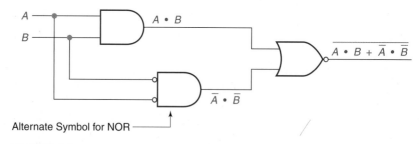

FIGURE 4-3 Equivalent exclusive-OR logic diagram

The pinouts for a 7486 and a 4070 quad two-input exclusive-OR are shown in Figure 4-4.

Some exclusive-OR gates and exclusive-NOR gates that are readily available are listed in Table 4-1.

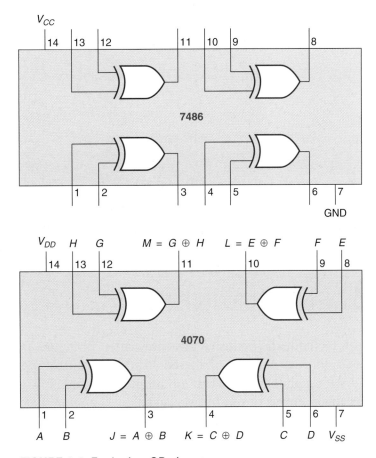

FIGURE 4-4 Exclusive-OR pinouts

TABLE 4-1 Exclusive-OR and Exclusive-NOR Gates

Number	Family	Description
74ACT86	Advanced CMOS-TTL compatible	Quad 2-input exclusive-OR
74ALS86	Advanced Low-Power Schottky TTL	Quad 2-input exclusive-OR
4030	CMOS	Quad 2-input exclusive-OR
4070	CMOS	Quad 2-input exclusive-OR
74HC7266	High-Speed CMOS	Quad 2-input exclusive-NOR

The IEC symbol for a 7486 quad two-input exclusive-OR gate integrated circuit is shown in Figure 4-5. The =1 sign indicates that exactly one input must be active. Since there are no triangles on inputs or outputs, all are active HIGH. Exactly one input HIGH makes the output HIGH.

The 1 sign was used to signify the inverter (Chapter 2), the ≥1 to signify the OR gate and NOR gate (Chapter 2), and =1 to signify the exclusive-OR gate.

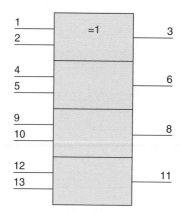

FIGURE 4-5 IEC-symbol—
7486 quad exclusive-OR gate

 # 4.2 ENABLE/INHIBIT

The truth table for the exclusive-OR is rewritten in Figure 4-6. When the control input is 0, data passes through the gate unaltered. When the control input is 1, the data passes through inverted. Although there is no real inhibit mode as there is with the basic gates, it is quite useful to be able to invert or not invert a signal by changing the control input. We will see applications of the exclusive-OR used in this mode in this and later chapters.

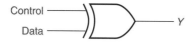

	Inputs		Output
	Control	Data	Y
Data passes	0	0	0
	0	1	1
Data passes inverted	1	0	1
	1	1	0

FIGURE 4-6 Enable/inhibit

EXAMPLE 4-1

Predict the output of each gate.

Solution Use the waveform as the data input and the static signal as the control. For the first gate the control input is 1, and the data passes through inverted. For the second gate the control signal is 0, and the data passes through unaltered.

4.3 WAVEFORM ANALYSIS

The output of an exclusive-OR is 1 when the inputs are unlike, and 0 when the inputs are the same. This makes it easy to predict the output waveform from given inputs. In Figure 4-7, the intervals in which A and B are unlike are shaded. The output, Y, is HIGH during these times and LOW elsewhere.

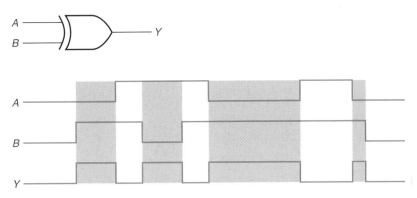

FIGURE 4-7 Waveform analysis

EXAMPLE 4-2

Predict the output of the exclusive-OR.

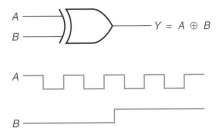

Solution Find the times when the inputs are different. The output is HIGH during those times. It is LOW at all other times.

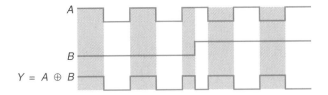

Another way to analyze this problem is to consider the B input as the control. In the first half of the waveform, B is 0 and the data passes to the output unaltered. When B goes HIGH, the data on A passes through inverted.

4.4 EXCLUSIVE-NOR

An **exclusive-NOR**, sometimes called **nonexclusive-OR**, has a truth table and symbol as shown in Figure 4-8. The output, Y, is HIGH when the inputs are alike and LOW when they are unlike.

$$Y = \overline{A \oplus B}$$

Inputs		Output
B	A	Y
0	0	1
0	1	0
1	0	0
1	1	1

FIGURE 4-8 Exclusive-NOR

EXAMPLE 4-3

Predict the output of each gate.

Solution The first two gates are exclusive-ORs. Since the inputs to the first gate are different, its output is 1. The last two gates are exclusive-NORs. Since the inputs to the third gate are different, its output is 0.

The 74HC7266 is an example of a quad two-input exclusive-NOR.

EXAMPLE 4-4

Predict the output.

Solution   When *C* is HIGH, the circuit functions as an exclusive-NOR. Find the times when *A* and *B* are different and draw the output LOW. For the rest of the time that *C* is HIGH, draw the output HIGH. When *C* is LOW, the circuit functions as an exclusive-OR. Find the times when *A* and *B* are different and draw the output HIGH. For the rest of the time that *C* is LOW, draw the output LOW.

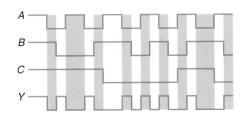

SELF-CHECK 1

Predict the output of each gate.

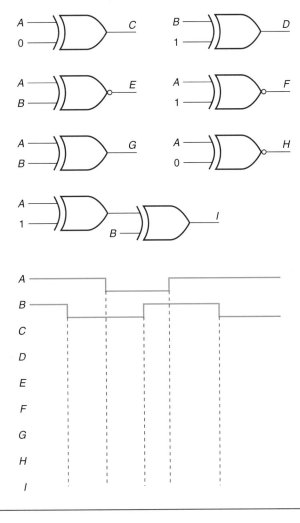

4.5 PARITY

When data bits are transferred from one circuit to another, an extra bit is sometimes added to ensure that the data is transferred correctly. The extra bit is called a **parity bit**. The system can work on an **even-parity** or **odd-parity** system. If the system is *even parity,* then the parity bit is chosen so that the total number of 1s in the word, including the parity bit, is even. For example, suppose that we have these seven data bits to transmit:

>   1011101

Let's add an even-parity bit as the most significant bit. There are five 1s in the data bits so the even-parity bit must be a 1 to make the total number of 1s even. The word would be transmitted as:

>   11011101

At the receiving end, the 1s would be counted. If the total number of 1s was not even, an alarm or flag would be set, notifying that an error has occurred.

EXAMPLE 4-5

Generate the even-parity bit for these data bits:

>   0000000

Solution   Zero 1s is considered even, so the even-parity bit must be a 0.

>   00000000

EXAMPLE 4-6

Generate the even-parity bit for these data bits:

>   1111111

Solution   Seven 1s is odd, so the even-parity bit must be a 1.

>   11111111

EXAMPLE 4-7

Generate the even-parity bit for these data bits:

>   1010101

Solution   Four 1s is even, so the even-parity bit must be a 0.

>   01010101

In an *odd-parity system,* the parity bit is chosen to make the total number of 1s odd.

EXAMPLE 4-8

Generate the odd-parity bit for these data bits:

1100110

Solution   Four 1s is even, so the odd-parity bit must be a 1.

11100110

EXAMPLE 4-9

Generate the odd-parity bit for these data bits:

1100111

Solution   Five 1s is odd, so the odd-parity bit must be a 0.

01100111

EXAMPLE 4-10

Generate the odd-parity bit for these data bits:

0000000

Solution   Zero 1s is considered even, so the odd-parity bit must be a 1.

10000000

4.6 EVEN-PARITY GENERATOR

A circuit that can determine whether the parity bit should be a 1 or 0 is called a **parity generator**. Exclusive-OR gates can be used to construct a parity generator. Figure 4-9 shows exclusive-OR gates being used as an even-parity generator. Each exclusive-OR gate checks for unlike inputs. Unlike inputs into the last exclusive-OR means that an odd number of 1s has been encountered. The 1 output would be used as a parity bit, making the total number of 1s even.

EXAMPLE 4-11

Use the even-parity generator in Figure 4-9 to generate the even-parity bit for 1011101.

Solution   Five 1s is odd. The generator supplies a 1 as the even-parity bit.

11011101

See Figure 4-10.

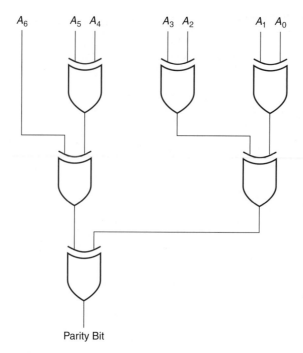

FIGURE 4-9 Even-parity generator

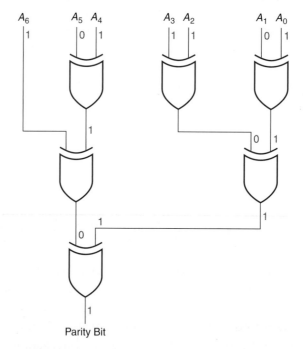

FIGURE 4-10

EXAMPLE 4-12

Use the even-parity generator in Figure 4-9 to supply the even-parity bit for 1000001.

Solution   Two 1s is even. The generator supplies an even-parity bit of 0.

01000001

See Figure 4-11.

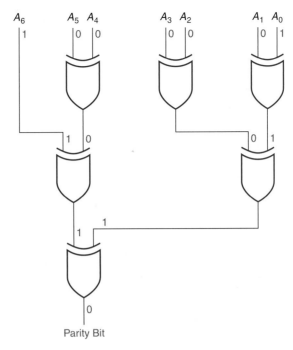

FIGURE 4-11

EXAMPLE 4-13

Use exclusive-OR gates to construct a 6-bit even-parity generator. The parity bit will become the seventh bit.

Solution   See Figure 4-12.

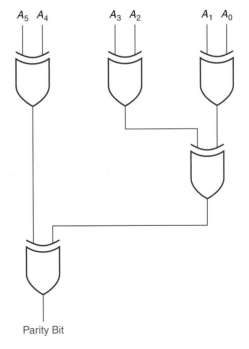

FIGURE 4-12

4.7 EVEN/ODD-PARITY GENERATOR

By adding an extra exclusive-OR gate, the circuit can be made more versatile. A 1 on the control input inverts the output and changes the circuit into an odd-parity generator. See Figure 4-13.

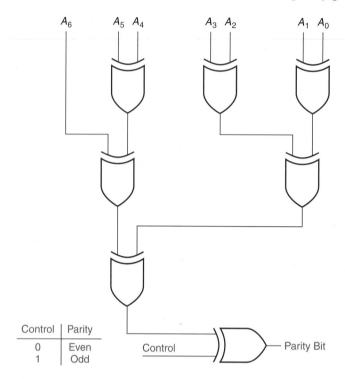

Control	Parity
0	Even
1	Odd

FIGURE 4-13 Even/odd-parity generator

EXAMPLE 4-14

Use the even/odd-parity generator in Figure 4-13 to supply the odd-parity bit for 0110101.

Solution   The control input must be 1 for an odd-parity generator. Four 1s is even, so the generator supplies an odd-parity bit of 1.

10110101

See Figure 4-14.

EXAMPLE 4-15

Use the parity generator in Figure 4-13 to supply the even-parity bit for 1111110.

Solution   The control input must be 0 for an even-parity generator. Six 1s is even, so the even-parity bit must be 0.

01111110

See Figure 4-15.

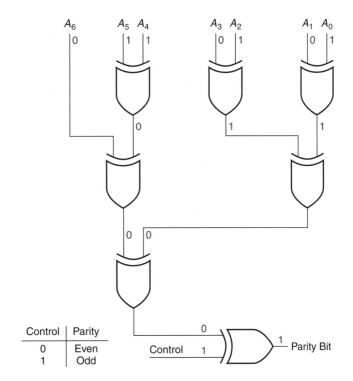

Control	Parity
0	Even
1	Odd

FIGURE 4-14

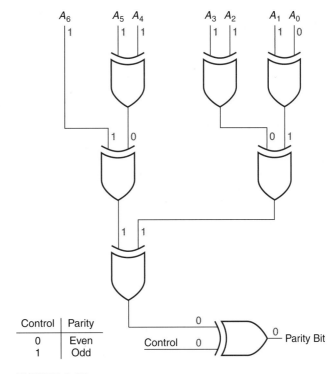

Control	Parity
0	Even
1	Odd

FIGURE 4-15

4.8 PARITY CHECKER

One way for a PC to communicate with the external world is through a communications program such as HyperTerminal®. Under a Windows operating system, HyperTerminal can be started with this sequence: Start Programs Accessories Communications HyperTerminal. After choosing which port to use (Com 1, Com 2), "Port Settings" are made. Port settings include the transmission rate in bits per second (baud), such as 9600, 1900, etc.; the number of data bits to send in each character, such as 5, 6, 7, or 8; and parity (none, even, odd). Suppose HyperTerminal is initialized to transmit out of Com 1 seven data bits with an even-parity bit. The receiving circuit checks to see that there is an even number of ones in the package. A circuit that can determine whether the total number of 1s is even or odd is called a **parity checker**. Figure 4-16 shows eight bits (seven data and one parity) fed into an exclusive-OR gate circuit.

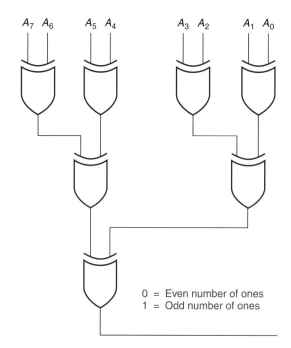

FIGURE 4-16 8-bit parity checker

EXAMPLE 4-16

01001101 has been received as seven data bits and an even-parity bit. Use the circuit in Figure 4-16 to check for parity errors.

Solution   The 0 out indicates that an even number of 1s has been received. No parity error has been detected.

See Figure 4-17.

EXAMPLE 4-17

11000110 has been received as seven data bits and an odd-parity bit. Use the circuit in Figure 4-16 to check for parity error.

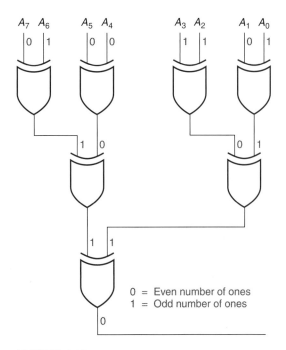

FIGURE 4-17

Solution   The 0 out indicates that an even number of 1s has been received. An error has occurred.

One of the bits has probably been inverted during transmission or reception.

See Figure 4-18.

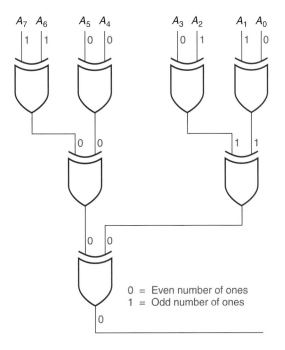

FIGURE 4-18

SELF-CHECK 2

1. Determine the even-parity bit for these data bits:

   0101110

2. Determine the odd-parity bit for these data bits:

   1011011

3. Use exclusive-OR gates to construct a 6-bit odd-parity generator. The parity bit will be the seventh bit.

4. 11010111 has been received as seven data bits and an odd-parity bit. Use the circuit in Figure 4-16 to check for parity error.

 4.9 9-BIT PARITY GENERATOR/CHECKER

The 74S280 is a medium-scale integrated circuit which functions as a 9-bit parity generator/checker. The pinout and truth table are shown in Figure 4-19. If the number of inputs (A through I) that are HIGH is even, then the Σ_{EVEN} output goes HIGH and the Σ_{ODD} output goes LOW (line 1 of the truth table). To use the IC as an even-parity generator, use the Σ_{ODD} output to generate the parity bit.

Number of Inputs (A–I) that are HIGH	Outputs	
	Σ EVEN	Σ ODD
0, 2, 4, 6, 8	H	L
1, 3, 5, 7, 9	L	H

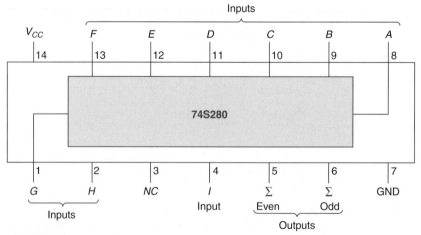

FIGURE 4-19 74S280 9-bit generator/checker

EXAMPLE 4-18

Use the 74S280 as an 8-bit even-parity generator (seven data bits and one parity bit). Generate the parity bit for 0101010.

Solution   *A* through *G* supply the seven data inputs. *H* and *I* are not used and are grounded. Pin 3 is not connected internally (NC).

0101010

Three inputs HIGH is the second line of the truth table.

Σ_{ODD} goes HIGH and sets the even-parity bit to 1.

10101010

The total number of 1s is even.

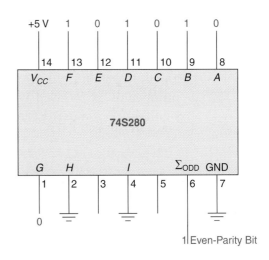

EXAMPLE 4-19

Use the 74S280 as an 8-bit even-parity generator (seven data bits and one parity bit). Generate the parity bit for 1111110.

Solution   Six inputs HIGH is the first line of the truth table. Σ_{ODD} goes LOW and resets the even-parity bit to 0.

01111110

The total number of 1s is even.

To use the 74S280 as an odd-parity generator, use the Σ_{EVEN} output to generate the parity bit.

EXAMPLE 4-20

Use the 74S280 as an 8-bit odd-parity generator (seven data bits and one parity bit). Generate the parity bit for 1010101.

Solution   Place the data bits on inputs *A–G*. Ground inputs *H* and *I* so that they will not affect the output. The output on pin 5 (Σ_{EVEN}) will be the odd-parity bit. In this case there are four 1s on the inputs. The first line of the truth table says that the Σ_{EVEN} output will be HIGH, as it should be for the odd-parity bit 11010101.

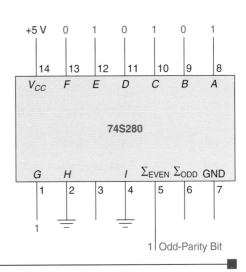

To use the 74S280 as an odd-parity checker, feed up to nine bits into inputs A through I. If the total number of inputs that are HIGH is odd, as it should be (bottom line of the truth table), then Σ_{ODD} goes HIGH and Σ_{EVEN} goes LOW. LOW-level outputs on this IC can sink (provide a path to ground) 20 mA so Σ_{ODD} could drive an LED as shown in Figure 4-20. If the total number of 1s received were even, then Σ_{ODD} would go LOW and the LED would light, signifying that a parity error had occurred.

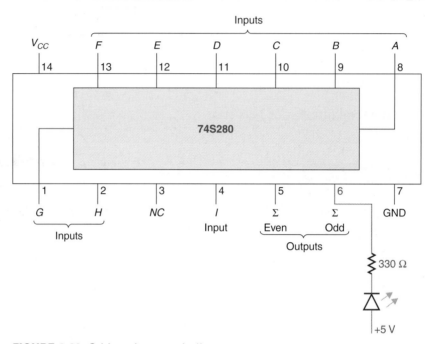

FIGURE 4-20 Odd-parity error indicator

EXAMPLE 4-21

Use the 74S280 as an 8-bit odd-parity checker (Figure 4-20) to check for parity error on the data 10111010.

Solution The eight bits are fed into the 74S280 on inputs A through H. I is grounded so that it will not influence the output. The second line of the truth table shows that five 1s in produce a HIGH on the Σ_{ODD} output, and the LED does not light. The LED off indicates lack of a parity error.

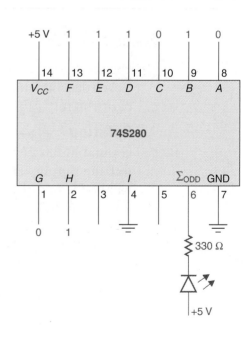

EXAMPLE 4-22

Use the 74S280 as an 8-bit even-parity checker. Have an LED light if there is no parity error. Test the circuit with 10111011 and 10111010.

Solution When an even number of 1s is input, the Σ_{ODD} output goes LOW (line 1 of the truth table). This output can be used to light an LED, indicating no parity error.

When 10111011 is input, Σ_{ODD} goes LOW (line 1 of the truth table), indicating an even number of inputs, and the LED lights (no parity error). When 10111010 is input, Σ_{ODD} goes HIGH (line 2 of the truth table), the LED does not light, indicating an even-parity error.

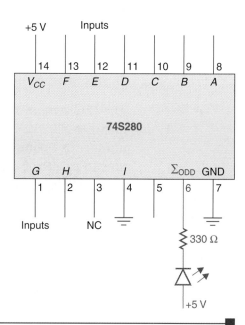

The 74S280 can be expanded by tying the Σ_{EVEN} output to a data input of a following stage.

EXAMPLE 4-23

Use two 74S280s to produce a 17-bit odd-parity checker. Have an LED light on parity error.

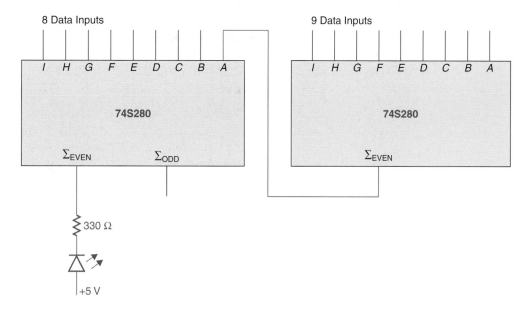

Solution Place all 1s into the first 74S280 (right) and all 0s into the second (left). The output on Σ_{EVEN} would be LOW. All 0s into the second 74S280 would cause its Σ_{EVEN} to go HIGH, and the LED would not light. This is correct since there is an odd number of 1s in.

In summary, in a parity generator, data bits are fed into the generator, and the output of the generator provides the parity bit. In a parity checker, the parity bit and data bits are fed into the checker, and the output indicates whether an error has been detected.

The IEC symbol for a 74S280 is shown in Figure 4-21. The 2k indicates that an even number of 1s must be present on the input pins to cause the output on pin 5 (Σ_{EVEN}) to go HIGH (no triangle) and the output on pin 6 (Σ_{ODD}) to go LOW (triangle).

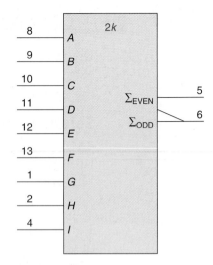

FIGURE 4-21

 # 4.10 COMPARATOR

The exclusive-OR can also be used to compare two numbers and decide if they are equal. Figure 4-22 shows a circuit that compares each bit of two numbers $A_3A_2A_1A_0$ and $B_3B_2B_1B_0$. If any corresponding bits are unequal, a 1 is fed into the NOR gate for a 0 out. Therefore, a 0 out indicates that the numbers are not equal, and a 1 out means that the numbers are equal.

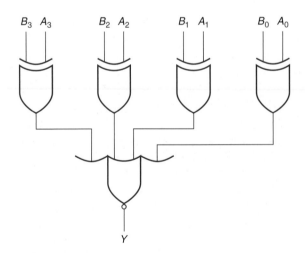

FIGURE 4-22 4-bit comparator

EXAMPLE 4-24

Compare 1010 and 1001.

Solution   0 out indicates that the numbers are not equal.

See Figure 4-23.

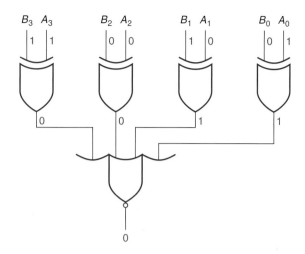

FIGURE 4-23

The 74LS85 and 74HC85 are 4-bit **magnitude comparators** whose outputs indicate whether $A = B, A < B$, or $A > B$. The comparator in Figure 4-22 indicates only whether $A = B$ or $A \neq B$.

The pinout and functional diagram for the 74LS85 4-bit magnitude comparator are shown in Figure 4-24. In the functional diagram, the pins have been grouped together according to function. Pins 2, 3, and 4 are expansion inputs that are used to expand the IC for use in circuits with more than four bits. When used as a 4-bit comparator, pins 2 and 4 ($I_{A<B}$ and $I_{A>B}$) must be grounded, and pin 3 ($I_{A=B}$) must be HIGH. Figure 4-25 shows the truth table of the '85. The IC compares the two 4-bit numbers on its inputs, $A_3A_2A_1A_0$ and $B_3B_2B_1B_0$, and sends output 5, 6, or 7 HIGH, depending on the relative magnitude of the 4-bit numbers' input. If the 4-bit number A is greater than B ($A > B$), then the output $Q_{A>B}$ goes HIGH. If A and B are equal ($A = B$), then output $Q_{A=B}$ goes HIGH. If A is less than B ($A < B$), then output $Q_{A<B}$ goes HIGH.

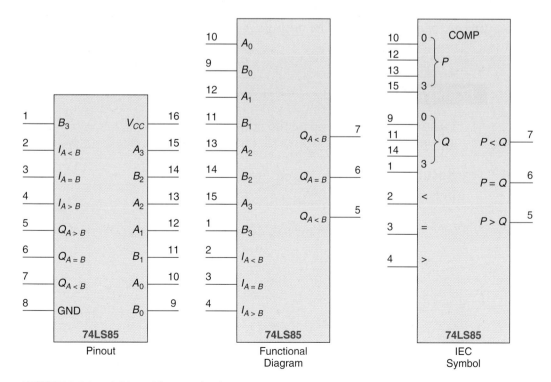

FIGURE 4-24 74LS85 4-bit magnitude comparator

Comparing Inputs				Cascading Inputs			Outputs		
A_3, B_3	A_2, B_2	A_1, B_1	A_0, B_0	$A > B$	$A < B$	$A = B$	$A > B$	$A < B$	$A = B$
$A_3 > B_3$	X	X	X	X	X	X	H	L	L
$A_3 < B_3$	X	X	X	X	X	X	L	H	L
$A_3 = B_3$	$A_2 > B_2$	X	X	X	X	X	H	L	L
$A_3 = B_3$	$A_2 < B_2$	X	X	X	X	X	L	H	L
$A_3 = B_3$	$A_2 = B_2$	$A_1 > B_1$	X	X	X	X	H	L	L
$A_3 = B_3$	$A_2 = B_2$	$A_1 < B_1$	X	X	X	X	L	H	L
$A_3 = B_3$	$A_2 = B_2$	$A_1 = B_1$	$A_0 > B_0$	X	X	X	H	L	L
$A_3 = B_3$	$A_2 = B_2$	$A_1 = B_1$	$A_0 < B_0$	X	X	X	H	H	L
$A_3 = B_3$	$A_2 = B_2$	$A_1 = B_1$	$A_0 = B_0$	H	L	L	H	L	L
$A_3 = B_3$	$A_2 = B_2$	$A_1 = B_1$	$A_0 = B_0$	L	H	L	L	H	L
$A_3 = B_3$	$A_2 = B_2$	$A_1 = B_1$	$A_0 = B_0$	L	L	H	L	L	H
		H = HIGH Level		L = LOW Level		X = Don't Care			

FIGURE 4-25   Truth table of the 74LS85

The truth table explains how the IC functions. The top line shows that if A_3 is greater than B_3, the other data inputs and cascading inputs are not considered, and A must be larger than B. The output $A > B$ goes HIGH. Likewise, if A_3 is less than B_3, A is less than B. The output $A < B$ goes HIGH. Lines 3 and 4 of the truth table show what happens if A_3 and B_3 are equal. A_2 and B_2 are compared, and the outputs are set accordingly. The bottom three lines show what happens when $A_3A_2A_1A_0$ is equal to $B_3B_2B_1B_0$.

Only then are the cascading inputs $I_{A>B}$, $I_{A<B}$, $I_{A=B}$ considered in the decision. These inputs represent the relative magnitudes of less significant digits when numbers of greater than four bits are being compared. The outputs $A > B$, $A < B$, $A = B$ follow the cascading inputs.

EXAMPLE 4-25

Use the 74LS85 to compare the numbers $A = 1011$ and $B = 1100$.

Solution   Since $A < B$, pin 7 will go HIGH, and pins 6 and 5 will go LOW.

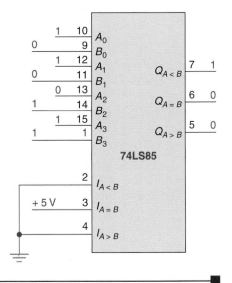

To expand the 74LS85 into an 8-bit comparator, the output $Q_{A<B}$ of the first (least significant) IC is wired to the input $I_{A<B}$ of the second (most significant) IC. Likewise, $Q_{A=B}$ is wired to $I_{A=B}$, and $Q_{A>B}$ is wired to $I_{A>B}$.

EXAMPLE 4-26

Use two 74LS85s to compare the 8-bit numbers $A = 9D_{16}$ and $B = B6_{16}$.

Solution $9D_{16} = 10011101_2$ and $B6_{16} = 10110110_2$

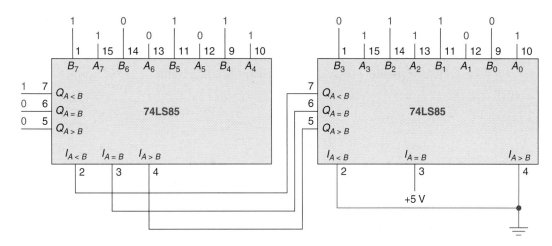

The outputs of the first IC (right) are wired into the expansion inputs of the second IC (left). Since A is less than B, pin 7 goes HIGH on the second IC and pins 6 and 5 go LOW.

The 74HC688 and 74LS682, 684, and 688 are additional examples of 8-bit magnitude comparators.

The 74ACT11521 and 74AC11521 are 8-bit **identity comparators** that compare two 8-bit binary or BCD numbers and produce a LOW output if the two numbers are identical. As shown in the truth table in Figure 4-26A, the enable input $\overline{E}$ must be LOW (first three lines of the truth table) for the IC to function. If E is HIGH, the IC ignores the inputs and the output is locked HIGH (fourth line of the truth table). Line one shows that the output $\overline{P = Q}$ goes LOW if input P = input Q. Figure 4-26B shows the pin configuration and logic symbols. The logic diagram of the internal circuitry in Figure 4-26C shows that the IC consists of inverters, exclusive NORs, and a 9-input NAND gate.

Inputs		Output
DATA P, Q	ENABLE $\overline{E}$	$\overline{P = Q}$
$P = Q$	L	L
$P > Q$	L	H
$P < Q$	L	H
X	H	H

FIGURE 4-26A 74ACT11521 truth table

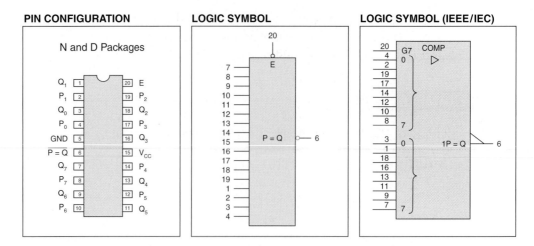

FIGURE 4-26B 74ACT11521 8-bit identity comparator

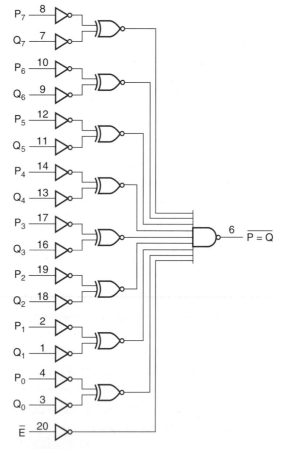

FIGURE 4-26C 74AC11521 8-bit identity
comparator

The 74ACT520 and 74FCT521 are other examples of 8-bit identity comparators.

 SELF-CHECK 3

1. Use the 74S280 as an 8-bit odd-parity generator (seven data bits and one parity bit). Generate the parity bit for 1011011.

2. Use the 74S280 as an 8-bit odd-parity checker. Check for parity error on the data 10101111.

3. Use the exclusive-OR comparator to compare the numbers 1110 and 1100.

4. Use the 7485 to compare 1010 and 1000.

5. Use the 8-bit identity comparator in Figure 4-26C to compare 11000110 to 11100110 by following the logic levels through the circuit. What logic level must be placed on $\bar{E}$, pin 20, to enable the output NAND gate?

4.11 PROGRAMMING A CPLD (Optional)

In this section, we will program a CPLD to perform exclusive-OR operations and to implement a 4-bit magnitude comparator.

Follow the procedure presented in Chapters 2 and 3:

Step 1. Create a new folder within the Altera folder and name it **ex_or**.

Step 2. Run QuartusII.

Step 3. Run "New Project Wizard."

Step 4. Create a new VHDL file and name it **ex_or.vhd**.

Step 5. Write a VHDL program that will implement these circuits:

> x is a exclusive-ORed with b.

> y is a, b, c, and d exclusive-ORed together.

> z is a AND b exclusive-ORed with c and d.

Step 6. Compile the program.

Steps 7–10. Create a .vwf (vector waveform file) named **ex_or.vwf** and use it to simulate the circuits. A sample vector waveform file for **ex_or.vwf** is shown in Figure 4-27. Notice the thin glitches that occur in the output waveforms each time multiple inputs change state at the same time. Note also the short delay between a change in the input signal and a corresponding change in the output signal. These propagation delays are discussed in Chapter 6.

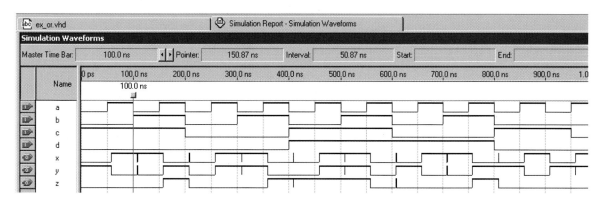

FIGURE 4-27 Simulation waveforms for ex_or.vwf

Step 11.   Assign pins to the input and output signals. Recompile your program.

Step 12.   Program your CPLD.

Step 13.   Test the operation of the CPLD using input switches and LEDs.

Ex_or.vhd looks like this:

```
ENTITY ex_or  IS
    PORT (a,b,c,d:IN BIT;
            x,y,z:OUT BIT);
END ex_or ;
ARCHITECTURE a OF ex_or IS
BEGIN
    x <= a XOR b;
    y <= a XOR b XOR c XOR d;
    z <= (a AND b) XOR (c AND d);
END a;
```

In the next project, we will use "IF" statements to make decisions about the input signals. An "IF . . . THEN" statement is used first. It is followed by one or more "ELSIF . . . THEN" statements. The last "ELSIF . . . THEN" statement is followed by "END IF." The set of "IF" statements is contained within a PROCESS. A PROCESS starts with a "sensitivity list" and "BEGIN" and ends with "END PROCESS." The sensitivity list is one or more signals listed within parentheses, for example, PROCESS (a,c,g). If one of the signals on the list changes value, then the PROCESS is executed to determine changes in outputs.

Signal names must begin with a letter. They can contain numbers and underscores (_) as well. There cannot be two underscores in succession, and a signal name cannot end with an underscore. A signal name cannot be a VHDL keyword such as PORT or BEGIN or END.

Let's create a magnitude comparer that will compare two 4-bit numbers, **a** (a_3 a_2 a_1 a_0) and **b** (b_3 b_2 b_1 b_0). The three outputs will tell whether a is equal to b, greater than b, or less than b. If a equals b, then aEQb will go HIGH and the other outputs will go LOW. Likewise, if a is greater than b, then aGTb will go HIGH and the other outputs will go LOW. If a is less than b, then aLTb will go HIGH and the other two outputs will go LOW. The project is named **magcomp_4**.

```
ENTITY magcomp_4  IS
    PORT(a,b:IN BIT_VECTOR (3 DOWNTO 0);
        aEQb, aGTb, aLTb:OUT BIT);
END magcomp_4;
ARCHITECTURE a OF magcomp_4 IS
BEGIN
    PROCESS (a,b)
    BEGIN
        IF a = b THEN
                aEQb <= '1';
                aGTb <= '0';
                aLTb <= '0';
```

```
          ELSIF a > b THEN
                aEQb <= '0';
                aGTb <= '1';
                aLTb <= '0';
          ELSIF a < b THEN
                aEQb <= '0';
                aGTb <= '0';
                aLTb <= '1';
          END IF;
     END PROCESS;
END a;
```

Notice these differences from the previous programs.

- **a** and **b** are declared as BIT_VECTORS, which means that a and b each consist of more than a single bit. (3 DOWNTO 0) means a is actually $a_3\ a_2\ a_1\ a_0$ and b is actually $b_3\ b_2\ b_1\ b_0$. BIT_VECTOR (0 TO 3) would define a and b as $a_0\ a_1\ a_2\ a_3$ and $b_0\ b_1\ b_2\ b_3$, respectively.

- IF statements are used to determine the relationship between a and b.

- IF statements must be contained within a PROCESS.

- There is an END IF statement.

- Both ARCHITECTURE and PROCESS have a BEGIN and an END.

- When the outputs are set to a logic level (1 or 0), the level is contained within single quotes and <= is used. E.g.: aLTb <= '1'.

There are two approaches to assigning input waveforms for vector (multi-bit) signals like a and b.

Process 1:   Click the "+" sign to the left of the signal names a and b to show waveforms for $a_3\ a_2\ a_1\ a_0$ and $b_3\ b_2\ b_1\ b_0$. Use the waveform editing tool to draw waveforms for each bit. Create inputs that will thoroughly test the circuit. Figure 4-28A shows those waveforms expanded so that each bit is shown.

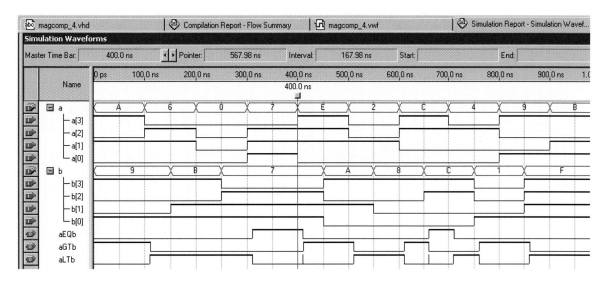

FIGURE 4-28A Simulation waveforms for magcomp_4.vwf expanded

Process 2: Click **Waveform Editing Tool**. On the a or b composite input signal (not one of the individual bits like a_3), click at the beginning of a time interval and drag to the end of the interval. This should highlight the interval to be defined. The "Arbitrary Window" appears. In the "Radix" window, choose "Binary" or "Hexadecimal." Enter the number to be assigned to that interval. The resulting patterns for this project's simulation waveforms are shown in Figure 4-28B with a and b in composite form.

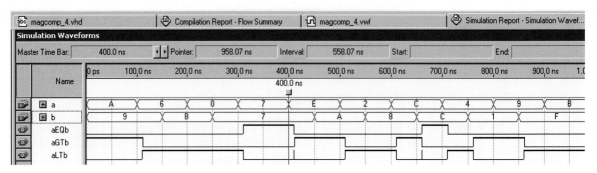

FIGURE 4-28B Simulation waveforms for magcomp_4.vwf

 SELF-CHECK 4

Write a VHDL program to implement these four circuits on the same CPLD. Call your program **circuit.vhd**.

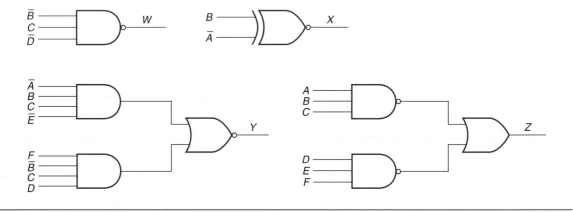

 4.12 TROUBLESHOOTING EXCLUSIVE-OR CIRCUITS

In the Chapter 4 labs you will be working with exclusive-OR gates and with a 9-bit parity checker/generator IC. These ICs are subject to several faults. If your IC is not behaving properly, check the power supply connections before worrying about anything else. Measure the voltages right on the pin of the IC itself, using an oscilloscope or voltmeter. On all TTL integrated circuits, V_{CC} should measure close to +5 volts, and the ground pin of the IC should measure 0 volts. If the supply voltages do not measure correctly, trace the wiring or printed circuit traces back to the power supply.

If the power pins measurement is correct, and the outputs are not correct, take a look at the voltage levels on the input pins. If the input levels are correct on the pins of the IC itself, and the outputs are not correct, either the IC is bad or something connected to the output pin is keeping it from operating correctly. Both possibilities must be considered.

ICs can have internal faults. An input pin can be shorted to (connected to) another input or to V_{CC} or to the ground pin. Or an input can be open, with the small internal conductor connecting the pin to the internal chip either broken or not connected. Likewise, an output pin can be shorted to another pin or open (not connected to anything). Decide that you can find the fault. With that attitude, your troubleshooting skills should improve.

In a circuit with several ICs, if the final output is not correct, and the power supply connections and input measurements are correct, divide the circuit into halves. Measure the voltages at test points in the center of the circuit. If they are correct, the fault lies in the second half of the circuit. If incorrect, the fault lies in the front half of the circuit.

EXAMPLE 4-27

Troubleshoot this circuit.

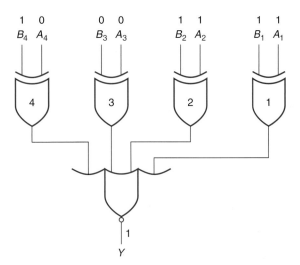

Solution This 4-bit comparator is comparing the numbers 1011 and 0011. Since the numbers are not equal, the output should be a 0. The power pins (V_{CC} and ground) measurements and the inputs are correct. Since Y is a 1, the NOR gate could be faulty, or it might be functioning correctly for the inputs it is receiving. Check the inputs to the NOR gate, particularly the output of exclusive-OR 4. Since B_4 and A_4 are different, the output of exclusive-OR 4 should be a 1. If it is a 1, the NOR gate output should be a 0 to indicate inequality. The NOR gate is probably faulty. If the output of exclusive-OR 4 is 0, either it is bad or the NOR gate input is shorted to ground and the exclusive-OR cannot go HIGH. If the ICs are in sockets, turn off power to the circuit and remove the NOR gate IC. If the output of exclusive-OR 4 goes HIGH when power is restored, the NOR gate was loading it down and the NOR gate is faulty. Otherwise, exclusive-OR 4 is probably the culprit.

Lab 4B has circuit files 4B-4.ms9, 4B-5.ms9, and 4B-6.ms9 to troubleshoot.

Parity

Single in-line memory modules (SIMM) and dual in-line memory modules (DIMM) are small printed circuit boards with several memory ICs soldered onto the boards. SIMMs or DIMMs are plugged into sockets on the motherboard to supply memory for the computer. SIMMs and DIMMs are available with and without parity bits. Those with parity bits are called ECC (Error Checking Code) versions; those without parity are called non-ECC versions. One parity bit is stored for each byte of data. A 72-pin parity type SIMM stores data in 32-bit chunks (4 bytes) with four parity bits. A 168-pin parity-type DIMM stores data in 64-bit chunks (8 bytes) with 8 parity bits.

When the parity checker on the motherboard detects a parity error, a non-maskable interrupt (NMI) is generated. The NMI subroutine (program) causes the screen to be cleared and an error message to be displayed. The message indicates whether the error occurred in memory ICs or in Input/Output cards plugged into the expansion slots. Responding to the error message by pressing S causes the parity checker to shut off, and the computer proceeds without parity checking. Pressing R causes the system to reboot. Pressing any other key causes the computer to proceed with parity checking enabled.

SUMMARY

- If the inputs to an exclusive-OR are like, the output is LOW.

- If the inputs to an exclusive-OR are different, the output is HIGH.

- When the control input of an exclusive-OR is LOW, data passes through unaltered.

- When the control input of an exclusive-OR is HIGH, data passes through inverted.

- If the inputs to an exclusive-NOR (nonexclusive-OR) are like, the output is HIGH.

- If the inputs to an exclusive-NOR are different, the output is LOW.

- To ensure that data is transmitted correctly, a parity bit is sent along with the data.

- In an even-parity system, the parity bit is generated so that the total number of 1s in a word, including the parity bit, is even.

- In an odd-parity system, the parity bit is generated so that the total number of 1s in a word, including the parity bit, is odd.

- A parity generator is a circuit that generates the parity bit.

- A parity checker is a circuit that checks the data bits and parity bit to determine whether an error has occurred during transmission.

- An identity comparator compares two numbers to determine whether they are equal.

- A magnitude comparator compares two numbers, a and b, and determines whether a = b, a < b, or a > b.

- In a VHDL program, "IF THEN" statements can be used to make decisions about input signals.

- "IF THEN" statements are contained within a PROCESS. They are followed by an "END IF" statement.

- A PROCESS has a BEGIN and an END PROCESS statement.

- A PROCESS has a "sensitivity list" that contains all the input signals used in the PROCESS.

- To troubleshoot an adder circuit, follow these steps:

  1. Establish the inputs.

  2. Determine whether the result is correct.

  3. If incorrect, confirm that power supply voltages, ground connections, and inputs are correct.

  4. Test voltages at a midpoint to divide the circuit and determine which part contains the fault.

  5. If the output of a device does not measure correctly, determine whether it is faulty or whether it is loaded down by a following device.

QUESTIONS AND PROBLEMS

1. Draw the symbol and write the truth table for an exclusive-OR.

2. Draw the symbol and write the truth table for an exclusive-NOR.

3. With a 1 on the control input of an exclusive-OR, will the data pass through unaltered or inverted?

4. Using Figure 4-29, sketch the output waveforms for *X* and *Y*.

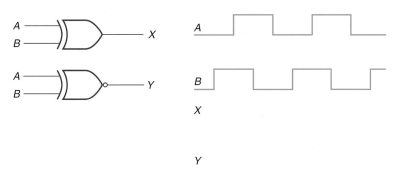

FIGURE 4-29

5. Predict the output of each gate.

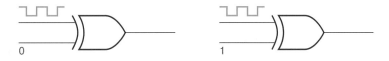

6. Predict the output of each gate.

7. Sketch the output of this exclusive-OR/NOR.

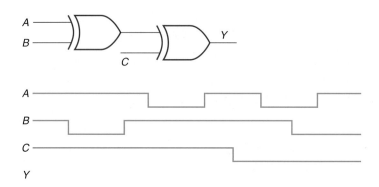

8. Sketch the output of this exclusive-OR/NOR.

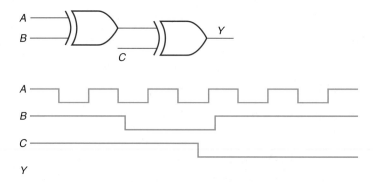

9. Draw the logic diagram of an exclusive-OR, using basic gates.

10. Sketch the logic diagram of an even-parity generator that uses 5 data bits and 1 parity bit.

11. Supply the parity bit.

    a. Even: 101101   **c.** Even: 000011

    b. Odd: 110000   **d.** Odd: 110010

12. Sketch the logic diagram of a 5-bit comparator and indicate the meaning of the output.

13. Sketch the pinout of a 7485 4-bit magnitude comparator. Show the expected logic levels on each pin when $A = 0110$ and $B = 1010$.

14. Describe, in your own words, the operation of a 4-bit magnitude comparator.

In problems 15 through 24, use test problems to verify your circuit.

15. Using two 7485s, draw the logic diagram of an 8-bit magnitude comparator.

16. Draw the logic diagram of a 74S280 used as an 8-bit (7 data bits plus 1 parity bit) even-parity generator.

17. Draw the logic diagram of a 74280 used as an 8-bit (7 data bits plus 1 parity bit) odd-parity generator.

18. Draw the logic diagram of a 74S280 used as an 8-bit (7 data bits plus 1 parity bit) even-parity checker. Use an LED to indicate that a parity error has occurred (on = error).

19. Draw the logic diagram of a 74280 used as an 8-bit (7 data bits plus 1 parity bit) odd-parity checker. Use an LED to indicate that an error has occurred (on = error).

20. Draw the logic diagram of two 74S280s being used as a 16-bit (15 data bits plus 1 parity bit) odd-parity generator.

21. Draw the logic diagram of two 74280s being used as a 16-bit (15 data bits plus 1 parity bit) even-parity generator.

22. Draw the logic diagram of two 74S280s being used as a 16-bit (15 data bits and 1 parity bit) odd-parity checker. Use an LED to indicate that a parity error has occurred (on = error).

23. Draw the logic diagram of two 74280s being used as a 16-bit (15 data bits plus 1 parity bit) even-parity checker. Use an LED to indicate that a parity error has occurred (on = error).

24. Draw the pinout of a CMOS comparator.

25. How do the IEC symbols differ for an inverter, OR gate, and an exclusive-OR gate?

26. Draw the IEC logic symbol for a 74S280 parity generator/checker.

27. Draw the IEC logic symbol for a 7485 4-bit magnitude comparator.

28. This 74AC11521 8-bit identity comparator (shown at the right) is comparing $C7_{16}$ and $D9_{16}$. The enable input on pin 20 is LOW. Place logic levels at the input to each inverter, the output of each inverter, the output of each exclusive-NOR, and the output of the NAND. Is the output correct?

29. Describe the process that should be followed in order to program a CPLD.

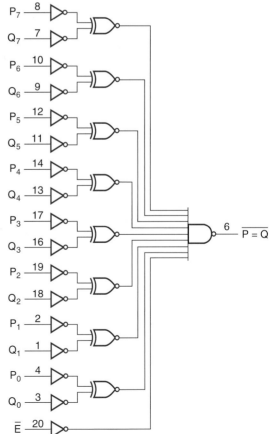

30. Write a VHDL program called prob30.vhd for these logic diagrams.

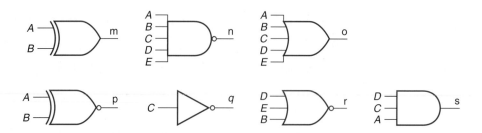

31. Write a VHDL program called prob31.vhd for these logic diagrams.

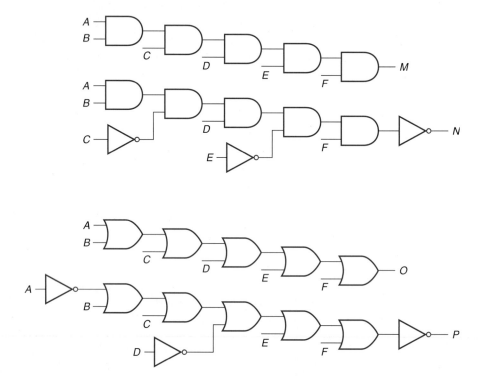

32. Write a VHDL program to implement an 8-bit even-parity generator (7 data bits, 1 parity bit). Name the inputs **A6–A0**, and name the output **pb**.

33. Write a VHDL program to implement an 8-bit odd-parity generator (7 data bits, 1 parity bit). Name the inputs **A6–A0**, and name the output **pb**.

34. Write a VHDL program to implement an 8-bit programmable parity generator. If the control input is a 1, the circuit will generate odd parity; if the control input is a 0, the circuit will generate even parity.

35. Write a VHDL program to implement an 8-bit even-parity checker (7 data bits, 1 parity bit). Name the inputs **A6–A0** and **pb**; name the output **err**. Make **err** go LOW to signify "no error" and HIGH to signify "parity error."

36. Write a VHDL program to implement an 8-bit odd-parity checker (7 data bits, 1 parity bit). Name the inputs **A6–A0** and **pb**; name the output **err**. Make **err** go LOW to signify "no error" and HIGH to signify "parity error."

37. Write a VHDL program to implement these functions:

$$x = a \oplus b \oplus c$$
$$y = \overline{a} \oplus \overline{b} \oplus \overline{c}$$
$$z = \overline{a \oplus b \oplus c}$$

38. Write a VHDL program to implement these functions:

six-input NOR:	$x = \overline{a + b + c + d + e + f}$
six-input NAND:	$y = \overline{a\,b\,c\,d\,e\,f}$
six-input OR:	$z = a + \overline{b} + c + \overline{d} + e + \overline{f}$
six-input AND:	$w = a\,b\,\overline{c}\,d\,\overline{e}\,f$
exclusive-OR:	$v = a\,b\,d \oplus d\,e\,f$
exclusive-NOR:	$u = \overline{a\,b \oplus c\,d \oplus e\,f}$

39. Find and correct four errors in this program:

```
INTITY new IS
      PORT (a,b,c,d IN BIT;
            x,y,z OUT BIT);
END new ;
ARCHITECTURE a OF new IS
BEGIN
      x = a XOR b;
      y = (a XOR b) AND (d XOR e);
      z = (a AND b) XOR (c AND d);
END a;
```

40. Find and correct four errors in this program:

```
ENTITY magnitude IS
      PORT(a,b:IN BIT_VECTOR (3 DOWNTO 0);
            aEQb, aGTb, aLTb:OUT BIT);
END magnitude;
ARCHITECTURE a OF magnitude IS
BEGIN
      PROCESS
      BEGIN
            IF a = b THEN
               aEQb <= 1;
               aGTb <= 0;
               aLTb <= 0;
            ELSIF a > b THEN
               aEQb <= '0';
               aGTb <= '1';
               aLTb <= '0';
            ELSIF a < b THEN
               aEQb <= '0';
               aGTb <= '0';
               aLTb <= '1';
      END a;
```

LAB 4A

EXCLUSIVE-OR

OBJECTIVES

After completing this lab, you should be able to:

- Use 7486s to construct a parity generator.
- Use a 74280 to generate parity bits.
- Use a 74280 to check for parity errors.
- Cascade two 74280s to make a 16-bit parity generator/checker.
- Use exclusive-OR gates to construct a 4-bit comparator.

COMPONENTS NEEDED

2	7486
2	74280 IC
1	4009
1	4012
1	4070
1	LED
1	330-Ω resistor

PREPARATION

In a parity generator, the data bits are input and the circuit generates the parity bit. The same circuit can be used as a parity checker by inputting the parity bit along with the data bits. The output becomes a signal or flag declaring whether there is a parity error.

Review the lab safety rules under SAFETY ADVICE in the PREPARATION section of Lab 1A.

PROCEDURE

1. Check each gate on a 7486 IC by determining its truth table.
2. Use 7486 ICs to construct an 8-bit parity generator (8th bit is parity bit).

   Input 136_8

   What is the output? Is this an even- or an odd-parity generator? How can it be converted to the other?

   Input 063_8

What is the output? Predict the output of your parity generator for the following numbers. Input these numbers and verify your conclusions.

135_8

056_8

060_8

177_8

3. Complete the truth table of a 74280 9-bit parity generator/checker.

Number of Inputs (A – I) that are HIGH	Outputs	
	Σ Even	Σ Odd
0, 2, 4, 6, 8		
1, 3, 5, 7, 9		

4. Use the 74280 as an odd-parity generator to determine the parity bits of the numbers in step 2 of this lab.

5. Use the 74280 as an odd-parity checker. Have an LED light if there is no parity error. Input your results from step 4.

6. Use two 74280s to make a 16-bit odd-parity generator. The 16th bit will be the parity bit. Use your circuit to determine the parity bit of the following words:

$2D6B_{16}$, $6F50_{16}$, $3BD4_{16}$

7. Use your circuit as a 16-bit even-parity checker by feeding all 16 bits into the inputs. Have an LED light if there is a parity error. Check these numbers for parity error:

$F809_{16}$, $400A_{16}$, $CD13_{16}$

8. Use a 4070, 4012, and 4009 to construct a 4-bit comparator. Input 0110 and 1010. What is the output? What does it indicate? Input 0110 and 0110. What is the output? What does it indicate? Try several other combinations of inputs.

LAB 4B

PARITY GENERATOR/CHECKER

OBJECTIVES

After completing this lab, you should be able to use Multisim to:

● Analyze a parity generator/checker.

● Analyze a 4-bit comparator.

● Troubleshoot a 4-bit comparator.

PROCEDURE

PART 1

Parity Generator

Open file 4B-1.ms9 and answer these questions.

1. The least significant bit of the word generator is connected to which input of the parity generator/checker?

2. The most significant bit of the word generator is connected to which input of the parity generator/checker?

3. What 2-digit hexadecimal number is supplied by the word generator at location 0000?

4. Convert the number at location 0000 to binary. Use the multimeter or voltage indicator to confirm that the inputs to the parity generator/checker are correct.

5. For the number at location 0000, what should the output Σ_{ODD} and Σ_{EVEN} be?

6. For the number at location 0000, what should the even-parity bit be? Odd-parity bit?

7. Which output should be used for the even-parity bit?

8. Does the circuit seem to be functioning?

9. Complete this chart for the remaining numbers.

Memory Location	Data Hex	Data Binary	Even-Parity Bit	Odd-Parity Bit	Outputs	
					EVEN	ODD
0000						
0001						
0002						
0003						
0004						
0005						

10. Summary of results:

PART 2

Parity Checker

Open file 4B-2.ms9. The nine bits being fed into the 74280 represent eight data bits and a parity bit. The 74280 is being used as an odd-parity checker. Complete the chart. Then determine which output, EVEN or ODD, should be used as an output. Wire a lamp indicator or light-emitting diode and current-limiting resistor to indicate parity error. The lamp should light to indicate parity error.

Memory Location	Data Hex	Data Binary	Odd-Parity Error (yes or no)	Outputs	
				EVEN	ODD
0000					
0001					
0002					
0003					
0004					
0005					

Conclusions:

PART 3

Open file 4B-3.ms9. Answer these questions.

1. Sketch the equivalent logic diagram of this circuit using individual gates.

2. What is the function of this circuit?

3. What does the light-emitting diode signify?

4. The word generator is being used to supply data to the circuit. The most significant four bits (nibble) supply data to 4A 3A 2A 1A, and the least significant nibble supplies data to 4B 3B 2B 1B. Enter the numbers to test the circuit. Try some of your own inputs.

5. Log your results.

PART 4

1. Open circuit file 4B-4.ms9. This circuit is the same as file 4B-3.ms9 except one fault has been introduced. Isolate the fault. Keep a log of your steps. What is your conclusion?

2. Open circuit file 4B-5.ms9. This circuit is the same as file 4B-3.ms9 except one fault has been introduced. Isolate the fault. Keep a log of your steps. What is your conclusion?

3. Open circuit file 4B-6.ms9. This circuit is the same as file 4B-3.ms9 except one fault has been introduced. Isolate the fault. Keep a log of your steps. What is your conclusion?

Adders

OUTLINE

KEY TERMS

arithmetic logic unit (ALU)

binary-coded decimal (BCD)

end-around carry (EAC)

fast carry

full adder

look ahead carry

1's complement

overflow

signed 2's complement

standard logic

standard logic vector

2's complement

OBJECTIVES

After completing this chapter, you should be able to:

- Define half adder and draw the block diagram and truth table.
- Develop the logic circuitry and construct a half adder.
- Define full adder and draw the block diagram and truth table.
- Develop the logic circuitry and construct a full adder.
- Subtract binary numbers using 1's complement method.
- Design the circuitry required to use a full adder as a 1's complement adder/subtractor.
- Subtract binary numbers using 2's complement method.
- Design the circuitry required to use a full adder as a 2's complement adder/subtractor.
- Convert from decimal to signed 2's complement and signed 2's complement to decimal.
- Add and subtract signed 2's complement numbers.
- Convert from decimal to BCD and BCD to decimal.
- Add BCD numbers.
- Design the circuitry required to use a full adder as a BCD adder.
- Combine numbers using the logic and arithmetic functions of an arithmetic logic unit.
- Program a CPLD to function as a 4-bit adder, 4-bit full adder, and BCD adder.
- Troubleshoot adder circuits.

5.1 HALF ADDER

A **half adder** is a circuit that has two inputs, *A* and *B*, and two outputs, sum and carry. It adds *A* and *B* according to the rules for binary addition and outputs the sum and carry. The block diagram and truth table for a half adder are shown in Figure 5-1. The truth table follows the rules for binary addition. The last line shows 1 plus 1 is 10, as it must be.

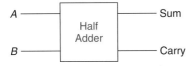

Inputs		Outputs	
B	*A*	*Carry*	*Sum*
0	0	0	0
0	1	0	1
1	0	0	1
1	1	1	0

FIGURE 5-1 Half adder

The sum output has a truth table identical to the exclusive-OR, and the carry output has a truth table identical to an AND gate. One way to construct a half adder is shown in Figure 5-2. Another way to construct a half adder is shown in Figure 5-3. In Figure 5-3, the exclusive-OR is constructed from one AND gate and two NOR gates, as discussed in Chapter 4. *A* is ANDed with *B* as part of the exclusive-OR and can be used as the carry signal.

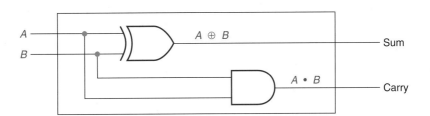

FIGURE 5-2 Logic diagram of a half adder

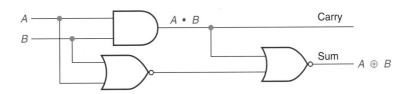

FIGURE 5-3 Using an AND gate and two NOR gates to construct a half adder

EXAMPLE 5-1

Add *A* = 1, *B* = 1.

Solution See Figure 5-4.

1 plus 1 has a sum of 0 and a carry of 1.

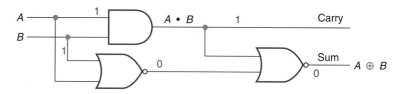

FIGURE 5-4

5.2 FULL ADDER

Whereas the half adder added two inputs, *A* and *B*, the **full adder** adds three inputs together, *A*, *B*, and a carry from a previous addition, and outputs a sum and carry. The truth table follows the rules for binary addition. The block diagram and truth table for a full adder are shown in Figure 5-5.

The sum is 1 each time the total number of 1s on inputs *A*, *B*, and carry-in is odd. This is analogous to an even-parity generator, as shown in Figure 5-6. The output is 1 for an odd number of 1s in.

Inputs			Outputs	
B	*A*	*Carry In*	*Carry Out*	*Sum*
0	0	0	0	0
0	0	1	0	1
0	1	0	0	1
0	1	1	1	0
1	0	0	0	1
1	0	1	1	0
1	1	0	1	0
1	1	1	1	1

FIGURE 5-5 Full adder

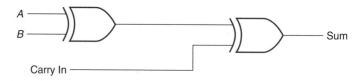

FIGURE 5-6 Full adder—sum

EXAMPLE 5-2

Add *A* = 1, *B* = 1, Carry-in = 1.

Solution   See Figure 5-7.

1 plus 1 plus 1 has a sum of 1.

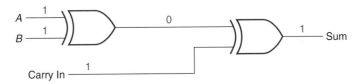

FIGURE 5-7

Each of the exclusive-OR gates in Figure 5-6 can be replaced with two NOR gates and an AND gate, Figure 5-8.

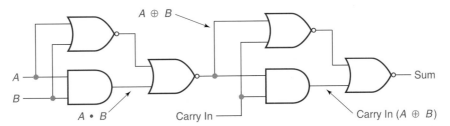

FIGURE 5-8 A second construction for a full adder—sum

EXAMPLE 5-3

Add $A = 1$, $B = 0$, Carry-in $= 1$.

Solution   See Figure 5-9.

1 plus 0 plus 1 has a sum of 0.

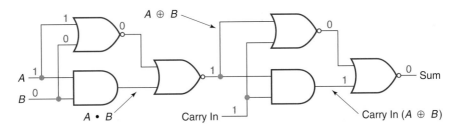

FIGURE 5-9

The carry output is 1 for the conditions on lines 4, 6, 7, and 8 of the truth table in Figure 5-5.

$$C_{OUT} = \overline{B}AC_{IN} + B\overline{A}C_{IN} + BA\overline{C}_{IN} + BAC_{IN}$$
$$= C_{IN}(\overline{B}A + B\overline{A}) + BA(\overline{C}_{IN} + C_{IN})$$
$$= C_{IN}(\overline{B}A + B\overline{A}) + BA$$
$$= C_{IN}(B \oplus A) + BA$$

In Figure 5-8, A has already been exclusive-ORed with B, and the result has been ANDed with C_{IN}. Also, A has been ANDed with B. A two-input OR gate will combine these two signals to produce the carry-out signal, as shown in Figure 5-10. The full adder in Figure 5-10 is constructed from two half adders and an OR gate. Each half adder is enclosed in broken lines.

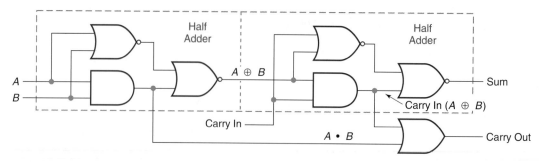

FIGURE 5-10 Full adder—sum and carry

EXAMPLE 5-4

Add $A = 1$, $B = 1$, Carry-in $= 0$.

Solution   See Figure 5-11.

Sum $= 0$; Carry $= 1$

1 plus 1 plus $0 = 10$

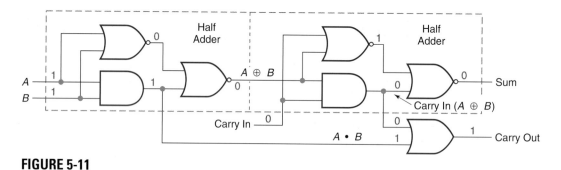

FIGURE 5-11

Table 5-1 lists some of the available 4-bit full adder ICs. The internal circuitry contains enough components for the IC to be classified as medium-scale integration. The 7483 was used in Lab 1A to add two 4-bit numbers, A and B, and a carry-in, called C_0. The outputs are a 4-bit sum and carry-out, C_4, as shown in Figure 5-12. Carries C_1, C_2, and C_3 are handled internally and do not appear on the pins of the IC.

TABLE 5-1   Medium Scale Integration Adder Circuits

Device No.	Family	Description
7483	TTL	4-bit binary adder with fast carry
74C83	CMOS	4-bit binary adder with fast carry
4008	CMOS	4-bit full adder with fast carry

$$
\begin{array}{r}
C_3\ C_2\ C_1\ C_0 \\
A_4\ A_3\ A_2\ A_1 \\
+\ B_4\ B_3\ B_2\ B_1 \\
\hline
C_4\quad \Sigma_4\ \Sigma_3\ \Sigma_2\ \Sigma_1
\end{array}
$$

FIGURE 5-12 4-bit full adder inputs and outputs

When working the addition problem longhand, C_4 is not determined until each of the columns has been added. The carry has to "ripple through" four stages of addition. The logic diagram in Figure 5-13 shows how a 7483 produces C_4 from the inputs without waiting for the "ripple effect" to take place. This results in a **fast carry** or **look-ahead carry**. The result is a faster operation; in fact, C_4 appears before the Σ outputs are established.

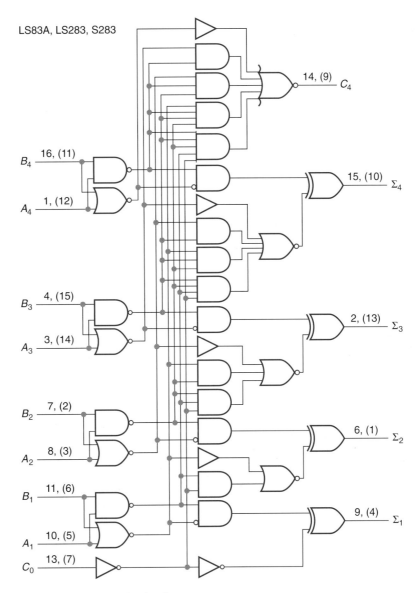

FIGURE 5-13 7483 logic diagram

One of the gates in the logic diagram shown in Figure 5-13 is not a basic gate, but is a combination of basic gates. The Boolean expression for the gate, its equivalent logic diagram, and its truth table are shown in Figures 5-14A and 5-14B. When A is 0 AND when B is 1 the output is 1.

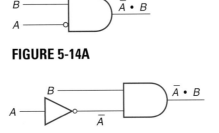

FIGURE 5-14A

FIGURE 5-14B

B	A	Y
0	0	0
0	1	0
1	0	1
1	1	0

EXAMPLE 5-5

For the gate shown in Figure 5-15A, write the Boolean expression, the equivalent logic diagram, and its truth table.

Solution   When A is 1 AND when B is 0 the output is 0. See Figure 5-15B.

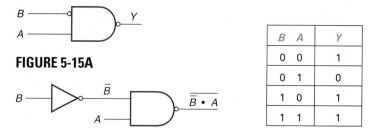

FIGURE 5-15A

FIGURE 5-15B

B	A	Y
0	0	1
0	1	0
1	0	1
1	1	1

EXAMPLE 5-6

Add these numbers using a 7483. Follow the logic levels through the logic diagram. $A = 1001$, $B = 1010$, and $C_0 = 1$

Solution   See Figure 5-16.

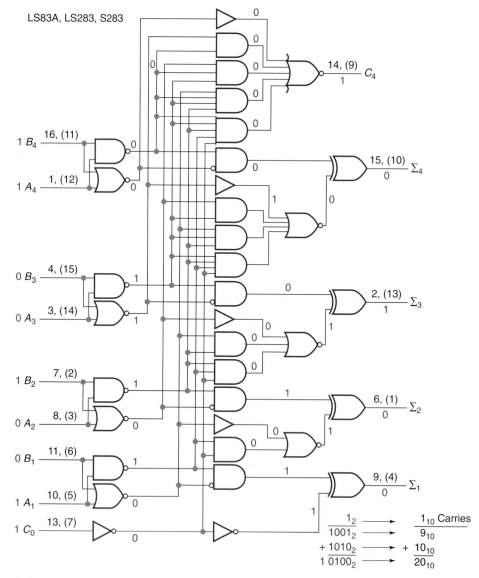

Note: Pin numbers shown in parentheses are for LS283, S283.

FIGURE 5-16

The truth table and connection diagram (pinout) for the 7483 and 74S283 are shown in Figure 5-17. Note that the 7483 is not "corner powered." V_{CC} is pin 5 and ground is on pin 12.

Writing a truth table for nine inputs creates a table of 512 lines (2^9). The truth table shown has been reduced to 16 lines. The note below the truth table explains that the table is used in two steps. A_1, B_1, A_2, B_2, and C_0 determine the outputs Σ_1, Σ_2, and C_2 that are internal to the IC. C_2 is then used with A_3, B_3, A_4, and B_4 to determine Σ_3, Σ_4, and C_4.

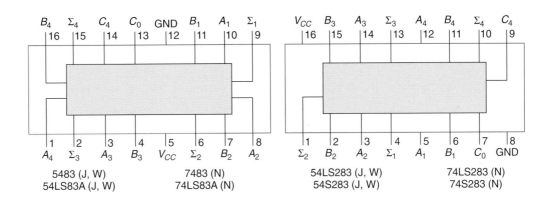

Input				Output When $C_0 = L$		When $C_2 = L$	When $C_0 = H$		When $C_2 = H$
A_1 / A_3	B_1 / B_3	A_2 / A_4	B_2 / B_4	Σ_1 / Σ_3	Σ_2 / Σ_4	C_2 / C_4	Σ_1 / Σ_3	Σ_2 / Σ_4	C_2 / C_4
L	L	L	L	L	L	L	H	L	L
H	L	L	L	H	L	L	L	H	L
L	H	L	L	H	L	L	L	H	L
H	H	L	L	L	H	L	H	H	L
L	L	H	L	L	H	L	H	H	L
H	L	H	L	H	H	L	L	L	H
L	H	H	L	H	H	L	L	L	H
H	H	H	L	L	L	H	H	L	H
L	L	L	H	L	H	L	H	H	L
H	L	L	H	H	H	L	L	L	H
L	H	L	H	H	H	L	L	L	H
H	H	L	H	L	L	H	H	L	H
L	L	H	H	L	L	H	H	L	H
H	L	H	H	H	L	H	L	H	H
L	H	H	H	H	L	H	L	H	H
H	H	H	H	L	H	H	H	H	H

H = HIGH Level, L = LOW Level

Note: Input conditions at A_1, B_1, A_2, B_2, and C_0 are used to determine outputs Σ_1 and Σ_2 and the value of the internal carry C_2. The values at C_2, A_3, B_3, A_4, and B_4 are then used to determine outputs Σ_3, Σ_4, and C_4.

FIGURE 5-17 Truth table and connection diagrams

EXAMPLE 5-7

Use the 7483 to add 0110 to 1101 with $C_0 = 1$.

A_4	A_3	A_2	A_1
0	1	1	0

B_4	B_3	B_2	B_1
1	1	0	1

C_0
1

Solution

Step 1.

A_1	B_1	A_2	B_2	=	L	H	H	L	(line 7)

with $C_0 = H$, $\Sigma_1 = L$, $\Sigma_2 = L$, $C_2 = H$

Step 2.

A_3	B_3	A_4	B_4	=	H	H	L	H	(line 12)

with $C_2 = H$, $\Sigma_3 = H$, $\Sigma_4 = L$, $C_4 = H$

Σ_4	Σ_3	Σ_2	Σ_1	=	L	H	L	L	= 0100

with $C_4 = H = 1$.
$0110 + 1101 + 1 = 10100$
$6 + 13 + 1 = 20$

The IEC logic symbol for a 74LS283 is shown in Figure 5-18. A capital sigma, Σ, is used to denote addition. This symbol uses P_3, P_2, P_1, P_0, and Q_3, Q_2, Q_1, Q_0 to represent the two 4-bit numbers to be added and Σ_3, Σ_2, Σ_1, Σ_0 to represent the result. Note that C_I is used for carry-in and C_O for carry-out.

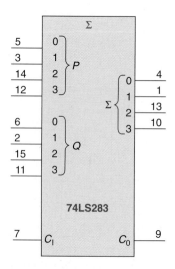

FIGURE 5-18 IEC logic symbol 74LS283

SELF-CHECK 1

1. How does a half adder differ from a full adder?

2. Draw the block diagram and truth table for a half adder.

3. Draw the block diagram and truth table for a full adder.

4. Add these numbers using a 7483. Follow the logic levels through Figure 5-13.

   a.   1 $(C_0 = 1)$ b.   0 $(C_0 = 0)$
       1001   0111
     + 0110   + 0110

Although circuits that subtract numbers can be designed and constructed, computers use a complement method of subtraction that converts the problem into addition. Then, adder circuits can be used to work subtraction problems. We will study two complement systems, 1's complement and 2's complement. Each system uses the concept of **overflow**. Overflow occurs when the sum of the most significant (left-most) column yields a carry. For example:

$$
\begin{array}{r} 872 \\ +345 \\ \hline 1\,217 \end{array}
\qquad\qquad
\begin{array}{r} 7326 \\ +0074 \\ \hline 0\,7400 \end{array}
$$

Overflow ⌐   No Overflow ⌐

In the second example, leading zeros are added to block in the two numbers.

5.3 BINARY 1'S COMPLEMENT SUBTRACTION

To take the **1's complement** of a binary number, simply change each bit. The 1's complement of 1 is 0 and vice versa. The 1's complement of 1001010 is 0110101. To subtract using 1's complement:

1. Take the 1's complement of the subtrahend (bottom number).

2. Add the 1's complement to the minuend (top number).

3. Overflow indicates that the answer is positive. Add the overflow to the least significant bit. This operation is called **end-around carry (EAC)**.

4. If there is no overflow then the answer is negative. Take the 1's complement of the original sum to obtain the true magnitude of the answer.

EXAMPLE 5-8

Subtract. $11001_2 - 10001_2$

Solution

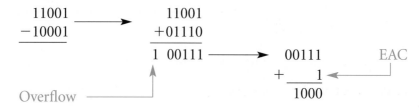

The answer is $+1000$.

Check.
$25_{10} - 17_{10} = 8_{10}$

EXAMPLE 5-9

Subtract. $1011_2 - 101_2$

Solution

$$
\begin{array}{r}
1011 \\
-\ 101 \\
\hline
\end{array}
\longrightarrow
\begin{array}{r}
1011 \\
+1010 \\
\hline
1\quad 0101
\end{array}
\longrightarrow
\begin{array}{r}
0101 \\
+\quad 1 \\
\hline
0110
\end{array}
\quad \text{EAC}
$$

Overflow

Note that the leading 0 becomes a 1. The answer is $+110$.

Check.
$11_{10} - 5_{10} = 6_{10}$

The same process is used when the subtrahend is larger than the minuend.

EXAMPLE 5-10

Subtract. $101_2 - 11000_2$

Solution

$$
\begin{array}{r}
101 \\
-11000 \\
\hline
\end{array}
\longrightarrow
\begin{array}{r}
101 \\
+00111 \\
\hline
01100
\end{array}
$$

No Overflow

The answer is negative. The true magnitude is the 1's complement of 01100 or 10011.
The answer is -10011.

Check.
$5_{10} - 24_{10} = -19_{10}$

EXAMPLE 5-11

Subtract. $10000_2 - 11101_2$

Solution

$$
\begin{array}{r}
10000 \\
-11101 \\
\hline
\end{array}
\qquad \longrightarrow \qquad
\begin{array}{r}
10000 \\
+00010 \\
\hline
10010
\end{array}
$$

No Overflow ⌐

The answer is negative. The true magnitude is the 1's complement of 10010 or 01101. The answer is -01101.

Check.
$16_{10} - 29_{10} = -13_{10}$

5.4 1'S COMPLEMENT ADDER/ SUBTRACTOR CIRCUIT

Design a circuit that will use a 7483 to add the 4-bit number B_4,B_3,B_2,B_1 to the 4-bit number A_4,A_3,A_2,A_1 or subtract B_4,B_3,B_2,B_1 from A_4,A_3,A_2,A_1. Use the 1's complement method for subtraction.

To use a 7483 4-bit full adder as a 1's complement adder/subtractor, the following details must be considered.

1. Refer to Figure 5-19. Leave the number B_4,B_3,B_2,B_1 unaltered for an addition problem, but take the 1's complement of the subtrahend for a subtraction problem. An exclusive-OR inverts data (1's complement) when the control input is HIGH. Exclusive-OR gates will be used to invert B_4,B_3,B_2,B_1 for subtraction. A control signal is needed that will be 1 for subtraction and 0 for addition. A_4,A_3,A_2,A_1 will be fed directly into the 7483.

2. Refer to Figure 5-20. If the problem is subtraction and if there is overflow ($C_4 = 1$), perform an EAC (end-around carry). To detect when subtraction and overflow occur, AND the control line with C_4. The output of the AND gate number 1 is 1 when an EAC results. But in this case, the output of AND gate number 1 can be fed directly into C_0.

3. Refer to Figures 5-21 and 5-22. If the problem is subtraction and if there is no overflow ($C_4 = 0$), indicate the answer is negative and take the 1's complement of the result to obtain the true magnitude of the answer. If C_4 is inverted, then an AND gate can be used to detect when a subtraction process is to be performed and when C_4 is 0. The control input and C_4 are fed into AND gate number 2. A HIGH output indicates a subtraction problem is being performed, and the answer is negative. This signal could be used to light an LED to indicate the answer is negative. The LED requires approximately 12 mA to burn brightly. As will be seen in Chapter 6, TTL can handle more current in the 0 mode than in the 1 mode. This signal will be inverted to drive the LED in the active LOW mode. A red

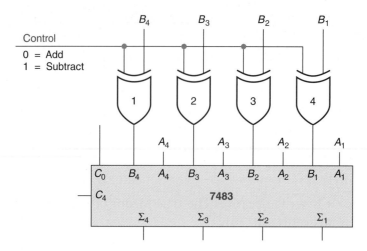

FIGURE 5-19

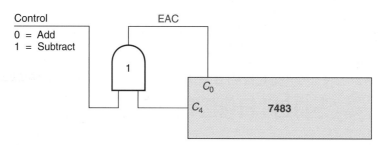

FIGURE 5-20

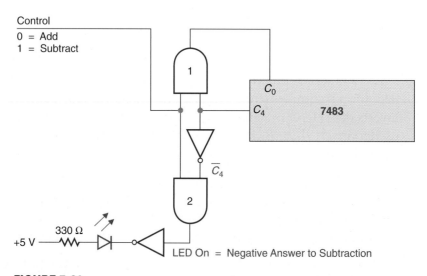

FIGURE 5-21

LED drops about 1.6 V or 1.7 V when lit (LED voltage drops vary greatly with different colors). This leaves 5 V − 1.7 V or 3.3 V to be dropped across the resistor. Ohm's law dictates the resistor should be about

$$\frac{3.3 \text{ V}}{12 \times 10^{-3} A} = 275\Omega$$

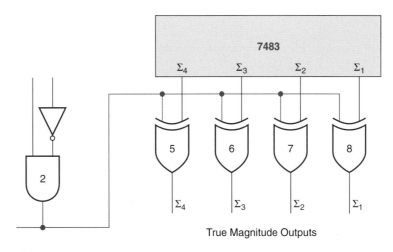

FIGURE 5-22

A 330-Ω resistor, the nearest standard size, will be used to limit the current through the LED. The 7404 inverter can handle 16 mA in the zero mode, which is more than enough to light the LED as designed. As shown in Figure 5-22, the output of AND gate number 2 can be used to control four exclusive-OR gates to invert (take the 1's complement) when the answer to a subtraction problem is negative. Refer to Figure 5-22. The full schematic is drawn in Figure 5-23.

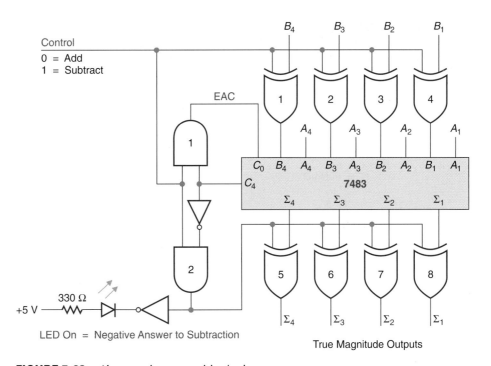

FIGURE 5-23   1's complement adder/subtractor

EXAMPLE 5-12

Add 1011 plus 0010.

Solution   See Figure 5-24.

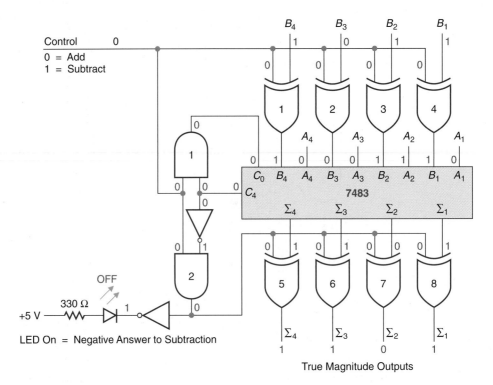

FIGURE 5-24

EXAMPLE 5-13

Subtract 0110 from 1001.

Solution   See Figure 5-25.

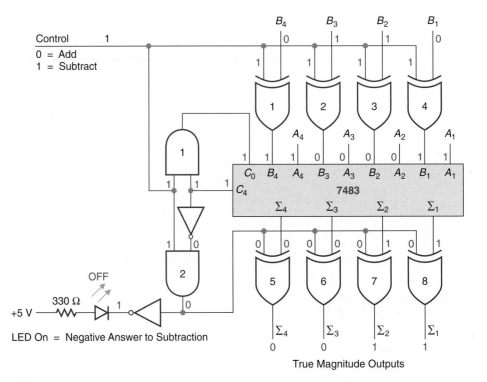

FIGURE 5-25

EXAMPLE 5-14

Subtract 1010 from 0011.

Solution See Figure 5-26.

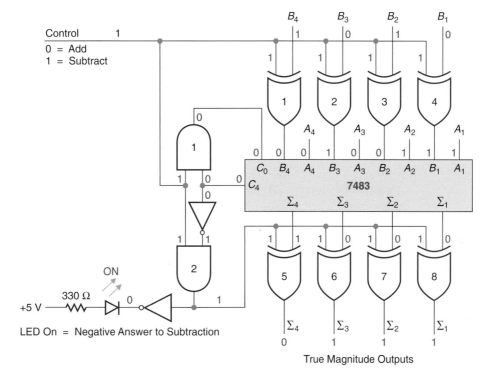

FIGURE 5-26

 SELF-CHECK 2

1. Subtract using the 1's complement method. Follow the logic levels through Figure 5-23.
   a.   0101   b.   11
       -1010 -1000

2. Subtract using the 1's complement method. Follow the logic levels through Figure 5-23.
   a.   1101   b.   1100
       -0110 -111

5.5 BINARY 2'S COMPLEMENT SUBTRACTION

To form the **2's complement** of a number, first take the 1's complement and then add 1. The 2's complement of 10110 is $01001 + 1 = 01010$. A shorter method is to start at the least significant bit and, moving to the left, leave each bit the same until the first 1 is passed. Then change each bit thereafter.

EXAMPLE 5-15

Find the 2's complement of 101101000.

Solution Change each bit to the left of the first 1.

Number = 101101000
2's complement = 010011000

EXAMPLE 5-16

Find the 2's complement of 1011011.

Solution
Method 1.

1's complement	0100100
Add 1	+ 1
2's complement	0100101

Method 2.
Change each bit to the left of the first 1.
1011011
0100101

EXAMPLE 5-17

Find the 2's complement of 101000000.

Solution
Method 1.

1's complement	010111111
Add 1	+ 1
2's complement	011000000

Method 2.
Change each bit to the left of the first 1.
101000000
011000000

To subtract using the 2's complement:

1. Take the 2's complement of the subtrahend (bottom number).
2. Add it to the minuend (top number).
3. Overflow indicates that the answer is positive. Ignore the overflow (no end-around carry).
4. No overflow indicates that the answer is negative. Take the 2's complement of the original sum to obtain the true magnitude of the answer.

EXAMPLE 5-18

Subtract. $1011_2 - 100_2$

Solution

$$
\begin{array}{r}
1011 \\
-\ 100 \\
\end{array}
\longrightarrow
\begin{array}{r}
1011 \\
+1100 \\
\hline
1\ 0111 \\
\end{array}
\qquad
\begin{array}{ll}
\text{1's complement} = & 1011 \\
& +\quad 1 \\
\text{2's complement} = & \overline{1100} \\
\end{array}
$$

Overflow

The answer is positive 111.

Check.

$11_{10} - 4_{10} = 7_{10}$

EXAMPLE 5-19

Subtract. $10011_2 - 10010_2$

Solution

$$
\begin{array}{r}
10011 \\
-10010 \\
\end{array}
\longrightarrow
\begin{array}{r}
10011 \\
+01110 \\
\hline
1\ 00001 \\
\end{array}
$$

Overflow

The answer is positive 1.

Check.

$19_{10} - 18_{10} = 1_{10}$

The process is the same when the subtrahend is larger than the minuend.

EXAMPLE 5-20

Subtract. $10010_2 - 11000_2$

Solution

$$
\begin{array}{r}
10010 \\
-11000 \\
\end{array}
\longrightarrow
\begin{array}{r}
10010 \\
+01000 \\
\hline
11010 \\
\end{array}
$$

No Overflow

The answer is negative. The true magnitude is the 2's complement of 11010 or 110. The answer is -110.

Check.

$18_{10} - 24_{10} = -6_{10}$

EXAMPLE 5-21

Subtract. $1001_2 - 10101_2$

Solution

$$
\begin{array}{r}
1001 \\
-10101 \\
\end{array}
\longrightarrow
\begin{array}{r}
01001 \\
+01011 \\
\hline
10100
\end{array}
$$

No Overflow

The answer is negative. The true magnitude is the 2's complement of 10100 or 1100. The answer is -1100.

Check.

$9_{10} - 21_{10} = -12_{10}$

Two advantages of subtraction by a complement system are:

1. The procedure is the same whether the subtrahend is larger or smaller than the minuend. This saves the extra time or circuitry for a digital machine to decide if one number is larger or smaller than another.

2. The subtraction problem is converted to an addition problem. The same circuitry could be used for both processes.

5.6 2'S COMPLEMENT ADDER/ SUBTRACTOR CIRCUIT

Design a circuit that will use a 7483 to add the 4-bit number B_4, B_3, B_2, B_1 to the 4-bit number A_4, A_3, A_2, A_1 and to subtract B_4, B_3, B_2, B_1 from A_4, A_3, A_2, A_1. Use the 2's complement method for subtraction.

To use the 7483 4-bit full adder as a 2's complement adder/subtractor, the following details must be considered.

1. Refer to Figure 5-27. Leave the number B_4, B_3, B_2, B_1 unaltered for an addition problem, but take the 2's complement of the subtrahend for a subtraction problem. The 2's complement can be formed by taking the 1's complement and adding 1. The 1's complement can be formed by using exclusive-OR gates as we did in the 1's complement subtractor. 1 can be added to form the 2's complement by wiring the control signal directly to C_0.

2. Refer to Figure 5-28. If the problem is subtraction and if there is no overflow ($C_4 = 0$), indicate the answer is negative and take the 2's complement of the result to obtain the true magnitude of the answer. As in the 1's complement subtraction circuit, C_4 can be inverted to form $\overline{C_4}$. $\overline{C_4}$ can be ANDed with the control signal. A HIGH out of the AND gate indicates that a subtraction problem is being performed and the answer is negative. This signal will be inverted to drive an LED in the active LOW mode. Refer to Figure 5-29. A HIGH

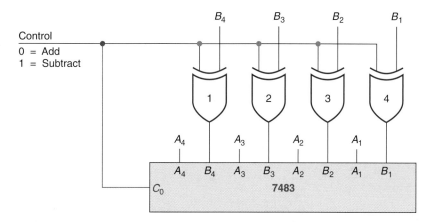

FIGURE 5-27

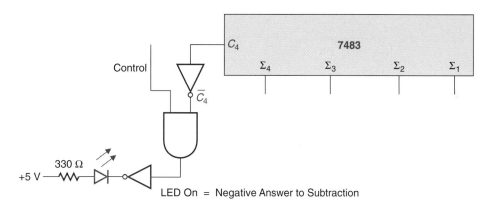

FIGURE 5-28

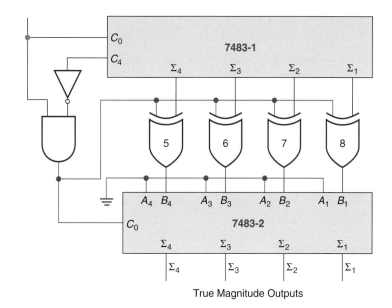

FIGURE 5-29

output of the AND gate also indicates the result of the addition should be 2's complemented to obtain the true magnitude of the answer. The 1's complement can be formed by exclusive-ORing the results with the output of the AND gate. To add 1 to form the 2's complement, another 7483 must be used. The output of the AND can be fed directly into C_0 of 7483-2 to complete the 2's complement process. The true magnitude outputs appear at $\Sigma_4, \Sigma_3, \Sigma_2, \Sigma_1$ of 7483-2.

3. If the problem is subtraction and if there is overflow ($C_4 = 1$), do not take the 2's complement of the result from 7483-1. The answer is already in true magnitude form and should not be altered by the following circuitry. In this situation the output of the AND gate will be 0. With a zero on the control inputs of exclusive-OR gates 5, 6, 7, and 8, the result from 7483-1 passes through 7483-2 unaltered, and the true magnitude outputs appear at Σ_4, Σ_3, Σ_2, Σ_1 of 7483-2. This detail has been taken care of by the circuitry developed in step 2. The full schematic is drawn in Figure 5-30.

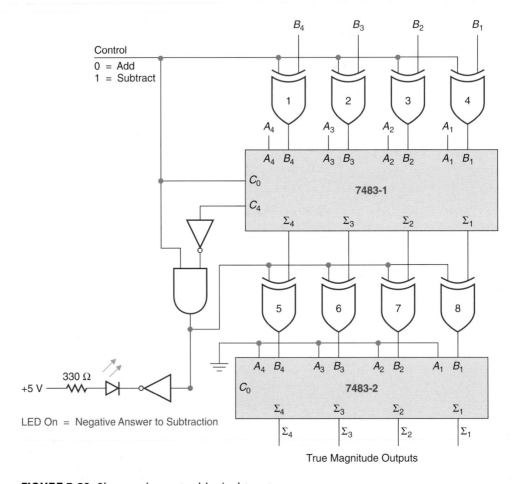

FIGURE 5-30 2's complement adder/subtractor

Try several addition and subtraction problems to fully understand the operation of the circuit.

EXAMPLE 5-22

Add 1001 to 0101.

Solution See Figure 5-31.

EXAMPLE 5-23

Subtract 0101 from 1001.

Solution See Figure 5-32.

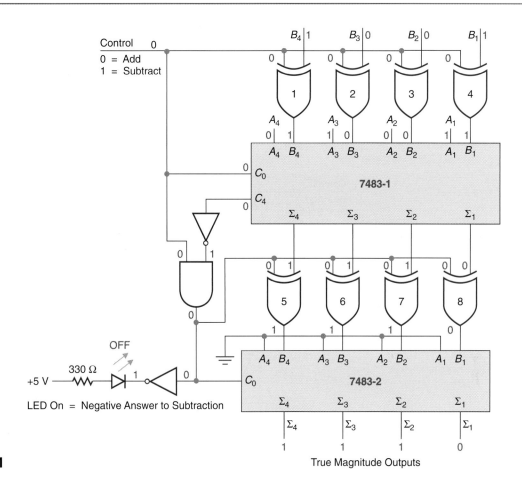

FIGURE 5-31

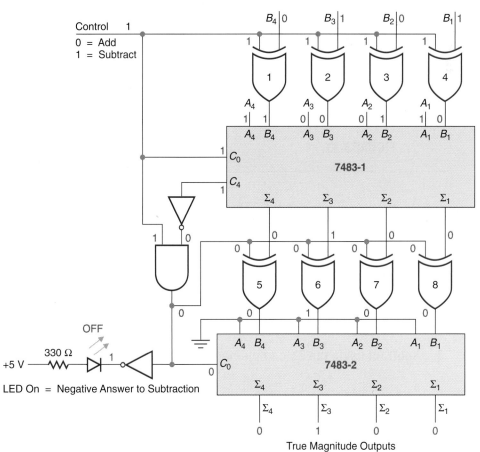

FIGURE 5-32

EXAMPLE 5-24

Subtract 1001 from 0101.

Solution   See Figure 5-33.

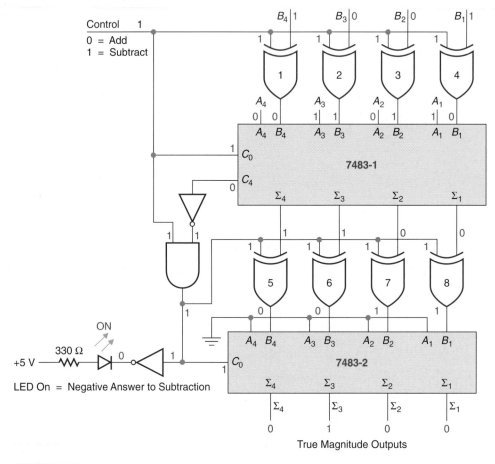

FIGURE 5-33

 SELF-CHECK 3

1. Subtract using the 1's complement method. Follow the logic levels through Figure 5-23.

    a. $\begin{array}{r} 0101 \\ -1010 \\ \hline \end{array}$ b. $\begin{array}{r} 11 \\ -1000 \\ \hline \end{array}$

2. Subtract using the 2's complement method. Follow the logic levels through Figure 5-30.

    a. $\begin{array}{r} 1001 \\ -0110 \\ \hline \end{array}$ b. $\begin{array}{r} 1100 \\ -\ 111 \\ \hline \end{array}$

5.7 SIGNED 2'S COMPLEMENT NUMBERS

Microcomputers sometimes use one bit of a binary number to indicate the sign of the number and the remaining bits to indicate the magnitude. Negative numbers are stored in memory in 2's complement form. This system is called **signed 2's complement**. In signed 2's complement numbers, the most significant bit is used as the sign bit. A zero in the sign bit usually indicates that the number is positive and the remaining bits express the number in true magnitude form.

EXAMPLE 5-25

Convert 00101101 in a signed 2's complement system to a decimal number.

Solution

0	0	1	0	1	1	0	1
	64	32	16	8	4	2	1

Sign Bit ⤴

The true magnitude is $32 + 8 + 4 + 1 = 45$. The number is positive. The answer is 45_{10}.

EXAMPLE 5-26

What is the highest positive number that can be represented in an 8-bit signed 2's complement system?

Solution The highest positive number that can be represented in an 8-bit signed 2's complement system is 01111111.

0	1	1	1	1	1	1	1
	64	32	16	8	4	2	1

Sign Bit ⤴

The true magnitude is $64 + 32 + 16 + 8 + 4 + 2 + 1 = 127$. The number is positive. The positive number 127_{10} is the highest decimal number that can be represented in this system.

A 1 in the sign bit usually indicates that the number is negative. The remaining bits express the number in 2's complement form.

EXAMPLE 5-27

Convert 10010011 in a signed 2's complement system to a decimal number.

Solution

10010011

Sign Bit ⤴ ⌐ Magnitude Bits
2's Complement of True Magnitude

To find the true magnitude, take the 2's complement of the complete number, including the sign bit.

$$10010011$$

2's complement $= 01101101$ True magnitude

0	1	1	0	1	1	0	1
128	64	32	16	8	4	2	1

$$64 + 32 + 8 + 4 + 1 = 109$$

The number is negative. The answer is -109_{10}.

EXAMPLE 5-28

Convert 11110000 in a signed 2's complement system to a decimal number.

Solution

```
        11110000
      ┌──┴────┐
Sign Bit ┘         └─ Magnitude Bits
                      2's Complement of True Magnitude
```

$$11110000$$

2's complement $= 00010000$ True magnitude

0	0	0	1	0	0	0	0
128	64	32	16	8	4	2	1

The number is negative. The answer is -16_{10}.

EXAMPLE 5-29

What is the most negative number that can be represented in an 8-bit signed 2's complement system?

Solution The most negative number that can be represented in an 8-bit signed 2's complement system is 10000000.

```
        10000000
      ┌──┴────┐
Sign Bit ┘         └─ Magnitude Bits
                      2's Complement of True Magnitude
```

2's complement $= 10000000$
True magnitude $= 10000000$

1	0	0	0	0	0	0	0
128	64	32	16	8	4	2	1

The negative number -128_{10} is the most negative number that can be represented in this system.

In an 8-bit signed 2's complement system, the numbers can range from -128_{10} to $+127_{10}$.

To express a negative decimal number in signed 2's complement form, convert the magnitude into binary and then take the 2's complement.

EXAMPLE 5-30

Express -78_{10} as an 8-bit signed 2's complement number.

Solution

$78_{10} =$

0	1	0	0	1	1	1	0
128	64	32	16	8	4	2	1

True magnitude $= 01001110$

2's complement $= 10110010$

$-78_{10} = 10110010$ (signed 2's complement)

Check.

10110010

Sign Bit ⌐ Magnitude Bits
2's Complement of True Magnitude

2's complement $= 10110010$
True magnitude $= 01001110$

0	1	0	0	1	1	1	0
128	64	32	16	8	4	2	1

The number is negative. The 8-bit signed 2's complement number 10110010 is equal to $-(64 + 8 + 4 + 2) = -78_{10}$.

Numbers in signed 2's complement form can be added using straight binary addition and subtracted by taking the 2's complement of the subtrahend and adding. The sign bit will indicate the sign of the answer. Positive answers will be in true magnitude form. Negative answers will be in 2's complement form.

EXAMPLE 5-31

Add these 8-bit signed 2's complement numbers.
$01011001 + 10101101$

Solution

$$
\begin{array}{ll}
01011001 & (+89) \\
+\ 10101101 & (-83) \\
\hline
100000110 & (+\ 6)
\end{array}
$$

Ignore Overflow ⌐

EXAMPLE 5-32

Add these 8-bit signed 2's complement numbers. Express the answer in decimal form.
$11011001 + 10101101$

Solution

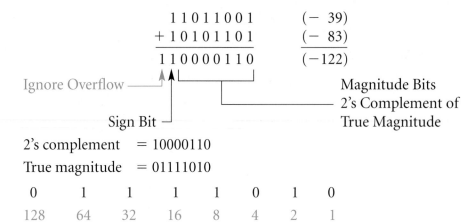

$$11011001 \qquad (-\ 39)$$
$$+\ 10101101 \qquad (-\ 83)$$
$$110000110 \qquad (-122)$$

Ignore Overflow

Sign Bit

Magnitude Bits
2's Complement of
True Magnitude

2's complement   = 10000110

True magnitude   = 01111010

0	1	1	1	1	0	1	0
128	64	32	16	8	4	2	1

The number is negative. The answer is

$-(64 + 32 + 16 + 8 + 2) = -122$

Check.

$-39 + (-83) = -122$

EXAMPLE 5-33

Subtract these 8-bit signed 2's complement numbers. Express the answer in decimal form.
01011011 − 11100101

Solution   To subtract, take the 2's complement of the subtrahend and add.

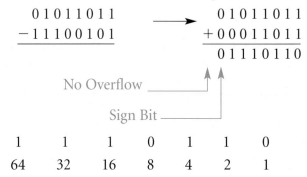

$$01011011 \qquad\qquad 01011011$$
$$-11100101 \longrightarrow\ +00011011$$
$$01110110$$

No Overflow

Sign Bit

1	1	1	0	1	1	0
64	32	16	8	4	2	1

The answer is positive. The answer is

$64 + 32 + 16 + 4 + 2 = 118$

Check.

$91 - (-27) = 118$

EXAMPLE 5-34

Subtract these 8-bit signed 2's complement numbers. Express the answer in decimal form.

10001010 − 11111100

Solution   To subtract, take the 2's complement of the subtrahend and add.

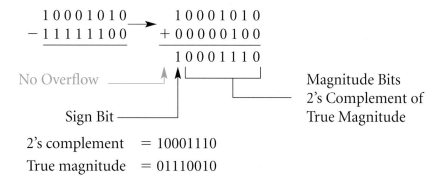

2's complement   = 10001110
True magnitude   = 01110010

0	1	1	1	0	0	1	0
128	64	32	16	8	4	2	1

The number is negative. The answer is

$-(64 + 32 + 16 + 2) = -114$

Check.

$-118 - (-4) = -114$

In each of the examples of signed 2's complement mathematics presented so far, the result has been correct. To ensure that the result is correct, the carry from column 7 into the sign bit and the overflow must be monitored and the following rules observed.

1. If there is a carry from column 7 into the sign bit and an overflow, the answer is correct.

2. If there is no carry from column 7 into the sign bit and no overflow, the answer is correct.

3. If there is no carry from column 7 into the sign bit and overflow occurs or vice versa, the answer is not correct.

Systems that use signed 2's complement mathematics must monitor the carry from column 7 into the sign bit and the overflow to signal whether or not an error has occurred.

In the following two examples, the results are not correct.

EXAMPLE 5-35

Subtract these 8-bit signed 2's complement numbers. Express the answer in decimal form.

10000101 − 01111111

Solution   To subtract, take the 2's complement of the subtrahend and add.

```
  1 0 0 0 0 1 0 1              1 0 0 0 0 1 0 1      (−123)
− 0 1 1 1 1 1 1 1     ──→    + 1 0 0 0 0 0 0 1     −(+127)
                            1 0 0 0 0 0 1 1 0      (−250)
```

Ignore Overflow

The result indicates that the answer is $+6$, but it should be -250. There was no carry into the sign bit, but there was overflow. Since these differ, the result is incorrect. The error has occurred because the true answer is too large to be expressed in an 8-bit signed 2's complement system.

EXAMPLE 5-36

Add these 8-bit signed 2's complement numbers. Express the answer in decimal form.
01111110 + 00111101

Solution

$$
\begin{array}{ll}
0\,1\,1\,1\,1\,1\,1\,0 & = +126 \\
+\,0\,0\,1\,1\,1\,1\,0\,1 & = +\ 61 \\
\hline
0\,1\,0\,1\,1\,1\,0\,1\,1 & \neq +187
\end{array}
$$

⎯⎯⎯⎯⎯ No Overflow

The result indicates that the answer is negative. A carry from bit 7 into the sign bit has changed the sign bit to a 1. No overflow occurred. Since they differ, the result is incorrect. The error occurred because $+187$ is too large to be represented in this 8-bit signed 2's complement system.

 SELF-CHECK 4

1. Convert from signed 2's complement to decimal.

   00001011   10110110

2. Convert from decimal to 8-bit signed 2's complement.

   100   −100

3. Add these signed 2's complement numbers. State whether the result is correct or incorrect.

   11100000   00111011
   10011101   10101011

4. Subtract these signed 2's complement numbers. State whether the result is correct or incorrect.

   00110110   10001111
   − 10101110   − 10101101

5.8 BINARY-CODED-DECIMAL ADDITION

Recall that BCD uses four bits to represent a decimal number, as shown in Figure 5-34. Although legitimate BCD numbers must stop at nine, there are six more counts before all four columns are full. These six steps are not legitimate BCD numbers. In BCD addition, care must be taken to compensate for these six forbidden states. If overflow occurs during an addition, or if one of the forbidden states occurs as a result of an addition, then six must be added to the result to "flip through" the unwanted states. In Figure 5-34, start at 7 and add 5. The result is 1100. To flip out of the forbidden states, count six more. The answer is 0010 or 2 with a carry to the next column. When you reach 1111 the next count is 0000 and a carry has occurred.

$$7 + 5 = 12$$

Legitimate BCD Numbers	0	0	0	0	0
	0	0	0	1	1
	0	0	1	0	2
	0	0	1	1	3
	0	1	0	0	4
	0	1	0	1	5
	0	1	1	0	6
	0	1	1	1	7
	1	0	0	0	8
	1	0	0	1	9
Forbidden Numbers	1	0	1	0	
	1	0	1	1	
	1	1	0	0	
	1	1	0	1	
	1	1	1	0	
	1	1	1	1	

FIGURE 5-34 BCD numbers

EXAMPLE 5-37

Add 3 plus 5.

Solution

```
  0011
+ 0101
  1000
```

There is no overflow, and the result is a legitimate BCD number, so it is correct. The answer is 8.

EXAMPLE 5-38

Add 8 plus 5.

Solution

```
  1000
+ 0101
  1101
```

There is no overflow, but the result is not a legitimate number. Six must be added to compensate for the six forbidden numbers.

$$
\begin{array}{r}
1101 \\
+0110 \\
\hline
10011
\end{array}
$$

The answer is 13.

EXAMPLE 5-39

Add 8 plus 9.

Solution

$$
\begin{array}{r}
1000 \\
+1001 \\
\hline
10001
\end{array}
$$

The result is a legitimate BCD number, but there is overflow. Six must be added to compensate for the forbidden states.

$$
\begin{array}{r}
10001 \\
+\ 0110 \\
\hline
10111
\end{array}
$$

The answer is 17.

EXAMPLE 5-40

Add 167 plus 396.

Solution

1	1		Carries
0001	0110	0111	
+0011	1001	0110	
0101	0000	1101	

Six is added to the least significant digit because the result is not a legitimate BCD number. Six must also be added to the middle digit because of the overflow. The most significant digit result produced no overflow, and it is a legitimate BCD number, so it is not necessary to add six.

0101	0000	1101
+ 0	0110	0110
0101	0110	0011

The answer is 563.

5.9 BINARY-CODED-DECIMAL ADDER CIRCUIT

To convert a binary adder into a BCD adder, logic must be provided that will produce a signal to indicate whether six should be added to the result of an addition. The carry out of the binary adder can be monitored to see if overflow resulted. But how do you distinguish a legitimate BCD number from a forbidden one? See Figure 5-34.

All 4-bit numbers above nine have 1s in the eight's column and a 1 in the four's column or two's column. Written in Boolean, this is $8(4 + 2)$. This signal can be produced with a two-input OR gate and a two-input AND gate. If this signal is 1, or if overflow occurs ($C_4 = 1$), six must be added to compensate for the six forbidden states. $8(2 + 4)$ is ORed with C_4 to produce the ADD 6 signal. Another 7483 will be used to add six to the result from 7483-1 when the ADD 6 signal is 1. Since 6 is 0110 in binary, A_4 and A_1 will be grounded, while the ADD 6 signal will be wired to A_3 and A_2. When ADD 6 is 0, A_4,A_3,A_2,A_1 will be 0000. When ADD 6 is 1, A_4,A_3,A_2,A_1 will be 0110 or 6. C_0 of 7483-2 must be grounded, or seven will be added instead of six. The complete schematic is shown in Figure 5-35.

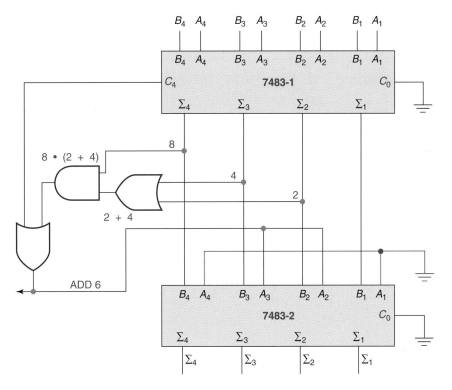

FIGURE 5-35 BCD adder

EXAMPLE 5-41

Use the BCD adder to add 9 and 3.

Solution   See Figure 5-36. The sum from the first adder is 1100, which generates an ADD 6 signal and a carry to the next stage. The second adder adds 1100 + 0110 for a sum of 0010 or 2. The answer is 12.

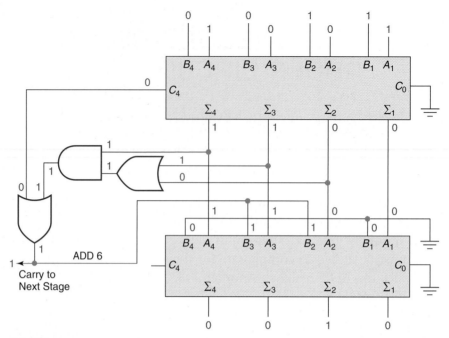

FIGURE 5-36

EXAMPLE 5-42

Use the BCD adder to add 9 and 7.

Solution   See Figure 5-37. The sum from the first adder is 0000 with a 1 out on C_4. This time C_4 generates the ADD 6 signal and the carry to the next stage. The answer is 16.

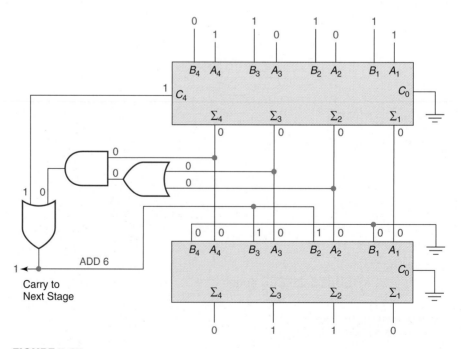

FIGURE 5-37

Follow several examples through until you understand the functioning of the ADD 6 circuit.

SELF-CHECK 5

1.   List the forbidden numbers in the BCD number system.

2.   List two conditions that cause six to be added to the preliminary sum in a BCD adder.

3.   Add these BCD numbers. Follow the logic levels through Figure 5-35.

   a.   0111
   +1000

   b.   1001
   + 1001

5.10 ARITHMETIC LOGIC UNIT (ALU)

An **arithmetic logic unit (ALU)** performs addition and subtraction as well as logical operations like AND and OR on the input data. The 74181 4-bit ALU has four Function Select inputs, S_3–S_0, that select sixteen different arithmetic operations or sixteen different logic operations. When the mode control input M is HIGH, all internal carries are inhibited, and the ALU performs logic operations, such as AND, OR, NAND, NOR, ex-OR, and ex-NOR. When M is LOW, the carries are enabled, and the ALU performs arithmetic operations, such as add, subtract, compare, and double. Table 5-2 defines the sixteen arithmetic functions and logic functions performed by the 74181 when using active HIGH inputs and outputs.

TABLE 5-2   74181 Function Table

S_3	S_2	S_1	S_0	$(M = 1)$ Logic	$(M = 0)$ C_n HIGH Arithmetic (inactive)
L	L	L	L	$\overline{A}$	A
L	L	L	H	$\overline{A + B}$	$A + B$
L	L	H	L	$\overline{A} B$	$A + \overline{B}$
L	L	H	H	0	0 minus 1
L	H	L	L	$\overline{A B}$	A plus $A \overline{B}$
L	H	L	H	$\overline{B}$	$A \overline{B}$ plus $(A + B)$
L	H	H	L	$A \oplus B$	A minus B minus 1
L	H	H	H	$A \overline{B}$	AB minus 1
H	L	L	L	$\overline{A} + B$	$A B$ plus A
H	L	L	H	$\overline{A \oplus B}$	A plus B
H	L	H	L	B	$A B$ plus $(A + \overline{B})$
H	L	H	H	$A B$	$A B$ minus 1
H	H	L	L	1	A plus A
H	H	L	H	$A + \overline{B}$	A plus $(A + B)$
H	H	H	L	$A + B$	A plus $(A + \overline{B})$
H	H	H	H	A	A minus 1

Arithmetic operations are performed on two 4-bit words, A_3, A_2, A_1, A_0 and B_3, B_2, B_1, B_0. The result appears on F_3, F_2, F_1, F_0. The 74181 is a full adder. The carry-in is called C_n and the carry-out is called C_{n+4}. C_n and C_{n+4} are active low. Place a LOW on C_n to represent a carry-in and a HIGH to represent no carry-in. Negative results are presented in 2's complement form. In the arithmetic column, "$A+B$" means A OR B (row 2). To add A to B go to "A plus B" (row 10).

EXAMPLE 5-43

If M is LOW, $S = 1001$, $A = 1011$, $B = 1000$, and C_n is HIGH, predict the outputs F and C_{n+4}.

Solution   With M LOW, arithmetic functions are selected. $S = 1001$ selects the function A plus B. C_n = HIGH means no carry-in. The 74181 will add A and B with no carry-in. C_{n+4} = LOW (signifying carry-out) and $F = 0011$.

Check.

$11 + 8 + 0 = 19$

EXAMPLE 5-44

If S is changed to 1100, predict the outputs.

Solution   1100 selects the function A plus A. C_{n+4} = LOW signifying carry-out and $F = 0110$.

Check.

11 plus 11 = 22.

Logic operations are performed on individual pairs of bits. The function $A + B$ ORs A_0 with B_0 and the result appears on F_0. Likewise, A_1 is ORed with B_1 and the result appears on F_1, and so on. For example, when 1010 is ORed with 1001, the result is 1011. The carries C_n and C_{n+4} are disabled during logical operations.

EXAMPLE 5-45

If M = HIGH, $S = 1011$, $A = 0110$, and $B = 1100$, predict the output F.

Solution M = HIGH selects the logic functions. $S = 1011$ selects AB (A AND B). C_n and C_{n+4} are disabled. Corresponding bits of A and B are ANDed. $F = 0100$.

EXAMPLE 5-46

If S is changed to 0001, predict the output.

Solution $S = 0001$ selects $\overline{A + B}$. Corresponding bits are NORed together.

$A = 0110$, $B = 1100$, and $F = 0001$.

■

Here are some other arithmetic logic unit ICs:

74381	4-bit ALU
74382	4-bit ALU with overflow output for 2's complement
74881	4-bit ALU
74582	4-bit BCD ALU
74583	4-bit BCD adder
74882	32-bit look ahead carry generator

 SELF-CHECK 6

Given these inputs, predict the outputs from a 74181.

1. $A = 0111$, $B = 1001$, $M =$ LOW. $S = 0110$, $C_n =$ LOW (1 has been added to A.)

2. $A = 1100$, $B = 0100$, $M =$ HIGH, $S = 0110$

 # 5.11 PROGRAMMING A CPLD (Optional)

In this section we will program a CPLD to function as a 4-bit adder, a 4-bit full adder, and a 4-bit BCD adder. Follow the procedure presented in Chapters 2 and 3:

Step 1.   Create a new folder within the Altera folder and name it **bin_add_4**.

Step 2.   Run Quartus II.

Step 3.   Run "New Project Wizard."

Step 4.   Create a new .vhd file and name it **bin_add_4.vhd**.

Step 5.   Write a VHDL program that will add two four-binary numbers ($a_3\ a_2\ a_1\ a_0$ and $b_3\ b_2\ b_1\ b_0$) and output the sum ($sum_3\ sum_2\ sum_1\ sum_0$).

In the ARCHITECTURE statement, all we have to do is define sum as a + b (sum <= a + b). For VHDL to perform that four-bit addition, it needs access to the ieee procedures for arithmetic operations for unsigned numbers. In order to invoke the ieee library, we need to begin this program with these three lines:

```
LIBRARY ieee;
USE ieee.std_logic_1164.all;
USE ieee.std_logic_unsigned.all;
```

It is the third line that defines the arithmetic operations. The second line defines standard logic. Whereas the BIT and BIT_VECTOR signals we have used so far can have the values of

1 or 0 only, standard logic and standard logic vectors can also have seven other values, including high impedance (covered in Chapter 6) and "don't care." Here we need to use the more complex standard logic signals with the ieee library of operations. So inputs **a** and **b** and output **sum** will be declared as standard logic vectors.

```
PORT(a,b:IN STD_LOGIC_VECTOR(3 DOWNTO 0);
    sum:OUT STD_LOGIC_VECTOR(3 DOWNTO 0));
```

bin_add_4.vhd looks like this:

```
LIBRARY ieee;
USE ieee.std_logic_1164.all;
USE ieee.std_logic_unsigned.all;
ENTITY bin_add_4 IS
    PORT(a,b:IN STD_LOGIC_VECTOR(3 DOWNTO 0);
        sum:OUT STD_LOGIC_VECTOR(3 DOWNTO 0));
END bin_add_4;
ARCHITECTURE a OF bin_add_4 IS
BEGIN
    sum <= a + b;
END a;
```

Step 6.   Compile the program.

Step 7.   Create a .vwf (vector waveform file) named **bin_add_4.vwf** to use in simulation.

Step 8.   Configure **bin_add_4.vwf** for use in this project.

Step 9.   Use one of the two approaches described in Chapter 4 to assign waveforms for signals a and b.

Create inputs that will thoroughly test the circuit. The resulting patterns for this project's simulation waveforms are shown in Figure 5-38A, with **a** and **b** and **sum** in composite form. Figure 5-38B shows those waveforms expanded so that each bit is shown.

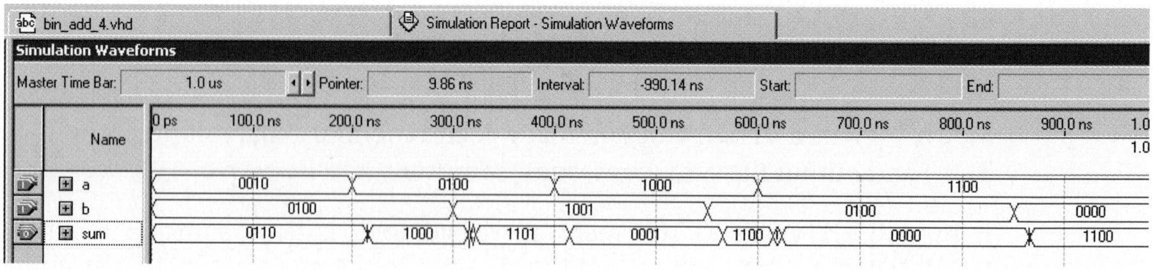

FIGURE 5-38A Simulation waveforms for bin_add_4.vwf

Step 10.   Simulate the circuit.

Step 11.   Assign pins to the input and output signals. Recompile your program.

Step 12.   Program your CPLD.

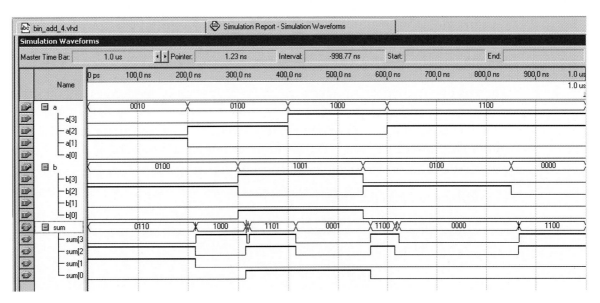

FIGURE 5-38B Simulation waveforms for bin_add_4.vwf expanded

Step 13.   Test the operation of the CPLD using input switches and LEDs.

Note in Figure 5-38A and Figure 5-38B that at the 500 nanosecond mark, $1000 + 1001 = 0001$ ($8 + 9 = 1$). No provision was made for a carry-out, so there is no indication that a carry into the 16's place has occurred.

Let's modify the program to make this a 4-bit full adder by adding a carry-in and a carry-out.

Name the project **bin_fulladd_4**. The outputs will be **Cout** and a 4-bit vector named **sum** (sum_3 sum_2 sum_1 sum_0).

In the ARCHITECTURE statement, we will declare and use a signal named **intsum** (internal sum) to keep track of the 5-bit sum (includes Cout). Intsum is declared as a 5-bit signal intsum ($intsum_4$ $intsum_3$ $intsum_2$ $intsum_1$ $intsum_0$). Then, intsum(4) is used as Cout, and intsum(3 DOWNTO 0) is used as the 4-bit sum output. Figure 5-39 summarizes these ideas.

				Cin
	a_3	a_2	a_1	a_0
	b_3	b_2	b_1	b_0
$intsum_4$	$intsum_3$	$intsum_2$	$intsum_1$	$intsum_0$
⇓	⇓	⇓	⇓	⇓
Cout	sum_3	sum_2	sum_1	sum_0

FIGURE 5-39 Internal sum (intsum)

After the end of the process and before the end of ARCHITECTURE **a**, Cout and sum are defined as part of intsum. Cout $<=$ intsum(4) declares Cout as the most significant bit of intsum. sum $\leq$ intsum(3 DOWNTO 0) declares sum as the four least significant bits of intsum. Note that the signal **intsum** is declared inside the ARCHITECTURE statement before BEGIN.

There is one more detail to take care of. Since intsum is declared as a 5-bit signal, one of the signals being added to produce intsum must also be a 5-bit signal (this is a requirement of VHDL language). Input **a** is extended to a 5-bit signal by concatenating (placing) a zero as the leading bit. Concatenation is denoted by the ampersand sign, **&**. So, ('0'& a) extends **a** into a 5-bit signal with zero as the most significant bit.

bin_fulladd_4 looks like this:

```
LIBRARY ieee;
USE ieee.std_logic_1164.ALL;
USE IEEE.STD_LOGIC_UNSIGNED.ALL;

ENTITY bin_fulladd_4 IS
    PORT(
        a,b:IN STD_LOGIC_VECTOR(3 DOWNTO 0);
        sum:OUT STD_LOGIC_VECTOR(3 DOWNTO 0);
        Cin:IN STD_LOGIC;
        Cout:OUT STD_LOGIC);
END bin_fulladd_4;

ARCHITECTURE a OF bin_fulladd_4 IS
    SIGNAL intsum:STD_LOGIC_VECTOR(4 DOWNTO 0);
BEGIN
    intsum <= ('0'& a) + b + Cin;
    Cout <= intsum(4);
    sum <= intsum(3 DOWNTO 0);
END a;
```

Figure 5-40A shows simulation waveforms for this project with **a** and **b** and **sum** in composite form, and Figure 5-40B shows those waveforms expanded so that waveforms for each bit are shown.

With those details taken care of, it is not difficult to modify the 4-bit full adder into a BCD full adder. Name the project **add_bcd_4**. The outputs will be **Cout** and a 4-bit vector named **sum**.

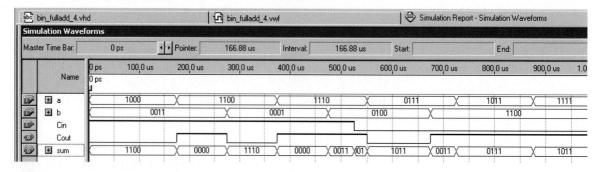

FIGURE 5-40A Simulation waveforms for bin_fulladd_4.vwf

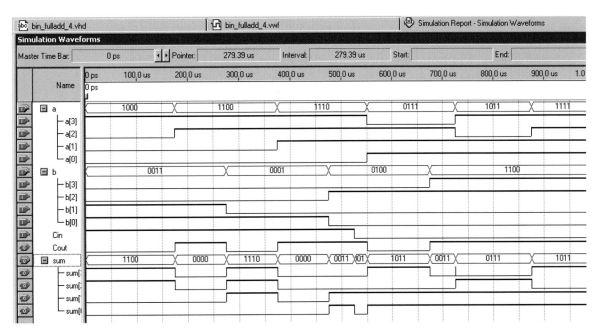

FIGURE 5-40B Simulation waveforms for bin_fulladd_4.vwf expanded

The inputs will be 4-bit vectors **a** and **b** and **Cin**. The addition process requires the ieee library. **a**, **b**, and **sum** will be declared as 4-bit standard logic vectors, and **Cin** and **Cout** will be declared as standard logic.

```
LIBRARY ieee;
USE ieee.std_logic_1164.ALL;
USE IEEE.STD_logic_unsigned.ALL;

ENTITY add_bcd_4 IS
     PORT(
          a,b:IN STD_LOGIC_VECTOR(3 DOWNTO 0);
          sum:OUT STD_LOGIC_VECTOR(3 DOWNTO 0);
          Cin:IN STD_LOGIC;
          Cout:OUT STD_LOGIC);
END add_bcd_4;
```

We will need a 5-bit internal sum as in the binary full adder project. It is named **intsum** and declared to be a 5-bit standard logic vector (4 DOWNTO 0). It is declared inside the ARCHITECTURE statement before BEGIN.

```
ARCHITECTURE a OF add_bcd_4 IS
     SIGNAL intsum:STD_LOGIC_VECTOR(4 DOWNTO 0);
BEGIN
```

IF THEN and ELSE statements are used to see whether intsum is greater than 9 (1001). If so, 6 (0110) must be added to correct for the six unused states. These statements must be contained within a PROCESS. The sensitivity list of the process contains intsum along with a, b, and Cin.

```
PROCESS (a,b,Cin,intsum)
BEGIN
```

Zero is concatenated with input signal a to produce a 5-bit signal compatible with the 5-bit preliminary result, intsum. a, b, and Cin are added to see whether the result is greater than nine. Multiple bits are enclosed in quotes, "01001." Note that 9 is written in binary as a 5-bit number to match the five bits of ('0'& a).

```
IF (('0'& a) + b + Cin)> "01001" THEN
```

If the preliminary sum is greater than nine, six is added to form the result.

```
intsum <= (('0'& a) + b + Cin)+ "0110";
```

Otherwise, the internal sum stays as is; six is not added.

```
ELSE intsum <= (('0'& a) + b + Cin);
    END IF;
END PROCESS;
```

After the end of the PROCESS and before the end of ARCHITECTURE a, **Cout** and **sum** are defined as part of **intsum**. Cout becomes the most significant bit, and sum becomes the four least significant bits.

```
Cout <= intsum(4);
sum <= intsum(3 DOWNTO 0);
END a;
```

add_bcd_4.vhd looks like this:

```
LIBRARY ieee;
USE ieee.std_logic_1164.ALL;
USE IEEE.STD_logic_unsigned.ALL;

ENTITY add_bcd_4 IS
    PORT(a,b:IN STD_LOGIC_VECTOR(3 DOWNTO 0);
        sum:OUT STD_LOGIC_VECTOR(3 DOWNTO 0);
        Cin:IN STD_LOGIC;
        Cout:OUT STD_LOGIC);
END add_bcd_4;

ARCHITECTURE a OF add_bcd_4 IS
    SIGNAL intsum:STD_LOGIC_VECTOR(4 DOWNTO 0);
BEGIN
    PROCESS (a,b,Cin,intsum)
    BEGIN
        IF (('0'& a) + b + Cin)> "01001" THEN
                intsum <= (('0'& a) + b + Cin)+ "0110";
            ELSE intsum <= (('0'& a) + b + Cin);
            END IF;
    END PROCESS;
    Cout <= intsum(4);
    sum <= intsum(3 DOWNTO 0);
END a;
```

Figure 5-41A shows the simulation waveforms for this project with **a** and **b** and **sum** in composite form; Figure 5-41B shows those waveforms expanded so that waveforms for each bit are shown.

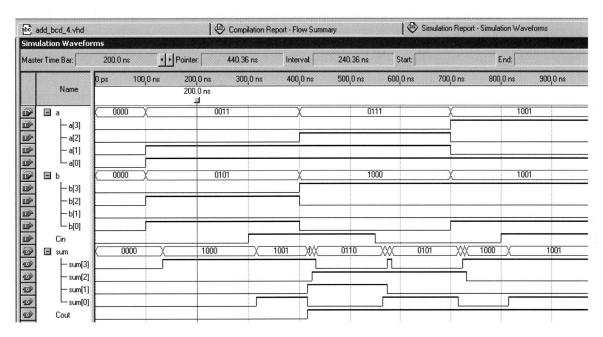

FIGURE 5-41A Simulation waveforms for add_bcd_4.vwf

FIGURE 5-41B Simulation waveforms for add_bcd_4.vwf expanded

5.12 TROUBLESHOOTING ADDER CIRCUITS

In the Chapter 5 labs, you will be working with the adder circuits discussed in this chapter. Chapter 1 discussed problems that can be encountered with a 4-bit full adder. Review those concepts. The adder circuits in this chapter involve several components. A systematic approach to troubleshooting these circuits needs to be taken.

1. Set the inputs for a problem that will test a particular aspect of the circuit. For example, in a BCD adder three test cases should be considered: a problem that requires no "add 6" correction, a problem that requires the "add 6" because an unused result occurred in the preliminary addition, and a problem that requires the "add 6" because C_4 occurred.

$0 + 0 = 0$ is a good first test. If the circuit gives an answer of 6, check out the "add 6" circuit to see how it is being enabled.

2. Determine whether the final output is correct.

3. If not, do these preliminary checks before spending time searching for a problem in the hardware.

   a. Check the power and ground connections on each IC.

   b. Check the inputs to the first adder IC to ensure that the adder circuit is working the correct problem.

4. If power connections and inputs are correct and the output is not, check voltages in the middle of the circuit to "divide and conquer." If the voltages in the middle are correct, the fault is in the second half of the circuit. If the voltages in the middle are not correct, the fault is in the first half of the circuit. Continue to divide the circuit until the general area of the fault is located.

5. If the inputs to a device are correct and the outputs are not, there are two possibilities.

   a. The device is bad.

   b. Something connected to the output of the device is loading it down (keeping it from assuming its proper voltage level). This condition is often overlooked.

EXAMPLE 5-47

Troubleshoot the BCD adder circuit in Figure 5-42.

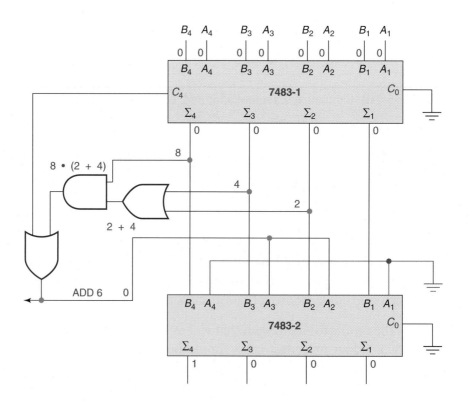

FIGURE 5-42 Troubleshooting a BCD adder

Solution

Step 1. A test problem has been entered. $0000 + 0000 = 1000$

Step 2. Check the preliminary sum from the first adder. 0000 is correct. The problem lies in the second half of the circuit. Since the result is 8 and not 6, the "ADD 6" circuit is probably not the culprit.

Step 3. Check the inputs to 7483-2. All inputs measure 0 except B_4. B_4 measures 1.7 V, which indicates a floating TTL input. A cold solder joint has left B_4 floating, and the 7483-2 is taking it as a 1.

◼

Lab 5B has circuit files 5B-4.ms9, 5B-5.ms9, and 5B-6.ms9 to troubleshoot.

DIGITAL APPLICATION

Floating-Point Unit (FPU)

Pentium central processing units (CPUs) have floating-point units (FPUs) built into them. FPUs are also called math coprocessors. The primary CPU is equipped to handle addition, subtraction, multiplication, and division of integer numbers. More complicated math problems such as logarithms, trigonometry and inverse trig functions, and floating-point math are handled by the FPU. Math-intensive computer application programs, such as computer-aided design (CAD) and other graphics programs and spreadsheet and database programs that run on the Pentium, use FPU instructions to handle their math needs efficiently. Intel processors prior to the 80486 used a separate math coprocessor IC to handle higher math functions. The 80386 processor worked with the 80387 math coprocessor. Moving the coprocessor onboard increased communication speed and lowered execution time.

SUMMARY

- A half adder adds two inputs and produces a sum and a carry.

- A full adder adds three inputs, two bits, and a carry-in, and produces a sum and a carry.

- A fast carry or look ahead carry is produced without waiting for the result to "ripple through" each stage of addition.

- To subtract using a complement method, add the complement of the subtrahend to the minuend.

- The 1's complement of a binary number is formed by changing each bit.

- To produce a 1's complement adder/subtractor from a full adder IC:

  1. Exclusive-ORs are used to complement the subtrahend for subtraction.

  2. An AND gate produces an EAC when overflow is produced during subtraction.

  3. An inverter and gate detect no overflow on subtraction.

4. When no overflow on subtraction is detected, exclusive-ORs take the 1's complement to produce true magnitude outputs.

- The 2's complement can be formed by two methods:

    1. Change every bit and add 1.

    2. Starting at the LSB and moving to the left, leave each bit unchanged until the first 1 is passed. Thereafter change each bit.

- To produce a 2's complement adder/subtractor from a full adder IC:

    1. Exclusive-ORs are used to take the 1's complement of the subtrahend for subtraction, and a 1 is input to C_0 to produce the 2's complement.

    2. An inverter and gate detect no overflow on subtraction.

    3. When no overflow is detected, the 2's complement must be taken to produce the true magnitude output. Exclusive-ORs take the 1's complement of the preliminary sum, and a second full adder is used to add 1 to produce the 2's complement.

- In a signed 2's complement system, the MSB indicates the sign of the number. 0 indicates positive, and 1 indicates negative.

- In a signed 2's complement system, negative numbers are written in 2's complement form.

- In a signed 2's complement system, the carry-in to the sign bit and the overflow are monitored to determine whether or not the result is correct.

- In binary-coded decimal (BCD), each decimal digit is represented with four bits according to the weighted 8, 4, 2, 1 system.

- In the BCD number system, there are six forbidden numbers: 1010, 1011, 1100, 1101, 1110, 1111.

- If one of the six forbidden numbers appears as a result of a BCD addition, six must be added to leap over the forbidden states.

- If a carry-out is generated as the result of a BCD addition, six must be added to compensate for the six forbidden numbers.

- To produce a BCD adder from a full adder IC:

    1. An OR gate and an AND gate watch for an 8 AND a 4 OR 2, which indicates that an invalid number has been produced as a preliminary sum.

    2. The invalid number OR a carry-out on C_4 of the adder indicates that six must be added to the original sum.

    3. A second full adder is used to add six when one of the above conditions is detected.

- When the ieee library is invoked, input and output signals are declared as standard logic or standard logic vectors.

- Bit signals and bit vector signals can have values of 0 and 1.

- Standard logic is more complex than bit logic and can have additional values such as high impedance (Chapter 6).

- To troubleshoot an adder circuit follow these steps:

    1. Establish the inputs.

2. Determine whether or not the result is correct.

3. Confirm that power supply voltages, ground connections, and inputs are correct.

4. Test voltages at a midpoint to divide the circuit and determine which part contains the fault.

5. If the output of a device does not measure correctly, determine whether it is faulty or if it is loaded down by a following device.

QUESTIONS AND PROBLEMS

1. Draw the truth table and logic diagram of a half adder. Show IC numbers and pin numbers.

2. Draw the truth table and logic diagram of a full adder. Show IC numbers and pin numbers.

3. Subtract using 1's complement.

    a. 1010_2 **b.** 10001_2

           -1000_2 -11101_2

4. Subtract using 1's complement

    a. 1101_2 **b.** 1001_2

           $-\ 100_2$ -1100_2

5. Work the following problems. Use the 1's complement method on the subtraction problems. Confirm Figure 5-23 by following these problems through the circuit.

    a.   0111   **b.**   1010   **c.**   0011

           $+1000$ -0111 -1000

6. Subtract using 2's complement.

    a. 11010_2 **b.** 10010_2

           $-\ 1100_2$ -11110_2

7. Subtract using 2's complement

    a. 100101_2 **b.** 10101_2

         $-\ 1001_2$ -11000_2

8. Work the following problems. Use the 1's complement method on the subtraction problems. Confirm Figure 5-23 by following these problems through the circuit.

    a.   0101   **b.**   1000   **c.**   0100

           $+1011$ $+0110$ -1100

9. Work the following problems. Use the 2's complement method on the subtraction problems. Confirm Figure 5-30 by following these problems through the circuit.

    a.   0101   **b.**   1001   **c.**   0011

           $+1000$ -0111 -1000

10. Work the following problems. Use the 2's complement method on the subtraction problems. Confirm Figure 5-30 by following these problems through the circuit.

    a.   0110
          +1000

    b.   1011
          −101

    c.   101
          −1010

11. Work the following BCD problems. Confirm Figure 5-35 by following these problems through the circuit.

    a.   0100
          +0101

    b.   1001
          +0110

    c.   1001
          +0111

12. Work the following BCD problems. Confirm Figure 5-35 by following these problems through the circuit.

    a.   101
          +1001

    b.   1001
          +1000

    c.   0101
          +0010

13. Draw the logic diagram of an 8-bit 1's complement adder/subtractor.

14. Draw the logic diagram of an 8-bit 2's complement adder/subtractor.

15. Draw the logic diagram of an 8-bit BCD adder.

16. Find a CMOS adder IC in a data book. Draw the logic diagram. Describe in your own words its function.

17. Work the following problems. Use the 1's complement method. Confirm your design by following the problem through the circuit that you designed in number 13.

    a.   01101100
          +00111010

    b.   10101011
          −01001100

    c.   00011100
          −10110101

18. Work the following problems. Use the 1's complement method. Confirm your design by following the problem through the circuit that you designed in number 13.

    a.   10000001
          +00111010

    b.   10001111
          −10000111

    c.   00111111
          −01000110

19. Work the following problems. Use the 2's complement method on the subtraction problems. Confirm your design by following the problem through the circuit that you designed in number 14.

    a.   01010001
          +01111010

    b.   11010011
          −00101101

    c.   11000000
          −11000001

20. Work the following problems. Use the 2's complement method on the subtraction problems. Confirm your design by following the problem through the circuit that you designed in number 14.

    a.   10011000
          +01100110

    b.   01111101
          −00110010

    c.   01101100
          −10010010

21. Work the following problems. Confirm your design by following the problem through the circuit that you designed in number 15. All the numbers are BCD.

    a.   10000010
          +00000111

    b.   00101000
          +01001001

    c.   10010101
          +01010001

22. Work the following problems. Confirm your design by following the problem through the circuit you designed in problem 15. All the numbers are BCD.

    a. 01110100
    +00111001

    b. 10000110
    +01111000

    c. 01010001
    +00110100

These questions refer to the 1's complement subtractor circuit in Figure 5-23.

23. How is the end-around carry accomplished?

24. Explain the function of exclusive-OR gates 1, 2, 3, and 4.

25. Explain the function of exclusive-OR gates 5, 6, 7, and 8.

26. What is the function of AND gate 2?

These questions refer to the 2's complement adder/subtractor in Figure 5-30.

27. What is the function of exclusive-OR gates 5, 6, 7, and 8?

28. When is the C_0 input on 7483-2 a 1 level?

29. What are the four pins that are grounded on 7483-2?

30. Why are they grounded?

31. When is the output of the AND gate a 1?

These questions refer to the BCD adder in Figure 5-35.

32. What is the function of the 7483-2?

33. Why does the C_4 output of 7483-1 need to be included in the development of the ADD 6 signal?

34. What would happen if C_0 on the 7483-2 were not grounded?

35. Draw the IEC symbol for a 4-bit full adder. Show the logic levels (1s and 0s) expected on each pin when 9 (input P) is added to C (input Q).

36. Express the following decimal numbers in 8-bit signed 2's complement form.

    a. -38
    b. $+57$

    c. -12
    d. $+12$

    e. -100
    f. $+60$

37. Express the following Decimal numbers in 8-bit signed 2's complement form.

    a. -50
    b. $+43$

    c. -2
    d. $+8$

    e. -120
    f. $+83$

38. Add these 8-bit signed 2's complement numbers. Use the carry from column 7 and the overflow to tell whether the answer is correct.

    a. 00011110 + 00111000
    b. 01011101 + 00111100

    c. 11100011 + 10000001
    d. 00110011 + 11001100

39. Add these 8-bit signed 2's complement numbers. Use the carry from column 7 and the overflow to tell whether the answer is correct.

    a. 00111101 + 11010110
    b. 01100111 + 11001001

    c. 10011100 + 10011011
    d. 01111111 + 01111111

40. Use the 2's complement method to subtract these 8-bit signed 2's complement numbers. Use the carry from 7 and the overflow to indicate whether the answer is correct.

   a. 00101010 − 01101101   **c.** 10001111 − 10100000
   b. 01111111 − 10000000   **d.** 10000000 − 10000000

41. Predict the output F of a 74181 ALU when $S = 1000$, $M = 1$, $A = 0111$, and $B = 1001$.

42. If, in problem 41, M is changed to 0 and C_n is inactive (HIGH), predict the outputs F and C_{n+4}.

43. Predict the output F of a 74181 ALU when $S = 0100$, $M = 1$, $A = 0011$, and $B = 1100$.

44. If, in problem 43, M is changed to 0 and C_n is active (LOW), predict the outputs F and C_{n+4}.

45. Configure a 74181 ALU to subtract B from A.

46. Configure a 74181 ALU to double-input A.

47. Write a VHDL program to implement an 8-bit adder.

48. Write a VHDL program to implement an 8-bit full adder.

49. Write a VHDL program to implement a 4-bit subtractor. Use the ieee library and the "−" operation. Output a negative answer in true magnitude form. To indicate a negative answer, program an output to go HIGH.

50. Write a VHDL program that will output the 1's complement of the 4-bit input signal.

51. Write a VHDL program that will output the 2's complement of the 4-bit input signal.

52. Write a VHDL program that will output the 1's complement of the 8-bit input signal.

53. Write a VHDL program that will output the 2's complement of the 8-bit input signal.

54. Write a VHDL program that will implement a 4-bit adder/subtractor. Output a negative answer in true magnitude form. To indicate a negative answer, program an output to go HIGH. If input signal **cntl** (control) is zero, the circuit will add. If **cntl** is one, the circuit will subtract.

55. Write a VHDL program that will implement an 8-bit adder/subtractor. Output a negative answer in true magnitude form. To indicate a negative answer, program an output to go HIGH. If input signal **cntl** (control) is zero, the circuit will add. If **cntl** is one, the circuit will subtract.

56. Find and correct four errors in this program.

```
LIBRARY ieee;
USE ieee.std_logic_1164.all;
USE ieee.std_logic_unsigned.all;
ENTITY bin_add IS
    BEGIN
    PORTS(a,b:IN STD_LOGIC_VECTOR(3 DOWNTO 0);
        sum:OUT STD_LOGIC_VECTOR(3 DOWNTO 0));
END bin_add_4;
ARCHITECTURE a OF bin_add_4 IS
BEGIN
    sum <= a + b;
END bin_add_4;
```

57. Find and correct four errors in this program.

```
LIBRARY ieee;
USE ieee.std_logic_1164.ALL;
ENTITY bin_fulladd_4 IS
    PORT(a,b:IN STD_LOGIC_VECTOR(3 DOWNTO 0);
        sum:OUT STD_LOGIC_VECTOR(3 DOWNTO 0);
        Cin:IN STD_LOGIC;
        Cout:OUT STD_LOGIC);
END bin_fulladd_4;
ARCHITECTURE OF bin_fulladd_4 IS
BEGIN
    intsum <= ('0'& a) + b + Cin;
    Cout <= intsum(4);
    sum <= intsum(3);
END;
```

OBJECTIVES

After completing this lab, you should be able to:

- Draw the logic diagram for a BCD adder.
- Construct and use a BCD adder.
- Draw the logic diagram for a 1's complement adder/subtractor.
- Construct and use a 1's complement adder/subtractor.

COMPONENTS NEEDED

1	7408 IC
1	7432 IC
2	7483 ICs
2	7486 ICs
5	LEDs
5	330-Ω resistors

PREPARATION

Test a BCD adder as follows:

1. Put 0000 and 0011 into the BCD adder input. Check to see if the output of the first 7438 IC is 0011. If it is, check the second 7483 IC for the sum of 0011.

2. If the adder works with sums of 9 or less but does not work with sums of 10 or more, then the ADD 6 part of the adder is not functioning.

3. If the sum of the two inputs 0000 and 0011 is 1001, then the ADD 6 part of the BCD adder is turned on when it should be off.

Troubleshoot the rest of Lab 5A if it is needed. It is a good idea to write the steps you use to troubleshoot a circuit in a notebook for reference during the troubleshooting procedure and for use at a later time.

Review the lab safety rules under SAFETY ADVICE in the PREPARATION section of Lab 1A, Chapter 1.

PROCEDURE

1. Draw the logic diagram of a BCD adder. Show pin numbers. Use two 7483s and additional gates as needed. Use LEDs to monitor the outputs. Have your instructor approve your drawing.

2. Construct the circuit. Let $A = 0101$ and $B = 0011$. What is the sum? Is there a carry-out of the first 7483?

3. Let $A = 0101$ and $B = 1001$. What is the sum? Is there a carry-out of the first 7483?

4. What is the purpose of the ADD 6 signal?

5. Add these combinations:

a.	0110	**b.**	1001	**c.**	1001
	$+0110$		$+1001$		$+0001$

6. Draw the logic diagram of a 1's complement adder/subtractor. Use one 7483 and additional gates as needed. Use LEDs to monitor the outputs. Have your instructor approve your drawing.

7. Construct the circuit. Add $A = 0101$ and $B = 0011$. What is the sum? Is there a carry-out? Does the first set of exclusive-OR gates complement B? Does the last set invert the answer? Is an EAC performed?

8. Let $A = 0101$ and $B = 1001$. What is the difference $(A - B)$? Is the result of the addition complemented by the second set of exclusive-OR gates? Is B complemented by the first set of exclusive-OR gates? Is an EAC performed?

9. Let $A = 1001$ and $B = 0101$. What is the difference $(A - B)$? Is B complemented by the first set of exclusive-OR gates? Is the result of the addition complemented by the second set of exclusive-OR gates? Is an EAC performed?

10. Try these combinations:

a.	0110	**b.**	1001	**c.**	1001
	-0110		-1010		-0001

If your circuit does not work properly, consider these points:

1. Check all power supply voltages and connections.

2. Check all input and output voltages for proper voltages.

3. Troubleshoot the circuit using the following basic steps. These steps can be used to troubleshoot any electronic circuit.

   a. First, understand the electronic circuit's operation theory.
   b. Determine or set the input value to the circuits.
   c. Measure the circuit's output value.
   d. Based on the input values, output values, and your understanding of how the circuits work, determine which part of the circuit is faulty. Then test the part you suspect.

OBJECTIVES

After completing this lab, you should be able to use Multisim to:

- Troubleshoot a BCD adder circuit.
- Troubleshoot a 1's complement circuit.
- Troubleshoot a 2's complement circuit.

PREPARATION

In each part of this lab, you are asked to troubleshoot one of the three adder/subtractor circuits covered in this chapter. In each circuit the word generator is used to supply inputs to the circuits. The first three addresses of the generator contain test problems. If you cannot determine the fault by studying the test problems, create some tests of your own. Use the lamp indicators, volt indicators, or voltmeters to check critical voltages. Keep a log of your procedure and write a paragraph discussing your conclusion.

PROCEDURE

PART 1

In Multisim, open circuit file 5B-1.ms9 and answer these questions.

1. What is the function of this circuit?
2. Which bits of the word generator are supplying the B inputs? A inputs?
3. The first three lines of the word generator contain test problems for the circuit. What are the three test problems?
4. Is the final result correct in each case?
5. Is the preliminary sum (output of the first 7483) correct in each case?
6. In each situation, should six be added to correct the preliminary result?
7. If so, is six being added correctly?
8. From your answers to questions 4 through 7, if you have detected a problem, what is the most probable fault? Write a paragraph explaining your decision.

PART 2

In Multisim, open circuit file 5B-2.ms9 and answer these questions.

1. What is the function of this circuit?
2. Which bits of the word generator are supplying the B inputs? A inputs? Add/subtract control signal?

3. The first three lines of the word generator contain test problems for the circuit. What are the three test problems?

4. Is the final result correct in each case?

5. Is the preliminary sum (output of the first 7483) correct in each case?

6. In each situation, should an end-around carry be added into the preliminary sum? If so, is it being added correctly?

7. In each situation, should the preliminary sum be complemented? If so, is it being complemented correctly?

8. From your answers to questions 3 through 7, if you have detected a problem, what is the most probable fault? Write a paragraph explaining your decision.

PART 3

In Multisim, open circuit file 5B-3.ms9 and answer these questions.

1. What is the function of this circuit?

2. Which bits of the word generator are supplying the B inputs? A inputs? Add/subtract control signal?

3. The first three lines of the word generator contain test problems for the circuit. What are the three test problems?

4. Is the final result correct in each case?

5. Is the preliminary sum (output of the first 7483) correct in each case?

6. In each situation, should the preliminary sum be complemented? If so, is it being complemented correctly?

. From your answers to questions 3 through 7, if you have detected a problem, what is the most probable fault? Write a paragraph explaining your decision.

PART 4

Troubleshoot circuit files 5B-4.ms9, 5B-5.ms9, and 5B-6.ms9. Keep a log of your discoveries and conclusions.

Specifications and Open-Collector Gates

OUTLINE

KEY TERMS

advanced CMOS (AC)

advanced CMOS-TTL compatible (ACT)

advanced high-speed CMOS (AHC)

advanced high-speed CMOS-TTL compatible (AHCT)

advanced low-voltage CMOS-TTL compatible (ALVT)

advanced low-power Schottky subfamily (ALS)

advanced Schottky subfamily (AS)

complementary metal-oxide semiconductor (CMOS)

dual in-line package (DIP)

emitter coupled logic (ECL)

FAST subfamily (F)

gullwing lead

high-speed CMOS (HC)

high-speed CMOS-TTL compatible (HCT)

J-lead

low-power Schottky subfamily (LS)

low-power subfamily (L)

low voltage (LV)

low-voltage CMOS (LVC)

low-voltage technology (LVT)

microminiaturization

open collector

open drain

plastic leaded chip carrier (PLCC)

Schottky subfamily (S)

shrink small outline package (SSOP)

small outline package (SOP)

small outline integrated circuit (SOIC)

staggered-pin grid array (SPGA)

surface mount technology (SMT)

thin shrink small outline package (TSSOP)

through-hole construction

Transistor Transistor Logic (TTL)

zero insertion force socket (ZIF)

OBJECTIVES

After completing this chapter, you should be able to:

■ Identify TTL commercial and military specification ICs.

■ Define TTL parameters and find their values in a spec book.

■ Define and calculate fan-out and noise margin.

■ List the TTL subfamilies according to speed and power consumption.

■ Describe the difference between totem-pole and open-collector outputs.

■ Use open-collector gates in applications.

■ Identify CMOS ICs.

■ Define CMOS parameters and find their values in a spec book.

■ Identify the 74CXX CMOS series ICs and discuss their characteristics.

■ Interface between logic families.

■ Identify the low-voltage CMOS subfamilies and their characteristics.

■ Identify the 10K and 100K ECL series ICs and describe their characteristics.

■ Describe surface mount IC packages.

■ Describe surface mount resistors.

6.1 TTL SUBFAMILIES

One popular, readily available family of digital ICs is **Transistor Transistor Logic (TTL)**. The TTL family is identified by the first two digits of the device number. 74XX denotes that the IC meets commercial specifications and has an operating range of 0°C to 70°C. 54XX denotes that the IC meets full military-range specifications and can operate between −55°C and +125°C. The 54XX series can withstand harsher environments than the 74XX series. The pinouts for corresponding ICs such as 5404 and 7404 are the same.

Any letters following 54 or 74 denote the subfamily of the IC. Common TTL subfamilies are listed as follows:

No letters	Standard TTL
LS	Low-power Schottky
S	Schottky
L	Low power
ALS	Advanced low-power Schottky
AS	Advanced Schottky
F	Fairchild advanced Schottky TTL (FAST)

The numbers following the subfamily designation indicate the function of the IC. The 54LS10 is a low-power Schottky triple three-input NAND gate that meets military specifications. The 74ALS32 is an advanced low-power Schottky quad two-input OR gate that meets commercial specifications.

Manufacturers guarantee that a 74 series can be operated with a supply voltage that ranges between 4.75 V and 5.25 V, and that a 54 series IC can be operated between 4.50 V and 5.50 V.

6.2 TTL ELECTRICAL CHARACTERISTICS

Figure 6-1 is reproduced from the National Semiconductor Data Book. It shows the recommended operating conditions, electrical characteristics, and switching characteristics for a 5400 and 7400 quad two-input NAND gate. These values are typical of all the standard TTL gates.

After these specifications have been studied, they will be compared to those of the L, S, AS, LS, F, and ALS subfamilies. In this table, conventional current flowing out of the gate is considered negative and conventional current flowing into the gate is positive. The units for each parameter are listed in the right-hand column.

V_{IH}, high-level input voltage, is listed as a minimum of 2 V. An input must be at least 2 V to be recognized as a 1. A 1-level input can range from 2 V to V_{CC}. V_{OH}, high-level output voltage, is listed as a minimum of 2.4 V. A 1 output can range from 2.4 V to V_{CC}.

If a gate supplies at least 2.4 V for a 1-level and a following gate can recognize down to 2.0 V as a 1, then there is a 0.4 V difference in levels. This safety margin is called the noise margin of

Recommended Operating Conditions

Symbol	Parameter	DM5400			DM7400			Units
		Min	Nom	Max	Min	Nom	Max	
V_{CC}	Supply Voltage	4.5	5	5.5	4.75	5	5.25	V
V_{IH}	High-Level Input Voltage	2			2			V
V_{IL}	Low-Level Input Voltage			0.8			0.8	V
I_{OH}	High-Level Output Current			−0.4			−0.4	mA
I_{OL}	Low-Level Output Current			16			16	mA
T_A	Free Air Operating Temperature	−55		125	0		70	°C

Electrical Characteristics over recommended operating free air temperature (unless otherwise noted)

Symbol	Parameter	Conditions		Min	Typ (Note 1)	Max	Units
V_I	Input Clamp Voltage	V_{CC} = Min, I_I = −12mA				−1.5	V
V_{OH}	High-Level Output Voltage	V_{CC} = Min, I_{OH} = Max V_{IL} = Max		2.4	3.4		V
V_{OL}	Low-Level Output Voltage	V_{CC} = Min, I_{OL} = Max V_{IH} = Min			0.2	0.4	V
I_I	Input Current @ Max Input Voltage	V_{CC} = Max, V_I = 5.5V				1	mA
I_{IH}	High-Level Input Current	V_{CC} = Max, V_I = 2.4V				40	μA
I_{IL}	Low-Level Input Current	V_{CC} = Max, V_I = 0.4V				−1.6	mA
I_{OS}	Short Circuit	V_{CC} = Max	DM54	−20		−55	mA
	Output Current	(Note 2)	DM74	−18		−55	mA
I_{CCH}	Supply Current With Outputs High	V_{CC} = Max			4	8	mA
I_{CCL}	Supply Current With Outputs Low	V_{CC} = Max			12	22	mA

Switching Characteristics at V_{CC} = 5V and T_A = 25°C (See Section 1 for Text Waveforms and Output Load)

Parameter	Conditions	C_L = 15 pF R_L = 400 Ω			Units
		Min	Typ	Max	
t_{PLH} Propagation Delay Time Low- to High-Level Output			12	22	ns
t_{PHL} Propagation Delay Time High- to Low-Level Output			7	15	ns

Note: All typicals are at V_{CC} = 5V, T_A = 25°C.
Note: Not more than one output should be shorted at a time.

FIGURE 6-1 TTL NAND gate specifications

the IC. As shown in Figure 6-2, 0.4 V of noise can be riding on a 1-level output, and the signal will still be recognized as a 1 by the following IC.

V_{IL}, low-level input voltage, is listed as 0.8 V maximum. The highest voltage that an IC will accept as a 0 is 0.8 V. A 0-level input can range from 0 to 0.8 V. V_{OL}, low-level output voltage, is listed as 0.4 V maximum. A 0-level output can range from 0 to 0.4 V. If the highest 0 level that an IC will supply is 0.4 V, but a following IC can recognize up to 0.8 V as a 0, then once again there is a noise margin of 0.4 V, as shown in Figure 6-3.

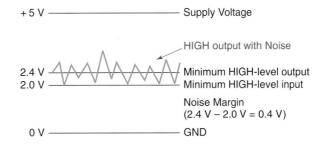

FIGURE 6-2 TTL 1-level noise margin

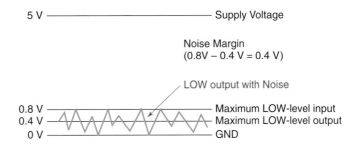

FIGURE 6-3 TTL 0-level noise margin

Standard TTL noise margin is summarized in Figure 6-4.

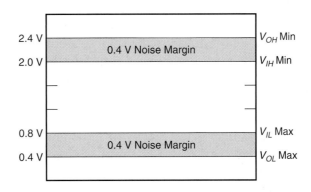

FIGURE 6-4 Voltage levels for TTL logic

Input and output voltages for each of the subfamilies are summarized in Figure 6-5.

DM5400/DM7400 2-INPUT NAND		TTL	L-TTL	LS	ALS	S	AS	F	Units
Military	V_{OH}	2.4	2.4	2.5	$V_{CC} - 2$	2.5	$V_{CC} - 2$		V
Commercial	V_{OH}	2.4	2.4	2.7	$V_{CC} - 2$	2.7	$V_{CC} - 2$	2.5	V
Military	V_{OL}	0.4	0.3	0.4	0.4	0.5	0.5		V
Commercial	V_{OL}	0.4	0.4	0.5	0.5	0.5	0.5	0.5	V
Mil. and Com.	V_{IH}	2	2	2	2	2	2	2	V
Military	V_{IL}	0.8	0.7	0.7	0.8	0.8	0.8		V
Commercial	V_{IL}	0.8	0.7	0.8	0.8	0.8	0.8	0.8	V

FIGURE 6-5 Input and output voltages: TTL subfamilies

EXAMPLE 6-1

Calculate the noise margin of a 54S00.

Solution   Figure 6-1 shows that $V_{OH} = 2.5$ V minimum, $V_{IL} = 0.8$ V maximum, $V_{OL} = 0.5$ V maximum, $V_{IH} = 2.0$ V minimum.

High-level noise margin $= V_{OH} - V_{IH} = 2.5$ V $- 2.0$ V $= 0.5$ V
Low-level noise margin $= V_{IL} - V_{OL} = 0.8$ V $- 0.5$ V $= 0.3$ V

EXAMPLE 6-2

Calculate the noise margin of a 74ALS00 operating at 5 V.

Solution

$V_{IL} = 0.8$ V maximum, $V_{OL} = 0.5$ V maximum,
$V_{IH} = 2$ V minimum, $V_{OH} = V_{CC} - 2$.

High-level noise margin $= V_{OH} - V_{IH} = 5 - 2 - 2 = 1$ V.
Low-level noise margin $= V_{IL} - V_{OL} = 0.8 - 0.5 = 0.3$ V.

I_{OL}, low-level output current, is listed as a maximum of 16 mA in Figure 6-1 (page 254). This is conventional current flowing into the gate, and the gate is said to be "sinking" current. The manufacturer guarantees that the 7400 can "sink" 16 mA of current without the zero-level output voltage rising above 0.4 V. I_{IL}, low-level input current, is listed as a maximum of -1.6 mA. This current is flowing out of the gate. Figure 6-6 shows a 7400 NAND gate sinking current from ten other gates, each with a low-level input current of -1.6 mA. The 1.6 mA is said to be "one standard TTL load." Fan-out is a measure of the number of loads that a gate can drive.

$$\boxed{\text{Fan-out} = \frac{I_{OL} \text{ of the driving gate}}{I_{IL} \text{ of the driven gate}}}$$

For a NAND gate driving other NAND gates or inverters,

$$\text{Fan-out} = \frac{I_{OL}}{I_{IL}} = \frac{16 \text{ mA}}{1.6 \text{ mA}} = 10 \text{ standard loads}$$

Each of the standard TTL gates can drive ten other standard gates.

Figure 6-7 shows the input and output currents for a NAND gate from each subfamily. I_{OL} is a measure of the drive capability of each subfamily. Schottky, advanced Schottky, and FAST are highest at 20 mA, followed by standard TTL at 16 mA, low-power Schottky and advanced low-power Schottky at 8 mA (4 mA for military), and finally 3.6 mA for low-power TTL (2 mA for military).

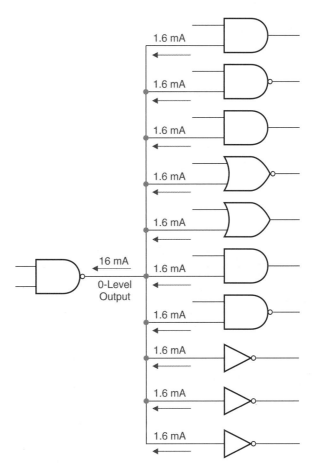

FIGURE 6-6 TTL fan-out

		TTL	L-TTL	LS	ALS	S	AS	F	Units
	I_{OH}	−400	−200	−400	−400	−1000	−2000	−1000	µA
Commercial	I_{OL}	16	3.6	8	8	20	20	20	mA
Military	I_{OL}	16	2	4	4	20	20		mA
	I_{IH}	40	10	20	20	50	20	20	µA
	I_{IL}	−1.6	−0.18	−0.36	−0.20	−2	−0.50	−0.6	mA

FIGURE 6-7 Input and output currents: TTL subfamilies

EXAMPLE 6-3

How many 54ALS00 gates can a 54L00 drive?

Solution   Figure 6-7 lists I_{OL} for a 54L00 as 2 mA and I_{IL} for a 54ALS00 as −0.2 mA.

$$\text{Fan-out} = \frac{I_{OL}}{I_{IL}} = \frac{2 \text{ mA}}{0.2 \text{ mA}} = 10$$

EXAMPLE 6-4

How many standard TTL loads can a 54LS00 drive?

Solution   Figure 6-7 lists I_{OL} for a 54LS00 as a 4 mA maximum.

$$\text{Fan-out} = \frac{I_{OL}}{I_{IL}} = \frac{4 \text{ mA}}{1.6 \text{ mA}} = 2.5$$

Since the 0.5 load is not a complete load, drop the 0.5. The 54LS00 has a fan-out of two standard TTL loads.

EXAMPLE 6-5

How many 74LS04 inverters can a 74LS00 drive?

Solution

I_{OL} for a 74LS00 is 8 mA.
I_{IL} for a 74LS04 is 0.36 mA.

$$\text{Fan-out} = \frac{I_{OL}}{I_{IL}} = \frac{8 \text{ mA}}{0.36 \text{ mA}} = 22.22$$

A 74LS00 can drive 22 74LS04 gates.

A gate from any of the TTL subfamilies can drive at least ten other gates from the same subfamily.

I_{OH}, high-level output current, is listed as -0.4 mA for the 7400. The negative sign implies that current is flowing out of the gate. The gate supplies or "sources" currents when outputting a 1. This group of gates can sink 16 mA on a low-level output, but can only source 0.4 mA on a high-level output. We can take advantage of the greater low-level current by using the gate in an active LOW mode, as shown in Figure 6-8. When A goes HIGH, B goes LOW. Since the 7404 can sink 16 mA when outputting a 0, the LED can burn brightly.

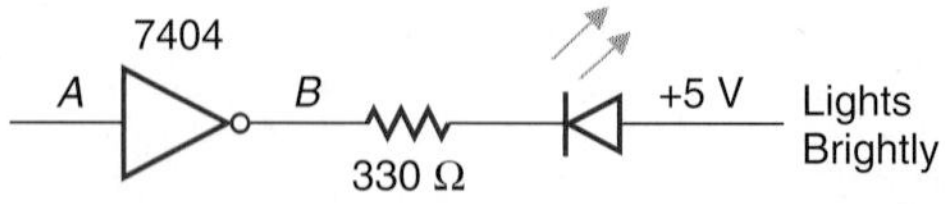

FIGURE 6-8 Active LOW mode

In the arrangement in Figure 6-9, a high-level output should light the LED. However, the data book lists I_{OH} for a 7408 as -800 µA maximum. The LED draws more than 800 µA and the 1 level could fall below 2.4 V. The output would no longer be a legitimate 1 level.

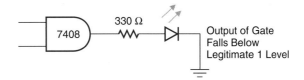

FIGURE 6-9 Active HIGH mode

I_{IH}, high-level input current, is listed as 40 μA for the standard gates in Figure 6-1. Since a standard NAND gate can source 400 μA on a high-level output, the gate can drive ten other gates. This result is compatible with what we learned from the low-level signals.

6.3 TTL SUPPLY CURRENTS

Figure 6-10, reproduced from the National Semiconductor Data Book, lists the current drawn from the power supply per IC for the 7400 NAND gates. I_{CCH} represents the total collector current drawn with all outputs HIGH, and I_{CCL} represents those with outputs LOW. The standard TTL 7400 draws a maximum of 8 mA from the supply with outputs HIGH and 22 mA maximum with outputs LOW. A Schottky 74S00 draws the most current at 36 mA (outputs LOW), followed by TTL, AS, F, LS, ALS, and finally the low-power 74L00 at 2.04 mA.

		S	TTL	AS	F	LS	ALS	L	Units
I_{CCH}	Supply current with outputs HIGH (maximum)	16	8	3.2	2.8	1.6	0.85	0.8	mA
I_{CCL}	Supply current with outputs LOW (maximum)	36	22	17.4	10.2	4.4	3.0	2.04	mA

FIGURE 6-10 Supply currents: TTL quad two-input NAND

To calculate the current that a power supply will need to provide for a circuit, follow these rules:

1. Total the worst-case currents for all the ICs in the circuit.

2. Add currents drawn by all other devices such as LEDs and displays.

3. As a rule of thumb, double the total and design your power supply accordingly.

6.4 TTL SWITCHING CHARACTERISTICS

Figure 6-11 lists the switching characteristics for TTL gates. The t_{PLH} (time propagation LOW-to-HIGH) is a measure of the time it takes for a change in the input to cause a LOW-to-HIGH change in the output. As shown in Figure 6-12, the t_{PLH} is measured from when the input reaches 1.5 V to when the output reaches 1.5 V on a LOW-to-HIGH transition. These

	L	TTL	LS	ALS	F	S	AS	TIME
t_{PLH}	60	22	15	11	5	7	4.5	ns
t_{PHL}	60	15	15	8	4.3	8	4	ns

FIGURE 6-11 Maximum switching characteristics: TTL quad two-input NAND

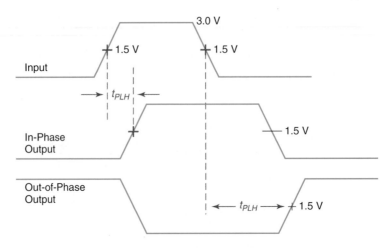

FIGURE 6-12 TTL propagation LOW-to-HIGH

parameters are measured with the gates driving a load equivalent to 10 gates of the same sub-family. Standard TTL has a maximum t_{PLH} of 22 nanoseconds, while advanced Schottky is the fastest with only 4.5 nanoseconds delay.

The t_{PHL} (time propagation HIGH-to-LOW) is a measure of the time it takes for a change in the input to cause a HIGH-to-LOW transition in the output. As shown in Figure 6-13, the t_{PHL} is measured from when the input reaches 1.5 V to when the output reaches 1.5 V on a HIGH-to-LOW transition.

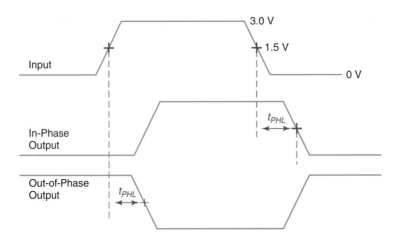

FIGURE 6-13 TTL propagation HIGH-to-LOW

Standard TTL has a maximum t_{PHL} of 15 nanoseconds. Note that standard TTL turns off more quickly than it turns on. Advanced Schottky is once again the fastest with a maximum t_{PHL} of 4 nanoseconds, and low-power TTL is slowest at 60-nanosecond delay.

The propagation delay limits the speed at which the IC can operate. When the propagation delay becomes a significant portion of the period of the applied signal, then the output levels and timing become distorted.

When a logic family is selected for use in a circuit, both speed and power consumption should be considered. Figure 6-14 summarizes these properties. Low-power TTL dissipates the least power but is the slowest. This is the classic trade-off, power consumption versus speed. Schottky is one of the fastest subfamilies, but it dissipates the most power. ALS is among the fastest, while dissipating less power than all subfamilies except L. These qualities make it a good choice for many applications.

Speed		Power Consumption	
Fastest	AS	Low	L
	F		ALS
	S		LS
	ALS		F
	LS		AS
	TTL		TTL
Slowest	L	High	S

FIGURE 6-14 TTL relative speed and power consumption

Because low-power Schottky is faster than standard TTL and draws less current from the power supply, it has been a popular subfamily.

Following is a brief synopsis of the evolution of the TTL subfamilies.

The **low-power subfamily L** was developed from TTL by increasing the resistances of the resistors in the internal circuitry by a factor of 10. The power dissipation of the L device is reduced by a factor of 10 but at a sacrifice of speed. L devices have a propagation delay that is three times longer than standard TTL.

The **Schottky subfamily S** uses Schottky diodes as clamps to prevent the transistors from becoming saturated and uses resistors whose resistances are about half those of standard TTL. This unsaturated logic switches three times more quickly than standard TTL but also consumes more power. Gates in this family dissipate about 20 milliwatts with typical propagation delays of 3 nanoseconds.

The **low-power Schottky subfamily LS** uses increased resistor values and diode inputs instead of the multi-emitter inputs seen in standard TTL circuitry. These diode inputs switch more quickly to yield typical propagation delays of 10 nanoseconds and a power dissipation of 2 milliwatts per gate.

The **advanced low-power Schottky subfamily, ALS**, uses refined fabricating techniques to increase switching speeds and lower power consumption over LS devices. ALS device gates have typical propagation delays of about 4 nanoseconds and a power dissipation of 1 mW per gate.

The **advanced Schottky subfamily AS** is designed for speed. It uses networks in the output circuits that reduce rise time. AS gates dissipate about 8 mW and have typical propagation delays of 1.5 ns. Fairchild's advanced Schottky gates, **FAST**, use the letter F to indicate the subfamily. 74F04 is a Fairchild advanced Schottky TTL quad two-input NAND gate. F gates have propagation delays of about 3 ns and dissipate about 5 mW.

Figure 6-15 shows the internal circuit of a TTL NAND gate. Although the gate can be used without knowledge of the internal circuitry, the TTL characteristics can be better understood by investigating the circuitry. A full description of this NAND gate circuit is included in Appendix D.

R_4, Q_3, D_3, and Q_4 constitute the output circuit as shown in Figure 6-16. Their configuration is called the "totem-pole" output. Normally either Q_3 or Q_4 is on, but not both. With Q_3 on, Y is pulled up to a one level, and with Q_4 on, Y is pulled down to a zero level. However, during switching, Q_3 and Q_4 are on simultaneously for a short time, and a heavy load is placed on the power supply. TTL circuits are designed to switch Q_3 and Q_4 quickly to minimize the effects on the power supply.

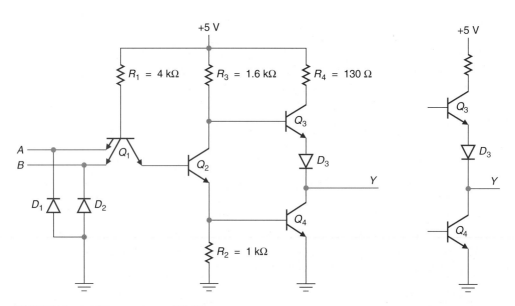

FIGURE 6-15 TTL two-input NAND gate

FIGURE 6-16 TTL totem-pole output

To filter the noise induced on the power supply by the switching action, wire a 0.01-μF ceramic capacitor across the power supply near the pins on the IC. One capacitor for every two ICs is a good rule of thumb to follow.

Two totem-pole outputs should not be connected directly together as shown in Figure 6-17. If the upper transistor of one totem-pole output turns on, and the lower transistor of the other turns on, then the power supply is shorted out, and heavy currents will flow. The result can be damage to the gates and power supply. Totem-pole outputs should be connected together through other gates.

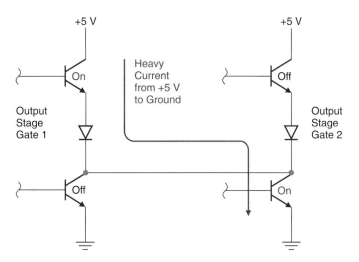

FIGURE 6-17 Danger from connecting totem-pole outputs together

 # 6.5 TTL OPEN-COLLECTOR GATES

A special type of gate called the **open-collector gate** has a modified output circuit. The upper transistor of the totem-pole pair has been omitted so that the output has no internal path to +5 V. When the output is driven LOW, the transistor turns on and connects the output Y to ground through a saturated transistor. When the output is driven into the HIGH state, Y is no longer connected to ground, nor is it connected to +5 V, since that path no longer exists. The output enters a high-impedance state, "HiZ," in which the gate has no influence on the output at all. The output is floating. Open-collector outputs can be connected together since there is no danger of the power supply being shorted. Figure 6-18 shows the output circuits of three open-collector gates with outputs connected together. The outputs are tied to a common resistor that is connected to +5 V. The resistor is called a "pull-up" resistor since it supplies the output with a path to +5 V and "pulls" the output up to a one level. If the output of any of the gates goes LOW, then the output Y goes LOW. Conventional current flows from +5 V through the pull-up resistor and through the saturated transistor to ground. Most of the 5 V is dropped across the pull-up resistor.

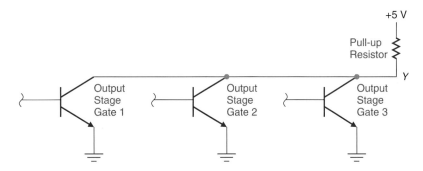

FIGURE 6-18 Open-collector gates

Figure 6-19 shows three open-collector inverters with outputs tied together. When A, B, and C are all LOW, the outputs are not connected to ground internally. The 1-kΩ resistor pulls the

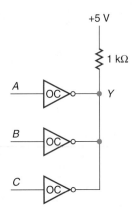

Inputs			Output
A	B	C	Y
0	0	0	1
0	0	1	0
0	1	0	0
0	1	1	0
1	0	0	0
1	0	1	0
1	1	0	0
1	1	1	0

FIGURE 6-19 Open-collector inverter and truth table

output *Y* up to a one level. If any of the inputs goes HIGH, the corresponding output transistor turns on and pulls *Y* down to a LOW level. The truth table for the circuit is also shown in Figure 6-19. The truth table is identical to that of a three-input NOR gate. This configuration is called a *wired-NOR* or a *dot-NOR* circuit.

Open-collector outputs are denoted in IEC and IEEE symbols by the symbol $\lozenge$ on each output.

Open-collector gates that are available for use are listed in Table 6-1.

TABLE 6-1 Open-collector gates

Device Number	Description	Device Number	Description
5401/7401 54ALS01/74ALS01 54L01/74L01 54LS01/74LS01 5403/7403 54L03/74L03 54LS03/74LS03 54ALS03/74ALS03 54S03/74S03	Quad 2-input NAND gates with open-collector outputs	54LS12/74LS12 54ALS12/74ALS12	Triple 3-input NAND with open-collector outputs
		54LS15/74LS15 54ALS15/74ALS15	Triple 3-input AND gates with open-collector outputs
		5416/7416	Hex inverter buffers with open-collector high-voltage outputs
5405/7405 54L05/74L05 54ALS05/74ALS05 54S05/74S05	Hex inverters with open-collector outputs	5417/7417	Hex buffers with open-collector high-voltage outputs
5406/7406 74F06	Hex inverter buffers with high-voltage open collectors	54ALS22/74ALS22 54LS22/74LS22 54S22/74S22	Dual 4-input NAND gates with open-collector outputs
5407/7407 74F07	Hex buffers with high-voltage open-collector outputs	5438/7438 54ALS38/74ALS38 54LS38/74LS38	Quad 2-input NAND buffers with open-collector outputs
5409/7409 54L09/74L09 54LS09/74LS09 54ALS09/74ALS09 54S09/74S09	Quad 2-input AND with open-collector outputs	54LS136/74LS136 54AS136/74AS136 54S136/74S136	Quad exclusive-OR with open-collector outputs
		54LS266/75LS266	Quad exclusive-NOR with open-collector outputs

EXAMPLE 6-6

Draw the IEC logic symbol for 74LS136 quad exclusive-OR with open-collector outputs.

Solution

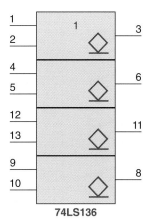

74LS136

The 1 indicates that exactly 1 input must be HIGH for the output to go into the high-impedance state.

 # 6.6 OPEN-COLLECTOR APPLICATIONS

The 7406, 7407, 7416, and 7417 are open-collector gates with high-voltage outputs. Although the ICs themselves operate at 5 V, the open-collector outputs can be pulled up to a higher voltage: 30 V for the 06 and 07, and 15 V for the 16 and 17. The 7406 and 7416 invert the input signal, and the 7407 and 7417 do not. Along with the higher voltages on the outputs, these gates can sink more low-level output current than the totem-pole output gates can. The 5406, 07, 16, and 17 gates can sink 30 mA and the 7406, 07, 16, and 17 can sink 40 mA. These open-collector gates are used in high-voltage applications and for tying outputs together.

Figure 6-20 shows a 7406 hex inverter with outputs pulled up to 30 V through a 1-kΩ resistor. When the output goes LOW, about 30 mA will be sinked by the gate. Since this is less than the 40 mA maximum listed by the specs, the output voltage will not rise above 0.4 V. This IC will be used in later chapters to step up to higher supply voltages, and in this chapter, to interface TTL with CMOS.

Figure 6-21 shows a 7406 open-collector high-voltage inverter driving a 12-V, 500-Ω relay coil. When the output of the inverter goes LOW, approximately 24 mA flows through the coil, and the contacts are activated. The 7406 can sink up to 40 mA. When input A goes LOW, the output of the gate goes HiZ and the current through the coil is interrupted. A reverse voltage is generated across the inductor. The diode limits the voltage across the inductor to approximately 0.7 V.

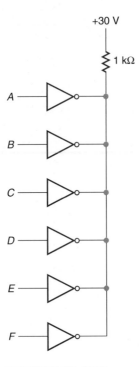

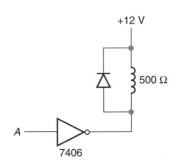

FIGURE 6-20 7406
open-collector inverter

FIGURE 6-21 Open collector
driving a relay coil

 SELF-CHECK 1

1. Identify these specifications and their values in a data book for a 7483.

    a. minimum 1-level output

    b. minimum 1-level input

    c. maximum 0-level output

    d. maximum 0-level input

    e. maximum current drawn from power supply

    f. maximum low-to-high propagation delay

    g. maximum high-to-low propagation delay

    h. maximum 0-level output current

    i. maximum 0-level input current

2. From these values calculate

    a. noise margin.

    b. fan-out to a 7432.

    c. how many 7486 inputs a 7432 can drive.

3. What is a pull-up resistor and why does an open-collector gate use one?

6.7 CMOS

CMOS stands for **complementary metal-oxide semiconductor** field effect transistor. Complementary means that a P-channel transistor and an N-channel transistor work together in a totem-pole arrangement, as shown in Figure 6-22. Metal-oxide refers to the silicon dioxide layer between the gate and channel. The gate, channel, and silicon dioxide insulator form a small capacitor. This capacitive input determines many of the characteristics of CMOS ICs.

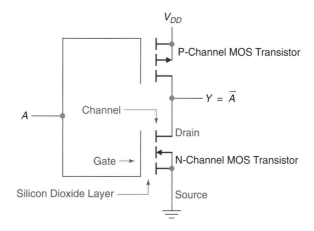

FIGURE 6-22 CMOS inverter

In Figure 6-22, when A is HIGH, the N-channel on the bottom is enhanced, and the output Y is connected to ground through a completed channel. The P-channel MOS on the top is turned off. When A is LOW, the P-channel is enhanced and the N-channel is turned off. Y is connected to V_{DD} through the P-channel. These two transistors produce an inverter.

The supply voltages for CMOS ICs are often called V_{DD}, for drain voltage, and V_{SS}, for source voltage. Sometimes supply voltages are called V_{CC}, for collector voltage and ground, as they were in TTL. V_{DD} varies for each subfamily. V_{SS} is usually 0 V.

A discussion of CMOS transistors and circuits is included in Appendix D.

6.8 CMOS SUBFAMILIES

One of the original series of CMOS ICs, the CD4000B metal gate series, has these advantages over TTL ICs:

1. wide operating voltage range (3–15 volts),

2. low power consumption (microwatts at low frequencies), and

3. high noise immunity (1 volt noise margin),

and these disadvantages:

1. long propagation delays (100 nanoseconds),

2. low current drive capabilities (1 LS-TTL load),

3. latch-up problems (gets stuck in low impedance state), and

4. sensitivity to electrostatic discharge (ESD).

Some examples of ICs in this family are:

4001	Quad 2-input NOR
4012	Dual 4-input NAND
4070	Quad 2-input exclusive-OR

The HE4000B silicon gate series retained the above mentioned advantages while decreasing propagation delays and doubling the drive current.

Another series of CMOS ICs can be identified by the letter C in the device number. CMOS subfamilies in this series are:

74Cxx or 54Cxx	CMOS
74HCxx	High-speed CMOS (HC)
74HCTxx	High-speed CMOS-TTL compatible (HCT)
74AHCxx	Advanced high-speed CMOS (AHC)
74AHCTxx	Advanced high-speed CMOS-TTL compatible (AHCT)
74ACxx	Advanced CMOS (AC)
74ACTxx	Advanced CMOS-TTL compatible (ACT)

The 74C and 54C ICs have the same pinout and function as the corresponding 74xx or 54xx IC. For example, a 7430 and 74C30 are both eight-input NAND gates with the same pinout. The 74HC, 74HCT, 74AHC, and 74AHCT are designed to be pin compatible with the 74LS series. Most of the TTL functions plus some of the functions of the 4000 series devices are duplicated in HC logic. For example, a 4060 is a 14-stage binary ripple counter with oscillator, as is a 74HC4060. A 7404 is a hex inverter, as is a 74HCT04. Also, propagation delays in 74HC and 74HCT devices have been reduced and are comparable to those of 74LS devices. Typical propagation delays are in the ten-nanosecond range. 74AHC and 74AHCT devices have propagation delays in the three-nanosecond range. 74HC devices can operate on a supply voltage from two volts to six volts. 74HCT devices are directly interfaced to LS-TTL devices, so their power supply range is limited to 4.5 to 5.5 volts.

To further increase the operating speed and output drive current of CMOS devices, advanced CMOS logic (ACL) was developed. Two subfamilies of ACL devices are available. 74AC and 54AC type devices are CMOS compatible and can operate on a supply voltage ranging from 3 to 5.5 volts. 74ACT and 54ACT devices are TTL compatible and operate on 5 volts +/− 10%.

The 74AC11xxx and 74ACT11xxx series adopted a "flowthrough architecture." In this architecture, the power supply pins are at the center of each side of the IC instead of the more common "corner power" arrangement. With the notch of the DIP package facing upward, V_{CC} is on the right side in the middle. The input pins are on the right, clustered around the V_{CC} pin. The ground pin is on the left side in the middle with output pins clustered around it. Control and enable pins are positioned at the corners. Figure 6-23 shows the pinout and logic diagram

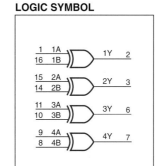

FIGURE 6-23 74AC11086 quad 2-input exclusive-OR

for a 74AC11086 quad 2-input exclusive-OR. Note that this is a 16-pin package, whereas the 7486 is a 14-pin package. 16-pin ACL devices with three or four outputs have two ground pins and two power supply pins. Note that six of the inputs are clustered around the V_{CC} supply pins, and the four outputs are clustered around the ground pins.

Figure 6-24 shows the pinout of a 74ACT11181 4-bit arithmetic logic unit. 20-, 24-, and 28-pin ACL devices with 3 or more outputs have four ground pins and two V_{CC} pins. The inputs A_0 through A_3 and B_0 through B_3 are on the right, clustered around the supply pins. The outputs F_0 through F_3, $A = B$, P, G, and C_{n+4} are on the left, clustered around the ground pins. S_0 through S_3 and M are control signals positioned toward the corners. C_n is the only input out of formation.

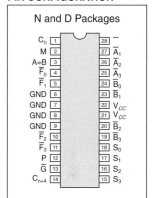

FIGURE 6-24
74ACT11181 4-bit ALU

This style of architecture provides two advantages. The internal inductance is minimized and system noise generated by high switching speeds is reduced. With this pin arrangement, propagation delays have been reduced to the five-nanosecond range. Secondly, having inputs on the right and outputs on the left simplifies printed circuit board design and helps reduce printed circuit board size. A disadvantage of this architecture is that AC11xxx and ACT11xxx devices are not pin compatible with TTL devices.

The 74HC and 74AC ICs have overcome the latch-up and ESD problems inherent in 4000 type devices. However, the handling procedures listed in the lab preparation for Chapter 2 are still recommended for all CMOS devices. Those guidelines are repeated here.

Care must be taken in the handling of CMOS ICs since they can be destroyed from excessive static buildup between pins. These guidelines should be followed:

1. Store CMOS ICs in antistatic tubes or in black conductive foam. Never push CMOS into styrofoam. They can be wrapped in aluminum foil.

2. In low-humidity environments where static buildup is a problem, avoid touching the pins of a CMOS IC when they are removed from storage unless precautions have been taken to bleed off the static charge. Conductive wrist straps connected through a resistor to ground is one method used.

3. Apply dc voltage to the CMOS circuit before signals are applied.

4. Remove signal sources before the dc supply is switched off.

5. Switch off supply voltages before inserting or removing devices from a circuit.

 # 6.9 CMOS SPECIFICATIONS

Table 6-2 contrasts the power dissipated by CMOS and TTL gates in milliwatts (load capacitance of 15 pF). The first line shows typical powers with the gates at rest, that is, not being switched (static). The subfamilies are listed in increasing order of power dissipation. Note that the CMOS gates draw the least power. The power dissipations are listed in milliwatts.

TABLE 6-2   Power Dissipation of CMOS and TTL Gates (in milliwatts)

	4000	74AC	74HC	74AHC	74ALS	74LS	74F	74AS	74	74S
Static	0.001	0.001	0.0025	0.15	1.2	2	5.5	8.5	10	19
100 kHz	0.1		0.075		1.2	2	5.5	8.5	10	19

As frequency of operation increases, the power consumed by the CMOS devices rises. This is because the small capacitors on the inputs of the MOS transistors must be charged and discharged. Power consumed by the TTL devices remains essentially the same. The second line of the table shows power dissipation at 100 kHz. At about 10 MHz, the power consumed by HC and HCT devices has risen to the levels consumed by TTL-LS devices.

Table 6-3 contrasts typical propagation delays and maximum clock frequencies in nanoseconds for CMOS and TTL gates (with a load capacitance of 15 pF). The AS, AC, F, and S subfamilies have the shortest delay times and highest clock frequencies. The HC and LS subfamilies operate at about the same speed. The 4000 series devices are the slowest.

TABLE 6-3   Propagation Delays and Maximum Clock Frequencies for CMOS and TTL Gates

| | | | | | | | | | | 4000 | |
	74AS	74AC	74F	74S	74AHC	74ALS	74HC	74LS	74	HE	CD
Delay (ns)	1.5	3	3	3	3.2	4	8	9.5	10	40	95
Max clk (MHz)	160	150	125	100	—	60	55	33	25	12	4

The 4000 series are the slowest gates but dissipate little power. This delay-power tradeoff often occurs. A single parameter to measure the delay and power consumed by a gate is formed by multiplying the delay in nanoseconds times the power consumed in milliwatts.

Power × Time = Energy

Watts times seconds is joules

milliwatts × nanoseconds = picoJoules

The delay/power product is measured in picoJoules.

Table 6-4 shows the delay/power product at 100 kHz for CMOS and TTL subfamilies. The low power consumption and relatively high speed of 74AC and HC devices puts them at the low end of the spectrum.

TABLE 6-4   Delay/Power Product in pJoules at 100 kHz

| | 74AC | 74HC | 74ALS | 4000 | | 74AS | 74F | 74LS | 74S | 74 |
				CD	HE					
Delay/Power (pJ)		0.6	4.8	9	4	13	16.5	19	57	100

Another way to view the power speed tradeoff is shown in Figure 6-25.

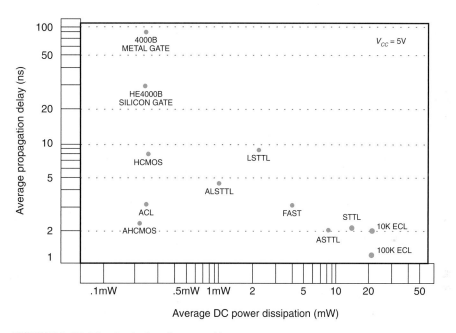

FIGURE 6-25 The logic family speed/power spectrum

Figure 6-25 shows the speed/power relationships for many of the subfamilies discussed in this chapter. The vertical axis is propagation delay in nanoseconds. Higher up on the chart implies longer propagation delays and slower speed of operation. 4000B series devices have the longest propagation delays and ECL devices (discussed later in the chapter) have the shortest. The horizontal axis is power dissipation in milliwatts. Devices to the right consume more power from the supply than those to the left. Notice that ACL (advanced CMOS logic) and AHCMOS are among the fastest of the families and among those that consume the least power from the supply.

Early CMOS ICs had low current drive capability. Recent advances in technology have increased CMOS current drive to TTL levels. Table 6-5 lists the current drive (I_{OL}) and fan-out capabilities of CMOS and TTL ICs.

4000 series devices do not have enough current drive to operate a TTL gate but are capable of driving a LS-TTL gate. The 74AC devices can sink 24 mA and drive 66 LS-TTL loads.

TABLE 6-5   Drive Current (I_{OL}) and Fan-out to LS-TTL Loads

	74AC	74S	74AS	74F	74	74LS	74ALS	74HC	4000 CD	4000 HE
Drive Current (mA)	24	20	20	20	16	8	8	4	.5	.8
Fan-out to LS-TTL	66	55	55	55	44	22	22	11	1	2

Fan-out calculations also require a knowledge of input current (I_{IL} and I_{IH}). Input currents to CMOS devices are quite small, 1 µA for the 74 HC and 74AC devices. Virtually an unlimited number of CMOS devices can be driven by other CMOS or TTL devices.

Table 6-6 summarizes the power supply requirements for the CMOS ICs. ACT, HCT, and 74AHCT devices are designed to directly interface to TTL ICs and must operate at five volts. The other CMOS ICs have a wider range of operation. The wide range of supply voltages does not eliminate the need for good supply voltage regulation. A fluctuating supply can cause errors in circuit operation.

TABLE 6-6   Power Supply Ranges for CMOS ICs

	4000	74C	74HC	74HCT	74AC	74ACT	74AHC	74AHCT
Power supply range (volts)	3–15	3–15	2–6	4.5–5.5	3–5.5	4.5–5.5	2–5.5	4.5–5.5

Table 6-7 lists the output voltages, high and low, and input voltages, high and low, of CMOS devices operating at 5 volts at +25°C. For ACT, HCT, and AHCT subfamilies, V_{IH} is 2.0 volts and V_{IL} is 0.8 volts. These values should look familiar. They are the same as those for TTL ICs. This means that ACT, HCT, and AHCT devices can recognize TTL output voltage levels without any special interfacing. Can TTL ICs recognize ACT, HCT, and AHCT outputs? The minimum HIGH-level output from a TTL device can range from 2.4 volts to 5 volts. The lowest HIGH level that a ACT, HCT, or AHCT device can output (V_{OH}) is about 4 volts, well within the legitimate HIGH level for TTL devices. ACT, HCT, and AHCT can drive and be driven by TTL devices without any special interfacing. However, AC and HC devices can only recognize HIGH-level inputs down to 3.5 volts (V_{IH}), and AHC can only recognize HIGHs down to 3.85 volts. A TTL output can range down to 2.4 volts. TTL does not directly interface with HC, AC, or AHC.

TABLE 6-7   Input and Output Voltage Levels for CMOS

		AC	ACT	HC	HCT	4000	74AHC	74AHCT
V_{OH}	(Vmin)	4.4	4.4	4.5	4.0	4.95	4.4	4.4
V_{OL}	(Vmax)	.36	.36	.26	.26	0.05	0.1	0.1
V_{IH}	(Vmin)	3.5	2.0	3.5	2.0	3.5	3.85	2.0
V_{IL}	(Vmax)	1.5	0.8	1.5	0.8	1.5	1.65	0.8

6.10 INTERFACING TTL TO CMOS

To interface TTL to CMOS operating at 5 V, care must be taken that high-level TTL outputs are high enough to be recognized by the following CMOS ICs. The TTL output should be pulled up through a 10-kΩ external resistor, as shown in Figure 6-26.

To interface TTL with CMOS that is operating at high voltage levels, one of the high-voltage open-collector gates can be used as shown in Figure 6-27. The open-collector output is pulled up to the operating voltage of the CMOS gate.

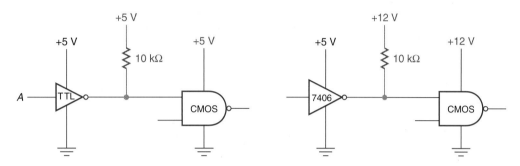

FIGURE 6-26 TTL to 5-volt CMOS

FIGURE 6-27 TTL to higher-voltage CMOS

In either case, TTL is capable of sinking sufficient current to drive an unlimited number of CMOS gates at low frequency.

Figure 6-28 shows a CMOS NAND gate that could be operated from 5 V to 18 V. The higher voltages on its output are not compatible with TTL inputs. These levels present no problem to the 4049 hex inverting buffer. The 4049 can operate at a supply voltage of 5 V and handle input voltages up to 15 V. The outputs are TTL compatible. The 4049 sinks sufficient current to drive two standard TTL gates; in this case a 7404 and a 7400. The 4050 hex non-inverting buffer can be used if inversion is not required.

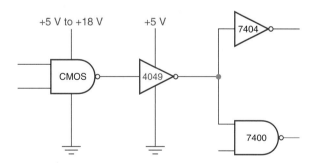

FIGURE 6-28 CMOS to TTL interface

Since CMOS HC operating at 5 V and HCT have outputs that are TTL compatible, HC and HCT devices can directly drive a TTL device. However, TTL HIGH-level outputs can range down to 2.4 V, and this is lower than the acceptable 3.5 V input that the HC devices recognize. In this case a pull-up resistor is required for the interface. The HCT devices can recognize TTL logic levels and interface directly to them. These interfaces are summarized in Figure 6-29.

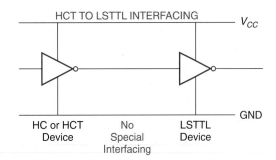

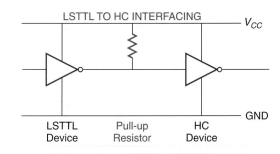

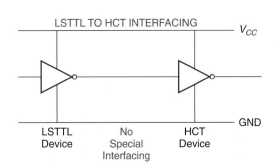

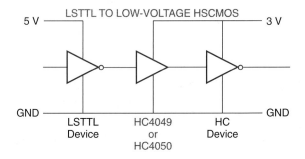

FIGURE 6-29 Interfacing TTL to HC and HCT devices (*Courtesy of Motorola, Inc.*)

The techniques shown for interfacing HC with TTL also apply to interfacing AC with TTL. ACT devices, like HCT, interface directly with TTL devices.

EXAMPLE 6-7

Drive as many 74LS00s as possible with a 74HCT00.

Solution   HCT outputs are directly compatible with TTL devices. No pull-up resistor is needed. A 74HCT00 has a fan-out of 10 LS gates.

EXAMPLE 6-8

Drive as many 74HC00 gates operating at 5 V as possible with a 74LS00.

Solution   A 74LS00 high-level output voltage, V_{OH}, can range down to 2.4 volts. A 74HC00 high-level input, V_{IH}, has minimum of 3.15 volts. The output of the 74LS00 needs to be "pulled up" with an external pull-up resistor. The maximum input current drawn by the 74HC00 is 1 μA. The low-level output current, I_{OL}, for 74LS00 is 8 mA, but the high-level output current is only 0.4 mA. Using 0.4 mA to calculate fan-out still allows us to drive 400 gates.

$$\text{Fan-out} = \frac{I_{OH}}{I_{IH}} = \frac{0.4 \text{ mA}}{1 \text{ μA}} = 400$$

By using a pull-up resistor (1 kΩ or so), 400 or more 74HC00s may be driven by a 74LS00.

6.11 LOW-VOLTAGE CMOS

A lower power supply voltage reduces the power consumption of the circuitry. Reduced power consumption enables portable equipment to be smaller, lighter, and run cooler. Systems can run longer before recharging or replacing the batteries. With lower supply voltages, equipment can run on one or two battery cells. To meet the need for reduced power consumption, CMOS low-voltage subfamilies LV, LVC, and LVT were developed. These families are manufactured by Phillips, Texas Instruments, and Hitachi.

Low Voltage (LV)

The LV family of ICs is designed to operate with a typical supply voltage of 3.3 volts. The supply can actually range from 1.0 to 5.5 volts. Operating at 3 volts, an LV gate can source or sink 6 mA and switches at the same speed as HCMOS. At 5 volts it can source or sink 12 mA and switches twice as fast as HCMOS. Table 6-8 lists some of the gates available in LV CMOS.

TABLE 6-8 Low-Voltage ICs

Device Number	Family	Description
74LV00	Low Voltage	Quad 2-input NAND gate
74LV32	Low Voltage	Quad 2-input OR gate
74LV86	Low Voltage	Quad 2-input Exclusive-OR gate
74LVC10	Low-Voltage CMOS	Triple 3-input NAND gate
74LVC27	Low-Voltage CMOS	Triple 3-input NOR gate
74LVT00	Low-Voltage Technology	Quad 2-input NAND gate
74LVT02	Low-Voltage Technology	Quad 2-input NOR gate

Low-Voltage CMOS (LVC)

LVC devices also operate at a typical voltage of 3.3 volts. LVC can operate at supply voltages from 1.2 volts to 3.6 volts. Operating at 3.3 volts, LVC is equivalent in switching speed to FAST, AST, AC, and ACT devices. The voltages on the inputs can rise to 5.5 volts, exceeding the supply voltage, with no damage to the IC. So LVC devices can be used to shift supply voltage levels in mixed 3-volt/5-volt systems. Since LV and LVC devices can operate at reduced speeds down to 1.2 volts, they can be operated by a single battery. Table 6-8 lists some of the available LVC devices.

Low-Voltage Technology (LVT)

LVT was designed to provide high output currents and fast switching speeds along with low voltage supply voltages and low power dissipation. Supply voltages can range from 3.0 to 3.6 volts. LVT devices can source 32 mA and sink 64 mA. Like HLL, LVT has propagation delays of less than 4 nanoseconds, making it twice as fast as AST devices. LVT inputs can rise above the supply voltage to 5.5 volts. This makes LVT compatible with 5-volt TTL logic. Some of the available LVT ICs are listed in Table 6-8.

Advanced Low-Voltage TTL Compatible (ALVT)

ALVT operates on a supply voltage of 3.3 V or 2.5 V. Operating at 3.3 V, ALVT has propagation delays of 2.5 ns; and at 2.5 V has propagation delays of 3.5 ns. ALVT can sink 64 mA. Inputs can rise above the supply voltage to TTL levels.

SELF-CHECK 2

1. Use a spec book to find the values of these parameters for a 74HC08 quad two-input AND operating at 4.5 volts.

    a. maximum low-level output voltage

    b. minimum high-level output voltage

    c. input current

    d. minimum output drive current N channel (LOW-level output current)

    e. minimum output drive current P channel (HIGH-level output current)

2. Calculate the fan-out to other 74HC08s.

3. Interface

    a. a 4070 operating at 5 volts to a 7408.

    b. a 74HCT00 to a 7400.

    c. a 74LS00 to a 74HC08 operating at 5 volts.

4. Which low-voltage subfamily is compatible with TTL?

5. Which low-voltage subfamilies can operate at 1.2 V.?

 # 6.12 EMITTER COUPLED LOGIC (ECL)

Two available **emitter coupled logic (ECL)** families are the 10K series and the 100K series. Like TTL, ECL is constructed of bipolar transistors. However, instead of using multi-emitter transistors for inputs, ECL uses a separate transistor for each input, connected in parallel. ECL transistor circuits are designed to not saturate when they turn on. Switching times are decreased, and for the 10K family, propagation delays of two nanoseconds result. 10K ECLs can operate at frequencies exceeding 200 MHz. ECL consumes more than twice as much power as TTL. Unused inputs on ECL gates can be left floating. ECL gates have high input impedance and low output impedance, which results in a fan-out of greater than 30. Unlike the TTL and CMOS gates, the basic ECL gate is an OR/NOR, a gate with complementary outputs. See Figure 6-30.

One member of the 10K family is the 10105, a triple 2-3-2 input OR/NOR gate. Its pinout is shown in Figure 6-31.

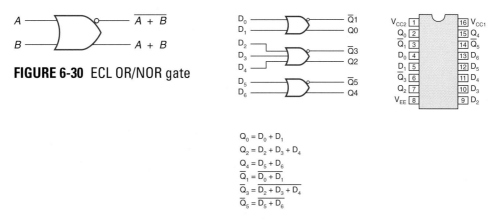

FIGURE 6-30 ECL OR/NOR gate

$$Q_0 = D_0 + D_1$$
$$Q_2 = D_2 + D_3 + D_4$$
$$Q_4 = D_5 + D_6$$
$$\overline{Q_1} = \overline{D_0 + D_1}$$
$$\overline{Q_3} = \overline{D_2 + D_3 + D_4}$$
$$\overline{Q_5} = \overline{D_5 + D_6}$$

FIGURE 6-31 10105 pinout

Notice that there are three power supply connections, V_{CC1}, V_{CC2}, and V_{EE}. 10K ECL operates best on two 5.2-volt power supplies, V_{CC1} and V_{CC2}. V_{CC1} supplies current to the switching transistors within the gate, and I_{CC2} supplies current to the output stage. This system reduces the amount of noise coupled between gates when the outputs are driving heavy loads. Both of the V_{CC} pins are connected to circuit ground, making V_{EE} −5.2 volts.

Figure 6-32 lists some of the other 10K series ICs. The strobe on a 10100 and 10101 is a single input wired to each of the gates on the IC. It is used to enable or inhibit all of the gates on the IC. If left unconnected, it does not influence the operation of the individual gates. The strobe input on the 10100 creates three-input gates, while the strobe input on the 10101 creates two-input gates. The enable input on the 10113 must be LOW for the gate to function as an exclusive-OR. If HIGH, the data inputs, D_0 to D_7, are ignored, and the outputs are locked LOW.

10100	Quad 2-input NOR gate with strobe
10101	Quad 2-input OR/NOR with strobe
10102	Quad 2-input NOR (3 NOR and 1 OR/NOR) gate
10103	Quad 2-input OR (3 OR and 1 OR/NOR) gate
10104	Quad 2-input AND gate
10106	Triple 4-3-3 input NOR
10107	Triple 2-input Exclusive-OR/Exclusive-NOR gate
10108	Dual 4-input AND/NAND gate
10109	Dual 4-5 input OR/NOR gate
10113	Quad Exclusive-OR gate with enable

FIGURE 6-32 10K ECL gate pinouts

A HIGH out of a 10K ECL gate, V_{OH}, ranges between −0.810 and −0.960 volts. A legitimate LOW, V_{OL}, ranges between −1.650 and −1.850 volts. The LOW-level noise margin on ECL is about 0.155 volts, and the HIGH-level noise margin is about 0.125 volts. Even though the ECL noise margin is considerably less than that of TTL and CMOS, circuit impedances make it possible for ECL to operate at high frequencies without noise-induced errors in the logic states.

Variations in the voltage and temperature compensation networks in the circuitry of the 10K family resulted in the 100K ECL family. The 100K family has propagation delays of 0.75 nanoseconds, which enable it to operate at frequencies in the GHz range. The optimum V_{EE} is −4.5 volts, although it can range down to −7 volts. Since the 100K family operates at a different V_{EE} from the 10K family, its output voltages are slightly different. V_{OH} ranges from −0.880 volts to −1.025 volts. V_{OL} ranges from −1.810 to −1.620 volts. V_{EE} can be increased to −5.2 volts to make 100K outputs compatible with 10K outputs.

Figure 6-33 lists some of the ICs available in the 100K family.

100101	Triple 5-input OR/NOR gate
100102	Quint 2-input OR/NOR gate with enable
100107	Quint Exclusive-OR/Exclusive-NOR gate with compare
100117	Triple 1-2-2 input OR-AND/OR-AND-INVERT gate
100118	Quint 2-4-4-5 input OR-AND gate
100166	9-bit comparator
100179	Carry look-ahead generator
100180	High-speed 6-bit adder
100181	4-bit binary/BCD ALU

FIGURE 6-33 100K ECL ICs

Positive emitter coupled logic (PECL) operates on a +5-volt power supply instead of a negative supply like traditional ECL. Low-voltage positive emitter coupled logic (LVPECL) operates on +3.3 volts.

6.13 INTERFACING ECL TO OTHER LOGIC FAMILIES

In systems where both ECL and TTL logic is used, there is a negative supply, V_{EE}, for ECL and a +5-volt supply, V_{CC} for TTL. The two supplies share a common ground. Translation circuits must be used to interface between the two logic systems. The following ICs are used to translate from one family to another, but not back. These are called unidirectional translators.

TTL to 10K	10124
10K to TTL	10125
TTL to 100K	100124
100K to TTL	100125

Figure 6-34 shows the pinout and logic diagram for a 10124 quad TTL-to-ECL translator.

Note that the 10124 has power supply connections for V_{CC}, V_{EE}, and ground. D_0 through D_3 and S are Schottky TTL inputs. The Q outputs are all 10K ECL outputs. Figure 6-35 shows a 74LS00 being interfaced to a 10101.

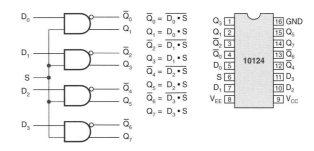

$$\overline{Q_0} = \overline{D_0 \cdot S}$$
$$Q_1 = D_0 \cdot S$$
$$\overline{Q_2} = \overline{D_1 \cdot S}$$
$$Q_3 = D_1 \cdot S$$
$$\overline{Q_4} = \overline{D_2 \cdot S}$$
$$Q_5 = D_2 \cdot S$$
$$\overline{Q_6} = \overline{D_3 \cdot S}$$
$$Q_7 = D_3 \cdot S$$

FIGURE 6-34 10124 pinout and logic diagram

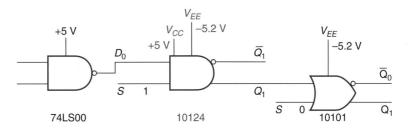

FIGURE 6-35 TTL to ECL interface

The 100255 is a 4-bit bidirectional translator. It can translate either from TTL to ECL or ECL to TTL. Figure 6-36 shows the logic diagram of a 100255. Inputs B_0 through B_4 are TTL compatible. Chip enable (CE), direction control (DIR), and A_0 through A_4 are ECL 100K compatible. CE must be HIGH to translate from ECL to TTL or vice versa. When CE goes LOW, all gates enter an HiZ state. The DIR pin controls the direction of transfer. A LOW on DIR enables the inverters that conduct from B to A (note the bubble on the control input on those inverters). With DIR LOW, the IC translates from TTL to ECL. A HIGH on DIR enables the inverters that conduct from A to B (no bubbles on the control inputs to those inverters). With DIR HIGH, the IC translates from ECL to TTL.

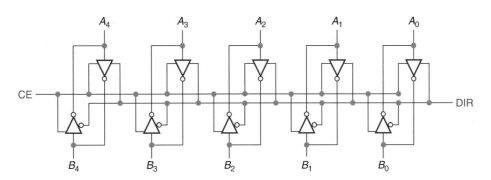

FIGURE 6-36 Bidirectional translation

The translators listed above for TTL also work for CMOS operating at five volts. To interface CMOS operating at other voltage levels, first translate to TTL and then use open-collector gates, as discussed earlier in the chapter, to interface to the desired voltage level.

6.14 SURFACE MOUNT TECHNOLOGY

Microminiaturization is a relatively new word which refers to the process of shrinking the size and power dissipation of electronic circuits. In our lab, we have used ICs contained in a dual in-line package (**DIP**). When mounted on a printed circuit board, the pins of the DIP IC pass through the board and are soldered to pads on the bottom side (component side, solder side). This is called **through-hole** construction. The DIP was a great improvement over equivalent circuits constructed from vacuum tubes or transistors. However, in pursuit of microminiaturization, the DIP package has given way to surface mount technology (**SMT**). SMT device leads are soldered to pads on the printed circuit board without passing through the board; hence the name "surface mount." SMT devices are often soldered to both sides of the board (primary side, secondary side).

The logic ICs we have studied in this chapter can be purchased in a variety of packages. As shown in Figure 6-37, the small outline package (**SOP**) and small outline integrated circuit (**SOIC**) are dual in-line style packages with leads bent down and out for easy mounting on printed circuit board pads. This shape lead is called **gullwing**. Whereas a DIP package has a lead spacing, or pin pitch, of 0.1″ (0.1″ from the center of one pin to the center of an adjacent pin), SOPs and SOICs have a pin pitch of half that—0.05″. Other available packages are the shrink small outline package (**SSOP**) and the thin shrink small outline package (**TSSOP**). The pin pitch of an SSOP or TSSOP package is 0.026″ or 0.02″, about half that of an SOIC.

More complex digital circuitry with more I/O pins is housed in packages that have pins on all four sides. Figure 6-37 shows a 160-pin Quad Flat Pack (**QFP**) with gullwing leads. QFPs range in size from 44 pins to 304.

The CPLD we have been programming comes in an 84-pin Plastic Leaded Chip Carrier package (**PLCC**). Its leads protrude down out of the plastic package and bend back under the package, forming a J shape; hence the name **J-lead**. See Figure 6-38(C). PLCCs are available in 20-, 28-, 44-, 52-, 68-, 84-, and 100-pin packages. Figure 6-38(B) shows the pinout for a 100180 ECL 8-bit full adder in a 28-pin PLCC package. As you can see, pin 1 is in the middle of one side and is marked by an indentation. Pins are numbered counterclockwise from pin 1. Note the index mark on the first corner counterclockwise from pin 1. PLCCs are often inserted into a socket that has been soldered to the PCB—see Figure 6-38(A) and Figure 6-38(C).

Passive components such as resistors and capacitors are also available in surface mount form. A surface mount resistor is stamped with a number that indicates its resistance, two significant digits and one multiplier. For example, 103 indicates 10 followed by three zeroes—10,000 Ω. A resistor stamped 151 would have a nominal resistance of 150 Ω. Figure 6-39 shows a 471-Ω resistor. The metal terminations on each end are soldered to pads on a printed circuit board.

Thick film chip resistors are referred to by their size in hundredths of an inch. For example, a 1210 surface mount resistor measures 0.12″ by 0.10″; an 0805 surface mount resistor measures 0.08″ by 0.05″; and an 0201 measures 0.02″ by 0.01″. Figure 6-39 shows the actual sizes of a series of surface mount resistors. They are listed here from largest to smallest along with their

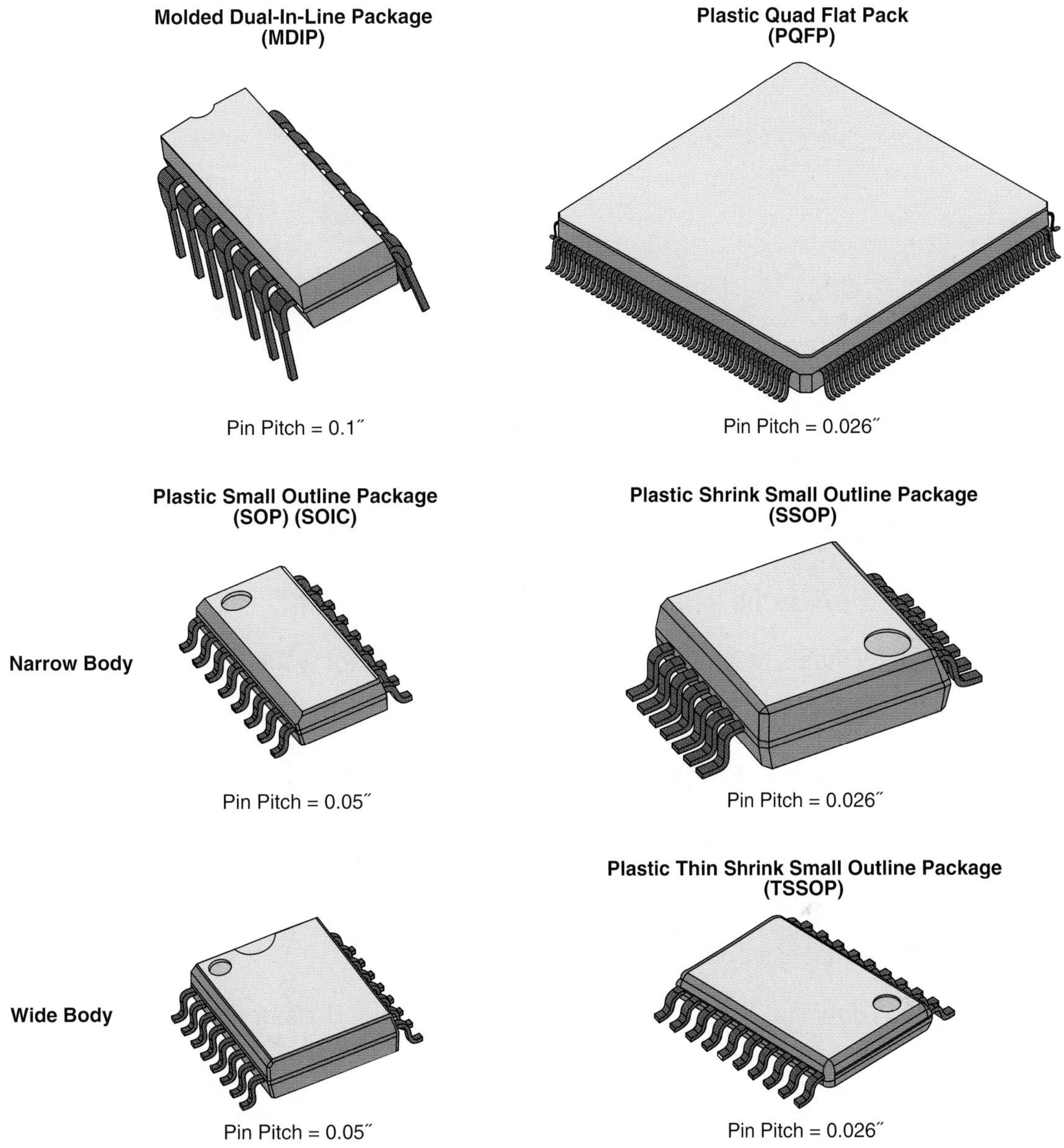

Molded Dual-In-Line Package
(MDIP)
Pin Pitch = 0.1″
Plastic Quad Flat Pack
(PQFP)
Pin Pitch = 0.026″
Plastic Small Outline Package
(SOP) (SOIC)
Narrow Body
Pin Pitch = 0.05″
Plastic Shrink Small Outline Package
(SSOP)
Pin Pitch = 0.026″
Plastic Thin Shrink Small Outline Package
(TSSOP)
Wide Body
Pin Pitch = 0.05″
Pin Pitch = 0.026″

FIGURE 6-37 Surface mount packages

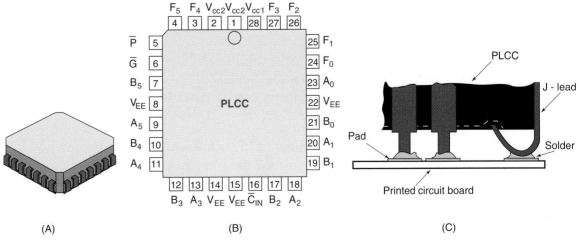

F5 F4 Vcc2 Vcc2 Vcc1 F3 F2
4 3 2 1 28 27 26
P̄ 5
Ḡ 6
B5 7
VEE 8
A5 9
B4 10
A4 11
PLCC
25 F1
24 F0
23 A0
22 VEE
21 B0
20 A1
19 B1
12 13 14 15 16 17 18
B3 A3 VEE VEE C̄IN B2 A2
PLCC
J - lead
Pad
Solder
Printed circuit board
(A)
(B)
(C)

FIGURE 6-38 Plastic leaded chip carrier (PLCC)

power rating: 1812 1/2 W; 1210 1/4 W; 1206 1/4 W; 0805 1/8 W; 0603 1/10 W; 0402 1/16 W; 0201 1/20 W.

470 Ω Resistor

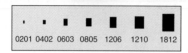

0201 0402 0603 0805 1206 1210 1812

Actual package sizes

FIGURE 6-39 Surface mount resistors

SELF-CHECK 3

1. Discuss the three power supply pins on an ECL IC.

2. What is the optimum power supply voltage for a 10K ECL IC? 100K?

3. Which translator is needed to interface a 10101 to a 74LS04?

4. How does ECL compare to the other logic families as far as speed is concerned?

5. How does ECL compare to other logic families as far as power consumption is concerned?

6. What is a gullwing lead?

7. What is a J-lead?

8. How much power can a 0201 surface mount resistor dissipate?

6.15 CPLD SPECIFICATIONS (Optional)

The CPLD we programmed in Chapters 2, 3, 4, and 5 is an Altera EPM7128SLC84-15 CPLD. The 128 in the IC number indicates 128 macro cells. The 84 indicates that the IC is in an 84-pin PLCC package. The "-15" suffix indicates that the delay time from an input pin to an output pin is on the order of 15 nanoseconds. This IC contains 2500 usable gates. It stores the program in Electrically Erasable Programmable Read-Only Memory (EEPROM). Each output pin can source or sink 25 mA.

Measure the propagation delay of the CPLD as follows: Write a program in VHDL that defines B as the complement of A. Compile the program and create and run simulation waveforms. Measure the propagation delay shown on the simulation waveforms by using the cursor to measure the time from when A goes HIGH to the time when B goes LOW. Assign pins to A and B. Assign pin 83 to A. Place a jumper on JP1; the onboard 4-MHz clock will function as input A. Observe A and B on a dual-trace oscilloscope to see if you can observe the propagation delay.

6.16 TROUBLESHOOTING TTL AND CMOS DEVICES

This chapter emphasized the relationship between output voltage and output current. When a LOW-level output sinks conventional current to ground, the output voltage tends to rise. If too much current is sinked, the output voltage will rise above the accepted LOW range. Following ICs might not be able to recognize the signal as a zero level. Likewise, when a HIGH-level output sources conventional current to a load, the output voltage tends to fall. If too much current is sourced, the output voltage will drop below the accepted HIGH range. Following ICs might not be able to recognize the signal as a one level. Be aware of outputs that have risen or fallen into the unused region between legitimate HIGH and LOW ranges.

Table 6-9 displays a summary of those limits. Refer to these limits if you have a concern about fan-out.

TABLE 6-9 Input/Output Voltage and Current Levels

Subfamily	V_{OL} (volts)	I_{OL} (mA)	V_{OH} (volts)	I_{OH} (mA)
74XX	.4	16	2.4	.4
LS	.4	8	2.7	.4
ALS	.4	8	3.0 (V_{CC} − 2)	.4
S	.5	20	2.7	1.0
AS	.5	20	3.0 (V_{CC} − 2)	2.0
AC	.36	24	4.4	24
ACT	.36	24	4.4	24
HC	.26	4	4.5	4
HCT	.26	4	4.0	4
4XXX	.05	.8	4.95	.8
AHC	0.1	25	4.4	25
AHCT	0.1	25	4.4	25

In Figure 6-40, open-drain inverters (CMOS version of open-collector inverters) are being used in a circuit called a partial decoder. The inputs A_{15} through A_8 are bits of the address bus of a microprocessor circuit. R_1 is the pull-up resistor for the "wired-NOR" configuration formed by the inverters. When A_{15} through A_{10} are all LOW, the outputs of the open-collector inverters are each in the HiZ state and R_1 pulls their common output up to +5 volts. If A_9 and A_8 are also HIGH, the output of the NAND gate is LOW, signifying that the bits on the upper part of the address bus are 00000011. The output of the partial decoder is used to enable a block of memory. The remaining address bus bits, A_7 through A_0, are decoded by the memory IC itself to select a single address within the memory block to be written to or read from. The

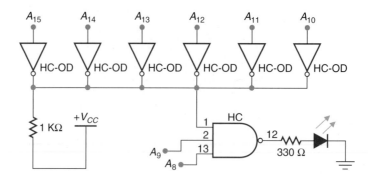

FIGURE 6-40 Partial decoder

partial decoder is used to indicate that A_{15} through A_{10} are being watched as a group to see whether they are all zeros or not. Decoders will be studied in detail in Chapter 14.

This circuit is easy to troubleshoot. If any of the inputs A_{15} through A_8 are HIGH, the corresponding output goes LOW, pulling the collective output of the wired-NOR configuration LOW. The output of the NAND gate goes HIGH, signifying that A_{15} through A_8 is no longer 00000011.

Suppose the partial decoder in Figure 6-40 is faulty. A_{15} through A_8 are confirmed to be 00000011, but the output of the NAND gate is slowly oscillating between an ON and OFF state. The power supply pins, including ground, check good. A check of the input pins of the NAND gate shows that the input voltage on pin 1 is unstable. It is floating when it should be a solid HIGH. It is not being pulled HIGH by R_1. The power supply is killed and R_1 is measured in circuit. It measures 3 MΩ, much higher than its upper bound (nominal value plus the tolerance). The resistor is bad. If a resistor measures lower than its lower bound (nominal value minus the tolerance) in circuit, it probably has a parallel path around it, causing it to read a lower value.

DIGITAL APPLICATION

Staggered-Pin Grid Array (SPGA)

Intel's Socket 5 configuration is shown in Figure 6-41. Columns are numbered 1 through 37 and rows are numbered A through AN. The pins are staggered for closer packing. This type of pinout is called staggered-pin grid array (**SPGA**). This socket was used for second-generation Pentium® processors that had 296 pins and plugged into this 320-pin IC socket. Since a processor draws considerable current, many pins are dedicated to V_{CC}, the positive voltage supply, and V_{SS}, the ground connection. V_{CC} was 3.3 V to 3.5 V. When the lever on the left is raised, the contacts are opened so that the 296 pins of the processor can be inserted with no force. Lowering the lever clamps the processor into the socket. This type of zero-insertion force (**ZIF**) socket is used by later-generation processors, including the Pentium 4®. Sockets are now named by the number of pins. AMD's Athlon 64 processor uses Socket 754, which has 754 pins, in a Micro Pin Grid Array (mPGA).

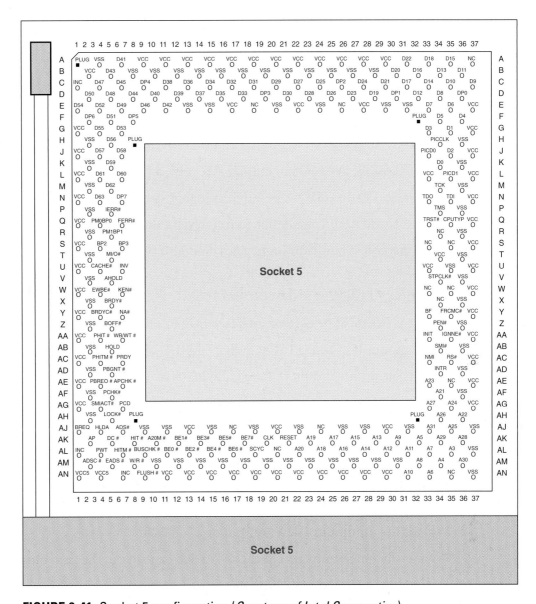

FIGURE 6-41 Socket 5 configuration (*Courtesy of Intel Corporation*)

SUMMARY

- In TTL, a high-level output can range from 2.4 volts to V_{CC}, but a high-level input can range from 2.0 volts to V_{CC}. The difference provides a 0.4-volt noise margin.

- 1.6 mA is called "one standard TTL load."

- Fan-out is the number of loads that a device can drive.

- From fastest to slowest, the TTL subfamilies are AS, F, S, ALS, LS, TTL, and L.

- From lowest power consumption to highest, the TTL subfamilies rank as L, ALS, LS, F, AS, TTL, S.

- An open-collector gate output circuit has no internal path to the power supply. An external pull-up resistor must be provided.

- Open-collector gate outputs can be tied together and share a common pull-up resistor.

- The CMOS equivalent to open collector is open drain.

- The 7406, 7407, 7416, and 7417 are open-collector gates with high-voltage outputs. Their pull-up resistors can be connected to voltages higher than V_{CC}.

- CMOS subfamilies include the 4000 series, 74C series, 74HC series, 74HCT series, 74AC series, 74ACT series, 74AHC series, and the 74AHCT series.

- To interface TTL to CMOS operating at 5 volts, a pull-up resistor is needed.

- To interface TTL to CMOS operating at higher voltages, a 4049 or 4050 is used.

- ACT, HCT, and AHCT are designed to interface directly with LS TTL.

- ECL is the fastest family of logic gates, and also draws the most current from the power supply.

- Two series of ECL ICs are the 10K and 100K.

- Translator ICs must be used to interface ECL to other logic families.

- Surface mount ICs come in SOP, SOIC, SSOP, TSSOP, QFP, and PLCC packages.

- The pins on surface mount ICs are gullwing-style or J-style.

- Surface mount resistors are named according to their size in hundredths of an inch.

QUESTIONS AND PROBLEMS

1. What is the minimum 1-level output on a 74LS08?

2. What is the maximum current that a 74LS08 can sink in the zero output state?

3. How many 5475s can a 7404 drive?

4. What is the maximum current drawn by a 74LS83A IC?

5. What is the maximum LOW-to-HIGH propagation delay time of a 5421? (Assume a load of 50 pF, 2 kΩ.)

6. At 15-V supply, how much current can a 4006BM sink?

7. What is the minimum 1-level output voltage for a 4006BM (5-V supply)?

8. Can a 4006BM (5-V supply) drive a 7400? Why or why not?

9. How much current can a 4049 sink on a LOW-level output ($V_{DD} = 5$ V)?

10. Interface a 7432 OR gate to a 4081 AND gate operating at 5 V.

11. Interface a 7404 inverter to a 4071 OR gate operating at 5 V.

12. Use a 7406 to interface a 7408 AND gate to a 4081 AND gate operating at 12 V.

13. Use a 7406 to interface a 7408 AND gate to a 74C08 AND gate operating at 12 V.

14. Use a 4049 to interface a 4081 AND gate operating at 12 V to a 7404 inverter.

15. Use a 4050 to interface a 74C00 NAND gate operating at 15 V to a 7400 NAND gate.

16. Use a 4049 to interface a 4001 operating at +12 V to as many 7404s as possible.

17. Use a 4049 to interface a 74LS00 to a 74HC06 operating at 3 V.

18. What is the acceptable range of power supply voltages for an HC device?

19. What is the acceptable range of power supply voltages for HCT devices?

20. Can a TTL device directly drive an HCT device?

21. Can a TTL device directly drive an HC device operating at 5 V?

22. What does AHCT stand for? What is the propagation delay of a 74AHCT device?

23. How many LS devices can a CMOS HC gate drive?

24. Rank these families according to power consumption, from least power to most power consumed: CMOS HC, CMOS, CMOS AC, ECL, TTL, LS, ALS, FAST, S, CMOS AHC.

25. Rank these families according to speed, in descending order: CMOS HC, CMOS AC, CMOS, TTL, LS, ALS, FAST, S, CMOS AHC.

26. What is the maximum propagation delay for a 4001 NOR gate?

27. At quiescence, how much current does a 4001 NOR gate draw?

28. What is the typical low-level drive current for a 4011 NAND gate?

29. What is the maximum low-level output voltage for a 4001 quad two-input NOR gate?

30. Draw the schematic of an open-collector inverter driving a relay coil connected to +20 V.

31. Draw the schematic diagram of three open-collector inverters, 7406s, tied to a pull-up resistor of 5 kΩ and a power supply of 20 V.

32. How does a totem-pole gate differ from an open-collector gate?

33. Draw the schematic symbol of an ECL OR/NOR gate.

34. Connect the gates of a 10105 to make a NOR gate with the maximum number of inputs (expand the NOR).

35. What logic levels must exist on CE and DIR of a 100255 for the IC to translate from ECL to TTL?

36. Define microminiaturization.

37. Name six styles of surface mount packages.

38. Name two styles of leads that are used on surface mount ICs.

39. List the advantages of operating ICs at lower supply voltages.

40. List four CMOS low-voltage subfamilies.

41. Which CMOS low-voltage subfamilies can sink 64 milliAmps?

42. Which CMOS low-voltage subfamilies are faster than advanced Schottky TTL devices?

43. Which CMOS low-voltage subfamilies can accept input signals from TTL devices?

44. Which CMOS low-voltage subfamilies can operate from a 1.2-volt cell?

45. Which CMOS low-voltage subfamilies offer basic gates?

46. An enhancement of the 74ACT subfamily is the 74ACTQ subfamily. Research this sub-family and report on its characteristics, including its power supply voltage, propagation delay, output drive current, and the meaning of "Q." List all the devices available within this subfamily.

47. Logic gates are available in single-gate packages. Research single-gate logic and report on: IC numbering, which gates are available in single-gate packages, pinouts, package size, and characteristics such as propagation delay, supply voltage, and fan-out.

48. Find a printed circuit board that contains surface mount components. Identify resistors, resistance and power rating, and surface mount IC packages.

49. If you were designing a new digital system, which subfamily would you use?

SPECIFICATIONS AND OPEN-COLLECTOR GATES

OBJECTIVES

After completing this lab, you should be able to:

- Measure and graph input current versus input voltage for a TTL gate.
- Measure and graph output current versus output voltage for a TTL gate.
- Measure and graph source current versus applied frequency for a CMOS IC.
- Construct a six-input wired NOR circuit.

COMPONENTS NEEDED

1	7404 IC
1	7406 IC
1	74C14 IC
1	4001 IC
1	1-kΩ resistor
1	1-kΩ potentiometer
1	100-Ω resistor

Review the lab safety rules under SAFETY ADVICE in the PREPARATION section of Lab 1A, Chapter 1.

PROCEDURE

1. Set up the circuit shown. Measure and record input voltages and currents. Then draw a graph of input voltage versus input current. Does the maximum input current measured fall within the IC specifications?

Input Voltage	Current
5.0 V	
4.5 V	
4.0 V	
3.5 V	
3.0 V	
2.5 V	
2.0 V	
1.5 V	
1.0 V	
0.5 V	
0.0 V	

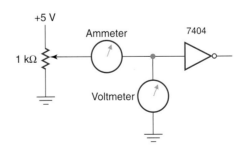

2. Connect the circuit shown. Measure and record output voltages and currents. Graph the output voltage versus the output current. Does the output voltage fall within the IC specifications?

Output Current	Output Voltage
4 mA	
6 mA	
8 mA	
10 mA	
12 mA	
14 mA	
16 mA	
18 mA	
20 mA	
25 mA	
30 mA	
40 mA	

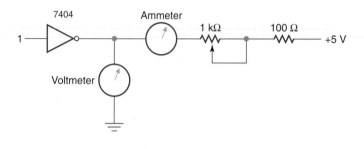

3. Construct the circuit shown. Measure and record output voltages and currents. Graph the output current versus the output voltage.

Output Current	Output Voltage
	5.0 V
	4.5 V
	4.0 V
	3.5 V
	3.0 V
	2.4 V
	2.0 V

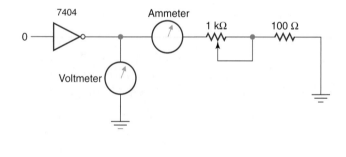

4. Construct the circuit shown. Measure and record frequency and current data. Graph the input frequency versus V_{DD} current. Use an oscilloscope or frequency counter to measure the input frequency. The 74C14 will be discussed in a later chapter. It is used here to convert the ac signal into a square wave that ranges from 0 to 10 volts. A signal generator that can be adjusted for a 0- to 10-volt square wave can be used instead.

Input Frequency	V_{DD} Current
1 kHz	
10 kHz	
100 kHz	
500 kHz	
1 MHz	
1.5 MHz	
2 MHz	
2.5 MHz	
3 MHz	

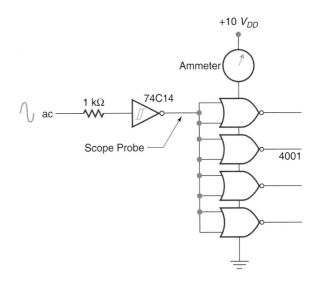

5. Use a 7406 open-collector gate to construct a six-input wired NOR gate.

LAB 6B SPECIFICATIONS AND OPEN-DRAIN INVERTERS

OBJECTIVES

After completing this lab, you should be able to use Multisim to:

- Define and measure fan-out.
- Measure t_{PHL}, high-to-low propagation delay.
- Troubleshoot open-drain inverters.

PREPARATION

Fan-out was defined in Chapter 6 as the measure of a digital device to supply current to following devices. In this lab you will determine whether circuits in Multisim exhibit the same limitations as actual ICs. Does the LOW output of a TTL gate rise as it sinks more current? In Part 2 you will measure the high-to-low propagation delay of an LS gate. Part 3 contains four partial decoder circuits to troubleshoot, files 6B-1.ms9 through 6B-4.ms9.

PROCEDURE

PART 1

LOW-level output characteristics

Determine the LOW-level fan-out characteristics of an LS TTL NAND gate as follows:

1. Select and place a 74LS00 2-input NAND gate in the circuit window. This will be the "gate under test."

2. Tie the inputs HIGH to force the output LOW.

3. Measure and record the output voltage.

4. Add a load of one 7400 NAND gate (connect the output of the 74LS00 gate to one input of a 7400 NAND gate). Measure and record the output voltage and current sinked by the "gate under test."

5. Continue to add gates to the output of the74LS00 and to measure and record the LOW-level output voltage and load current. Do this until the output voltage rises to 0.4 V, or until it is determined that the gate is ideal and its model is not limited in fan-out.

6. Describe the results.

PART 2

High-to-low propagation delay

Use Multisim to measure t_{PHL} high-to-low propagation delay of a 74LS00 NAND gate. Describe your procedure and results.

PART 3

Troubleshooting partial decoder circuits

Troubleshoot each circuit file 6B-1.ms9 through 6B-4.ms9. Use the word generator to supply test inputs. Log the steps taken to isolate each fault. Describe your conclusions.

Flip-Flops

OUTLINE

KEY TERMS

$\overline{\text{CLEAR}}$

crossed NAND gated SET-RESET flip-flop

crossed NAND $\overline{\text{SET}}$-$\overline{\text{RESET}}$ flip-flop

crossed NOR SET-RESET flip-flop

debounce

edge triggered

flip-flop

master-slave D flip-flop

$\overline{\text{PRESET}}$

$\overline{\text{SET}}$-$\overline{\text{RESET}}$ flip-flop

transparent D flip-flop

OBJECTIVES

After completing this chapter, you should be able to:

- Explain the operation of a SET-RESET flip-flop.

- Use a crossed NAND or crossed NOR flip-flop as a debounce switch.

- Explain the operation of a gated SET-RESET flip-flop.

- Explain the operation of a transparent D flip-flop.

- Explain the operation of a D flip-flop used as a latch.

- Explain the operation of a master-slave D flip-flop.

- Describe typical IC flip-flops.

- Understand the use of programmable logic devices to create $\overline{\text{SET}}$-$\overline{\text{RESET}}$ flip-flops.

- Troubleshoot a typical flip-flop circuit.

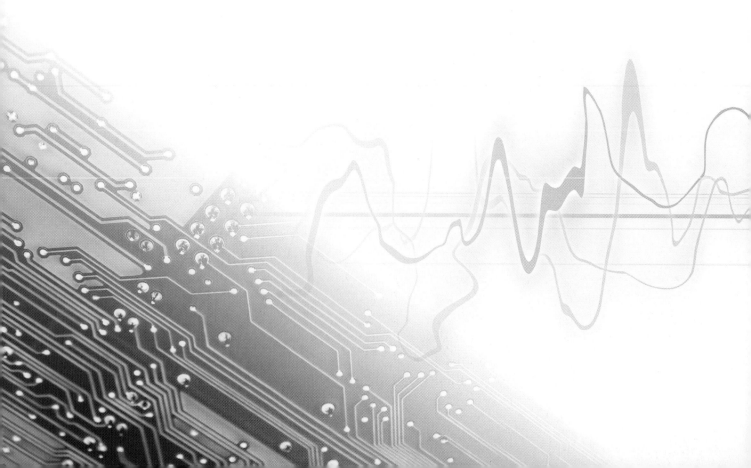

7.1 INTRODUCTION TO FLIP-FLOPS

A **flip-flop** is a digital circuit that has two outputs Q and $\overline{Q}$, which are always in the opposite states. If Q is 1 then $\overline{Q}$ is 0, and the flip-flop is said to be set, on, or preset. If Q is 0 then $\overline{Q}$ is 1, and the flip-flop is said to be reset, off, or cleared. There are several types of flip-flops, and the control inputs vary with each type. The logic levels on the flip-flop's inputs will determine the state of the Q and $\overline{Q}$ outputs according to the truth table for that type of flip-flop.

Unlike the gates studied up to this point, the flip-flop can in some states maintain its output state (on or off) after the input signals which produced the output state change. Thus, the flip-flop can store a bit of information or one place of a larger binary number. There are many other uses for flip-flops as we will see in the next few chapters.

7.2 CROSSED NAND $\overline{\text{SET}}$-$\overline{\text{RESET}}$ FLIP-FLOPS

A **$\overline{\text{SET}}$-$\overline{\text{RESET}}$ flip-flop** is a digital circuit whose output is set by the $\overline{\text{SET}}$ input but can only be reset by the $\overline{\text{RESET}}$ input. The two crossed NAND gates in Figure 7-1 form a $\overline{\text{SET}}$-$\overline{\text{RESET}}$ flip-flop.

The inputs $\overline{\text{SET}}$ and $\overline{\text{RESET}}$ are active LOW. The $\overline{\text{SET}}$ input must be a 0 to set the Q output to a 1. Notice the complement bar over the $\overline{\text{SET}}$ and $\overline{\text{RESET}}$ inputs. This means they are active LOW inputs. The outputs of a flip-flop are usually labeled Q and $\overline{Q}$, meaning that if Q is a 1, $\overline{Q}$ is a 0 and vice versa.

When the $\overline{\text{SET}}$ input goes to a 0 and the $\overline{\text{RESET}}$ input is held at 1, the output of the crossed NAND flip-flop will have the configuration shown in Figure 7-2. This is because any 0 into a NAND gate makes its output a 1. This will set the Q output to 1 and the $\overline{Q}$ output to a 0.

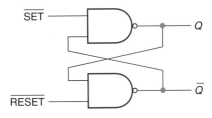

FIGURE 7-1 Crossed NAND $\overline{\text{SET}}$-$\overline{\text{RESET}}$ flip-flop

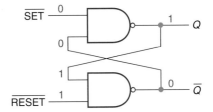

FIGURE 7-2 Setting the Q output for a crossed NAND $\overline{\text{SET}}$-$\overline{\text{RESET}}$ flip-flop

If the $\overline{\text{SET}}$ input goes to a 1 and the $\overline{\text{RESET}}$ remains at 1, the output does not change, as shown in Figure 7-3. This is because the outputs are fed back to the input of the opposite gate, which makes them retain their original output configuration.

To reset the flip-flop, you must bring the $\overline{\text{RESET}}$ input to a 0 and keep the SET at 1, as shown in Figure 7-4. As can be seen, the $\overline{\text{SET}}$ input cannot reset the Q output to a 0. This can only be done by bringing the $\overline{\text{RESET}}$ input to a 0 while keeping the $\overline{\text{SET}}$ input at 1. The same thing is true for the $\overline{\text{RESET}}$ input. It cannot set or bring the Q output to a 1; only reset it or bring the Q output to a 0.

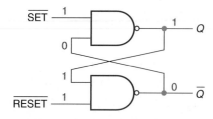

FIGURE 7-3 The unchanged state for a crossed NAND $\overline{\text{SET}}$-$\overline{\text{RESET}}$ flip-flop

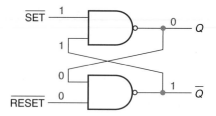

FIGURE 7-4 Resetting the Q output for a crossed NAND $\overline{\text{SET}}$-$\overline{\text{RESET}}$ flip-flop

The only other possible input state not yet covered for the crossed NAND $\overline{\text{SET}}$-$\overline{\text{RESET}}$ flip-flop is 0 on both inputs, as shown in Figure 7-5. This is an unused state. We never want a flip-flop to have Q and $\overline{Q}$ with the same value.

The input that returns to 1 first will determine the resulting state of the flip-flop. The truth table for a **crossed NAND $\overline{\text{SET}}$-$\overline{\text{RESET}}$ flip-flop** is shown in Figure 7-6.

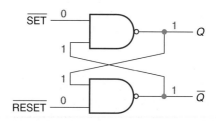

FIGURE 7-5 The unused state for a crossed NAND $\overline{\text{SET}}$-$\overline{\text{RESET}}$ flip-flop

SET	RESET	$\overline{\text{SET}}$	$\overline{\text{RESET}}$	Q	$\overline{Q}$	STATE
1	1	0	0	1	1	Unused state
1	0	0	1	1	0	SET
0	1	1	0	0	1	RESET
0	0	1	1	Q	$\overline{Q}$	SET

FIGURE 7-6 Truth table for a crossed NAND $\overline{\text{SET}}$ $\overline{\text{RESET}}$ flip-flop

EXAMPLE 7-1

Construct two $\overline{\text{SET}}$-$\overline{\text{RESET}}$ flip-flops from one 74LS00 IC.

Solution   Refer to Figure 7-7.

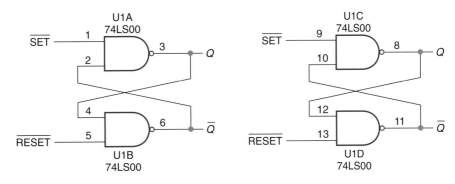

FIGURE 7-7

7.3 CROSSED NOR SET-RESET FLIP-FLOPS

Figure 7-8 shows a **crossed NOR SET-RESET flip-flop**. Note that the inputs are not complemented; therefore they are active HIGH.

When the SET input goes to a 1 and the RESET remains at 0, then the Q output goes to a 1 state, as shown in Figure 7-9. Any 1 into a NOR gate will produce a 0 output.

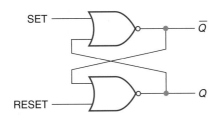

FIGURE 7-8 Crossed NOR SET-RESET flip-flop

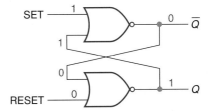

FIGURE 7-9 Setting the Q output for a crossed NOR SET-RESET flip-flop

When the SET input returns to 0 and the RESET is also 0, the outputs Q and $\overline{Q}$ do not change, as shown in Figure 7-10. This is because the outputs of the NOR gates are tied back to the opposite gates' input. This keeps the gates from changing states.

To bring Q back to a 0, the RESET input must be made 1 while the SET input is held at 0. This is shown in Figure 7-11.

The unused state for the crossed NOR SET-RESET flip-flop is where the SET is 1 and the RESET is 1. This is shown in Figure 7-12.

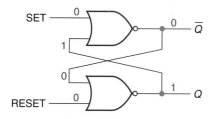

FIGURE 7-10 Unchanged state for a crossed NOR SET-RESET flip-flop

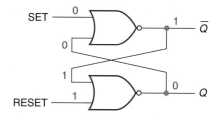

FIGURE 7-11 Resetting the Q output for a crossed NOR SET-RESET flip-flop

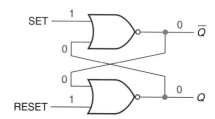

FIGURE 7-12 The unused state for a crossed NOR SET-RESET flip-flop

The first input to return to the 0 state will determine the output state of the Q and $\overline{Q}$ outputs. The truth table for the crossed NOR SET-RESET flip-flop is shown in Figure 7-13.

SET	RESET	$\overline{SET}$	$\overline{RESET}$	Q	$\overline{Q}$	STATE
1	1	0	0	0	0	Unused state
1	0	0	1	1	0	RESET
0	1	1	0	0	1	SET
0	0	1	1	Q	$\overline{Q}$	Unchanged state

FIGURE 7-13 Truth table for a crossed NOR SET-RESET flip-flop

SELF-CHECK 1

1. What is the logic state of the Q and $\overline{Q}$ outputs of a $\overline{SET}$-$\overline{RESET}$ crossed NAND flip-flop if the $\overline{SET}$ is 1 and the $\overline{RESET}$ is 0?

2. What would be the Q and $\overline{Q}$ outputs logic states of the flip-flop in question 1 if the $\overline{RESET}$ were moved to a logic 1 state and the $\overline{SET}$ remained a logic 1?

3. What would be the logic state of the Q and $\overline{Q}$ outputs of a SET-RESET crossed NOR flip-flop if the SET were 1 and the RESET were 0?

4. What would be the Q and $\overline{Q}$ logic states for the flip-flop in question 3 if the RESET were moved to logic 1 and the SET remained 1?

7.4 COMPARISON OF THE CROSSED NAND AND THE CROSSED NOR SET-RESET FLIP-FLOPS

When you compare the truth table for the crossed NAND $\overline{\text{SET}}$-$\overline{\text{RESET}}$ flip-flop in Figure 7-6 with the table for the crossed NOR SET-RESET flip-flop in Figure 7-13, you'll find that they differ in the values of Q and $\overline{Q}$ in the unused state.

Another major difference between the two flip-flops is the SET and RESET inputs. The crossed NOR inputs are active HIGH and the $\overline{\text{SET}}$-$\overline{\text{RESET}}$ inputs on the crossed NAND gates are active LOW. This means that the crossed NOR flip-flop will change state when an input goes HIGH or 1, and the crossed NAND will change state when an input goes LOW or 0.

EXAMPLE 7-2

Draw the waveform for the Q output of a crossed NAND flip-flop and a crossed NOR flip-flop from the given inputs in Figure 7-14.

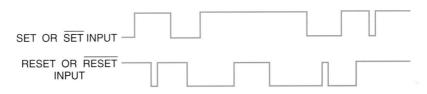

FIGURE 7-14

Solution   Refer to Figure 7-15.

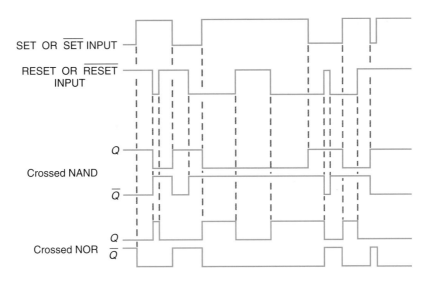

FIGURE 7-15

7.5 USING A $\overline{\text{SET}}$-$\overline{\text{RESET}}$ FLIP-FLOP AS A DEBOUNCE SWITCH

When a normal metal contact on a single-pole switch closes or opens, the contacts do not make and break the circuit smoothly. Instead they bounce, making and breaking contact many times before they finally come to rest. This happens very fast and can cause havoc with digital circuits which may be counting the number of switch closings that occur. Figure 7-16 shows a typical circuit with a single-pole switch and the corresponding waveforms generated when the switch is closed. If a counter had been clocked by the output of the inverter in Figure 7-16, it would have counted 3 pulses instead of 1.

Figure 7-17 shows how a single-pole double-throw switch can be used in conjunction with a $\overline{\text{SET}}$-$\overline{\text{RESET}}$ flip-flop to prevent switch bounce.

These circuits work because when the switch is moved the flip-flop outputs will remain unchanged until the center pole of the switch contacts the opposite pole of the switch, at which time the flip-flop changes state and remains there even though the switch bounces. Figure 7-18 shows the switch movement using a crossed NAND $\overline{\text{SET}}$-$\overline{\text{RESET}}$ flip-flop.

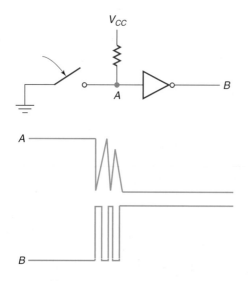

FIGURE 7-16 Switch bounce

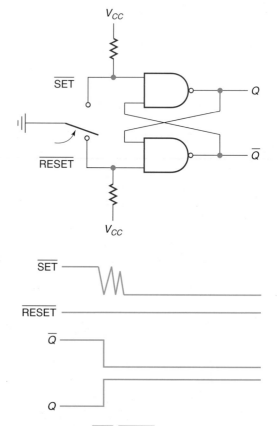

FIGURE 7-17 $\overline{\text{SET}}$-$\overline{\text{RESET}}$ flip-flop used as a debounce switch

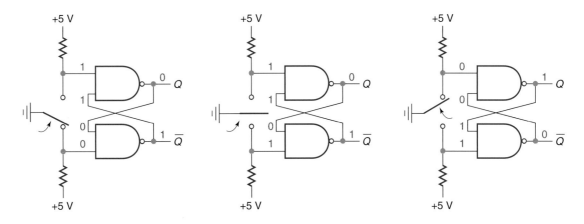

FIGURE 7-18 A crossed NAND SET-RESET flip-flop changing states

EXAMPLE 7-3

Draw the circuit for a debounce switch using a SET-RESET flip-flop made from a set of NOR gates.

Solution   Refer to Figure 7-19.

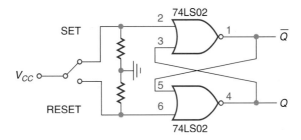

FIGURE 7-19

7.6 THE GATED SET-RESET FLIP-FLOP

Figure 7-20 shows a **crossed NAND gated SET-RESET flip-flop** and its truth table. There are two NAND gates which are used to gate the SET-RESET inputs to the $\overline{\text{SET}}$-$\overline{\text{RESET}}$ flip-flop. The clock input is used to enable or inhibit the two gates. If a 0 is put on the clock input, the output of the two NAND gates will be forced to a 1. This places the crossed NAND $\overline{\text{SET}}$-$\overline{\text{RESET}}$ flip-flop in its remembering or unchanged state. Therefore, when the clock is 0, the flip-flop outputs cannot be changed. When the clock is made 1, the gates are enabled, or turned on, and the values of the inputs are passed through as their complements. This is shown in Figure 7-21.

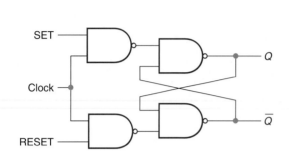

Clock	SET	RESET	Q	$\overline{Q}$	STATE
0	0	0	Q	$\overline{Q}$	
0	0	1	Q	$\overline{Q}$	
0	1	0	Q	$\overline{Q}$	Unchanged state
0	1	1	Q	$\overline{Q}$	
1	0	0	Q	$\overline{Q}$	
1	0	1	0	1	RESET
1	1	0	1	0	SET
1	1	1	1	1	Unused state

FIGURE 7-20 Gated SET-RESET flip-flop

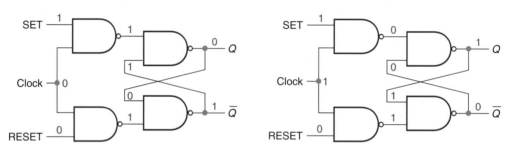

FIGURE 7-21 Enabling a gated SET-RESET flip-flop

Because the NAND gates invert the inputs, when the SET is 1 and RESET 0, Q is 1 and $\overline{Q}$ is 0. Also, when the SET is 0 and RESET is 1, Q is 0 and $\overline{Q}$ is 1. This means that when the clock is 1, the Q and $\overline{Q}$ outputs follow the values of the SET and RESET respectively.

EXAMPLE 7-4

Draw the waveform of the Q output for a gated SET RESET flip-flop from the inputs given in Figure 7-22.

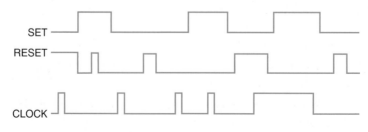

FIGURE 7-22

Solution   Refer to Figure 7-23.

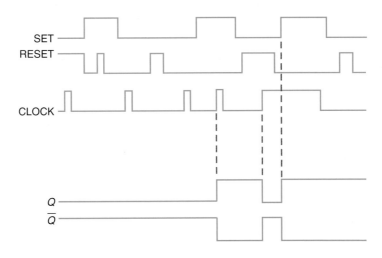

FIGURE 7-23

 7.7 THE TRANSPARENT *D* FLIP-FLOP

One problem with the gated NAND SET-RESET flip-flop is that there can be a 1 on the Q and a 1 on the $\overline{Q}$ when the SET and RESET inputs are both 1. This is the unused state, which should be avoided if possible. Also, it would be much more convenient if one input could SET and RESET the flip-flop. Both of these problems can be alleviated by placing an inverter between the SET and RESET inputs as shown in Figure 7-24. This makes a new input which we call the *D* input. Notice that the SET and RESET inputs can never be the same value because of the inverter. This means that the unused state can never exist. Also, there is one *D* or data input to SET or RESET the flip-flop.

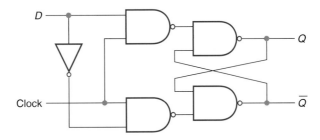

FIGURE 7-24 *D* flip-flop

It can be readily seen that when the clock is 1, which enables the gates to the $\overline{\text{SET}}$-$\overline{\text{RESET}}$ flip-flop, the value of *D* (1 or 0) is transferred to the *Q* output. When the clock is 0, the *Q* and $\overline{Q}$ outputs cannot be changed by the *D* input.

This type of D flip-flop is called a **transparent D flip-flop** because when the clock is 1, Q changes when D changes. The flip-flop appears transparent until the clock falls to 0, at which time the flip-flop becomes opaque. Figure 7-25 shows the truth table for the transparent D flip-flop.

D	Clock	Q	$\overline{Q}$	
0	0	Q	$\overline{Q}$	Unchanged
1	0	Q	$\overline{Q}$	state
0	1	0	1	
1	1	1	0	

FIGURE 7-25 Truth table for transparent D flip-flop

EXAMPLE 7-5

Draw a D flip-flop using only NOR gates. See Figure 7-26.

Solution   Notice that the Q and $\overline{Q}$ have been reversed to make the flip-flop work correctly and that the clock is active low.

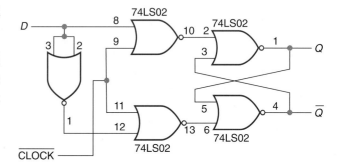

FIGURE 7-26

The D flip-flop is used to store bits of binary numbers. Because it can be turned on or off by the clock, it is also used to catch or latch a binary number present on the D input for a short time and store it on the Q and $\overline{Q}$ outputs. A D flip-flop can be used as the output port of a microcomputer, as shown in Figure 7-27. When the computer wants to output an 8-bit binary number to the printer, it places the binary number on the data bus and then strobes the clocks of the D flip-flops. This causes the Q outputs to take on the value of the data bus. The number is now latched on the Q outputs and will not change even though the data bus changes. A typical TTL IC, which contains four transparent D flip-flops, is the 7475 quad latch.

The symbol for the D flip-flop is also shown in Figure 7-27.

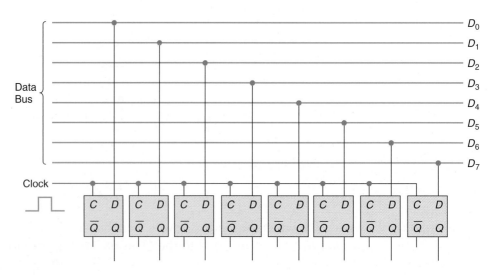

FIGURE 7-27 Computer-output port

SELF-CHECK 2

1. Draw a gated SET-RESET flip-flop using NOR gates.

2. Draw the output waveforms from the following inputs to a transparent *D* flip-flop with an active HIGH clock. Refer to Figure 7-28.

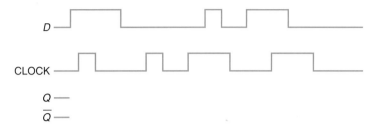

FIGURE 7-28

3. Using the following waveforms as the inputs to a $\overline{\text{SET}}$-$\overline{\text{RESET}}$ flip-flop made from NAND gates, draw the output waveforms. Refer to Figure 7-29.

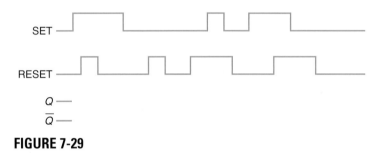

FIGURE 7-29

 ## 7.8 THE MASTER-SLAVE *D* FLIP-FLOP

Figure 7-30 shows a **master-slave *D* flip-flop** made from NAND gates. The master section is a transparent *D* flip-flop while the slave section is a gated SET-RESET flip-flop. The clock is fed to an inverter which is connected to the slave clock.

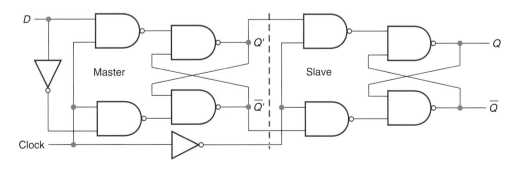

FIGURE 7-30 Master-slave *D* flip-flop

This type of master-slave *D* flip-flop is called a negative **edge-triggered** *D* flip-flop because the *Q* outputs will take on the value of the *D* input only on the falling edge of the clock pulse.

As shown in Figure 7-31, when the clock is 1, the master part of the flip-flop, which is a transparent D flip-flop, is turned on. The Q' output will follow the D input. The slave part, which is a gated SET-RESET flip-flop, is turned off because the inverter on the clock made the clock a 0. Because the slave is turned off, the Q outputs cannot change.

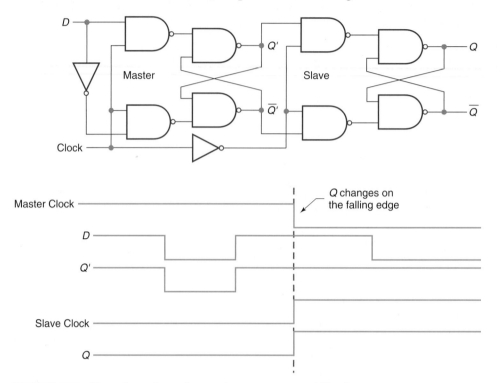

FIGURE 7-31 Negative edge-triggered master-slave D flip-flop

When the clock falls from 1 to 0, the master is turned off and cannot change; but the slave transfers the values of Q' and $\overline{Q}'$ to Q and $\overline{Q}$ because the slave clock goes to 1. The slave will not change if the D input changes because the master is turned off by the 0 on the clock. Therefore, the Q outputs can only change on the falling edge of the clock and will take on the value of the D input at the instant the falling edge happens.

If the inverter is reversed as shown in Figure 7-32, the flip-flop will change states on the positive or rising edge of the clock. Figure 7-32 also shows **CLEAR** and **PRESET** inputs, which can

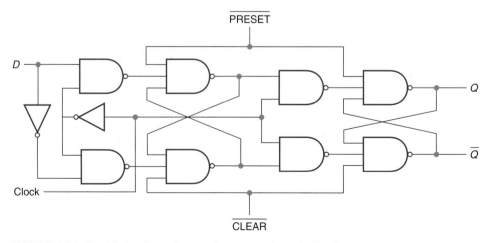

FIGURE 7-32 Positive edge-triggered master-slave D flip-flop

be used to force the output of the flip-flop to 0 or 1, regardless of the clock or D input values. When the $\overline{\text{PRESET}}$ goes LOW, the Q output is set or forced to 1. When the $\overline{\text{CLEAR}}$ goes LOW, the Q output is cleared or forced to 0. Notice that an invalid state will result if both $\overline{\text{PRESET}}$ and $\overline{\text{CLEAR}}$ are LOW or 0 at the same time. The $\overline{\text{PRESET}}$ and $\overline{\text{CLEAR}}$ inputs could be labeled $\overline{\text{SET}}$ and $\overline{\text{RESET}}$ also, because they make the flip-flop behave like a simple $\overline{\text{SET}}$-$\overline{\text{RESET}}$ flip-flop.

Figure 7-33 shows the symbol for a positive edge-triggered D flip-flop and a negative edge-triggered flip-flop. Notice the bubbled input on the active LOW inputs and the > mark which indicates the flip-flop is edge triggered.

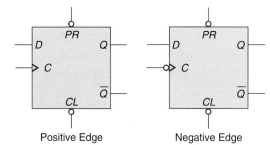

Positive Edge   Negative Edge

FIGURE 7-33 Flip-flop symbols

A typical positive edge-triggered flip-flop is the TTL 7474, which has two flip-flops on one IC. The 4013 is a CMOS dual, edge-triggered flip-flop with active high, SET, and RESET inputs. Figure 7-34 shows a truth table for a positive edge-triggered flip-flop such as the 7474 IC. Two commonly used symbols for rising edge are the ⌐ or ↑.

X = 1 or 0

CLEAR	PRESET	Clock	D	Q	$\overline{Q}$	STATE
0	1	X	X	0	1	CLEAR
1	0	X	X	1	0	PRESET
0	0	X	X	1	1	Unused state
1	1	⌐	1	1	0	SET
1	1	⌐	0	0	1	RESET

FIGURE 7-34 Truth table for a positive edge-triggered D flip-flop

As will be seen in the next chapter, master-slave flip-flops are used for many different types of digital circuits, such as shift registers, counters, and frequency dividers, because they are edge triggered.

Figure 7-35 shows some of the commonly used D flip-flops and $\overline{\text{SET}}$-$\overline{\text{RESET}}$ flip-flops which are manufactured by several integrated circuit manufacturers. The 7475 is a quad transparent D flip-flop which can be used as a latch or storage register. The 74LS74 is a positive edge-triggered flip-flop with active LOW presets and clears. The 74LS174 and 74LS175 are often used as an output port latches for microcomputers because they have an active LOW clear; but the 74LS273 has eight D flip-flops on one 20-pin chip, which means it can store 1 byte. That makes it a better choice for a microcomputer output port. The 74LS279 has four crossed NAND $\overline{\text{SET}}$-$\overline{\text{RESET}}$ flip-flops in one 16-pin dual in-line package.

An interesting IC is the MC14042B from Motorola. This chip is a 4-bit CMOS transparent latch that has an exclusive-OR gate on the clock. This allows the user to change the clock from active HIGH to active LOW by the logic value placed on pin 6 of the IC.

The MC14043B and MC14044B ICs are SET-RESET and $\overline{\text{SET}}$-$\overline{\text{RESET}}$ respectively with outputs that are gated with tri-state gates. The tri-state gate is a gate that can pass the 1 or 0 of the logic signal or have a third state of very high impedance. This makes the gate output look as if it is not connected to anything when it is in this HiZ state. This type of gate will be studied later in this book.

Very useful ICs for microcontroller output ports are the MC14514B and the MC14515B. These chips are 4-bit latches made of transparent D flip-flops that have the output of the latch feed to the input of a 4-to-16 decoder. They can be used by a microcontroller as an output port to select one of 16 possible things to enable.

At first glance the MC74F803 looks just like a simple 4-bit latch with inverted outputs, and this is true except this chip is designed to have matched propagation delays for the four D flip flops. This IC is used in a very high-speed synchronous logic system where the outputs must change at the same time on each transition of the clock.

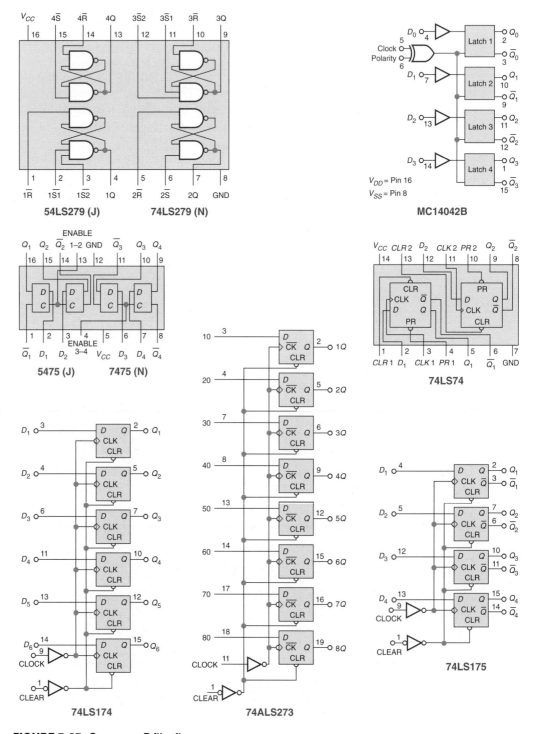

FIGURE 7-35 Common D flip-flops

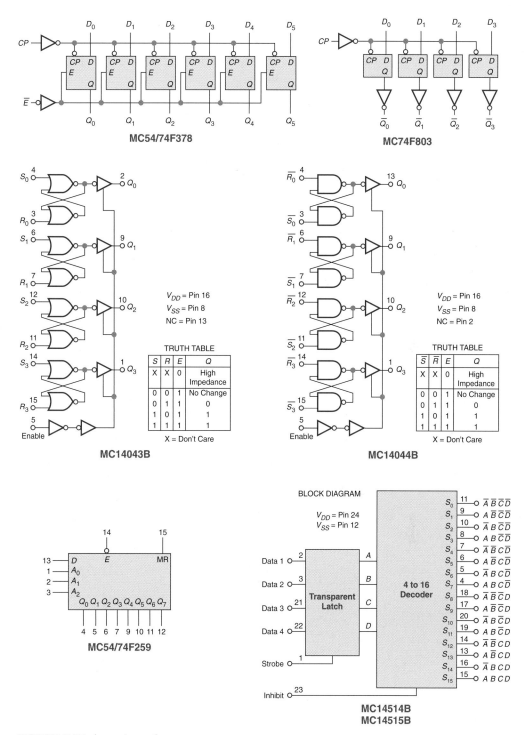

FIGURE 7-35 (*continued*)

The outputs of the MC54/74F378 can be placed in the "no change" or "remembering" state by an ENABLE signal. This allows an external signal to stop the flip-flops from being controlled by the clock and D input.

The MC54/74F259 also has an ENABLE control, but it only has one D input for the seven Q outputs of the flip-flops. Which flip-flop gets the value placed on the D input fed to it is determined by the binary number placed on the select inputs A_0, A_1 and A_2. This chip can also be used by a microcontroller as an output port. The microcontroller can choose the Q output which is to receive the incoming data on the D input by placing a binary number on the A select inputs.

Draw the Q and $\overline{Q}$ outputs for the given inputs to the flip-flop shown in Figure 7-36.

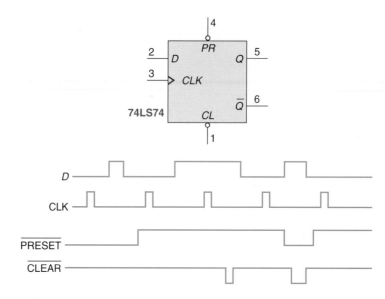

FIGURE 7-36

Solution   Refer to Figure 7-37.

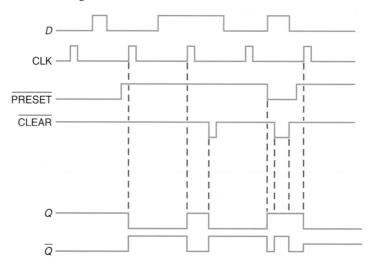

FIGURE 7-37

7.9 THE PULSE EDGE-TRIGGERED
D FLIP-FLOP

The master-slave D flip-flop is not the only way to make an edge-triggered D flip-flop. Figure 7-38 shows an edge-triggered D flip-flop which uses a pulse generator on the clock input to enable and inhibit the clock of the transparent D flip-flop very quickly. Because this short pulse

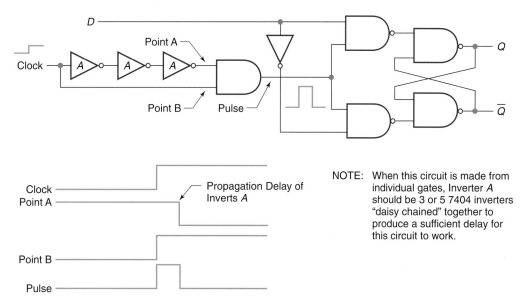

FIGURE 7-38 Edge-triggered *D* flip-flop

happens only on the rising edge of the input clock, the *D* flip-flop is an edge-triggered *D* flip-flop.

The circuit which produces the short pulse on the rising edge of the clock is called an edge-triggered one-shot. We will study one-shots more in a later chapter.

7.10 PROGRAMMING A CPLD (Optional)

In VHDL, signals are named according to these rules:

> A signal name begins with a letter.
>
> The name can contain any number of letters, numbers, and underscores.
>
> There cannot be two underscores in a row.
>
> The name cannot end with an underscore.
>
> The name cannot be the same as a VHDL key word like END.

Comments in a VHDL program begin with a double hyphen: "--." The compiler ignores all text on a line following a double hyphen.

In this section, we will program a CPLD to implement a transparent flip-flop and a set-reset flip-flop. Refer to Section 2.19 and Section 3.10—"Programming a PLD"—as needed.

Step 1. Create a new folder within the Altera folder and name it **transD**.

Step 2. Run Quartus II.

Step 3. Run "New Project Wizard."

Step 4. Create a new .vhd file and name it **transD.vhd**.

Step 5.   Write a VHDL program that will produce a transparent data latch. When the clock is high, the output Q will follow the data input D. The input data will be latched on the high-to-low transition of the clock. While the clock is low, the input data will be ignored.

- Invoke the ieee library by beginning your program with these two lines:

  LIBRARY ieee;

  USE ieee.std_logic_1164.all;

- Declare D and the clock to be standard logic inputs.

  D,clk : IN STD_LOGIC

- Declare Q1 and Q2 to be buffers. With Q1 as a buffer, we can program its value and then declare Q2 to be its complement. If Q1 is declared OUT STD_LOGIC we cannot read its value and therefore cannot assign its complement to Q2. Q1 will function as Q and Q2 will function as Q complement.

  Q1,Q2:BUFFER STD_LOGIC

Step 6.   Compile the program.

Steps 7–10.   Create a .vwf (vector waveform file) named **transD.vwf** and use it to simulate the circuits. Sample simulation waveforms are shown in Figure 7-39.

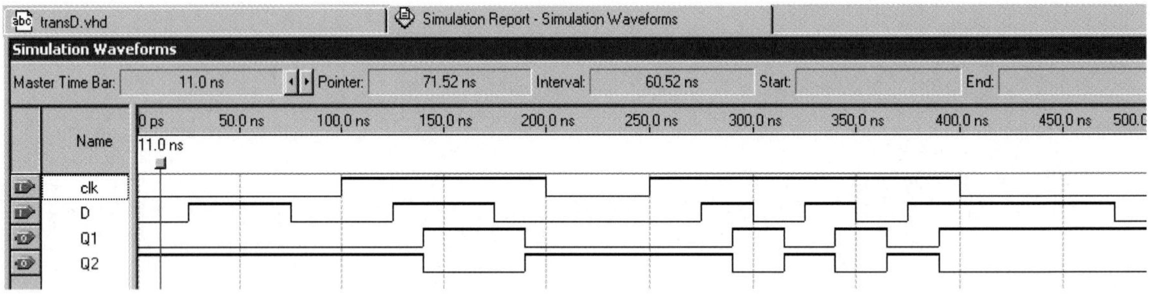

FIGURE 7-39 Simulation waveforms for transD.vwf

Note that Q1 follows D while the **clk** input is HIGH. At time 200 nanoseconds, **clk** goes LOW and latches the "0" on D. At time 400 ns, **clk** goes LOW and latches the "1" on D. The data change on D at 500 ns is ignored because **clk** is LOW.

Step 11.   Assign pins to the input and output signals. On the RSR PLDT-2 trainer, S5 consists of four debounced toggle switches with LEDs for indicating the HIGH or LOW state of each. S3 and S4 are momentary push-button switches, not debounced switches. Any of the six switches can be used as our clock input. Assign **clk** to pin 75 or to another pin that is physically close to the switches. Recompile the program.

Step 12.   Program your CPLD.

Step 13.   Test the operation of the CPLD using input switches and LEDs.

transD.vhd looks like this:

```
Library ieee;
USE ieee.std_logic_1164.all;
```

```
ENTITY transD IS
    PORT (D,clk:IN STD_LOGIC;
          Q1,Q2:BUFFER STD_LOGIC);
END transD;

ARCHITECTURE a OF transD IS
BEGIN
    PROCESS (D,clk,Q1) --IF statements must be contained in a PROCESS
    BEGIN
        IF clk = '0' THEN  --When clk is LOW, ignore data.
                Q1 <= Q1;
                Q2 <= NOT Q1;
        ELSIF clk = '1' THEN  --When clk is HIGH, data passes.
                Q1 <= D;
                Q2 <= NOT D;
        END IF;
    END PROCESS;
END a;
```

Note these features in the program:

- IF and ELSIF (ELSE IF) statements are used to test the clk input.

- These statements are housed in a PROCESS.

- The PROCESS statement ends with an END PROCESS statement.

- The "=" sign is used when checking a value in an IF statement or ELSIF statement; for example, IF **clk** = '0'.

- The <= sign is used when a signal is being defined; for example, Q1 <= D.

- The value of a single bit is enclosed in single quotes: '0' or '1'.

EXAMPLE 7-7

Write a VHDL program that will describe the operation of a crossed-NAND $\overline{set}$-$\overline{res}$ flip-flop. Name the inputs **set** and **res**. Name the outputs **Q1** and **Q2**.

Solution   Name the new project and folder **set_reset**. Create a new .vhd file and name it **set_reset.vhd**.

We will use IF THEN statements to test the set and reset inputs. IF THEN statements are contained in a PROCESS. To include a PROCESS in the ARCHITECTURE body, we need the ieee library. Invoke the ieee library by beginning your program with these two lines:

LIBRARY ieee;
USE ieee.std_logic_1164.all;

Declare **set** and **res** to be standard logic inputs, and declare **Q1** and **Q2** to be buffers so that Q1 can be used as an output—and so that its output value can be used to define Q2.

```
ENTITY set_reset IS
     PORT (set,res : IN STD_LOGIC;
           Q1,Q2 : BUFFER STD_LOGIC);
END set_reset;
ARCHITECTURE a OF set_reset IS
BEGIN
```

We have to BEGIN the ARCHITECTURE statement and the PROCESS.

Use IF THEN statements to test the states of **set** and **res**. In order to define **Q2**, in our PROCESS we are watching for changes in **set**, **res**, or **Q1**.

```
PROCESS (set,res, Q1)
BEGIN
    Q2 <= NOT Q1;
```

A zero on the set input should cause the flip-flop to turn on: Q1 goes HIGH and the Q2 output goes LOW.

```
IF (set = '0' AND res = '1') THEN
    Q1 <= '1';
    Q2 <= '0';
```

A zero on the reset input should cause the flip-flop to turn off: Q1 goes LOW and Q2 goes HIGH.

```
ELSIF (set = '1' AND res = '0') THEN
    Q1 <= '0';
    Q2 <= '1';
```

When both inputs are HIGH, the flip-flop should retain its present value. No action is taking place, so set = 1 and res = 1 is not tested for. If **set** and **res** are active at the same time, both Q1 and Q2 go HIGH. This is the unused state of this flip-flop, but it is included here as a possibility.

```
ELSIF (set = '0' AND res = '0') THEN
    Q1 <= '1';
    Q2 <= '1';
```

Now we have to END the IF statements, the PROCESS, and the ARCHITECTURE statement.

```
        END IF;
    END PROCESS;
END a;
```

set_reset.vhd looks like this:

```
Library ieee;
USE ieee.std_logic_1164.all;
ENTITY set_reset IS
     PORT (set,res : IN STD_LOGIC;
           Q1,Q2 : BUFFER STD_LOGIC);
```

```
END set_reset;
ARCHITECTURE a OF set_reset IS
BEGIN
     PROCESS (set, res, Q1)
        BEGIN
          Q2 <= NOT Q1;
            IF (set = '0' AND res = '1') THEN --Turn off flip-flop
                 Q1 <= '1';
                 Q2 <= '0';
             ELSIF (set = '1' AND res = '0') THEN --Turn on flip-flop
                 Q1 <= '0';
                 Q2 <= '1';
             ELSIF (set = '0' AND res = '0') THEN --Unused state
                 Q1 <= '1';
                 Q2 <= '1';
             END IF;
        END PROCESS;
END a;
```

Simulation waveforms for set_reset are shown in Figure 7-40.

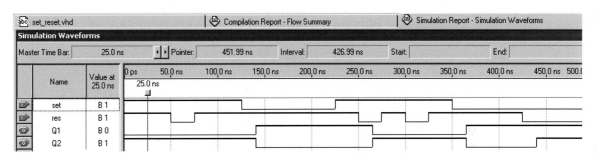

FIGURE 7-40 Simulation waveforms for set_reset.vwf.

At 50 nanoseconds, **res** goes LOW. **Q1** and **Q2** do not respond because the flip-flop is already off. At 125 ns, **set** goes LOW; the flip-flop turns on at 140 ns, a propagation delay of 15 ns. At 425 ns, both **set** and **reset** go LOW, causing both **Q1** and **Q2** to go HIGH (unused state).

7.11 TROUBLESHOOTING A DIGITAL CIRCUIT

Figure 7-41 is the schematic of a digital circuit designed to control the speed of a stepper motor used in the mechanical feed system for a film developing machine. The motor does not run and a technician has been asked to fix it.

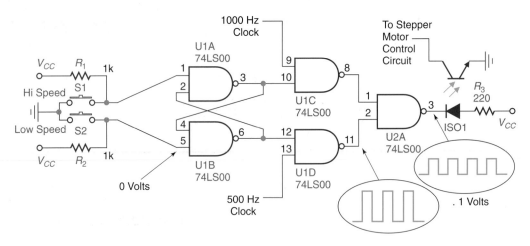

FIGURE 7-41 Stepper motor speed control circuit

The first step in troubleshooting anything is to gain as much knowledge as possible about how the device works. With electronic devices this means getting the schematic, if possible, and then studying it to determine how it works. The circuit in Figure 7-41 is made from two push-button switches and some NAND gates which feed an optocoupler. The two buttons should toggle the $\overline{\text{SET}}$ $\overline{\text{RESET}}$ flip-flop made from NAND gates and pass either the 1000-Hz signal or the 500-Hz signal to the optocoupler and then to the stepper motor control circuit.

The next step in troubleshooting a device is to determine the value of the inputs and the outputs of the circuit. Armed with this information you can make some conclusions about the probable cause of the problem. To determine the input and output values you need to select a test instrument which will give you the proper information you need. With this circuit you could use a simple volt-ohm-millammeter (VOM) to measure the voltage values of the IC pins, but that will not work well for the inputs with fast on/off signals such as the 1000-Hz input to pin 9 of U1. Therefore, the selection of an oscilloscope would be a better choice. The oscilloscope will show you the true waveform at the pins and not an average voltage as the VOM.

The first thing that does not seem right is that the voltage at pin 3 of U2 is at .1 volts. This pin should have a clock signal of either 1000 Hz or 500 Hz. Checking the two inputs of the U2A NAND gate shows that pin 2 of U2 has a 500-Hz signal on it and pin 1 of U2 has a good logic one of 3.8 volts on it. This shows that the U2A gate is not working correctly; if it were, the signal on pin 2 of U2 would be seen on the output of the gate (pin 3 U2). A novice technician will assume that the U2 (the 74LS00 IC) is bad and will spend time and effort removing it from the PC board and replacing it. The truth of the matter is, bad ICs are not often the problem.

The problem may be that the output pin 3 of U2 is being pulled down by something that has gone to ground and is connected to the pin. The technician should first inspect the board for solder splashes and cold solder joints.

The next thing to try is lifting pin 3 of U2 and disconnecting it from the circuit. This can easily be done if the IC is socketed by removing the IC and bending out the pin so it is no longer in the socket. This may not be practical if the IC is soldered to the board or if it is a surface mount IC. If the pin can be isolated from the circuit easily, the technician can check it with the oscilloscope for the proper signal. If there is a signal, the problem is not in the IC but in the net which is connected to the pin of the IC. Sometimes it is possible to tell that something is shorting the output pin to ground by using an oscilloscope to display the pin's voltage. Place the oscilloscope on ac coupling. Then crank up the voltage setting to something like 50 mV per cm. If the thing which is shorting the output pin to ground has some resistance such as a shorted diode or IC input, you will see a very small indication of the 500-Hz signal on the pin. This is because the top transistor of the totem-pole in the IC's output is trying to pull the output up to V_{CC} but can't raise it above a very small voltage.

After inspecting the board, the technician can not find any solder splashes so he replaces the optocoupler. This restores the signal at pin 3 of U2 to a good 500-Hz TTL level signal, but the stepper motor still does not run. The next step is to see if the 500-Hz signal has gotten through the optocoupler to the collector of the transistor. After looking at the collector pin of the optocoupler and finding no signal, the technician may conclude that the new optocoupler he just installed is also bad. The probability of this is quite low. The real problem is more than likely in the current-limiting resistor to the diode of the optocoupler (R_3).

When resistors go bad they do not get smaller in resistance; they get larger and larger until they burn up and open. To test the circuit for a possible bad resistor, simply place a known good resistor in parallel with the suspected bad resistor to lower the resistance to a value which will make the circuit work or show a change that indicates that the resistor has risen to a higher value than it should. Our technician selects a 330-ohm resistor and holds it across R_3. This causes the 500-Hz signal to appear at the collector of the transistor on the optocoupler and the stepper motor starts to run. The technician replaces R_3. When he flips over the board to do this he notices a burn mark on the bottom of the board where the lead of the R_3 resistor has touched the back of a transistor on the board mounted just beneath the one he has been working on. He quickly measures the voltage to ground of the transistor back and finds that it is −45 volts. Now it all makes sense. Someone had pressed the board down causing resistor R_3 to touch the back of the transistor on the board below. This caused a large current in resistor R_3 and the diode of the optocoupler resulting in the failure. Our technician places a piece of heat-shrink tubing over the transistor and a piece of electrical tape on the bottom of the board to prevent this from happening again.

After competing the repairs, the technician notices that the stepper motor does not change speeds when the high-speed button is pressed. Instead of switching to the higher speed, the motor is erratic and returns to the lower speed when the high-speed button is released. Measuring the voltages on the pins of the $\overline{\text{SET}}$ $\overline{\text{RESET}}$ flip-flop, he finds that pin 5 of U1 is always low. This pin should only be low when the low-speed button is pressed. After further examination the technician finds that the button, S2, is jammed closed, causing pin 5 of U1 to be at ground level all the time. The technician realizes that this is why the board was pressed so hard that the resistor lead of R_3 was forced to touch the back of the transistor on the board below. A very good conclusion on his part. The technician replaces the switch and the circuit works correctly in all ways. Our technician fills out an engineering repair report that is sent to the design engineer recommending that a center-mounted standoff be added to prevent the board from flexing when the buttons are pressed hard.

Mechanical Flip-Flop

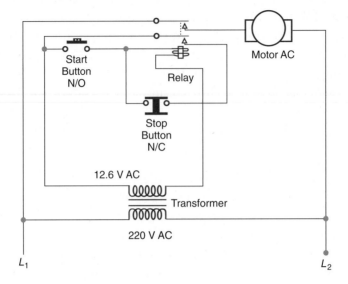

Have you ever seen a control box for a large machine with a start button and a stop button to start and stop the machine? The electrical circuit that controls the machine is an electro-mechanical SET-RESET flip-flop. The schematic shown is a typical start/stop circuit for a large machine, such as a large compressor for an air conditioner.

The circuit uses a relay which can be latched on by pushing the start button (SET input) and delatched by pushing the stop button (RESET). The relay contacts can handle the large ac currents needed to turn the large machine on and off, but the current needed to run the relay coil is small. do you think you could design a digital circuit using NAND gates to run the relay?

SUMMARY

- Flip-flops are logic circuits designed to store one bit of a binary number.

  The outputs of a flip-flop are called Q and $\overline{Q}$ and should always have different logic states from one another. There are several types of flip-flops which are used for different things.

- The crossed NAND has active LOW inputs called the $\overline{\text{SET}}$ and $\overline{\text{RESET}}$, while the crossed NOR has active HIGH inputs called SET-RESET.

  These flip-flops are called SET-RESET flip-flops and contain an unused state in their truth tables. Flip-flops of this type are often used for debouncing switches and storing logic states in more complex circuits.

- Gated flip-flops are SET-RESET flip-flops that have a pair of gates such as NAND or NOR gates to gate the SET and RESET inputs.

  This gives the gated flip-flop a new input called a clock. The clock will enable the flip-flop or inhibit it. When enabled, the flip-flop can change states, but when inhibited the Q and $\overline{Q}$ can not change states.

- The transparent D flip-flop is a gated flip-flop with an inverter between the SET and RESET inputs to prevent them from ever being the same logic state.

  This type of flip-flop is used quite often as a storage register or latch in microprocessor circuits. These types of flip-flops often have $\overline{PRESET}$ and $\overline{CLEAR}$ inputs to force the Q output to a 1 or 0 respectively.

- A master-slave D flip-flop is made of a transparent D flip-flop with its output tied to the inputs of a gated SET-RESET flip-flop.

  The D flip-flop is the master and the gated flip-flop is the slave. This makes a flip-flop that will only change states on the edge of the clock. These flip-flops can be rising or falling edge flip-flops. Figure 7-42 shows the common symbols used for flip-flop clocks.

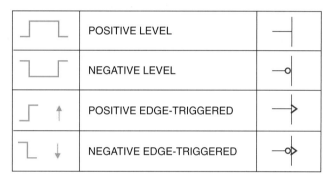

FIGURE 7-42 Table of flip-flop clock types

- Edge-triggered flip-flops can also be made using delay circuits, which will be studied later in this text.

  Edge-triggered flip-flops are also used as frequency dividers and counters, as we will see in the next chapter.

QUESTIONS AND PROBLEMS

1. Draw a $\overline{SET}$-$\overline{RESET}$ flip-flop logic diagram using NAND gates.

2. Draw the logic diagram of a gated $\overline{SET}$-$\overline{RESET}$ flip-flop using NAND gates.

3. Draw a logic diagram of a transparent D flip-flop using NAND gates.

4. Draw a logic diagram of a negative edge-triggered master-slave D flip-flop using NAND gates or NOR gates.

5. Complete the waveform for the negative edge-triggered master-slave flip-flop.

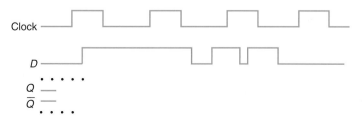

6. Draw a SET-RESET flip-flop logic diagram using NOR gates.

7. Draw a gated SET-RESET flip-flop logic diagram using NOR gates.

8. List two CMOS D flip-flop ICs. Draw the pinouts for these ICs.

9. Using the waveforms in number 5, draw the output waveforms for a positive edge-triggered D flip-flop.

10. Using the waveforms in number 5, draw the output waveforms for a transparent D flip-flop.

11. Draw the logic diagram for a positive edge-triggered D flip-flop using NOR gates.

12. Draw two commonly used symbols for indicating positive edge triggering in flip-flop truth tables.

13. Draw the logic symbol for a negative edge-triggered D flip-flop with active LOW, $\overline{\text{CLEAR}}$ and $\overline{\text{PRESET}}$.

14. What would be the value (1 or 0) of the Q output of a $\overline{\text{SET}}$-$\overline{\text{RESET}}$ flip-flop made from NAND gates if the $\overline{\text{SET}}$ input and the $\overline{\text{RESET}}$ input were LOW or zero?

15. Draw the logic diagram for a debounce switch using NOR gates.

16. Use a 74LS279 IC to design four debounce switches. Show the pin numbers.

17. Use a 74LS08, 74LS00, and a 74LS04 IC to design a pulse edge-triggered D flip-flop. Show the pin numbers.

18. Draw the logic diagram of a transparent D flip-flop using NOR gates on two 74LS02 ICs. Show the pin numbers.

19. What would be the logic value of the Q output for a negative edge-triggered D flip-flop if the D input were 1 and the CLOCK input made a transition from 1 to 0?

20. Draw the truth table for a transparent D flip-flop.

21. Why is an oscilloscope a better tool to troubleshoot a digital circuit than a volt-meter or logic probe?

22. What are the voltage ranges for a good TTL logic one signal?

23. What are the voltage ranges for a good TTL logic zero signal?

24. If you read a voltage reading of 1.5 volts using a volt-ohm meter on a TTL output pin, does it mean the IC is bad. If not, why?

25. What would be the value of Q and $\overline{Q}$ of a SET-RESET flip-flop made from crossed NAND gates if both inputs (SET and RESET) were held low?

26. Would the condition in question 25 damage the IC?

27. Could the pulsed edge-triggered D flip-flop shown in Figure 7-38 be made from only NOR gates?

28. Draw the logic diagram for a transparent D flip-flop made from OR gates and inverters.

29. What would happen to the operation of the circuit in Figure 7-30 if the inverter between the master and the slave were to have its output shorted to ground?

30. What would happen if the inverter in question 29 did not invert but passed the input to the output unchanged?

31. Write a program using VHDL that implements a crossed-NOR set-reset flip-flop.

32. Write a program using VHDL that implements a gated $\overline{\text{SET}}$-$\overline{\text{RESET}}$ flip-flop.

33. Find and correct four errors in this program:

```
ENTITY set_reset IS
     PORT (set,reset : IN STD_LOGIC;
           Q1,Q2 : BUFFER STD_LOGIC);
END set_reset;
ARCHITECTURE a OF set_reset IS
BEGIN
     PROCESS (set, res, Q1)
     BEGIN
          Q2 = NOT Q1;
          IF (set = '0' AND res = '1') THEN
                 Q1 <= '1';
                 Q2 <=  '0';
          ELSIF (set = '1' AND res = '0') THEN
                 Q1 <= '0';
                 Q2 <= '1';
          ELSIF (set = '0' AND res = '0') THEN
                 Q1 <= '1';
                 Q2 <= '1';
          END IF;
     END PROCESS;
END a;
```

34. Find and correct four errors in this program:

```
Library ieee;
USE ieee.std_logic_1164.all;

ENTITY transp IS
BEGIN
     PORT (D,clk : IN STD_LOGIC;
           Q1,Q2 : OUT STD_LOGIC);
END transp;

ARCHITECTURE a OF trans IS
BEGIN
     IF clk = '0' THEN
             Q1 <= Q1;
             Q2 <= NOT Q1;
     ELSIF clk = '1' THEN
             Q1 <= D;
             Q2 <=  NOT D;
     END IF;
END a;
```

OBJECTIVES

After completing this lab, you should be able to:

- Construct a debounce switch.
- Explain the operation of a gated SET-RESET flip-flop.
- Explain the operation of a master-slave D flip-flop.

COMPONENTS NEEDED

3	7400 quad NAND gate ICs
1	7408 quad AND gate IC
1	single-pole, double-throw switch

PROCEDURE

1. Use a 74LS00 or 74HCT00 quad NAND IC and a single-pole double-throw switch to make a debounce switch.

2. Draw the logic diagram for a negative edge-triggered D master-slave flip-flop with a $\overline{\text{CLEAR}}$ and $\overline{\text{PRESET}}$. Use one 74LS08 or 74HCT08 IC and three 74LS00 or 74HTC00 ICs. Show pin numbers on your drawing.

3. Construct the flip-flop in number 2 and have the instructor check the operation.

4. Write the truth table for the flip-flop in number 3 and verify its operation.

5. When does the flip-flop change states?

6. Bring both $\overline{\text{CLEAR}}$ and $\overline{\text{PRESET}}$ HIGH and place a wire from $\overline{Q}$ to the D input. Next put a 1-kHz TTL square wave signal on the clock of the master-slave D flip-flop and compare it to the Q output with the oscilloscope. What is the frequency of the Q output? The toggling of a master-slave flip-flop is the subject of the next chapter. You may want to read the next few pages of Chapter 8 to better understand the flip-flop's action.

If your circuit does not work properly, consider these points:

1. Test all power supply connectors to the ICs.

2. Check all input and output for proper voltage levels.

3. Disconnect the outputs of the first two NAND gates from the master part of the flip-flop. Now test their inputs and outputs for proper operation.

4. If the operation in step 3 of the circuit check does not correct the problem, remove the output of the crossed NAND $\overline{\text{SET}}$-$\overline{\text{RESET}}$ flip-flop from the input NAND gate of the slave and test it for proper operation.

OBJECTIVES

After completing this lab, you should be able to use Multisim to:

- Construct a transparent D flip-flop.
- Troubleshoot a master-slave flip-flop.

PROCEDURES

1. Using the SET-RESET flip-flop in the Multisim toolbox and three NAND gates, construct a transparent D flip-flop. Use the word generator to drive the flip-flop input and clock to check the flip-flop's operation.

2. Open File 7B-1.ms9. This file contains a master-slave flip-flop that is wired to toggle.

3. Run the simulation and observe the input and output of the flip-flop on the oscilloscope. It is not working correctly.

4. Use the on/off indicator and your knowledge of the circuit to debug the flip-flop and make it work correctly. There are three problems within the circuit.

Master-Slave *D* and *JK* Flip-Flops

OUTLINE

KEY TERMS

JK flip-flop   shift counter
propagation delay   toggle

OBJECTIVES

After completing this chapter, you should be able to:

- Make a master-slave D and JK flip-flop toggle.
- Explain the operation of a JK flip-flop.
- Explain the operation of a 2-phase nonoverlapping clock.
- Explain the operation of a shift counter.
- Describe typical JK flip-flop ICs.
- Troubleshoot JK flip-flops.

8.1 TOGGLING A MASTER-SLAVE *D* FLIP-FLOP

Figure 8-1 shows a master-slave *D* flip-flop, constructed from NAND gates, that is wired to toggle. A flip-flop is said to **toggle** when the *Q* outputs change states on each clock pulse. This means if *Q* is 1, after the next clock pulse, *Q* would toggle or change to 0, and after the next clock pulse, *Q* would toggle back to 1. This would continue as long as the clock pulses continued. Notice the $\overline{Q}$ output is tied back to the *D* input. This is why the flip-flop toggles.

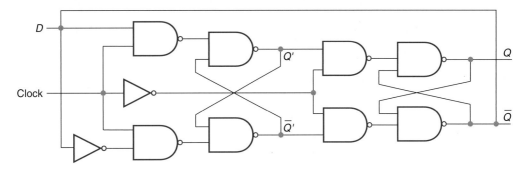

FIGURE 8-1 Master-slave *D* flip-flop wired to toggle

Figure 8-2 shows the waveforms for a toggling of the master-slave *D* flip-flop in Figure 8-1. When the clock goes HIGH, the master part of the flip-flop, which is a transparent *D* flip-flop, transfers the value of the *D* input, which is 0, to the *Q'* outputs of the master. The slave part of the flip-flop cannot change its output yet, because the inverter between the master and slave makes its clock 0 and turns off the slave. When the clock falls to 0, the master is turned off first, preventing *Q'* and $\overline{Q}'$ from changing. A few nanoseconds later, the *Q* and $\overline{Q}$ outputs change states to match the values of *Q'* and $\overline{Q}'$. This changes the value of the *D* input to 1. The master is turned off by the LOW clock. Therefore, *Q'* and $\overline{Q}'$ cannot change until the clock returns to a 1.

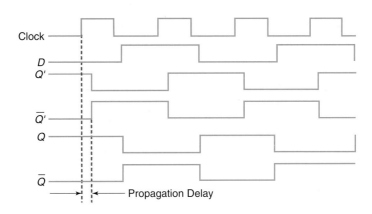

FIGURE 8-2 Toggling a master-slave *D* flip-flop

When the clock returns to a 1, the slave is turned off by the 0 on its clock. A few nanoseconds later, the master passes the 1 on the *D* input to the *Q'* outputs, and the whole cycle starts over again.

This circuit will work because of the **propagation delay** of the NAND gates used. The slave will turn off before the outputs of the master change states, and the master will turn off before the outputs of the slave change states. On the falling edge of every clock pulse to the flip-flop, the Q output switches states. Notice in Figure 8-2 that the Q output has one positive-going pulse for two positive-going pulses of the clock.

When a flip-flop toggles, the Q output frequency is one-half of the clock input to the flip-flop. This will be studied more in Chapter 10.

EXAMPLE 8-1

Draw the logic drawing for a master-slave D flip-flop wired to toggle using NOR gates.

Solution   Refer to Figure 8-3.

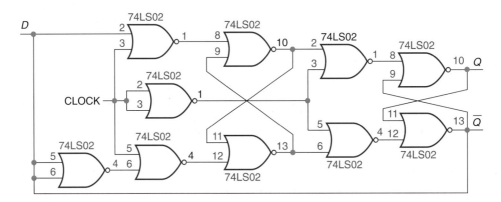

FIGURE 8-3

 8.2 THE *JK* FLIP-FLOP

The **JK flip-flop** is a special kind of master-slave flip-flop. Figure 8-4 shows a *JK* flip-flop constructed with NAND gates, the logic symbol used for a negative edge-triggered *JK* flip-flop, and its truth table.

This type of flip-flop can be wired or programmed to do the job of any type of flip-flop. The Q and $\overline{Q}$ outputs are wired back to K and J gates, respectively. This will allow the flip-flop to toggle when the J and K inputs are 1s. The J and K inputs are used to steer the Q outputs. There are 2 inputs, called $\overline{\text{PRESET}}$ and $\overline{\text{CLEAR}}$, which force the Q outputs to 1 and 0, respectively, when they are brought LOW.

Notice that the $\overline{\text{PRESET}}$ and $\overline{\text{CLEAR}}$, which are active LOW inputs, go to the master and slave parts of the *JK* flip-flop. When the $\overline{\text{PRESET}}$ goes LOW, this forces the Q outputs of the master and slave to 1 and keeps them at 1 until the $\overline{\text{PRESET}}$ returns to 1. The same thing applies to the $\overline{\text{CLEAR}}$ input, except it forces the Q output to 0. $\overline{\text{PRESET}}$ works just like a $\overline{\text{SET}}$ input, and $\overline{\text{CLEAR}}$ is the $\overline{\text{RESET}}$ input for a simple crossed NAND $\overline{\text{SET}}$-$\overline{\text{RESET}}$ flip-flop. The thing to remember about the $\overline{\text{CLEAR}}$ and $\overline{\text{PRESET}}$ inputs is that they override all other inputs to a *JK* flip-flop, as can be seen in the truth table. It should be noted that, as in the

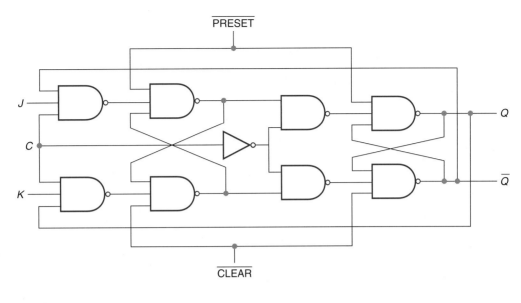

PRESET	CLEAR	J	K	C	Q	Q̄	
0	1	X	X	X	1	0	
1	0	X	X	X	0	1	
0	0	X	X	X	1	1	Unused state
1	1	0	1	⌐_	0	1	
1	1	1	0	⌐_	1	0	
1	1	0	0	X	Q	Q̄	Unchanged state
1	1	1	1	⌐_	Toggle		

FIGURE 8-4 *JK* flip-flop

crossed NAND $\overline{\text{SET}}$-$\overline{\text{RESET}}$ flip-flop, the *JK* has an unused state where $\overline{\text{CLEAR}}$ and $\overline{\text{PRESET}}$ are both 0.

If both the *J* and *K* inputs are 0, then the master is turned off just as if the clock was 0, because any 0 into a NAND gate produces a 1 on its output. This places the crossed NAND gates of the master in the unchanged state. This keeps the output of the slave from changing. Therefore, the *JK* flip-flop is in its unchanged or remembering state when the *J* and *K* inputs are 0.

When the *JK* inputs are not the same—that is, *J* is 1 or 0 and *K* is the opposite value—the *Q* outputs will change to the same value on the falling edge of the clock.

The only other possible input configuration for the *J* and *K* inputs is where both inputs are 1s. When this happens and the clock is HIGH, the *Q* and $\overline{Q}$ outputs, which are tied back to input NAND gates, control the master's outputs. This will cause the flip-flop to toggle or change states every falling edge of the clock because the *Q* output is tied back to the opposite *K* gate, and the $\overline{Q}$ is tied back to the opposite *J* gate.

Facts to remember about negative edge-triggered *JK* flip-flops include the following:

- The *Q* output only changes on the falling edge of the clock except when the $\overline{\text{CLEAR}}$ or $\overline{\text{PRESET}}$ go LOW.

- $\overline{\text{CLEAR}}$ and $\overline{\text{PRESET}}$ override all other inputs on the *JK* flip-flop.

- When the *J* and *K* are both 1, the flip-flop will toggle on the falling edge of the clock.

- When *J* and *K* are not equal, the output will follow the *J* and *K* on the falling edge of the clock.

- When *J* and *K* are both 0, the *Q* outputs will not change their values.

Figure 8-5 shows the waveform for the Q and $\overline{Q}$ outputs of a negative edge-triggered *JK* flip-flop for a given set of input waveforms on the *J*, *K*, $\overline{\text{CLEAR}}$, and $\overline{\text{PRESET}}$. Notice that, except for when the $\overline{\text{CLEAR}}$ or $\overline{\text{PRESET}}$ go active, the Q and $\overline{Q}$ outputs do not change states except on the falling edge of the CLK (clock) signal. The value of the Q and $\overline{Q}$ outputs is determined by the value of the *J* and *K* inputs before the falling edge of the CLK input to the flip-flop. They follow the truth table in Figure 8-4.

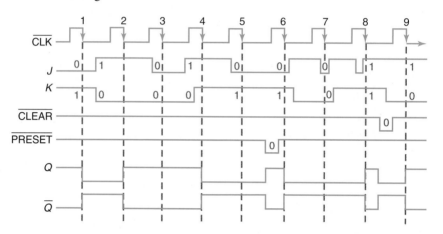

FIGURE 8-5 Output waveforms for a *JK* flip-flop

EXAMPLE 8-2

Draw the logic diagram for a *JK* master-slave flip-flop using NOR gates.

Solution   Refer to Figure 8-6.

Note: The toggling state for this *JK* flip-flop is 0 on the *JK* inputs, and the remembering state is 1 on the *JK* inputs. This is not the standard for *JK* flip-flops. Also, the $\overline{\text{CLEAR}}$ and $\overline{\text{PRESET}}$ are active HIGH.

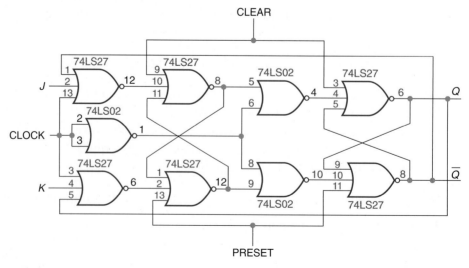

FIGURE 8-6

SELF-CHECK 1

1. How do you wire an edge-triggered *D* flip-flop to toggle?

2. What must be on the *JK* inputs of a *JK* flip-flop to make it toggle?

3. What would be the state of the *Q* and $\overline{Q}$ outputs of the *JK* flip-flop in Figure 8-4 if the $\overline{\text{CLEAR}}$ were LOW and the $\overline{\text{PRESET}}$ were HIGH?

 8.3 THE NONOVERLAPPING CLOCK

Figure 8-7 shows a *JK* flip-flop used to make a nonoverlapping clock. Notice that *CP* and *CP'* are one-half the frequency of the clock and are 180° out-of-phase. They are said to be nonoverlapping because the leading or rising edges and trailing or falling edges of *CP* and *CP'* will never occur at the same time.

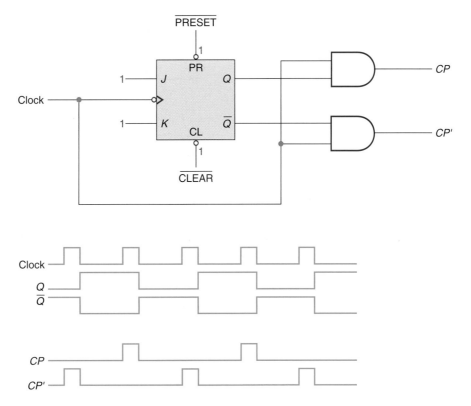

FIGURE 8-7 Nonoverlapping clock

The *JK* flip-flop is wired to toggle on the falling edge of the clock. This enables the *CP* AND gate and then the *CP'* AND gate on the next falling edge of the clock. Each time an AND gate is enabled by the flip-flop's *Q* or $\overline{Q}$ output, the next positive clock pulse is passed through the gate. On the falling edge of that clock, the flip-flop toggles, inhibiting one AND gate and enabling the other AND gate. This process can be seen in the wave diagram in Figure 8-7. This type of nonoverlapping clock is used for generating strobe signals and various waveforms for digital devices.

EXAMPLE 8-3

Draw the logic diagram and waveforms for a nonoverlapping clock and the waveforms for CP and CP' using a 74LS76 and a 74LS02 IC.

Solution   Refer to Figure 8-8.

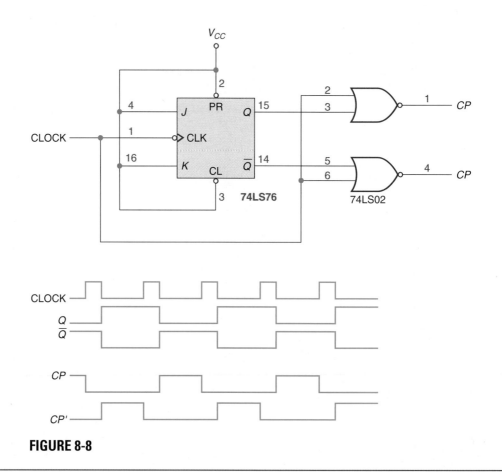

FIGURE 8-8

8.4 THE SHIFT COUNTER

Figure 8-9 is a **shift counter** made from three JK flip-flops. The Q and $\overline{Q}$ outputs of the A flip-flop are connected to the J and K inputs of the B flip-flop, respectively; and the B flip-flop's Q and $\overline{Q}$ outputs are connected to the J and K inputs of the C flip-flop in the same way.

Because all the clocks of the three flip-flops are tied together when the CP falls, the value of the A, Q, and $\overline{Q}$ will be passed on to the Q and $\overline{Q}$ outputs of the B flip-flop. The value of the B, Q,

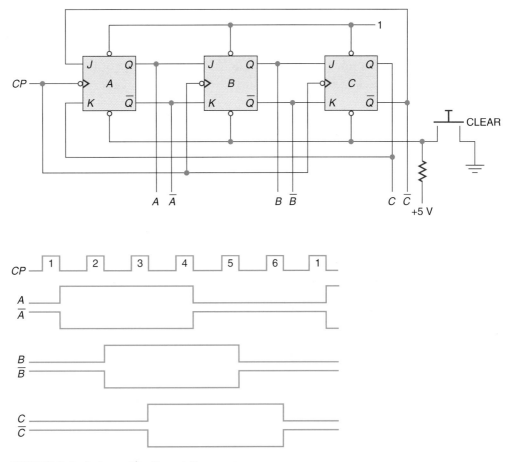

FIGURE 8-9 A three-flip-flop shift counter

and $\overline{Q}$ outputs before the *CP* falling edge will be passed on to the *Q* and $\overline{Q}$ outputs of the *C* flip-flop.

The *Q* and $\overline{Q}$ outputs of the *C* flip-flop are tied back to the *J* and *K* inputs of the *A* flip-flop in reverse order. *Q* is connected to *K*, and $\overline{Q}$ is connected to *J*. This means that after the *CP* falls, the *Q* and $\overline{Q}$ outputs of the *A* flip-flop will have the opposite value from the *Q* and $\overline{Q}$ outputs of the *C* flip-flop before the *CP* falling edge.

If the shift counter is cleared by bringing all of the $\overline{\text{CLEAR}}$ inputs LOW, all of the *Q* outputs will be 0, and the $\overline{Q}$ outputs will be 1. This state will exist as long as the $\overline{\text{CLEAR}}$ inputs are LOW. When they are brought back HIGH and after the falling edge of the next *CP*, the *A* flip-flop's *Q* output will go to the opposite of the *C* flip-flop's *Q* output, which was 0. Therefore, after the falling edge of the first *CP*, the output of the *A* flip-flop will be 1. After the next *CP* falling edge, the 1 on the *A* flip-flop's *Q* output will be passed on to the *B* flip-flop's *Q* output. When the third *CP* falling edge occurs, the 1 on the *Q* output of the *B* flip-flop will be passed on to the output of the *C* flip-flop.

The fourth *CP* causes the *A* flip-flop's *Q* output to go to 0 because of the reversed output feedback to the *A* flip-flop's *JK* inputs. After the fifth *CP*, the *B* flip-flop's *Q* output will be 0, and after the sixth *CP*, the entire counter will be 0 on all of the *Q* outputs.

If you examine the waveforms in Figure 8-9 for the three-flip-flop shift counter shown, you will notice that each flip-flop is 1 for three *CP* and then 0 for three *CP*. The outputs of the three flip-flops are 120° out-of-phase; that is, after *A* goes to 1 on the next *CP*, *B* goes to 1, and so on.

With the use of a few simple gates, *CP*, *CP'*, and outputs of the counter, it is possible to make any waveform needed for a digital device that will repeat every six *CP*. Figure 8-10 shows some examples of this.

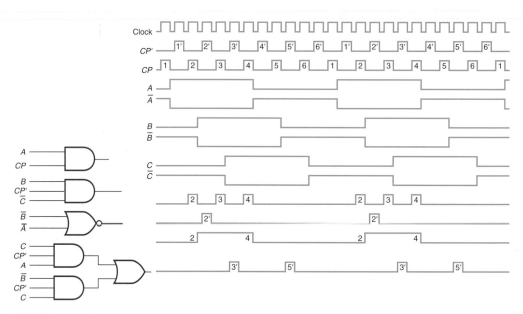

FIGURE 8-10 Nonoverlapping clock and three-flip-flop shift-counter waveforms

EXAMPLE 8-4

Design a nonoverlapping clock and 2-bit shift counter using two 74LS76 ICs and one 74LS08 IC. Show the pin numbers.

Solution   Refer to Figure 8-11.

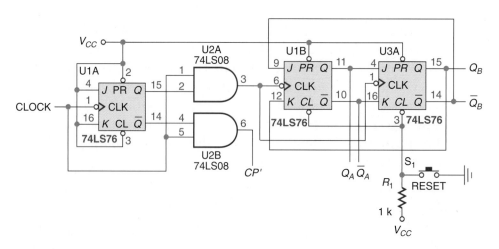

FIGURE 8-11

8.5 TYPICAL *JK* FLIP-FLOP ICs

Figure 8-12 shows four commonly used *JK* flip-flops and their pinouts. The 74LS73 is a negative edge-triggered master-slave *JK* flip-flop with only $\overline{\text{CLEAR}}$ inputs. The 74LS76 and 7476 are full-blown negative edge-triggered master-slave *JK* flip-flops with both $\overline{\text{CLEAR}}$ and $\overline{\text{PRESET}}$ inputs. This costs two more pins on the IC package, though. If the designer wishes to have a 14-pin IC package and to retain both $\overline{\text{CLEAR}}$ and $\overline{\text{PRESET}}$, then he or she can choose the 74LS78, which has them but also has a common $\overline{\text{CLOCK}}$ and $\overline{\text{CLEAR}}$ pin for both flip-flops. A typical positive edge-triggered *JK* flip-flop is the 74LS109.

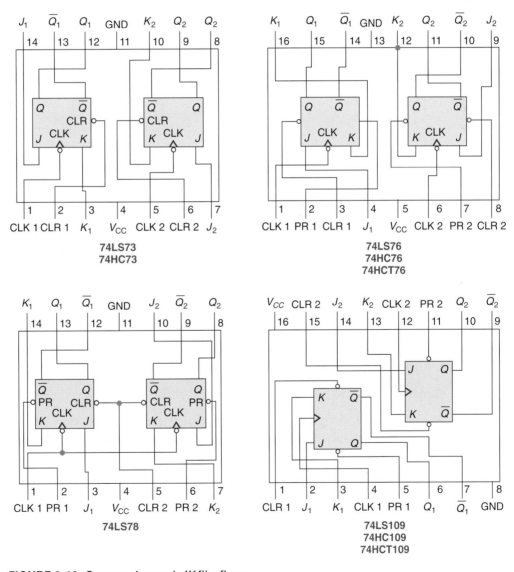

FIGURE 8-12 Commonly used *JK* flip-flops

All these flip-flops will work for the examples in this textbook as well as others not mentioned here. The student should refer to a manufacturer's IC specification manual to get a feel for the number and diversity of *JK* flip-flop ICs available today.

✓ SELF-CHECK 2

1. Draw the waveforms for the nonoverlapping clock and 2-bit shift counter in the example on page 334. Refer to Figure 8-13.

CLOCK
CP —
CP' —
A —
$\overline{A}$ —
B —
$\overline{B}$ —

FIGURE 8-13

2. Draw the output of the gate shown in Figure 8-14 for the given inputs in question 1.

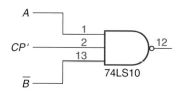

FIGURE 8-14

8.6 PROGRAMMING A CPLD (Optional)

In this section we will program a CPLD to implement both a negative edge-triggered *JK* flip-flop and a positive edge-triggered *JK* flip-flop with preset and clear.

- Create a new folder within the Altera folder and name it **JK_FF**.

- Create a new .vhd file and name it **JK_FF.vhd**.

- Write a VHDL program that will describe the operation of a negative edge-triggered *JK* flip-flop. Invoke the ieee library by beginning your program with these two lines:

```
LIBRARY ieee;
USE ieee.std_logic_1164.all;
```

In this program, the output state of Q1 and Q2 (HIGH or LOW) will be used. For instance, when it is time to toggle the flip-flop, the state of Q2 becomes the state of Q1. Also, the complement of the state of Q1 is assigned to Q2. So Q1 and Q2 will be declared as standard logic buffers.

```
ENTITY JK_FF IS
      PORT (J,K,clk:IN STD_LOGIC;
            Q1,Q2:BUFFER STD_LOGIC);
END JK_FF;
```

(HIGH), then the second set of IF statements looks for a LOW-to-HIGH transition of the clock (positive edge-triggered):

IF (clk'EVENT AND clk = '1') THEN

When the clock occurs, the third set of IF statements check the states of *J* and *K*. Three END IF statements are needed.

JK_pre_clr.vhd looks like this.

```
Library ieee;
USE ieee.std_logic_1164.all;
ENTITY JK_pre_clr IS
    PORT (J,K,clk,pre,clr:IN STD_LOGIC;
          Q1,Q2:BUFFER STD_LOGIC);
END JK_pre_clr;

ARCHITECTURE a OF JK_pre_clr IS
BEGIN
    PROCESS (J,K,clk,Q1,Q2,pre,clr)
    BEGIN
        IF pre <='0' THEN         --turn on
                Q1<='1';
                Q2<='0';
        ELSIF clr <='0' THEN      --turn off
                Q1<='0';
                Q2<='1';
        ELSIF (PRE<='1' AND clr<='1') THEN
            IF (clk'EVENT AND clk = '1') THEN    --positive clock edge
                IF (J='0' AND K='0') THEN   --no change
                        Q1<=Q1;
                ELSIF (J='1' AND K='0') THEN       --turn on
                        Q1<='1';
                ELSIF (J='0' AND K='1') THEN       --turn off
                        Q1<='0';
                ELSIF (J='1' AND K='1') THEN       --toggle
                        Q1<=Q2;
                END IF;
            END IF;
        END IF;
        Q2<= NOT Q1;     --Q2 is the complement of Q1
    END PROCESS;
END a;
```

Sample vector waveform files for **JK_pre_clr.vwf** are shown in Figure 8-16. Note that changes in Q1 occur following a LOW-to-HIGH transition of the clock (positive edge-triggered), after a short propagation delay. Q2 becomes the complement of Q1 after another short propagation delay.

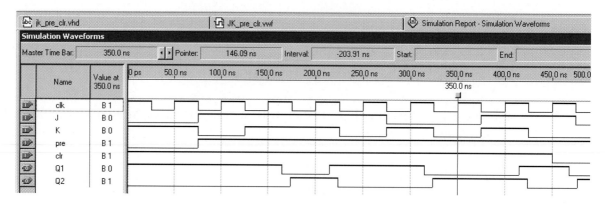

FIGURE 8-16 Simulation waveforms for JK_pre_clr.vwf

8.7 TROUBLESHOOTING *JK* FLIP-FLOPS

The circuit shown in Figure 8-17 is the control logic for a film advance to a photo enlarger. It does not work properly and once again the technician has been called in to get it back on-line and working.

The technician has been told by the operator of the enlarger that when the film advance button is pressed the film continues to advance until he presses the button again and then it stops. The film should advance only one frame when the button is pressed. With this information in mind, the technician moves the enlarger to his workbench and looks for the schematics of the machine. He knows that the first step in troubleshooting and repairing a device is to understand how it works.

After studying the control logic for the film advance portion of the device, he concludes that when the film advance button is pressed the SET-RESET flip-flop made from the two NAND gates, U6B and U6A, debounces the button and causes the *JK* flip-flop (U7A) to toggle. He understands how a *JK* flip-flop works and can see that the *J* and *K* inputs are tied high. This will cause the flip-flop to toggle if the $\overline{\text{CLEAR}}$ is not active. The $\overline{\text{CLEAR}}$ is fed from an AND gate. The inputs of this AND gate come from a shift counter made from three flip-flops (U7B, U9A, and U9B). The clock for the shift counter comes from a NAND gate which is gated on and off by the *Q* output of the *JK* flip-flop, U7A. Now thoroughly intrigued by the design, the technician draws a wave diagram of the expected waveforms coming from the shift counter (U7B, U9A, and U9B). These waveforms are shown in Figure 8-17. After doing this, he realizes how the whole film advance circuit works. When the *JK* flip-flop, U7A, toggles to make the *Q* output a 1, the CLOCK on pin 9 of U6C is gated to the output of the NAND gate on pin 8 of the same IC. This starts the shift counter running and produces the two waveforms which cause the film hold-down to be lifted and the film to be advanced under it. After the film is advanced, the film hold-down is lowered and a short reset pulse is generated to clear the *JK* flip-flop, thus gating off the CLOCK to the shift counter and stopping the whole process.

Now things start to make sense to the technician. The operator of the enlarger had told him that when he pressed the film advance button the film kept advancing instead of stopping at the next frame, but when he pressed it again, it would stop. The technician concluded that the

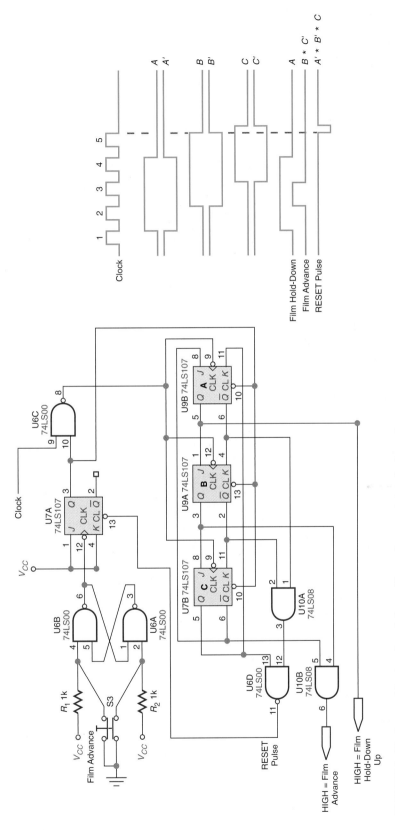

FIGURE 8-17 Control logic circuit for a film advance to a photo enlarger

JK flip-flop (U7A) was not getting cleared and, therefore, did not stop the CLOCK which was being fed to the shift counter.

With this information in mind, the technician got an oscilloscope and went to test the circuit to see if he could find the reset pulse. He did not just grab any oscilloscope, but chose a high-speed digital storage oscilloscope. He made this decision because he realized that the reset pulse would be short in duration and would only happen once each time he pressed the film advance button. He knew that the reset pulse would be only about six propagation delays long. He set the time sweep of the oscilloscope to "50 nano sec per cm" and set it on "storage single sweep." Placing the oscilloscope probe on pin 11 of U6D, he pressed the film advance button and the LOW-going reset pulse popped up on the oscilloscope screen. He reasoned that if the reset pulse is on pin 11 of U6D, maybe it is also on the clear pin of the *JK* flip-flop (pin 13 of U7A). After resetting the oscilloscope and pressing the button again, he got the reset pulse again on the oscilloscope. This puzzled the technician. If the reset pulse was reaching the *JK* flip-flop, then why did it not clear? He concluded that the chip must be bad.

Just as he was reaching for the solder sucker to remove the IC, he noticed that the film advance had worked correctly that time and only advanced one frame. This puzzled him even more, so he picked up the printed circuit board and began to examine it closely. This is something he should have done earlier, he thought. Pressing on the clear pin of the *JK* flip-flop he noticed that it moved and, flipping over the board, he found that the pin had a cold solder joint. The joint was so bad that when he had placed the oscilloscope probe on the pin, it moved the pin enough to make contact with the pad on the board. This is why the film advance worked correctly that time. The technician quickly fixed the cold solder joint with his soldering iron and tested the board again. It worked flawlessly. He could capture the reset pulse on the oscilloscope and see all the signals he expected to see.

Before the technician replaced the board in the enlarger, he inspected it very carefully and found two more cold solder joints, which he fixed. These joints had not come loose yet but very well might have later. This, he thought, will save me work later.

DIGITAL APPLICATION — ## Circuit for an IGBT Bridge

This is a portion of a digital circuit used to turn on and off an IGBT (insulated gate bipolar transistor) bridge. The current flowing in the IGBT bridge is used to change the polarity of the magnets in the rotor of a large generator (40 to 100 KVA). The current being controlled by this circuit is about 30 to 80 amps. By changing the polarity of the magnets as the rotor turns in the generator, the output voltage, phase, and frequency of the generator can be controlled even if the rotor rpm slows down. This is a patented technology called written-pole. The circuit is courtesy of Precise Power Inc. Notice the use of two *D* flip-flops (U38A and U38B) to sync the two signals POLARITY_1CD and DISABLE_1CD to the LATCH_1CD signal. Also notice the SET-RESET flip-flop made from two NAND gates (U10A and U10B) used to feed the signal from the window detector made from the two op amps (U13A and U13B).

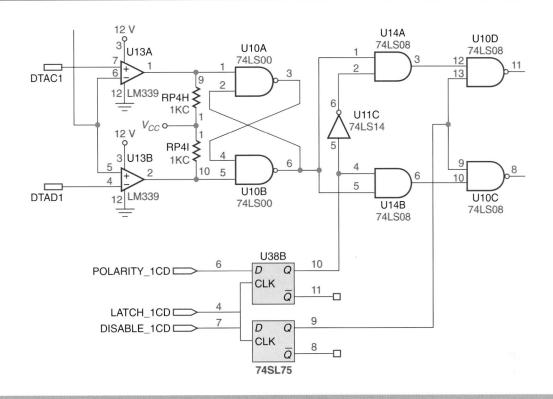

SUMMARY

- A master-slave flip-flop has the ability to toggle because it is edge triggered.

  To make an edge-triggered flip-flop toggle, the value of the Q output must be placed on the flip-flop's input as an inverted value. This can be done in the D master-slave flip-flop by connecting the $\overline{Q}$ to the D input. This will have the effect of changing the logic value on the D input each time the flip-flop is clocked.

- The *JK* flip-flop is a universal type of flip-flop which can be programmed or controlled by the logic values placed on the *J* and *K* inputs to the flip-flop.

  The basic rules for the use of a *JK* flip-flop are as follows:

  1. Depending on the type of *JK* flip-flop, the Q and $\overline{Q}$ outputs will change on the falling or rising edge of the clock, except when the $\overline{\text{CLEAR}}$ or $\overline{\text{PRESET}}$ are active.

  2. The $\overline{\text{CLEAR}}$ and $\overline{\text{PRESET}}$ override all other inputs to the *JK* flip-flop.

  3. When the *J* and *K* inputs are held at a logic 1, the flip-flop will toggle.

  4. When the *J* and *K* inputs are not the same logic value, the Q and $\overline{Q}$ outputs will follow the *JK* inputs, respectively, on the edge of the clock.

  5. When the *J* and *K* inputs are logic 0, the flip-flop will not change the Q and $\overline{Q}$ values on the edge of the clock. This is the unchanged or remembering state.

- When a flip-flop is wired to toggle, the Q output will be one-half the frequency of the clock input to the flip-flop.

This property of a toggling flip-flop can be used to make counters and frequency dividers. This is the subject of Chapter 10.

- The shift counter and the nonoverlapping clock can be used to generate complicated waveforms which are quite often used in microprocessor applications and control signals for more complex digital circuits, such as real-time clocks and frequency counters.

- A nonoverlapping clock can be made from one D flip-flop. The flip-flop is set to toggle by connecting the $\overline{Q}$ complement output to the D input.

  The flip-flop is positive edge triggered and will toggle on the rising edge of the clock input.

QUESTIONS AND PROBLEMS

1. Draw a logic diagram of a delayed clock and shift counter which will repeat every 10 *CP*. Include the IC pinout and pin number on the logic diagram.

2. Using the TTL and CMOS data manuals, locate two ICs which have negative clocks and two ICs which have positive clocks for both TTL and CMOS. Draw their pinouts.

3. Draw the logic diagram for a 2-phase nonoverlapping clock using one *JK* flip-flop and three NOR gates. Label the IC used and put pin numbers on the drawing.

4. Using Figure 8-13, draw the waveforms for the following gates.

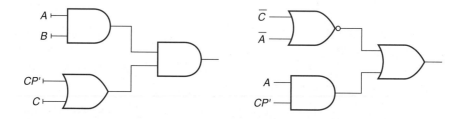

5. Using Figure 8-13, draw the logic diagram of the gates which produce the following waveforms.

6. Draw the logic diagram for a negative edge-triggered *JK* flip-flop with active LOW CLEAR and PRESET using NOR gates.

7. Draw the logic symbol for a negative edge-triggered *JK* flip-flop with active LOW $\overline{\text{CLEAR}}$ and $\overline{\text{PRESET}}$.

8. Draw the output waveforms for the *Q* output of a negative edge-triggered *JK* flip-flop with active LOW $\overline{\text{CLEAR}}$ and $\overline{\text{PRESET}}$. Use the following input waveforms.

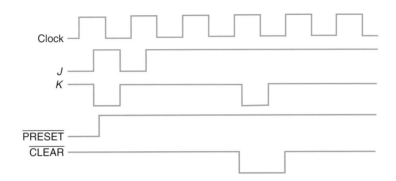

9. Repeat number 8 with a positive edge-triggered flip-flop with active HIGH $\overline{\text{CLEAR}}$ and $\overline{\text{PRESET}}$.

10. If the *J* input was 0 and the *K* input was 0 before the falling edge of the clock, what would be the value of the *Q* output of a negative edge-triggered *JK* flip-flop after the falling edge of the clock?

11. Draw the logic symbol for a flip-flop and put the correct values on the inputs to cause it to toggle.

12. Draw the waveforms for the flip-flop shown using the waveforms in Figure 8-10 as the inputs.

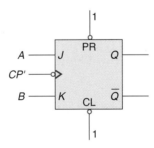

13. Using the waveforms in Figure 8-10 and a single *JK* flip-flop, draw the logic diagram of a circuit which will produce an even-duty clock which has the same frequency as *CP*.

14. What would have to be put on the inputs of the flip-flop in Figure 8-4 for *Q* and $\overline{Q}$ to be HIGH?

15. What would be the value of *CP* in Figure 8-7 if the $\overline{\text{PRESET}}$ input was tied LOW on the *JK* flip-flop?

16. Can a 74LS75 *D* flip-flop be made to toggle? Why?

17. Draw the logic diagram to make a 74LS74 IC toggle. Show the pin numbers.

18. Use a TTL data book and list five different edge-triggered flip-flop ICs.

19. Draw the waveform for the nonoverlapping clock in Figure 8-7 if the $\overline{\text{PRESET}}$ were tied LOW.

20. Draw the Q output for the JK flip-flop shown here.

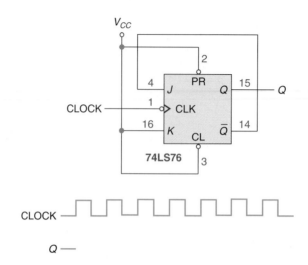

21. What would be the effect on the flip-flop in Figure 8-1 if the inverter between the master and slave parts of the flop-flop was reversed?

22. Draw the expected waveforms for the flip-flop in Figure 8-7 if the CLEAR input has been grounded.

23. What must be the values of the J and K inputs of a flip-flop if the Q outputs do not change on the transition of the clock?

24. Could a nonoverlapping clock such as the one shown in Figure 8-7 be made from only NOR gates?

25. How would the circuit in Figure 8-19 work if the J and K inputs to flip-flop U7A had been grounded?

26. Draw a nonoverlapping clock circuit using an edge-triggered D flip-flop and two NAND gates.

27. Draw the waveforms for the circuit shown in Figure 8-11.

28. What would happen to the operation of the circuit in Figure 8-11 if the clock signal became disconnected from the clock input of the JK flip-flop, U1A?

29. Using the waveforms in Figure 8-10, construct a circuit which will produce a $1'$, $2'$, and $5'$ pulse.

30. What would be the effect on a JK flip-flop if the J or K input was pulled up to a voltage of 1.5 volts?

31. Write a program using VHDL that will implement a flip-flop that will toggle its outputs on each LOW-to-HIGH transition of the clock.

32. Write a program using VHDL that will implement a flip-flop with asynchronous inputs **preset** and **clear** that will toggle its outputs on each LOW-to-HIGH transition of the clock. Make **preset** and **clear** active LOW.

33. Write a program using VHDL that will implement a delayed-clock system. Name the input clock **clk** and the nonoverlapping outputs **cp1** and **cp2**.

34. To your delayed-clock system in problem 33, add an active LOW signal **preset** that will make **cp1** HIGH and **cp2** LOW, and an active low signal **clear** that will make **cp1** LOW and **cp2** HIGH.

35. Identify and correct four errors in this VHDL program.

```
Library ieee;
USE ieee.std_logic_1164.all;

ENTITY FF IS
      PORT (J,K,clk:IN STD_LOGIC;
            Q1,Q2:OUT STD_LOGIC);
END FF;

ARCHITECTURE OF FF IS
BEGIN
      PROCESS (J,K,clk,Q1,Q2)
      BEGIN
            IF (clk'EVENT AND clk = '0') THEN
                  IF J<='1' AND K<='1' THEN
                        Q1<=Q1;
                  ELSIF J<='1' AND K<='0' THEN
                        Q1<='1';
                  ELSIF J<='0' AND K<='1' THEN
                        Q1<='0';
                  ELSIF J<='0' AND K<='0' THEN
                        Q1<=Q2;
            END IF;
            Q2 <= NOT Q1;
      END PROCESS;
END a;
```

36. Identify and correct four errors in this VHDL program.

```
Library ieee;
USE ieee.std_logic_1164.all;
ENTITY _ff_pre_clr IS
      PORT (J,K,clk,pre,clr:IN STD_LOGIC;
            Q1,Q2:BUFFER STD_LOGIC);
END _ff_pre_clr;

ARCHITECTURE able OF _ff_pre_clr IS
BEGIN
      PROCESS (J,K,clk,Q1,Q2,pre,clr);
      BEGIN
            IF pre <='0' THEN      --turn on
                  Q1<='1';
                  Q2<='0';
            ELSIF clr <='0' THEN  --turn off
                  Q1<='0';
                  Q2<='1';
```

```
                ELSIF (PRE<='1' AND clr<='1' )THEN
                   IF (clk'EVENT AND clk = '1') THEN --positive clock edge
                          IF (J='0' AND K='0') THEN    --no change
                                Q1<=Q1;
                          ELSIF (J='1' AND K='0') THEN    --turn on
                                Q1<='1';
                          ELSIF (J='0' AND K='1') THEN    --turn off
                                Q1<='0';
                          ELSIF (J='1' AND K='1') THEN    --toggle
                                Q1<=Q2;
                          END IF;
                   END IF;
                END IF;
          END PROCESS;
        END able;
```

SHIFT COUNTER AND DELAYED CLOCK

OBJECTIVES

After completing this lab, you should be able to:

- Use a *JK* flip-flop to make a nonoverlapping clock.
- Use *JK* flip-flops to make a shift counter.
- Use simple gates to make various waveforms.

COMPONENTS NEEDED

2	7476 *JK* flip-flop ICs
1	7408 quad AND gate IC
1	10-kΩ, ¼-W resistor
1	7404 hex inverter IC
1	7432 quad OR gate IC
1	7410 triple three-input NAND gate IC
1	7427 triple three-input NOR gate IC

PROCEDURE

1. Use a *JK* flip-flop and two AND gates to generate a clock, *CP*, and delayed clock, *CP'*.

2. On a sheet of graph paper, plot the waveforms that you would expect to see for clock, *Q*, $\overline{Q}$, *CP*, and *CP'*.

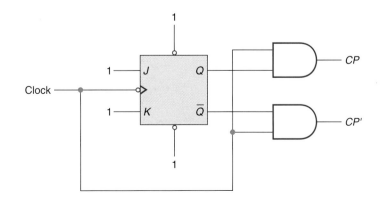

3. Use an oscilloscope to observe the actual waveforms of the following: clock, Q, $\overline{Q}$, CP, and CP'. Graph the actual waveforms.

4. Construct the shift counter shown and connect to it the delayed clock from number 1.

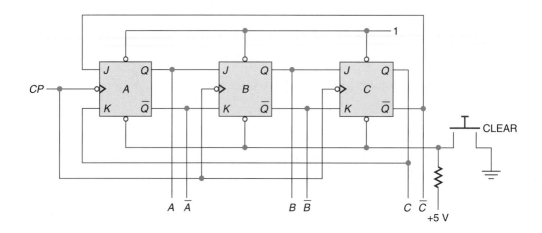

5. Plot the waveforms that you would expect to see for CP, CP', A, $\overline{A}$, B, $\overline{B}$, C, and $\overline{C}$.

6. Use an oscilloscope to observe the actual waveforms CP, CP', A, $\overline{A}$, B, $\overline{B}$, C, and $\overline{C}$. Graph the actual waveforms.

7. Connect the gates necessary to produce the following waveforms and observe them on the oscilloscope.

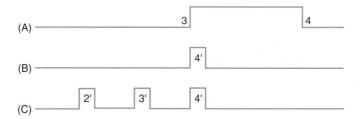

8. Predict the outputs of these gates.

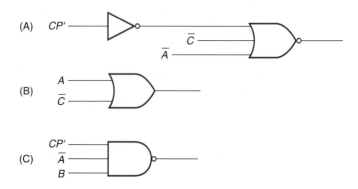

9. Connect the gates in number 8 and observe the actual outputs.

10. Make up a waveform not covered in this lab and create it.

If your circuit does not work properly, consider these points:

1. Check the power supply connection to each IC.

2. Check all the inputs and outputs of each gate for proper voltage levels.

3. The nonoverlapping clock:

    a. Make sure that you have tied the *J*, *K*, $\overline{CLEAR}$, and $\overline{PRESET}$ to +5 V. *JK* flip-flops do not always consider an unconnected input as a logic 1.

    b. Use your oscilloscope to trace the clock signal to the *JK* flip-flop's clock input pin and to the inputs of the two NAND gates.

    c. Use one channel of the oscilloscope to display the input clock and the other channel to display the *Q* output of the flip-flop. This will tell you if the flip-flop is toggling or not.

4. Shift counter:

    a. Check to see if you left a $\overline{PRESET}$ or $\overline{CLEAR}$ unconnected.

    b. Be sure you have cleared all the flip-flops to properly sequence them.

    c. Use the oscilloscope to trace the *CP* signal to each flip-flop's clock input pin.

LAB 8B

JK FLIP-FLOPS

OBJECTIVES

After completing this lab, you should be able to use Multisim to:

● Construct a delayed clock from NAND gates and a *JK* flip-flop.

● Troubleshoot a delayed clock and shift counter.

PROCEDURES

1. Using the *JK* flip-flop and NAND gates in Multisim, construct a delayed clock and use the clock generator and logic analyzer to drive the clock and show the three waveforms. Print out the circuit and the waveforms.

2. Open the file named 8B-1.ms9. This is a delayed clock with a 3-bit shift counter. The circuit has three faults. Find the faults and correct them.

3. Using the fixed delayed clock and shift counter in file 8B-1.ms9, add the gates shown below and print out the waveforms for each gate that the logic analyzer shows.

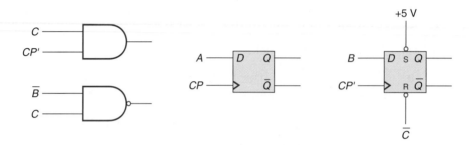

Shift Registers

OUTLINE

KEY TERMS

ASCII
asynchronous
baud rate
DB-25 connector
framing bits

RS-232C
synchronous
UART
universal serial bus (USB)

OBJECTIVES

After completing this chapter, you should be able to:

- Explain the operation of a shift register.
- Describe how to parallel load a shift register and serially shift the data out and serial-in, parallel-out.
- Describe typical serial digital data transmission methods.
- Describe typical IC shift registers.
- Describe the RS-232C standard.
- Describe the ASCII code.
- Understand how to use a programmable logic device to make a shift register.
- Troubleshoot the RS-232C.

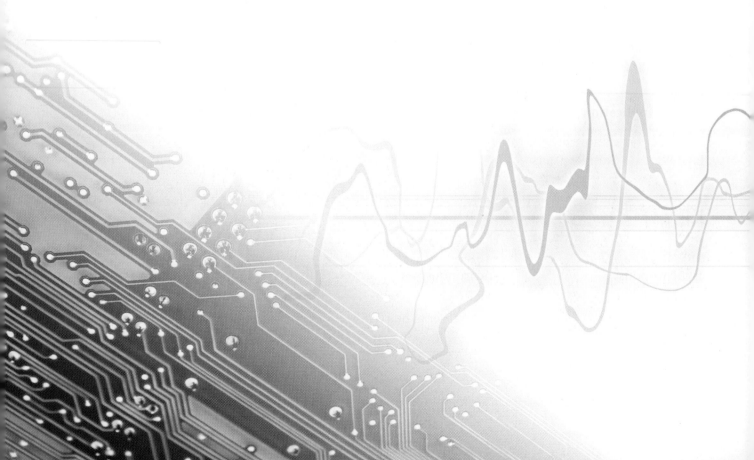

9.1 SHIFT REGISTER CONSTRUCTED FROM *JK* FLIP-FLOPS

Figure 9-1 shows a 4-bit serial-in, parallel-out shift register made from *JK* flip-flops. Notice the inverter between the *J* and *K* inputs of the *A* or first flip-flop. This means that the *J* and *K* inputs can never be the same. When the *A* flip-flop is clocked with the falling edge of the clock, the Q and $\overline{Q}$ outputs will take on the values of the *J* and *K* inputs. The Q and $\overline{Q}$ outputs of the *A* flip-flop are connected to the *JK* inputs of the *B* flip-flop. The *B* flip-flop is connected to the *C* flip-flop in the same way. This method of connecting the flip-flops would continue until you obtained the desired number of binary bits for your shift register.

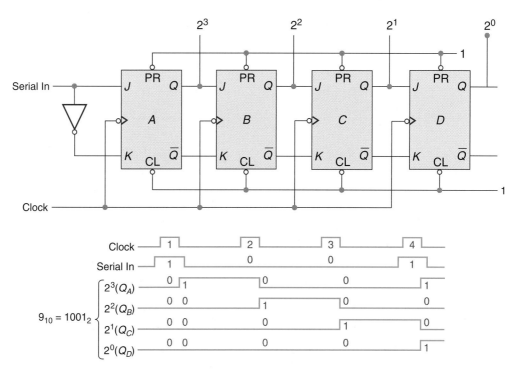

FIGURE 9-1 A 4-bit shift register

Because the clock inputs of all the flip-flops are tied together, the flip-flops will all change states at the same time, and their Q outputs will reflect the *J* input before the falling edge of the clock.

At the falling edge of the first clock pulse in Figure 9-1, the Q output of the *A* flip-flop (Q_A) is set to 1 because the serial-in was 1 before the falling edge of the clock. The Q output of the *B* flip-flop (Q_B) is still 0 after the falling edge of clock pulse 1 because Q_A was 0 before the falling edge of the first clock pulse. The outputs of the *C* and *D* flip-flops (Q_C and Q_D) are 0 for the same reason.

At the falling edge of the second clock pulse, Q_A changes to 0 because the serial-in input was 0 before the falling edge. Q_B now changes to 1 because Q_A was 1 before the falling edge of the second clock pulse.

On the third falling edge, the 1 on the output of the *B* flip-flop would be transferred to the output of the *C* flip-flop, and after the fourth clock pulse the register would be full. In Figure 9-1, the binary number equivalent to 9 was clocked into the shift register in serial form and was converted to parallel form after the fourth clock pulse.

EXAMPLE 9-1

What would be the waveforms for the outputs of the shift register in Figure 9-1 if the number 5 were shifted into the shift register?

Solution   Refer to Figure 9-2.

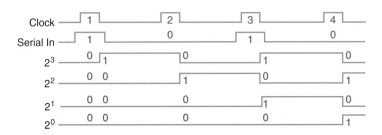

FIGURE 9-2

9.2 PARALLEL AND SERIAL DATA

Data in serial form is fed one bit at a time over only one line or wire at a rate which is constant and in phase with a clock reference. Parallel data, on the other hand, has one line or wire for each bit in the binary number or data word and does not have to be referenced to a clock to transfer it from one register to another.

The frequency of the reference clock or bits per second of the serial-in input is usually called the **baud rate** of the serial transfer. A typical baud rate for a teletype machine is 110 bits per second. At this rate you can send ten 11-bit binary numbers in one second.

If the same reference clock is used to clock a parallel register with one line for each bit of the 11-bit binary number, you can transfer one hundred ten 11-bit numbers in one second.

As can be seen, parallel data transmission is much faster but takes many more lines or wires than serial transmission. When digital data is transferred over long distances or is placed on a long magnetic tape for storage, serial methods are used because it only takes one line or wire to do the job.

Computers transfer data in 8-bit binary numbers called bytes. These bytes are moved to and from memory by the CPU (central processing unit) in parallel form. The wires or electrical connections which do the parallel transfer are called a bus and one computer may have several of these buses to move data. The parallel form is used in a computer because it is much faster and the distance of the transfer is small. However, a serial format is used to move data to and from the terminal, which has the keyboard and the display monitor. This is because the terminal may be some distance from the computer, and the rate of data transfer does not have to be great. A parallel-to-serial shift register is used to convert from the parallel format to the serial format when going from the computer to the terminal, and a serial-to-parallel shift register is used to receive data from the terminal to the computer.

 # 9.3 PARALLEL-IN SERIAL-OUT

Figure 9-3 shows a parallel-in serial-out shift register and the waveform for loading the binary number equivalent to 9 and shifting it out to the right. The A flip-flop has a 0 on the J input and a 1 on the K input. This will cause the Q output of the A flip-flop to go to 0 after the falling edge of the clock input. If the shift register was clocked four times, the Q outputs of all the flip-flops would be 0 and stay 0 until the shift register was broadside loaded or parallel loaded with a new binary number.

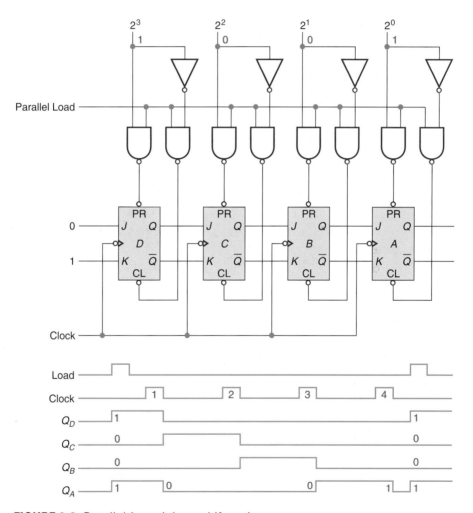

FIGURE 9-3 Parallel-in serial-out shift register

This is done by placing a binary number on the parallel inputs and raising the parallel load input or control to 1. This enables the NAND gates which feed the $\overline{PRESET}$ and $\overline{CLEAR}$ inputs of each flip-flop. Because of the inverter between the two NAND gate inputs, one NAND gate output will be 1 while the other will be 0. They can never be the same value while the NAND gates are enabled. This will cause the Q output of the flip-flop to be set or reset depending on the value of the parallel input to the NAND gates feeding the flip-flop.

Because $\overline{\text{CLEAR}}$ and $\overline{\text{PRESET}}$ take precedence over all other inputs of the *JK* flip-flop, the *Q* outputs of the flip-flops will not change as long as the parallel load input is 1, since this enables the NAND gates. When the parallel load input falls to 0, the NAND gates are inhibited, and their outputs go to 1 because any 0 into a NAND gate produces a 1 on the output.

The shift register is now loaded with the binary number desired. In the case of Figure 9-3, the binary number is equivalent to 9. The $\overline{\text{PRESET}}$ and $\overline{\text{CLEAR}}$ are 1, which means when the falling edge of the clock occurs, the shift register will shift each bit right one place, shift in a 0 on the left, and shift out a 1 on the right. After four clock pulses, the number will be shifted out to the right, and the shift register will be empty or 0 and ready for a new number.

The *Q* output of the *D* flip-flop could be connected to a serial-in parallel-out shift register which was clocked by the same clock, and the binary number equivalent to 9 would be transferred to the other shift register in four clock pulses. This idea is represented in Figure 9-4.

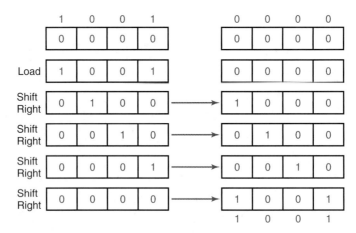

FIGURE 9-4 Serial data transfer

What would be the waveforms in Figure 9-3 if the binary number 1010 (decimal 10) were parallel loaded into the shift register?

Solution   Refer to Figure 9-5.

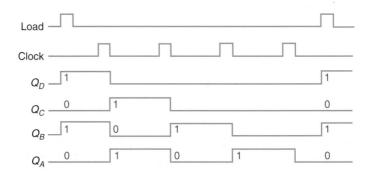

FIGURE 9-5

9.4 SERIAL DATA TRANSMISSION FORMATS

There are several standardized serial data transmission formats. Two of these are the RS-232 serial interface used in computers and the 20 mA current loop used on older teletypes. Figure 9-6 shows a typical word format for these serial interfaces. This is called asynchronous serial transfer.

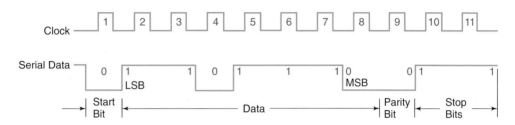

FIGURE 9-6 Asynchronous serial format

Each word is started with a LOW-going start bit which starts the shift register clocking in the data. The next seven bits are data, usually in the form of ASCII code for letters of the alphabet. These are followed by a parity bit and two stop bits, which are 1s. Notice that it takes eleven clock pulses to shift in one word. A typical teletype runs at a 110 baud rate. This means that if it used the format in Figure 9-6, it could send or receive ten words or characters per second. This is quite slow compared to today's computer, which runs at 9600 baud rate or higher.

EXAMPLE 9-3

What would the waveform be in Figure 9-6 if the binary number sent were 43 Hex?

Solution The binary number for 43 Hex is 1000011. The binary number in Figure 9-5 is 111011. This number has an odd number of 1s, and the parity bit is 0. This means that the parity system being used is odd. The number 43 Hex is also an odd parity, so the parity bit to be sent will be a 0. By adding a 0 for a start bit and two 1s for stop bits we get 01100001011, as shown in Figure 9-7.

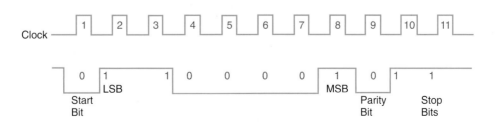

FIGURE 9-7

The start bit and the stop bits are called **framing bits** and are used to start and stop the serial shift register which is receiving the data. This type of serial data transfer is called **asynchronous** serial data transfer because the data comes in neat little packages and at any time interval between these packages. **Synchronous** serial data does not use framing bits, and the data must come one word after the other at a constant rate. Synchronous serial data transfer is usually done in blocks of binary numbers or words. A good example of synchronous serial data transfer is the way most disk drives store data. The data is stored on the disk in sectors. These sectors usually comprise 256, 512, or more bytes. A *byte* is an 8-bit binary number.

To store one byte on a disk using asynchronous serial methods would take a minimum of ten bits: one start bit, eight data bits, and one stop bit. To store one byte on a disk using the synchronous methods would take only eight bits for the actual data. Therefore, the synchronous method would let us store 20 percent more data on the disk than the asynchronous method. This is good for disk storage, but for serial data transfer over long distances, synchronous methods are not easy to use. The clock must be carried with the data stream, in some method, to run the serial shift register which converts the data from serial form to parallel form. Because the clock is started with the start bit and stopped with the stop bit, in asynchronous serial data transfer, the clock can be derived from a very accurate crystal clock at the receiver end of the serial transfer and does not need to be carried with the data stream.

Figure 9-8 shows a serial input logic system designed to input serial data in the format of Figure 9-6. This is a typical system which would input serial data and latch it out into a set

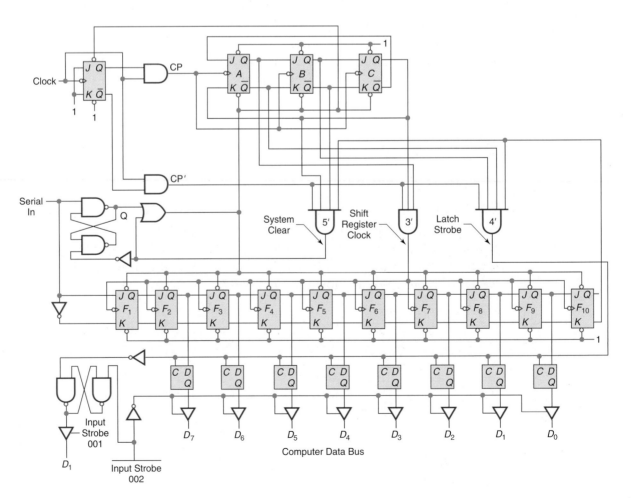

FIGURE 9-8 Asynchronous serial receiver

of D flip-flops for a microprocessor to input into its memory. Today, such logic circuits are not usually constructed from basic gates and flip-flops because they are put on LSI circuits as one integrated circuit. The LSI circuits are called **UART** ICs and include a parallel-to-serial transmitter logic system and all the necessary logic to check the parity, stop bits, and framing of the incoming data. Most of the LSI UART ICs are programmable by the microprocessor they are feeding. The word UART is an acronym for universal asynchronous receiver transmitter. Figure 9-8 is an example of how the receiver of a typical UART might be constructed.

Figure 9-9 shows the waveforms for the eleventh clock pulse of the shift register clock. The shift register clock is obtained by decoding the $3'$ pulse from the delayed clock and shift counter. This means the clock which is driving the delayed clock must be twelve times the baud rate of the incoming serial data.

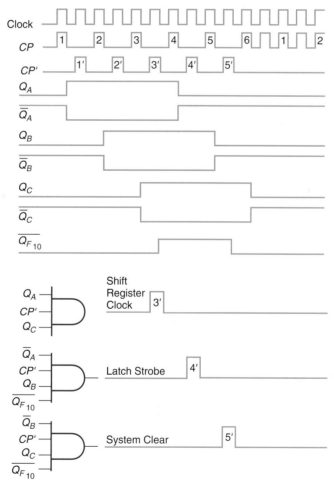

FIGURE 9-9 The last clock cycle of the asynchronous serial receiver shown in Figure 9-8

When there is no data being sent to the serial input, the Q output of the crossed NAND gates is 0. This causes the JK flip-flops of the shift counter to be cleared, and their Q outputs will be

0 as long as the Q of the crossed NAND gates is 0. Also, the Q outputs of the shift register will be 1 because the Q of the crossed NAND gates is tied through an OR gate to the $\overline{\text{PRESET}}$s of the ten JK flip-flops of the shift register.

This condition will exist until the serial input goes LOW. This causes the crossed NAND gates to change states producing a 1 on the Q output of the crossed NAND gates. Three clock pulses of CP' later, the shift register will be clocked. This makes F_1 go LOW and shifts a 1 out of the end of the shift register.

The shifting of the shift register will continue for ten 3′ clocks, after which the output $\overline{Q}_{F1}$ will go to a 1, enabling the 4′ and 5′ AND gates. When the 4′ pulse goes HIGH, it latches the value of the eight bits of data into the D latches. When the 5′ pulse goes HIGH, the shift counter is cleared and the shift register is preset because the crossed NAND gates are reset by the 5′ pulse. The receiver is now ready to shift in another serial word when another HIGH-to-LOW transition of the serial input occurs. When the 4′ pulse goes HIGH, latching the data into the D flip-flops, it also sets a crossed NAND $\overline{\text{SET}}$-$\overline{\text{RESET}}$ flip-flop, which the computer will scan to see if any data has been received. This serial receiver could run with one stop bit; but two stop bits ensure that the serial input remains 1 during the last clock, even if the shift-counter clock and data rate are not quite synchronous.

SELF-CHECK 1

1. What would be the waveforms for the shift register outputs in Figure 9-1 if the binary number 1010 were shifted into the shift register?

2. Draw a parallel-in serial-out shift register that could shift out a 7-bit number. Use JK flip-flops.

3. Draw the waveform for a binary number of 11 Hex using the serial format in Figure 9-6.

9.5 IC SHIFT REGISTERS

Figure 9-10 shows a logic diagram of a 7495 4-bit shift register. This shift register can be wired to shift right or left, parallel-in, serial-out or serial-in, parallel-out. Each of the JK inputs on the four flip-flops is controlled by a set of two AND gates fed into a NOR gate. The AND gates are enabled or inhibited by the mode control (pin 6). When the mode control goes HIGH, the AND gates which have the A, B, C, and D parallel inputs are enabled and the outputs are inhibited. This means when the mode is HIGH, the value of the JK inputs of the flip-flops is controlled by the $ABCD$ parallel inputs. This is a parallel load or broadside load.

When the mode control is made LOW, the parallel inputs are inhibited and the serial inputs to the JK flip-flops are enabled. Notice the A flip-flop is fed from the serial input (pin 1) and the B flip-flop is fed from Q_A, and so on. In this configuration, when clock 1 is brought LOW,

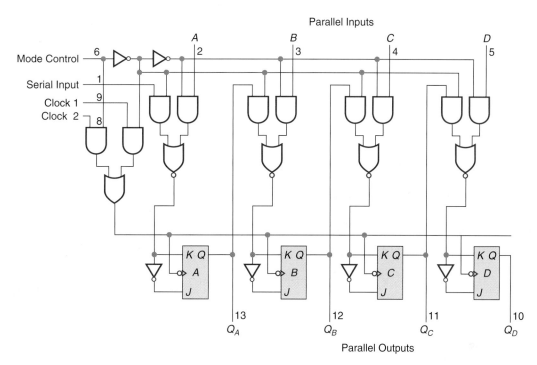

FIGURE 9-10 A logic drawing of the 7495 4-bit shift register

the flip-flops will shift right one place. When the mode control is HIGH, the shift register can be broadside loaded, and when the mode control is LOW the shift register can be shifted right.

To get the 7495 to shift left, you must wire Q_D back to the C input, Q_C back to the B input, and Q_B back to the A input. Then place the mode control HIGH to enable the $ABCD$ inputs.

The 7495 shift register is only four bits wide but is quite versatile. Figure 9-11 shows four 8-bit shift registers, which are not as versatile but have the advantage of shifting a whole byte in one IC.

The 74LS164 is an 8-bit serial-in, parallel-out shift register with an active LOW CLEAR. The counterpart to this IC is the 74LS165, which is a parallel-in, serial-out IC with eight bits also. The 8-bit IC which most resembles the old 7495 is the 74LS166. As you can see, it has the same logic configuration as the 7495, but the parallel outputs are missing and not brought out of the IC.

The 74198 is an 8-bit shift register IC which can shift right, left, serial-in parallel-out, parallel-in serial-out, and clear the register. This register is similar to the 7495 except that it does not need to have the outputs wired back to the inputs to shift left. The direction of the shift and the parallel load are controlled by the inputs called S_1 and S_0. When S_1 and S_0 are both HIGH, the shift register will be loaded from the parallel inputs when the clock goes from a LOW to a HIGH state. This is called a broadside or parallel load of the shift register. When S_1 is LOW and S_0 is HIGH, the shift register will shift Q_A to Q_B and the rest of the outputs to the left when the clock makes a LOW-to-HIGH transition. The shift register will shift right when S_1 and S_0 are both HIGH; and when S_1 and S_0 are both LOW, the shift register will not shift in either direction when the clock makes its transition. The CLEAR input overrides all the other control inputs and brings all the outputs to a 0 value.

This shift register is very flexible and has the added feature of being eight bits wide.

It is recommended that the student refer to a good IC manufacturer's specification manual for other shift register ICs.

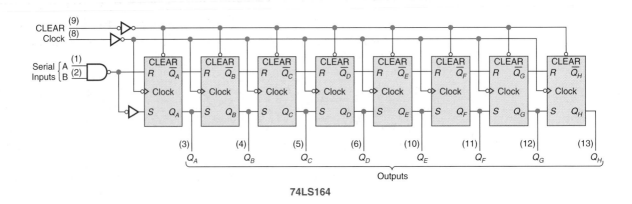

74LS164

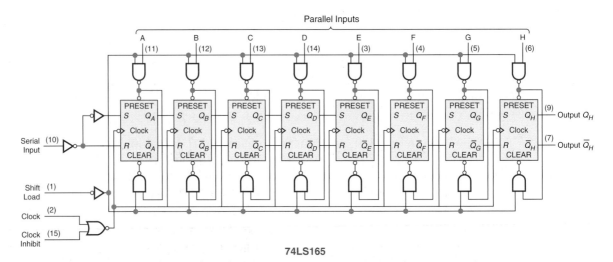

74LS165

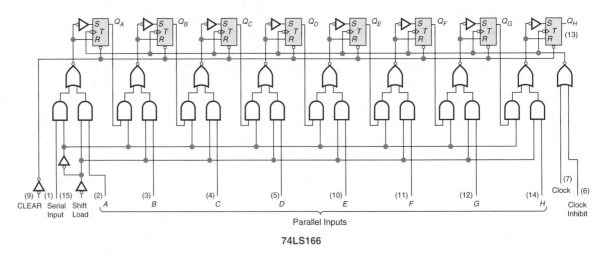

74LS166

FIGURE 9-11 Common shift registers

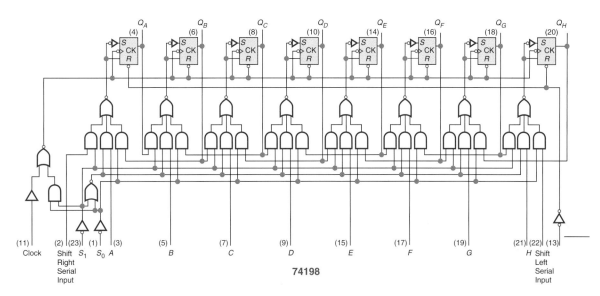

FIGURE 9-11 (*continued*)

9.6 SERIAL DATA STANDARDS

There are several standards published by several different standards organizations which define the logic levels and protocol to be used for serial data communication. The one which has become the most widely used in the computer world is the **RS-232C** standard published by the Electronics Industry Association (EIA). Late in the 1960s it became necessary to standardize the serial data transfer used by modems. The RS-232C standard was first published in 1969. This standard defines the voltage levels for a logic 1 and 0, as well as many other things needed to make one piece of computer equipment communicate with another. The RS stands for Recommended Standard, the 232 is the standard identification number, and the letter C means that the standard has been revised three times.

The RS-232C standard is not as complete as it could be in defining all the parameters needed for proper serial communication. For instance, the RS-232C standard does not specify the use of a standard connector. This means that a piece of equipment can be fully RS-232C compatible but have a connector that will not fit any other piece of RS-232C equipment. This has caused the EIA organization to release several newer serial data standards such as the RS-422 and the RS-249. These standards are used, but not to the extent of RS-232C.

While there is no standard connector defined in the RS-232C, there have been two connectors which have become the de facto standard. These connectors are the DB-25 and the DB-9. Figure 9-12 shows the pin definition used for each of these connectors. The DB-25 connector was used long before the DB-9 came into use. When IBM developed their first personal computer, they used the DB-9 connector for the RS-232C port. The reason for this was that the standard 34-pin connector used for the Centronics printer port was too wide to fit in the I/O slots, so they used the DB-25 connector for the printer port. The DB-25 connector was thinner and would fit on the back of the printer card. To prevent the computer from having two of the same type of connector with different functions, they used the DB-9 for the RS-232C port.

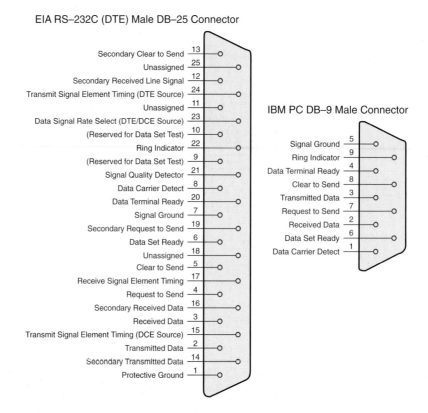

FIGURE 9-12 De facto RS-232C pinouts

The old IBM PC was so successful that many devices that require an RS-232C interface—including most personal computers—now use the DB-9 connector.

The RS-232C standard divides serial equipment into two types, called DCE and DTE. The DCE uses the female connector, and the DTE uses the male connector. Originally, DCE equipment was data-connecting equipment, such as modems, and DTE equipment was data-terminating equipment. Today, however, the distinction between the two is not that clear and is based solely on the type of connector used with the equipment.

The voltage levels used for defining a logic 1 and a logic 0 are shown in Figure 9-13. Notice that the voltage level for a 1 is lower than the voltage level for a 0. This makes the RS-232C standard a negative logic system. Also, the 1 voltage is a negative voltage while the 0 voltage is a positive voltage. This means that the current traveling across the serial conductor is an ac voltage. The noise margin is two volts, and the voltage spread for a valid 1 or 0 is 10 volts. Converting this negative logic system to a TTL logic system can be done with some simple transistors and resistors, but it is most often done with two integrated circuits called RS-232C interface drivers. Figure 9-14 shows these two chips and how they would be used to interface to TTL logic.

There are 25 pins on the DB-25 connector, but most of the time only eight of them are used and quite often only three are used. Pin 1 is the equipment ground and is connected to the chassis ground. This is sometimes used as the signal ground, but it is not a good practice. Pins 2 and 3 are the serial data transmitter and receiver pins; which one is which depends on the type of connector, DCE (female) or DTE (male). Pin 7 is the signal ground and must be connected to the circuit ground of the serial transmitter or receiver. Only these three pins (2, 3, 7) are needed to establish serial communication if the serial equipment uses software

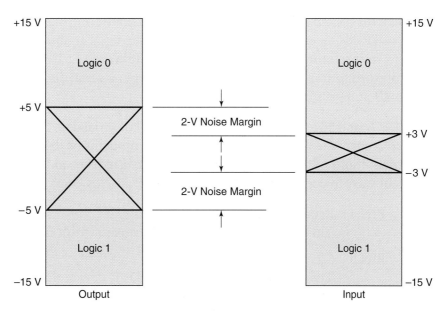

FIGURE 9-13 RS-232C minimum and maximum logic voltages

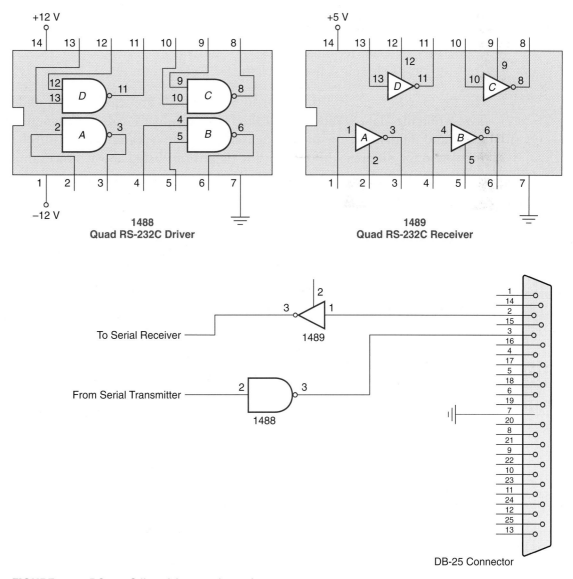

FIGURE 9-14 RS-232C line driver and receiver

protocol for controlling the serial data flow. Software control of the data flow is called X-ON and X-OFF. X-ON is a data word (13 Hex) that the serial equipment will recognize as a control code to start sending serial data. X-OFF is a data word (11 Hex) that will be recognized as a control word to stop sending serial data. This type of serial data control is quite often used by terminals and printers.

Pins 4, 5, 6, and 20 are used for hardware control of serial data flow. These pins were originally used to control the data flow from terminals to modems or data sets, as they were called. The exact definitions of these four control signals are listed, but today they are often used in much different ways to control the serial data flow.

PIN 4   Request to Send
This pin is used to request to transmit from the DTE device to DCE device.

PIN 5   Clear to Send
This pin is used to signal the DTE device that the DCE device is ready to receive serial data.

PIN 6   Data Set Ready
This pin is used to signal the DTE device that the DCE device is on and connected to the communication channel, usually the phone line.

PIN 20   Data Terminal Ready
This pin is used to tell the DCE device that the DTE device is ready to operate.

The Request to Send and the Clear to Send signals are used to start and stop serial data flow, while the Data Set Ready and the Data Terminal Ready signals indicate that the terminal and modem are on and configured to communicate. These control signals are quite often used for different things than the definitions listed above. For example, many printers use the Data Terminal Ready as the Busy indicator to stop serial data flow to the printer from the computer. Because of the many ways these signals are used, a technician may have some trouble constructing a serial cable for many serial devices if he does not have the wiring diagram for the cable. A good device for the technician to use when debugging a serial device is an RS-232C break-out box. This device plugs in-line with the serial cable and allows the technician to make the connections between the two devices that are communicating over the serial cable. He can also use a oscilloscope to monitor the serial signal as it is transmitted.

 # 9.7 THE ASCII CODE

ASCII is an acronym for American Standard Code for Information Interchange. This code is used to represent printable letters, numbers, some punctuation marks, and control codes. This 7-bit code is supported by the American National Standards Institute (ANSI) and has become the international de facto standard for such codes. It is not the only code for this purpose, but it is the most widely used by far. One other code which saw some use was IBM's EBCDIC. EBCDIC was used on IBM's main frames and terminals, but even IBM has now started using ASCII.

The ASCII code is broken into two parts: control codes and printable characters. The printable characters are upper and lower case letters (A to Z) and some punctuation marks such as ? < > @ # $ % ^ & * ( ) _ [ ] { } + = " : ; ' ~ \ /. The control codes are used to control the terminal, printer, or whatever device is being sent serial data. If you send a 13 Hex to an ASCII terminal, the cursor on the screen will move back to the left side of the screen, and if you send a 0A Hex, the cursor will move down one line. With the control codes, the operator can control display of the characters and many of the operational characteristics of the terminal. Figure 9-15 shows the ASCII code. You can use this table to determine the binary number for each letter and control code.

Least significant bits (3, 2, 1, 0)	Most significant bits (6, 5, 4)							
	000	001	010	011	100	101	110	111
0000	NUL	DLE	SP	0	@	P	`	p
0001	SOH	DC1	!	1	A	Q	a	q
0010	STX	DC2	"	2	B	R	b	r
0011	ETX	DC3	#	3	C	S	c	s
0100	EOT	DC4	$	4	D	T	d	t
0101	ENQ	NAK	%	5	E	U	e	u
0110	ACK	SYN	&	6	F	V	f	v
0111	BEL	ETB	'	7	G	W	g	w
1000	BS	CAN	(	8	H	X	h	x
1001	HT	EM	)	9	I	Y	i	y
1010	LF	SUB	*	:	J	Z	j	z
1011	VT	ESC	+	;	K	[	k	{
1100	FF	FS	,	<	L	\	l	|
1101	CR	GS	-	=	M	]	m	}
1110	SO	RS	.	>	N	^	n	~
1111	SI	US	/	?	O	—	o	DEL

Control characters			Control characters	
NUL	Null		DC1	Device Control 1
SOH	Start of Heading		DC2	Device Control 2
STX	Start of Text		DC3	Device Control 3
ETX	End of Text		DC4	Device Control 4
EOT	End of Transmission		NAK	Negative Acknowledge
ENQ	Enquiry		SYN	Synchronous Idle
ACK	Acknowledge			End of Transmission
BEL	Bell		ETB	Block
BS	Backspace		CAN	Cancel
HT	Horizontal Tabulation		EM	End of Medium
LF	Line Feed		SUB	Substitute
VT	Vertical Tabulation		ESC	Escape
FF	Form Feed		FS	File Separator
CR	Carriage Return		GS	Group Separator
SO	Shift Out		RS	Record Separator
SI	Shift In		US	Unit Separator
DLE	Data Link Escape		DEL	Delete

FIGURE 9-15 ASCII code

EXAMPLE 9-4

What is the ASCII code for the letter C?

Solution　First find the capital letter C in the table of Figure 9-15. The most significant bits of the binary code are at the top of the column in which you find the letter C (100). These are bits 6, 5, 4 of the binary number. Next, get the least significant four bits of the binary code from the row where you find the letter C (0011). These are bits 3, 2, 1, 0 of the binary code. Put the two parts of the binary code together and you will have the 7-bit binary code for the letter C (1000011). This is 43 Hex or 67 decimal.

Notice that all but one of the control codes are in the first two columns of the ASCII table. This means that except for DEL (delete), all the control codes are in the first 0 to 31 binary numbers. A standard ASCII keyboard (such as those used on most computers) can send the first 26 of these codes by using the control key and then pressing a letter. The control code LF (line feed) can be sent by using the control key and the J key. The ESC (escape) key is code 27 decimal and is one code past the control Z key combination. This key (ESC) is used by most ANSI terminals to send a string of control commands for formatting the display of text on the terminal. A thorough knowledge of the ASCII code is very useful when communicating to a standard ANSI terminal.

SELF-CHECK 2

1. What does the RS stand for in the RS-232C standard?

2. What would be the waveform for the number 41 Hex using the RS-232C standard? No parity bit and two stop bits.

3. What would be the waveform for the ASCII capital letter C if it were sent using the RS-232C serial standard? Use odd parity bit and two stop bits.

9.8 PROGRAMMING A CPLD (Optional)

In this section we will write programs in VHDL to program a CPLD to implement an 8-bit serial-in, parallel-out, shift-right shift register and an 8-bit parallel-in, serial-out, shift-right shift register.

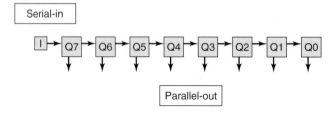

FIGURE 9-16 Serial-in, parallel-out shift register

Follow the procedure presented in previous chapters:

- Create a new folder within the Altera folder and name it **s_in_p_out_sr**.

- Create a new .vhd file and name it **s_in_p_out_sr.vhd**.

- Write a VHDL program that describes the operation of an 8-bit serial-in, parallel-out shift register. Refer to Figure 9-16. The level that is present on the data input should be shifted into Q7 on the falling edge of the clock and then through the shift register to Q0. So the least significant bit of the data should appear first on the data input.

  The usual ieee LIBRARY statements are not invoked because no special operations are used and the signals are declared as bit types.

  Declare the **D** and **clk** as input bits.

      D,clk:IN BIT

  Declare **Q** as an 8-bit vector, Q7–Q0.

      Q:OUT BIT_VECTOR(7 DOWNTO 0)

  Within a PROCESS, each output signal is evaluated, but none of the values actually change until the end of the process. Therefore, the shifting process is easy to describe. Q(7) becomes what was on D; Q(6) becomes what was on Q(7); Q(5) becomes what was on Q(6); and so on down to Q(0). These statements do not even need to be in descending order. Any order will do.

      Q(7)<=D;

      Q(6)<=Q(7);

      Q(5)<=Q(6);

- Compile the program.

- Create a .vwf (vector waveform file) named **s_in_p_out_sr.vwf** and use it to simulate the circuits. Sample simulation waveforms are shown in Figure 9-17. Study each HIGH-to-LOW transition of **clk**. The data at each clock edge is in this sequence: 1 1 0 0 1 0 0 0. After 8 clocks this data has been shifted into the flip-flops. The data can be read at time 800 ns from bottom to top, Q(0) to Q(7): 11001000. The byte sent—from most significant bit to least significant bit—is 00010011.

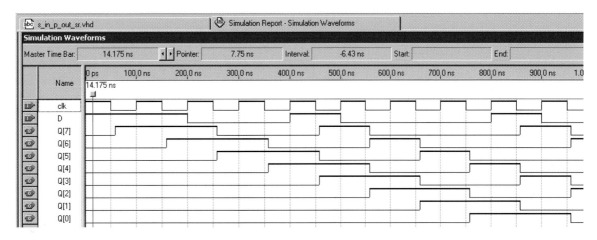

FIGURE 9-17 Simulation waveforms for s_in_p_out_sr .vwf

- Assign pin numbers to each of the input and output signals. Use switches S5, S4, S3 to simulate Data and clock. Use LEDs 1 through 7 to display the states of Q7 to Q0. Recompile your program.

- Program your CPLD.

- Test the operation of the CPLD using input switches and LEDs. Watch the data created shift across the LEDs.

s_in_p_out_sr.vhd looks like this:

```
ENTITY s_in_p_out_sr IS
    PORT(D,clk:IN BIT;
         Q:BUFFER BIT_VECTOR(7 DOWNTO 0));
END s_in_p_out_sr;

ARCHITECTURE a OF s_in_p_out_sr IS
BEGIN
    PROCESS (clk)
    BEGIN
        IF (clk'event AND clk = '0') THEN
            Q(7) <= D;
            Q(6) <= Q(7);
            Q(5) <= Q(6);
            Q(4) <= Q(5);
            Q(3) <= Q(4);
            Q(2) <= Q(3);
            Q(1) <= Q(2);
            Q(0) <= Q(1);
        END IF;
    END PROCESS;
END a;
```

Now create a project that implements a parallel-in, serial-out, shift-right shift register. Refer to Figure 9-18. Name the folder, project, and vhd and vwf files **p_in_s_out_sr**.

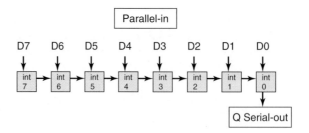

FIGURE 9-18 Parallel-in, serial-out, shift-right shift register

We will use an input named **load** as a signal to broadside-load 8 bits into the shift register. There is only one output bit, **Q**. We need to keep track of the shifting in the shift register, but those latches do not show up as outputs. Therefore, they are not declared in the ENTITY statement. Instead, we will declare an 8-bit bit_vector signal (called **internal**) inside the ARCHITECTURE statement. Data will be shifted within the CPLD from **internal(7)** down to **internal(0)**. Just before ending the ARCHITECTURE statement, output **Q** will be equated to **internal(0)**.

Data will shift on the leading edge of **clk**, IF (clk'event AND clk = '1') THEN. As data is shifted to the right, zeroes are shifted into **internal(7)**. Nested IF THEN statements are needed. The first set checks **load**, and the second set checks for the leading edge of **clk**.

```
ENTITY p_in_s_out_sr IS
    PORT(
        D:IN BIT_VECTOR (7 DOWNTO 0);
        Q:OUT BIT;
        clk,load:IN BIT);
END p_in_s_out_sr;


ARCHITECTURE a OF p_in_s_out_sr IS
    SIGNAL internal:BIT_VECTOR (7 DOWNTO 0);
BEGIN
    PROCESS (clk,load,D)
    BEGIN
        IF load = '1' THEN      --broadside load
            internal<=D;
        ELSIF load = '0' THEN
            IF (clk'event AND clk='1') THEN
                --low-to-high transition
            internal(7)<= '0';      --shift 0 into int(7)
            internal(6)<= internal(7);
            internal(5)<= internal(6);
            internal(4)<= internal(5);
            internal(3)<= internal(4);
            internal(2)<= internal(3);
            internal(1)<= internal(2);
            internal(0)<= internal(1);
            END IF;
        END IF;
    END PROCESS;
    Q <= internal(0);  --define Q as lsb of internal
END a;
```

Note these features in the program:

- When checking the value of "load" in the "IF" statement, the "=" sign is used (IF load = '1' THEN).

- When assigning a relationship in the following line, "<=" is used (internal <= D).

Sample simulation waveforms are shown in Figure 9-19. At time 0, the data on the 8 Data inputs is AB_{16} (10101011_2). Load is HIGH, so AB_{16} is loaded into the 8-bit register named **internal**. **Load** goes LOW at 25 ns. Then, at each LOW-to-HIGH transition of **clk**, one bit of data is shifted into **Q**. **Q** should contain AB_{16}, beginning at the least significant bit. In order to confirm proper operation, find the value of Q at each HIGH-to-LOW transition of the clock. When **load** goes HIGH again at 625 ns, **D** has changed to 57_{16}. 01010111_2 is loaded into **internal**.

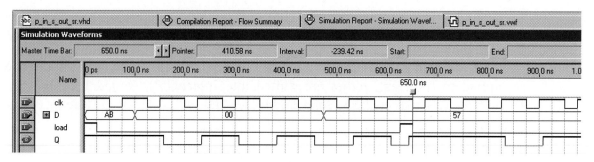

FIGURE 9-19 Simulation waveforms for p_in_s_out_sr.vwf

9.9 TROUBLESHOOTING AN RS-232C SYSTEM

Many devices communicate to computers and general purpose terminals with the RS-232C serial communication standard. One of the problems with this standard is that no standard connector was ever set. Therefore, many devices will have their own type of connector and special cable to connect them to a computer or terminal. If the cable is lost and no documentation exists to tell the technician what the pinout of the connector is and how to make a new cable, the device is often rendered useless. This places a burden on the technician to understand how an RS-232C device works; he has to use common sense and be willing to take extra care in solving the problem.

Such a problem presented itself to our technician one afternoon. It seems that the CAD (computer aided design) department had placed a plotter in storage some time ago and, when they decided to get it out and use it, the serial cable which came with it was nowhere to be found. They called on the technician to make a new one; it seems the old one would have been too short anyway. The technician thought about it and decided all he had to do was find the documentation on this plotter, get a few connectors, and solder up a new, longer cable. But to his dismay, no documentation could be found. This did not deter the technician. He reasoned that if he got his oscilloscope and took a look at the voltages on the connector, he should be able to identify transmit, receive, and ground.

When the technician got to the plotter and looked at the connector, he noticed that it had only three terminals and was just a simple 3-post connector. He had several of these connectors in his toolbox. Using his soldering iron, he quickly constructed a connector to fit the plotter's RS-232C output. He made the wires about two feet long and plugged them into a small protoboard which he always carried in his toolbox. He then got out his IBM compatible serial cable. This cable had a DB-9 connector on one end to plug into the host computer and wires on the other end which the technician had labeled. He had the receive, transmit, ground, clear to send, request to send, and all the rest of the connector pins clearly labeled. He put the wire ends of this cable on the protoboard opposite the three wires from the plotter serial port.

The technician got his oscilloscope out and looked on the plotter's circuit board for a place to put the ground of the oscilloscope. Without any documentation to help him, he started by checking the part numbers of the ICs on the board. He soon found a 74LS08, which he knew had ground on pin 7. He could also see that this pin was connected to a ¼-watt resistor which he could easily connect the oscilloscope ground to. Now he had good signal ground to measure the voltages with. He found that pin 1 had no voltage, pin 2 had −12 volts, and pin 3 had no voltage. This meant that pin 2 must be the transmit line coming from the plotter because the idle state of a standard RS-232C transmit line is one which is −5 to −15 volts. The other two pins were ground and receive, but which was which? To answer this he turned off the plotter and got out his volt-ohm meter to measure the resistance between pin 1 and the signal ground which the oscilloscope was attached to. This measured .01 Ω and pin 3 measured 1 MΩ. This meant that pin 1 was signal ground. He knew pin 2 was the transmit, so pin 3 must be the receive line. Because the connector only had three pins on the plotter, he assumed correctly that the plotter must use X-ON X-OFF codes to control the flow of serial data. He felt lucky at this because he knew that many serial devices use "Request to Send" and "Clear to Send" or "Data Terminal Ready" to control the serial data flow by hardware handshaking. If this were the case, it would have been much harder to figure out the plotter's serial port connector.

Now that he had the pinout of the plotter connector, he used some jumper wires to connect the plotter's ground signal and transmit line to the ground and receive line of the computer, and the computer transmit line to the receive line of the plotter. The technician turned on the computer and the plotter and proceeded to send a test plot to the plotter. The plotter started and then stopped; it did not plot even the first line. This was not what should have happened at all, so the technician hooked up the oscilloscope to the plotter transmit line and placed it on storage. He then repeated the test. This time he could see serial data coming from the plotter. Quickly he pressed the hold button on the oscilloscope and saw the waveform of an asynchronous serial data transmission. He could not make much sense of the waveform, so he tried latching several more and soon got a waveform—he could see the start bit, data, and the stop bits. This is shown in Figure 9-20. Looking at the sweep frequency he had used to get one frame of data, he concluded that the baud rate must be 9600. To verify this he set the measurement cursors for the oscilloscope on 960 Hz, and they fit across the serial frame exactly. He knew that it should be correct because there are ten bits to one serial frame (one start bit, eight data bits and one stop bit). He then went to the host computer and set the baud rate for serial transmission to 9600, and the plotter came to life and plotted the test. With all the necessary information at hand to make the cable, the technician soldered up a cable and put covers on both connectors.

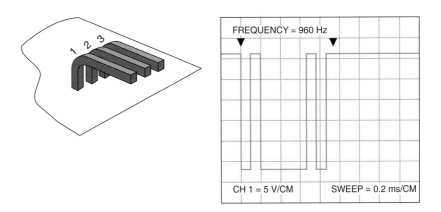

FIGURE 9-20 RS-232C connector and waveform for a 9600-baud asynchronous signal

9.10 THE USB PORT

USB is an acronym for **Universal Serial Bus**. This is a new standard, which is a synchronous high-speed serial port used in today's computers. This standard does not have a start and stop bit for each byte of information sent, as does the RS-232C standard. Instead, it uses a packet of many bytes which have a start and end signal in combination with synchronous bytes that define the packet. There are three speeds at which the USB Rev. 2 standard can run: 1.5 MB/sec., 12 MB/sec., and 480 MB/sec. As you can see, this is much faster than the older RS-232C system. The problem of nonstandard connectors has also been addressed in this standard. There are two connectors used, which are not the same: one for upstream use, and one for downstream use. This prevents those devices that use the serial system from being connected up wrong. The two connectors are shown in Figure 9-21. The cable contains four wires: two to carry the differential serial signal (D+ and D−), and two for power and ground. This gives the system the ability to supply power to the device that is connected to it.

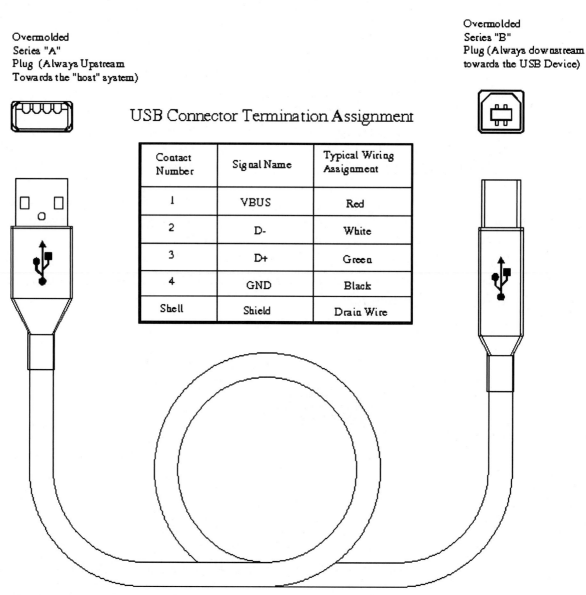

FIGURE 9-21 USB connectors

The serial signal is carried by two conductors that switch the polarity of the current flowing through them. These conductors are called a differential pair and are much better at noise reduction at high speeds. The current flow is typically 17 mA, and the voltage drop across the two data lines is typically 400 mV when a device is plugged into the host computer. When the device is unplugged, this voltage will change and the system can then detect the disconnected device. This means that devices can be placed online and removed at will. The system has a complete standardized hierarchy of programming that allows it to communicate with more than one device on the same pair of conductors. The system is a half-duplex system, using the two differential conductors to both send and receive information.

The host computer controls who is being talked to and which way the information is being sent. When the host computer detects a new device placed on the serial bus, it will assign a number to it for identification. When it sends information out to this device, the packet will contain this number. All devices connected to the bus will then decode the packet header to determine if the device is the one to respond to the packet sent. In this way, the system can control up to 127 devices on one bus. The actual number of devices on the bus at any given time is determined by the cumulative bandwidth of each device that is on the bus at that time.

Any synchronous serial system must have the clock carried with the signal, which is used to clock the shift register, or the data cannot be reproduced at the other end of the cable. There is no conductor for a clock signal in the USB cable. The clock is encoded in the data stream in much the same manner as was done with data stored on old tape recorders. This system is called NRZI data encoding. A **1** in the data is represented by no change in the level of the output on the cable. A **0** is indicated by a change in the level of the output on the cable. Figure 9-22 shows the waveforms for this system of encoding. The clock is derived by using the rising and falling edges of the incoming data stream on the cable to sync a free-running clock which is used to shift in each bit. This will work well if the data stream does not have a great number of ones in a row. If this happens, the clock can get out of sync with the data it is to shift in at the other end of the cable. To prevent this, a method called **bit stuffing** is used. If there are more than 6 ones in a row in the data stream, a zero is placed in the data stream in order to produce a rising or falling edge in the output stream on the cable. This insures at least one edge transition every 6 bits of data to sync the free-running clock.

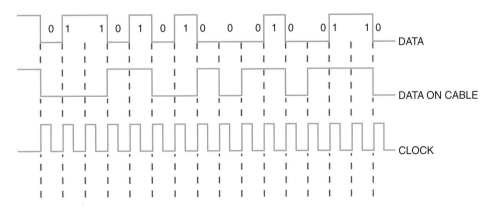

FIGURE 9-22 Clock encoding used in USB serial transmission

USB is a very powerful serial communication system, one that is slowly replacing the older RS-232C standard used for communication between computers and devices such as printers, modems, cameras, and other devices. There are several manufacturers now making LSI chips that implement the USB serial system. This is making it easier to use this system in the design of computer products.

Converting Voltage by Using a 1489 Chip

This circuit shows an unusual use for the 1489 chip (see Figure 9-14). This chip was designed to convert the plus and minus voltage of an RS-232C signal to a TTL logic level voltage. In the following schematic it is being used to convert the $+12$ volt to -12 volt output of the LM348 op amp to TTL logic voltages. The op amp is used as a zero crossing detector with hysteresis. This particular circuit is used to convert the voltage pulse of a proximity detector to TTL logic voltage levels. The signal is used to measure the period and frequency of the rotor of large ac motors. (Schematic courtesy of Precise Power Inc.)

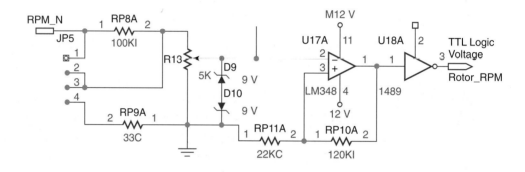

SUMMARY

- A shift register is a set of edge-triggered flip-flops that have their outputs connected to the input of the next flip-flop and have all the clocks tied together.

  This will shift the contents of the preceding flip-flop to the next flip-flop at the transition of the clock.

- Serial data can be asynchronous or synchronous.

  Asynchronous serial format uses a set of framing bits to start and stop the serial bit flow, while the synchronous data format must have a common clock and external sync method to keep the transmitter and receiver together. Synchronous serial data format can send more data in the same number of clocks than asynchronous because it has no framing bits.

- The RS-232C standard for serial data transmission is one of the most widely used serial data transfer standards in use today.

  It is a negative logic system and uses plus and minus voltages for the logic levels.

- The ASCII code is the international de facto standard used for coding letters, numbers, and control codes.

It is used to send and store text and data of all kinds on all types of data transmission and storage devices.

- Today, most shift registers are made on single chips and many are part of complete systems of serial data transmission on a single integrated circuit.

- In VHDL programming, the signal assignments made inside a PROCESS are not applied until the END of the PROCESS. This means that the order of assignments inside the PROCESS does not alter the outcome.

QUESTIONS AND PROBLEMS

1. Draw the logic diagram for a 5-bit shift register using *JK* flip-flops.

2. Draw a logic diagram for a 4-bit shift register using 4027 CMOS *JK* flip-flops. Show pin numbers on the drawing.

3. Draw the wave diagram for Figure 9-3 if the binary number equivalent to 6 is on the parallel inputs.

4. What is the purpose of the start bit in the serial format of Figure 9-6?

5. Draw a logic diagram for an 8-bit shift register using two 7495 ICs.

6. Draw the waveforms for the 4-bit shift register in Figure 9-1 if the binary number equivalent to 6 is shifted into the shift register.

7. Draw the waveforms for the shift register in Figure 9-3 if the binary number equivalent to 7 is on the parallel inputs.

8. Using the clock pulse given in Figure 9-6, draw the waveform for an asynchronous serial word which has no parity bit, one stop bit, and data of the binary number equivalent to 62.

9. Using Figure 9-8 as an example, design an asynchronous serial receiver that uses a clock sixteen times the incoming baud rate of the serial data.

10. Draw the logic diagram of a 7495 4-bit shift register wired to shift left.

11. Draw the logic diagram of a 7495 4-bit shift register wired to be a parallel-in, serial-out shift register.

12. Which serial method of storing data on a magnetic device will store the most data?

13. Write the code for the words DIGITAL ELECTRONICS in ASCII code.

14. Look up the 74164 IC in your data book and write a description of it.

15. Draw the logic diagram of a 16-bit shift register using the 74164. Show pin numbers.

16. What is the voltage level for a logic 1 in the RS-232C standard?

17. How many codes are in the ASCII code?

18. What is the ASCII hex number to move the cursor on an ANSI terminal down one line?

19. Does the RS-232C standard have a standard connector?

20. What do DCE and DTE stand for in the RS-232C standard?

21. How much time is used to transmit one frame (start bit, eight data bits, and one stop bit) at 9600 bits per second?

22. What would happen if the output of the AND gate 4' in Figure 9-8 were to go HIGH and never go LOW?

23. Draw the logic diagram for a 4-bit shift register using negative edge-triggered *D* flip-flops.

24. How can you find which pin on an RS-232C connector is the transmit pin?

25. Which pin of a DB-25 connector is usually used for signal ground for RS-232C?

26. What does the ASCII code 0D Hex do?

27. Which ASCII code will do nothing?

28. What would happen if the clock of the *JK* flip-flop F_{10} in Figure 9-8 became disconnected?

29. Will a plus and minus 9-volt RS-232C signal work correctly?

30. What is the idle voltage for a typical RS-232C signal?

31. Write a program using VHDL that will implement the 3-bit shift counter used in Chapter 8 and Chapter 9. Name the outputs **A1**, **A2**, **B1**, **B2**, **C1**, **C2**, where **A2** is the complement of **A1**, **B2** is the complement of **B1**, and **C2** is the complement of **C1**. Include an asynchronous input named **clear** that will reset all three flip-flops for proper operation.

32. Write a program using VHDL that will implement an 8-bit ring counter that will shift a single '1' from left to right on the leading edge of the clock. When the '1' reaches the flip-flop on the right, the next clock will cause the flip-flop on the left to turn on, so that the '1' continuously travels around the ring.

33. Write a program using VHDL that will implement an 8-bit "Kit" shift register that will cause a single '1' to shift from left to right. When the '1' reaches the flip-flop on the right, the '1' will begin to shift from right to left. When the '1' reaches the flip-flop on the left, the '1' will begin to shift again from left to right.

34. Write a program using VHDL that will implement a circuit that will parallel-load an 8-bit binary number. At each positive edge of an input signal named **double**, the number will double in value.

35. Write a program using VHDL that will implement a circuit that will parallel-load an 8-bit binary number. At each positive edge of an input signal named **half**, the number will be divided by two.

36. Find and correct four errors in this VHDL program.

```
ENTITY s_in_p_out  IS
    PORT(D,clk:IN STD_LOGIC;
        Q:BUFFER STD_LOGIC_VECTOR(7 DOWNTO 0));
END s_in_p_out;
ARCHITECTURE new OF s_in_p_out IS
```

```
BEGIN
    PROCESS (clk)
        IF (clk'event AND clk = '0') THEN
                Q(7) = D;
                Q(6) = Q(7);
                Q(5) = Q(6);
                Q(4) = Q(5);
                Q(3) = Q(4);
                Q(2) = Q(3);
                Q(1) = Q(2);
                Q(0) = Q(1);
        END IF;
    END PROCESS;
END a;
```

37. Find and correct four errors in this VHDL program.

```
ENTITY p_in_s_out IS
    PORT(
        D:IN BIT_VECTOR (7 DOWNTO 0);
        Q:OUT BIT_VECTOR (7 DOWNTO 0);
        clk,load:IN BIT);
END p_in_S_out;

ARCHITECTURE a OF p_in_s_out IS
    SIGNAL internal:BIT_VECTOR;
BEGIN
    PROCESS (clk,load,D)
    BEGIN
        IF load = '1' THEN  --broadside load
                Internal <= D;
        ELSIF load = '0' THEN
            IF (clk event AND clk = '1') THEN
                --low-to-high transition
                internal(7)<= '0';  --shift 0 into int(7)
                internal(6)<= internal(7);
                internal(5)<= internal(6);
                internal(4)<= internal(5);
                internal(3)<= internal(4);
                internal(2)<= internal(3);
                internal(1)<= internal(2);
                internal(0)<= internal(1);
            END IF;
        END IF;
    END PROCESS;
    Q <= internal;    --define Q as lsb of internal
END a;
```

SHIFT REGISTERS

OBJECTIVES

After completing this lab, you should be able to:

- Use the 7495 shift register.
- Use the ASCII code.
- Use a team approach to solving a problem.

COMPONENTS NEEDED

PART 1

1	7476 *JK* flip-flop IC
1	7414 hex Schmitt-trigger input inverter IC
1	7408 quad AND gate IC

PART 2

2	7476 *JK* flip-flop ICs
1	7420 dual four-input NAND gate IC
1	7410 triple three-input NAND gate IC

PART 3

1	74LS164 shift register IC
1	7400 quad NAND gate IC
1	74LS32 quad OR gate IC
1	74LS74 D flip-flop

PART 4

2	7475 quad transparent *D* latch ICs
8	Red LEDs
8	330-Ω, ¼-W resistors

PROCEDURE

1. The instructor will assign each lab group one part of the serial receiver to construct.
2. Each lab group will construct its part of the serial receiver and test it to be sure it works.

3. After construction, the separate lab groups will combine their parts of the serial receiver to make a complete serial receiver and connect the input to the output of an ASCII terminal provided by the instructor.

4. There will be a message sent in serial ASCII form. The characters will be spaced about five seconds apart to allow you to read them from the LEDs on your serial receiver.

PART 1

1. Construct the nonoverlapping clock shown using one 7476 *JK* flip-flop, one 7408 AND gate IC, and one 7414 IC.

2. Feed the clock with the signal from the sine wave frequency generator* which has been squared up by the 7414 Schmitt-trigger inverter.

3. Set the signal generator at 1320 Hz and measure the frequency on the oscilloscope to get it correct. This is very important, or the serial receiver will not work.

4. Put the inputs and outputs of the clock on a blank spot of the protoboard to make it easy to connect to the next part of the serial receiver.

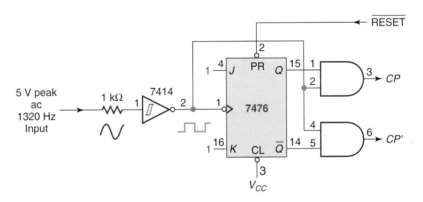

If your circuit does not work properly, consider these points:

1. Check all power supply connections.

2. Check to be sure there are no unconnected inputs on the *JK* flip-flop.

3. Make sure you have connected the signal generator ground to the circuit ground.

4. Use the oscilloscope to trace the input clock signal from the signal generator to all points in the circuit.

PART 2

1. Construct the shift counter shown and the 3′, 4′, and 5′ outputs. Use two 7476 *JK* flip-flops, one 7420 four-input NAND gate, and one 7410 three-input NAND gate.

2. Feed the shift counter with the clock from your trainer and check the output waveforms on the oscilloscope to be sure they are correct.

3. Place the inputs and outputs of the shift counter on a blank spot on your protoboard to make it easy to connect to the next part of the serial receiver.

---

* If an RS-232C equipped terminal is used for the serial output device, be sure to use a buffer IC such as LM1489 to convert the RS-232C voltage levels to standard TTL voltage levels.

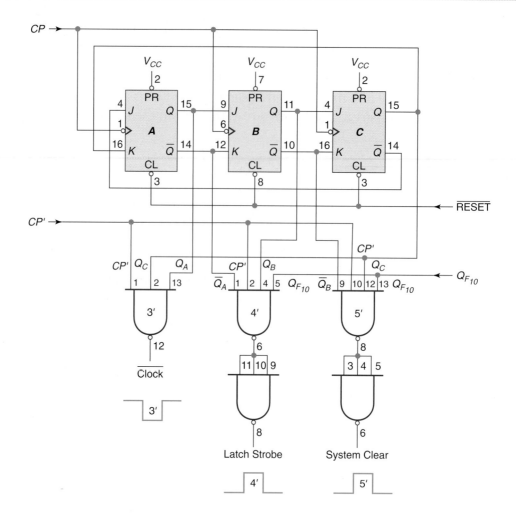

If your circuit does not work properly, consider these points:

1. Check all the power supply connections to the ICs.

2. Connect a clock of about 1 kHz to the *CP* input and reset the shift counter by momentarily bringing the $\overline{\text{RESET}}$ input LOW.

    a. Make sure that the PRESET inputs to the *JK* flip-flops are HIGH.

    b. Use the oscilloscope to test the *Q* outputs for their proper signal. Q_A should be one clock pulse out-of-phase with Q_B and Q_C should be two clocks out-of-phase with Q_A. If these are not the waveforms you see, then check your wiring.

    c. Check your wiring by putting channel one of the oscilloscope on a pin and channel two on a pin which should be tied to the first pin. If the signals are not the same, they are not connected.

3. Tie *CP'* and $\overline{Q}_{F11}$ HIGH.

    a. Use the oscilloscope to test the "clock," "latch strobe," and "system clear" for proper operation. These should be one clock pulse out-of-phase.

    b. Bringing $\overline{Q}_{F11}$ LOW and *CP'* HIGH should cause only "latch strobe" and "systems clear" to be LOW.

    c. If the circuit does not work in the manner described, use the method described in 2C to trace your wiring.

PART 3

Description of the circuit operation

This portion of the serial receiver contains the starting $\overline{\text{SET}}$-$\overline{\text{RESET}}$ flip-flop and the shift register. The starting $\overline{\text{SET}}$-$\overline{\text{RESET}}$ flip-flop is constructed from two crossed NAND gates. The shift register is constructed from one 74LS164 8-bit serial-in parallel-out shift register and two 74LS74 edge-triggered D flip-flops. The two extra flip-flops are used to store the start bit and one of the two stop bits.

Before the first LOW-going start bit reaches the SERIAL INPUT of the starting flip-flop (IC 8A and IC 8B), the shift register (IC 11) and the two D flip-flops (IC 10A and IC 10B) are held in the cleared state by the RESET output from the OR gate (IC 9A). When the LOW-going start bit reaches the starting flip-flop (IC 8A and IC 8B), the flip-flop is SET, which causes a logic 1 on pin 1 of the RESET OR gate (IC 9A). This logic 1 is then passed through the OR gate because any 1 into an OR gate produces a 1 on the output. A logic 1 on the RESET will allow the shift counter to start counting, which will produce the clock and shift in the rest of the serial data. When the LOW-going start bit is clocked into the 74LS164 (IC 11), it is inverted to a logic 1. This is done by feeding the shift register input from the $\overline{Q}$ of the D flip-flop (IC 10A). The 74LS164 (IC 11) had been cleared and will shift out 0s to the last D flip-flop (IC 10B) until the inverted start bit reaches the end of the shift register, and a logic 1 will be shifted into the last D flip-flop (IC 10B). This will make Q_{F10} logic 1 and cause the shift counter to produce the LATCH STROBE and then the SYSTEM CLEAR pulses. When the LATCH STROBE goes HIGH, the content of the 74LS164 is latched into the parallel output register, and when the SYSTEM CLEAR goes HIGH, the starting flip-flop will be RESET. This will stop the serial receiver and make it ready for the next LOW-going start bit. It should be noted that a logic 1 on the SYSTEM CLEAR keeps the logic 1 on the RESET for the full period of the HIGH pulse because it is the second input to the OR gate that produces the RESET signal. This makes sure that the system is reset completely before the next serial start bit comes along. The output of the shift register is the complement value of the serial data. This data will be complemented again by the latch output in part 4.

Construction and debug procedures

1. Construct the serial receiver shift register from one 74LS00 quad NAND gate, one 74LS32 quad OR gate IC, one 74LS74 dual D flip-flop IC, and one 74LS164 shift-register IC.

2. Bring all the control inputs and outputs which come from or go to other parts of the receiver out to the edge of the protoboard to allow for easy connection.

3. Connect the clock input to a debounce button on your trainer or construct a debounce button from a pair of NAND gates. Be sure that the debounce button is at a logic 1 except when pressed. Bring the SYSTEM CLEAR, HIGH, and then LOW. This should clear all the D flip-flops and the 74LS164 shift register. You can use an oscilloscope or voltmeter to check the logic levels on the Q outputs and D_0 through D_7. The first D flip-flop (IC 10A) should be preset to a logic 1.

4. Bring the serial input to a logic 0. Now press the debounce button connected to the clock input and monitor the D_0 to D_7 outputs with a oscilloscope or voltmeter.

Watch for the inverted 0 (1) to be clocked across the shift register. Be sure that the last *D* flip-flop (IC 10B) clocks in the 1.

5. Connect this part of the serial receiver to the other parts. Be sure to connect a common ground to the other parts of the receiver.

If your circuit does not work properly, consider these points:

1. Check all power supply connections.

2. Check all inputs and outputs for proper voltage levels.

3. Supply a 1-Hz or slower clock to the clock input, tie the serial input HIGH, and pulse SYSTEM CLEAR, HIGH, and then LOW.

   a. Pin 3 of the 7400 AND gate should be LOW. If it is HIGH, there is something wrong with the crossed NAND $\overline{\text{SET}}$-$\overline{\text{RESET}}$ flip-flop or the system CLEAR input.

   b. Using the oscilloscope or voltmeter, check all the pins for their expected logic levels. The D_0 through D_7 should be HIGH.

4. Put channel one of the oscilloscope on the slow 1-Hz clock input to the circuit. Put channel two of the oscilloscope on pin 9 of the 74LS74, which supplies the signal for $\overline{Q}_{F10}$. Momentarily bring the serial input LOW for one clock pulse of the 1-Hz clock.

   a. If the shift registers are working correctly, pin 9 should go HIGH ten clock pulses later.

   b. If pin 9 never goes HIGH, then back up to D_2 and repeat the procedure.

   c. If D_2 never goes LOW, then repeat the procedure with pin 6 of the 74LS74.

   d. Use the oscilloscope to trace wiring to find any faults discovered.

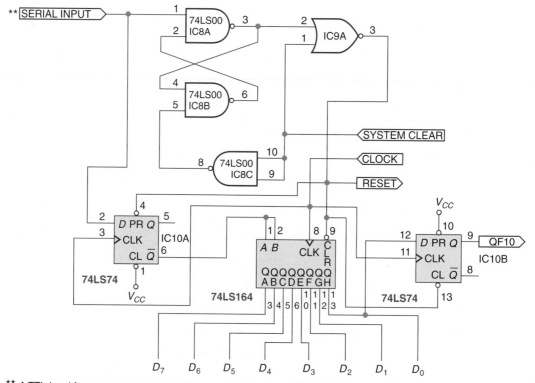

** A TTL-level frequency generator could be used in the place of the 7414 IC.

PART 4

1. Construct the parallel register shown. Use two 7475 *D* latches and eight LEDs.

2. Test your register to make sure it works properly.

3. Put the inputs to your register on the protoboard in a logical order to make it easy to connect to the next part of the serial receiver.

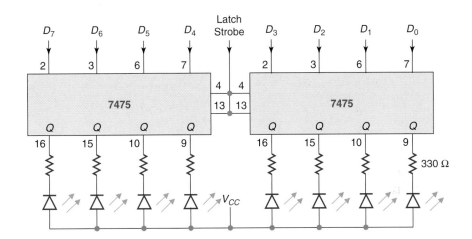

If your circuit does not work properly, consider these points:

1. Check all power supply connections.

2. Bring D_0 through D_7 HIGH and pulse the latch strobe HIGH. This should turn on all the LEDs.

3. Bring D_0 through D_7 LOW while the latch strobe is LOW. This should not change the state of the LEDs until you pulse the clock HIGH again, at which time they should turn off.

4. If the current-limiting resistors are too low, then the 7475 IC will not work properly because of the heavy I_{CC}.

PART 5

If the total circuit does not work properly after you have combined all the parts, consider these points:

1. Is each circuit connected to a common ground?

2. Put an oscilloscope probe on the serial input to detect the incoming data word from the terminal.

3. Use the other oscilloscope channel to trace the signal back from the serial input.

The ASCII code is a standard used by almost all computers in the world to represent letters, numbers, and control commands for input/output terminals. This code is used to store text in computer files, to send text over phone wires, and to perform similar tasks.

ASCII CODE CONVERSION TABLE

BITS 4 THRU 6 —	0	1	2	3	4	5	6	7
0	NUL	DLE	SP	0	@	P	'	p
1	SOH	DC1	!	1	A	Q	a	q
2	STX	DC2	"	2	B	R	b	r
3	ETX	DC3	#	3	C	S	c	s
4	EOT	DC4	$	4	D	T	d	t
5	ENQ	NAK	%	5	E	U	e	u
6	ACK	SYN	&	6	F	V	f	v
7	BEL	ETB	'	7	G	W	g	w
8	BS	CAN	(	8	H	X	h	x
9	HT	EM	)	9	I	Y	i	y
A	LF	SUB	*	:	J	Z	j	z
B	VT	ESC	+	;	K	[	k	{
C	FF	FS	,	<	L	\	l	|
D	CR	GS	-	=	m	]	m	}
E	SO	RS	.	>	N	^	n	~
F	SI	US	/	?	O	—	o	DEL

BITS 0 thru 3 (row labels)

The first 32 ASCII codes are commands for the computer terminal, and the rest, with the exception of 7F Hex, are printable characters.

The binary numbers that your instructor will send to your newly constructed serial receiver will be in ASCII code. You must write the binary code in hex form and use the ASCII conversion table to translate the message to letters and numbers. Do this by using the most significant hex digit to find the column and the least significant hex digit for the row in the table. For instance, the hex number 4A would be the ASCII code for the capital letter J.

If the ASCII characters being sent to your serial receiver are coming from a standard computer terminal with an RS-232C interface, hold the control key down and press the letters starting with A. You will find that this will give you the first 26 control codes in the ASCII table.

LAB 9B   SHIFT REGISTERS

OBJECTIVES

After completing this lab, you should be able to use Multisim to:

- Construct a shift register.
- Troubleshoot a serial communication system using shift registers.

PROCEDURES

1. Using Multisim, construct a 4-bit parallel-in serial-out shift register. Use the word generator to produce the CLOCK and the LOAD signals. Use the logic analyzer to display the CLOCK, LOAD, Q_A, Q_B, Q_C, and SDO (serial data out) signals.

2. Open file 9B-1.ms9. This is a 4-bit parallel-in serial-out shift register connected to a serial-in parallel-out shift register. Run the simulation and change the parallel data being fed to the parallel-in serial-out shift register. Work with this working design until you feel you have become familiar with its operation.

3. Open file 9B-2.ms9. This is the same circuit you just worked with, but it has one fault. By looking at the waveforms and the logic indicators, determine the problem with the circuit and fix it. When you have finished with this problem, open files 9B-3.ms9 and 9B-4.ms9. They also each have one fault that prevents them from working correctly.

Counters

OUTLINE

KEY TERMS

down counter

presettable counter

ripple counter

synchronous counter

OBJECTIVES

After completing this chapter, you should be able to:

- Explain the operation of a ripple counter.
- Describe the decode-and-clear method of making a divide-by-*N* counter.
- Explain how to design a divide-by-*N* synchronous counter.
- Explain the use of a presettable counter.
- Describe the up-down counter.
- Use typical MSI counter ICs.
- Understand how to use a typical PLD to create a synchronous counter.
- Troubleshoot counters.

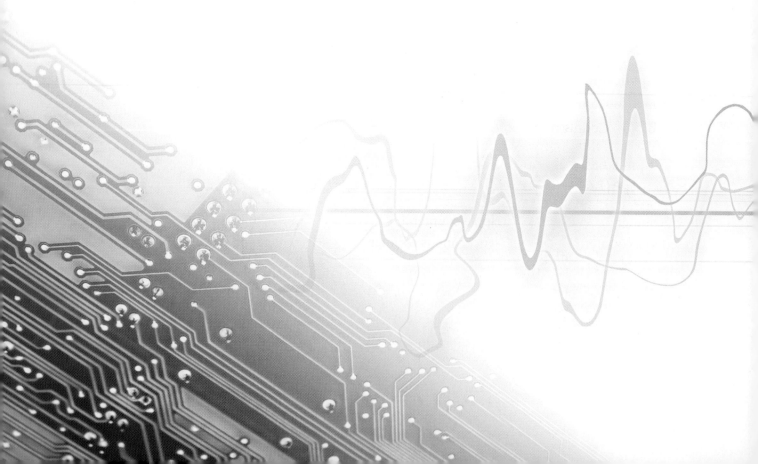

◐ **10.1** THE RIPPLE COUNTER

Figure 10-1 shows a 4-bit ripple counter and the waveform it generates. The negative edge-triggered *JK* flip-flops are set to toggle. The clock is tied to the *Q* output of the previous flip-flop. This means the first flip-flop (*A*) must change states from a *Q* of HIGH-to-LOW in order for the next flip-flop (*B*) to toggle. Notice that flip-flop *A* changes states on every trailing edge of the input clock, flip-flop *B* changes states on the trailing edge of flip-flop *A*, and flip-flop *C* changes states on the trailing edge of flip-flop *B*. This procedure continues for as many flip-flops as are in the counter.

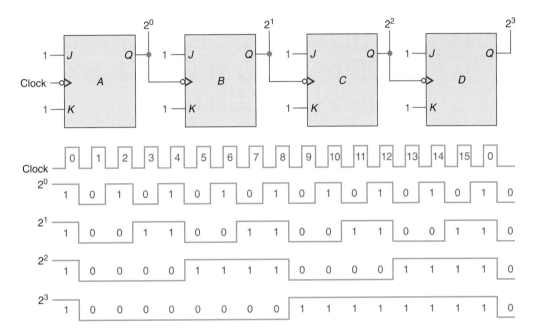

FIGURE 10-1 A 4-bit ripple counter

Notice also that the output frequency of each flip-flop is one-half the frequency of the previous flip-flop. This means the output frequency of flip-flop *A* is one-half the clock frequency, and the output frequency of flip-flop *B* is one-half the frequency of flip-flop *A* or one-fourth the clock frequency. This divide-by-2 for each succeeding flip-flop continues for as many flip-flops as are in the counter.

Because the counter divides the frequency of the previous flip-flop by 2, the value of the outputs for each flip-flop at any count of the input clock will be the binary number of that clock pulse. This means the counter in Figure 10-1 will count in binary from 0000 to 1111, and then start over again with 0000. The largest number the counter will display is a function of the number of flip-flops in the counter, because each flip-flop produces one bit in the binary number. Therefore, the formula for the largest binary number that can be displayed for a given number of bits will determine the largest number that can be displayed for a given counter of *N* flip-flops.

Largest binary number $= 2^N - 1$

For the counter in Figure 10-1, the largest number that can be displayed is $2^4 - 1$ or 15. Since the D output divides the input frequency by 2^4 or 16, we would call this counter a divide-by-16 ripple counter. If you were to add one more flip-flop to the counter in Figure 10-1, it would be a divide-by-2^5 or divide-by-32 counter.

The last flip-flop in a **ripple counter** must wait for the input signal to ripple down through each preceding flip-flop before it can change. Because of this cumulative propagation delay of a ripple counter, the larger it is, the slower it is. If each flip-flop in Figure 10-1 has a propagation delay of 25 nanoseconds, it would take 4×25 nanoseconds for the counter to change from 0111 to 1000. The output does not change together, but one after the other.

 10.2 THE DECODE-AND-CLEAR METHOD OF MAKING A DIVIDE-BY-N RIPPLE COUNTER

The divide-by-5 counter in Figure 10-2 is a ripple counter which uses the decode-and-clear method of resetting the counter after the fifth clock pulse. The number 5 is decoded by the NAND gate, which produces a LOW output when the number 5 appears on the output of the counter. As the waveforms show, the number 5 is not present for very long before the counter is reset to 000, causing the LOW output of the NAND gate to go back HIGH and starting the

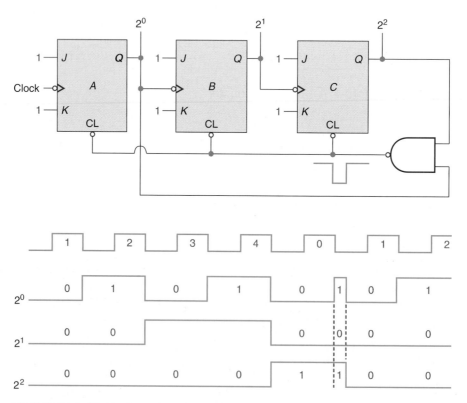

FIGURE 10-2 Divide-by-5 decode-and-clear counter

counter over again at 000 for the next pulse of the input clock. This effectively resets the counter at the number 5, but it produces a small spike at the 5 count on the *A* output. Also, the *C* output is a bit longer than it should be. Because they are counted as an extra pulse, these spikes can cause problems if the output which contains them is used for input clocks to other counters. A counter of divide-by-any number you wish can be constructed by this method.

EXAMPLE 10-1

Design a divide-by-10 decode-and-clear counter.

Solution A divide-by-10 counter will count from 0 to 9 and then start over again at 0. To make the counter reset to 0 after the 9 count, the designer must decode the 10 count to clear the flip-flops. To count to 10 in binary you will need four flip-flops. Notice the table of binary numbers from 0000 to 1010 (10 decimal). Only one binary number has a 1 in the most significant bit (2^3) and a 1 in the 2^1 bit. That binary number is 1010 or decimal 10. Using a NAND gate to decode the number 10 will produce the divide-by-10 counter as shown in Figure 10-3.

Decimal	Binary
0	0000
1	0001
2	0010
3	0011
4	0100
5	0101
6	0110
7	0111
8	1000
9	1001
10	1010

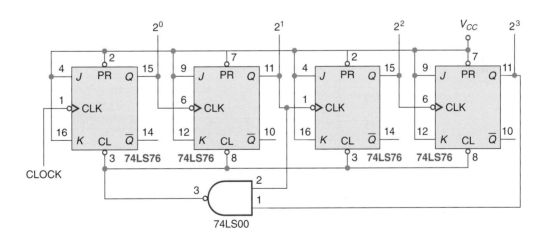

FIGURE 10-3

The divide-by-*N* ripple counter has many problems which have been previously discussed. To correct these problems, you can use a synchronous counter. A **synchronous counter** has all the clock inputs tied together so that each flip-flop changes state at the same time. This means that the propagation for the whole counter is the same as one flip-flop, no matter how many flip-flops are in the counter. To do this, each flip-flop must be steered by using the *JK* input so it will change to its proper state when the next clock comes along. The counter in Figure 10-4 does this by allowing the flip-flop to toggle or not to toggle by using AND gates to produce a

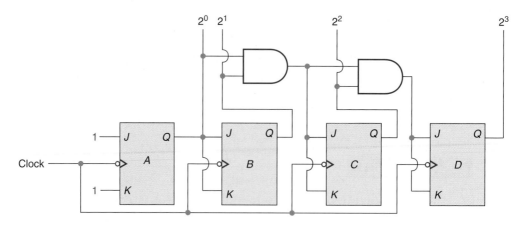

FIGURE 10-4 Synchronous counter

1 on the *JK* input only when there is 1 on all the outputs before the flip-flop. This counter will produce the same waveforms as the ripple counter in Figure 10-1, but it is much faster.

 10.3 THE DIVIDE-BY-*N* SYNCHRONOUS COUNTER

To design a divide-by-*N* synchronous counter, the first thing you must do is define the values of the *JK* inputs before the clock pulse to get a desired change in *Q* after the clock pulse. The truth table in Figure 10-5 shows this for a negative edge-triggered *JK* flip-flop. Notice that in all lines of the truth table there is an *X* for one of the values of *J* or *K*. This means that the *J* or *K* input can be either a 1 or a 0. Take the first line in the table, for instance. If the *Q* output is

PRESET	CLEAR	J	K	C	Q	$\overline{Q}$	STATE
0	1	X	X	X	1	0	Set
1	0	X	X	X	0	1	Reset
0	0	X	X	X	1	1	Unused state
1	1	0	1	⌐L	0	1	
1	1	1	0	⌐L	1	0	
1	1	0	0	X	Q	$\overline{Q}$	Unchanged state
1	1	1	1	⌐L	Toggle		

Before Clock	After Clock	Before Clock	
Q	Q	J	K
0	0	0	X
0	1	1	X
1	0	X	1
1	1	X	0

X = 1 or 0

FIGURE 10-5 Truth table for a negative edge-triggered *JK* flip-flop

0, and we want it to be 0 after the falling edge of the clock to the flip-flop, then J must equal 0, but the K can be 1 or 0. If $J = 0$ and $K = 1$, the Q output would be forced to 0, and if $J = 0$ and $K = 0$, the Q output would not change from its present 0 state.

The second step in the design of a divide-by-N synchronous counter is to define the JK input for each flip-flop in the counter before the clock to obtain the desired Q output after the clock. Figure 10-6 shows this for a divide-by-5 synchronous counter. The desired Q output in this case is a binary count from 0 to 101. The output for the first Q values before the first clock would be 000. The desired Q output after the LOW-going edge of the first clock would be 001. This output can be achieved by making $J = 0$ and $K = X$ (0 or 1) on the two most significant flip-flops, C and B, and by making $J = 1$ and $K = X$ (1 or 0) on the A flip-flop. After the first clock, the Q output will be 001 and will be ready to be set up for the second LOW-going clock, which will change it to 010.

Q Before the Clock			Q After the Clock			C		B		A		
C	B	A	C	B	A	J	K	J	K	J	K	
0	0	0	0	0	1	0	X	0	X	1	X	$X = 1$ or 0
0	0	1	0	1	0	0	X	1	X	X	1	
0	1	0	0	1	1	0	X	X	0	1	X	
0	1	1	1	0	0	1	X	X	1	X	1	
1	0	0	0	0	0	X	1	0	X	0	X	

FIGURE 10-6 *JK* input for a divide-by-5 synchronous counter

Once this is done, a Boolean expression needs to be made for each J and K input which will express the J or K input with respect to the Q outputs before the clock. This can be done by expressing each 1 in the J and K columns of the truth table as an AND gate whose inputs are the Q values before the clock, then OR each AND expression together as shown in Figure 10-7. Reduce the expression to its lowest Boolean terms.

$$K_A = A\,\overline{B}\,\overline{C} + A\,B\,\overline{C}$$
$$K_A = A\,\overline{C}\,(\overline{B} + B)$$
$$K_A = A\,\overline{C}$$

FIGURE 10-7 Reducing to minimal terms

This method will give a correct Boolean statement, but it will not always be the simplest form which will work for the truth table because it presumes that all the other combinations of the truth table must be 0. This is not always the case, as can be seen if you examine the K_A column of the truth table in Figure 10-6. Notice that two input combinations must be a 1. All the other input combinations can be 1 or 0 as shown by the X. Therefore, K_A could be made 1 and fulfill the requirements for the truth table.

Figure 10-8 shows the simplest Boolean expressions for the truth table in Figure 10-6. Examine these expressions and confirm that there is no simpler form.

$J_A = \overline{C}$	$J_B = A$	$J_C = AB$
$K_A = 1$	$K_B = A$	$K_C = 1$

FIGURE 10-8 Minimal terms for divide-by-5 synchronous counter

All that is needed to complete the design of our divide-by-5 synchronous counter is to draw the logic diagram from the Boolean expressions, as shown in Figure 10-9.

This design method could be used to design a string of JK flip-flops that would step through any sequence of output you desired.

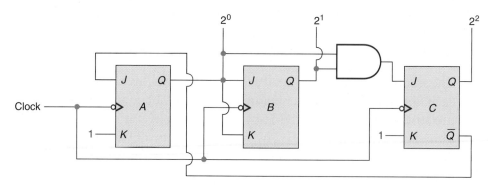

FIGURE 10-9 Divide-by-5 synchronous counter

EXAMPLE 10-2

Design a synchronous divide-by-6 counter using negative edge *JK* flip-flops.

Solution   The counter must count from 000 to 101 and start over again. This means that you will need three flip-flops. Define the states of the *J* and *K* inputs for each of the three flip-flops before the clock to obtain the desired state after the clock, as shown. Next determine the simplest Boolean expression for each *JK* set using the *Q* states before the clock. Finally, draw the logic diagram using the Boolean expressions, as shown in Figure 10-10.

Before Clock	After Clock	C		B		A		
		J	*K*	*J*	*K*	*J*	*K*	
000	001	0	X	0	X	1	X	$J_A = 1$
001	010	0	X	1	X	X	1	$K_A = 1$
010	011	0	X	X	0	1	X	$J_B = \overline{C} A$
011	100	1	X	X	1	X	1	$K_B = A$
100	101	X	0	0	X	1	X	$J_C = A B$
101	000	X	1	0	X	X	1	$K_C = A$

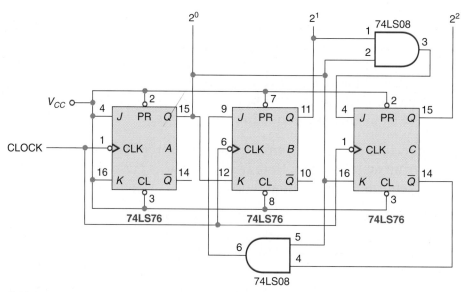

FIGURE 10-10

SELF-CHECK 1

1. Draw the logic diagram for a ripple counter that will count to 64 before starting over again at 0.

2. Draw the logic diagram for a synchronous counter that will count to 32 before starting over again at 0.

3. Design and draw the logic diagram for a synchronous counter which will count to 11 and start over at 0.

4. Use the decode-and-clear method to make a divide-by-3 counter.

 10.4 PRESETTABLE COUNTERS

The **presettable counter** shown in Figure 10-11 uses a set of NAND gates to supply a $\overline{\text{CLEAR}}$ or $\overline{\text{PRESET}}$ signal to each flip-flop in the ripple counter. A 1 on the PRESET CONTROL will enable the NAND gates, allowing the data on the PRESET INPUT to pass and thus setting the counter's outputs to the value on the PRESET INPUTS.

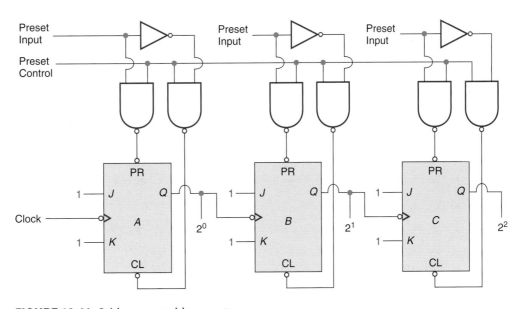

FIGURE 10-11 3-bit presettable counter

While the PRESET CONTROL is HIGH, the counter will hold the value of the PRESET input because the $\overline{\text{CLEAR}}$ or $\overline{\text{PRESET}}$ of a *JK* flip-flop overrides the clock of the flip-flop; but when the PRESET CONTROL falls LOW, the $\overline{\text{PRESET}}$ and $\overline{\text{CLEAR}}$ will both be HIGH, allowing the counter to start counting from the PRESET value on the next falling edge of the clock.

The ability of this counter to preset to a predetermined value before counting begins allows us to use it as a programmable divide-by-N counter, as shown in Figure 10-12. Notice the decode 0 and preset routine produces a short spike just as in the decode-and-clear method used for the divide-by-N counter. The advantage here is we can control the divisor by the binary number put on the PRESET INPUTS. The counter in Figure 10-12, PRESET to the binary number equivalent to 5, starts counting up to the binary number equivalent to 7 and then is RESET on the next falling edge of the clock to 5 again. Therefore, this is a divide-by-3 counter.

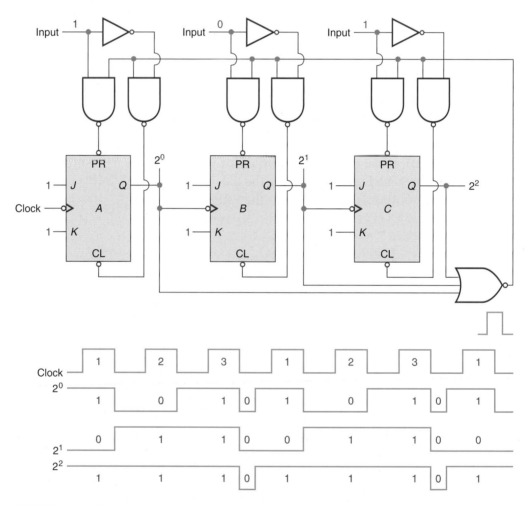

FIGURE 10-12 Presettable counter set to divide by 3

EXAMPLE 10-3 ──

Design a 4-bit presettable counter like the one in Figure 10-11 using two 74LS00 ICs, two 74LS76 ICs, and one 74LS04 IC.

Solution   Refer to Figure 10-13.

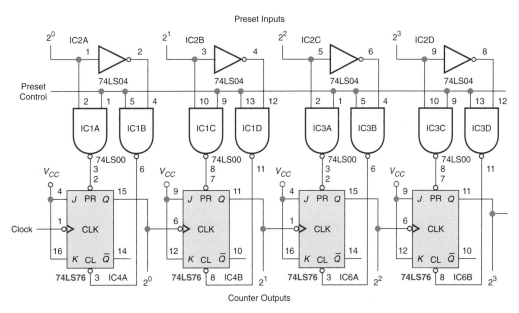

FIGURE 10-13 4-bit presettable counter

10.5 THE UP-DOWN COUNTER

If you construct a ripple counter by using the $\overline{Q}$ of each flip-flop as the clock to the next flip-flop, the counter will count down from its maximum count to 0 and then start over again. Figure 10-14 shows this for a divide-by-8 **down counter**. Figure 10-15 shows a synchronous down counter. Notice that the steering logic which controls the flip-flops is derived from the $\overline{Q}$ outputs of the flip-flops. This causes the counter to count down.

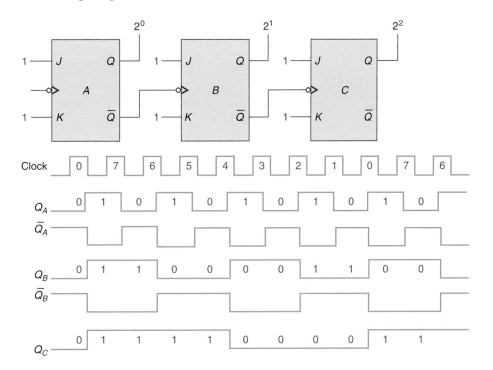

FIGURE 10-14 Ripple down counter

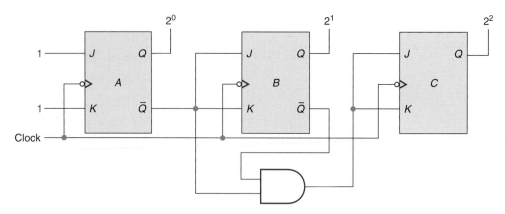

FIGURE 10-15 Synchronous down counter

An up-down counter can be made by adding steering logic derived from the Q outputs to the counter, as in Figure 10-15, and a control input to control which steering logic will be enabled. As shown in Figure 10-16, this gives us a counter that will count up or down depending on the level of the up-down input. Such a counter could be used to keep a count of the number of people in a room. When the sensor at the door senses a person entering, it counts up one pulse. When the sensor senses a person leaving the room, it counts down one count.

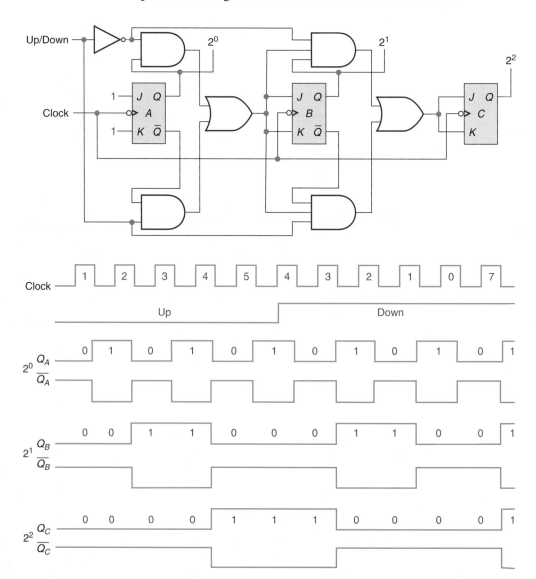

FIGURE 10-16
Synchronous up-down
counter

SELF-CHECK 2

1. Use three 74LS00 ICs, three 74LS76 ICs, and one 74LS04 IC to design a 6-bit presettable counter.
2. Design a 3-bit presettable up-down counter using AND gates, OR gates, inverters, and *JK* flip-flops.

10.6 TYPICAL MSI COUNTER ICS

Figure 10-17 shows the logic diagram for three medium-scale integration TTL counters. The 7490 is a combination divide-by-2 and divide-by-5 counter which can be configured to be a

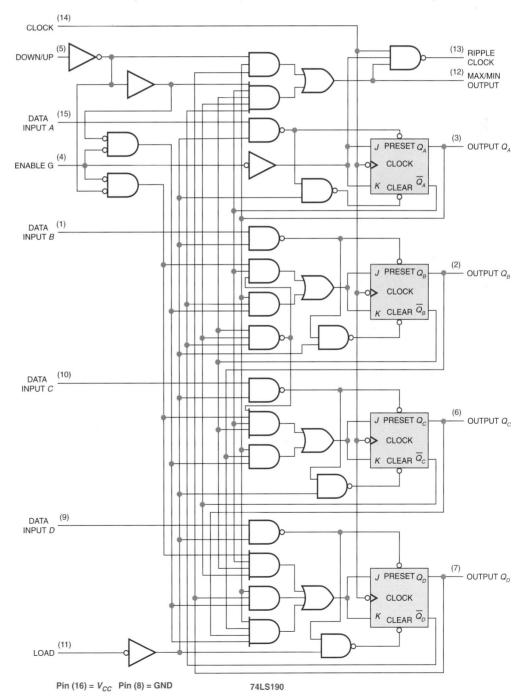

FIGURE 10-17
Medium-scale integration TTL counters

Pin (16) = V_{CC} Pin (8) = GND   74LS190

divide-by-10 counter by connecting the output of the divide-by-2 counter to the input of the divide-by-5 counter. Notice that the divide-by-5 part of the 7490 is part ripple and part synchronous.

The 7492 is a divide-by-2 and divide-by-6 counter. A divide-by-12 counter can be made by connecting these two counters. Also, the divide-by-6 part of the 7492 is part ripple and part synchronous.

The 7493 is a straight divide-by-2 and divide-by-8 ripple counter which can be connected to produce a divide-by-16 counter.

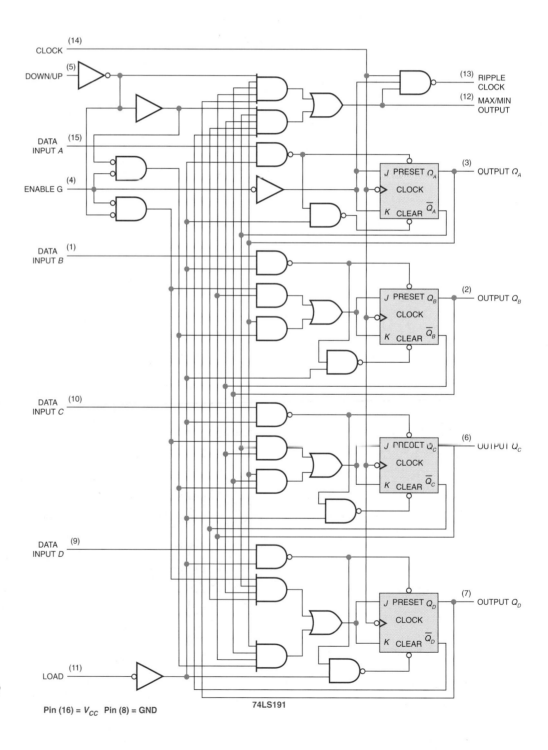

FIGURE 10-17
(*continued*)

Pin (16) = V_{CC} Pin (8) = GND

74LS191

Two typical up-down counters are the 74LS190 and 74LS191. The 74LS190 is a BCD counter and the 74LS191 is a binary counter. Both are synchronous and use *JK* master-slave flip-flops. If parallel clocking is used, they can be cascaded by feeding the ripple clock output to the enable input of the succeeding counter. The ripple clock will produce a LOW when the counter turns to 0000, which will allow the next counter to count for one count. The MAX/MIN output will produce a HIGH output for one count when the counter turns to 0000.

The 74HC4020 is a 14-bit ripple counter in a 16-pin package. Because this counter is in a 16-pin package, not all the bits of the counter are available.

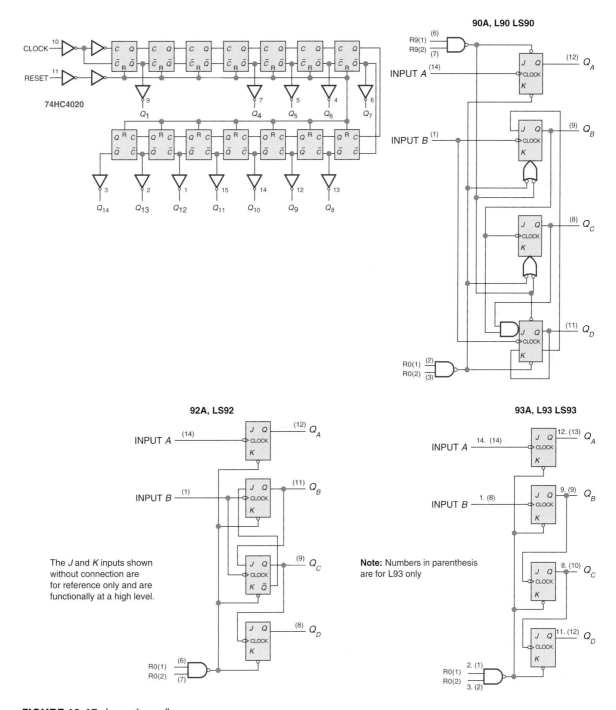

FIGURE 10-17 (*continued*)

EXAMPLE 10-4

Design a divide-by-60 BCD counter using one 74LS90 and one 74LS92 IC.

Solution   The divide-by-6 part of the 74LS92 does not count from 0 to 5 but counts 0, 1, 2, 4, 5, 6. This is so the most significant bit Q_D will have an even-duty cycle. A divide-by-6 counter that will count from 0 to 5 (0,1,2,3,4,5) can be made by connecting the output of the divide-by-2 counter Q_A to the B input of the divide-by-6 counter. This will cause the Q_A, Q_B, and Q_C outputs to count from 0 to 5 in the normal counting order (0,1,2,3,4,5,0,1,2, etc.). Refer to Figure 10-18.

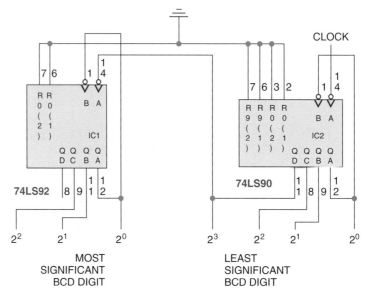

FIGURE 10-18 Divide-by-60 counter

Figure 10-19A and Figure 10-19B show these counters used to make a digital clock. The 7447 IC is a binary-coded decimal to seven-segment decoder/driver, used to display the time on the common-anode LEDs. We will study these ICs and LEDs in a later chapter.

10.7 THE DIVIDE-BY-N ½ COUNTER

All the counters thus far have divided the incoming signal by a whole number. There is a clever method of using the decode-and-clear method to produce a divisor of $N - ½$. This is done by using a flip-flop and an XOR gate to change the edge on which the counter will change states. Figure 10-20 shows how this works.

When the counter reaches the desired count, the decoder will clear the counter and toggle the edge control flip-flop (IC 1B). This causes the clock signal to be inverted on one pass of the counter and not on the next. This causes the counter to miss ½ of a clock cycle on the 0 count, thereby making the count ½ clock cycle shorter. The missing ½ clock cycle can be seen in the waveforms of Figure 10-20 at the 0 count.

Notice the decode 6 spike that clears the counter and toggles the inverting flip-flop (IC 1B). A similar synchronous counter can be made by applying the same methods if the decode spike is a problem.

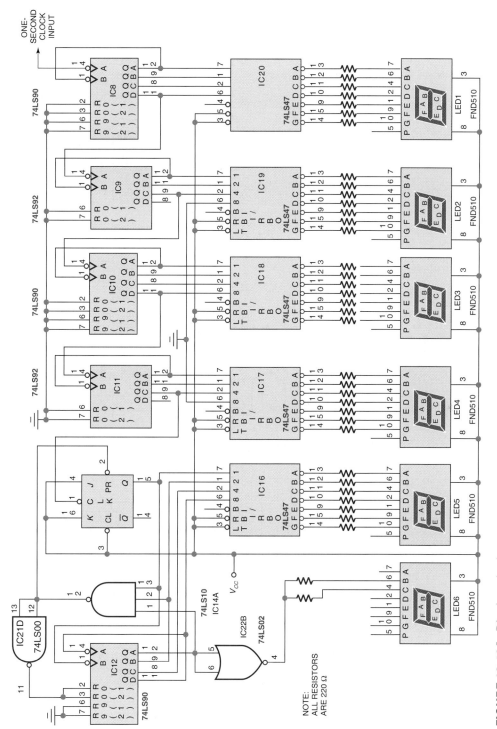

FIGURE 10-19A Display clock

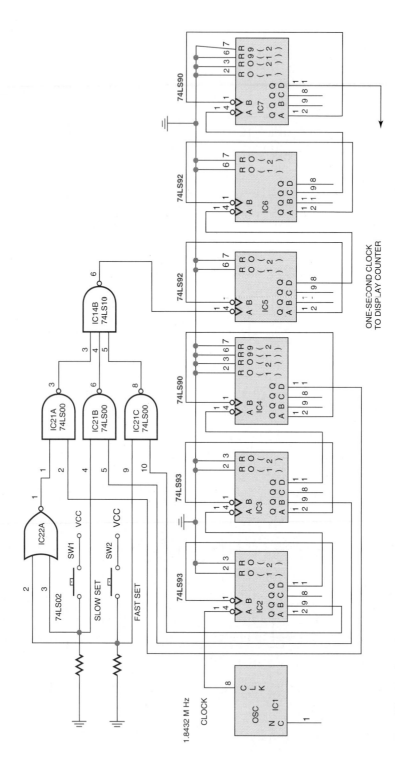

FIGURE 10-19B One-second clock

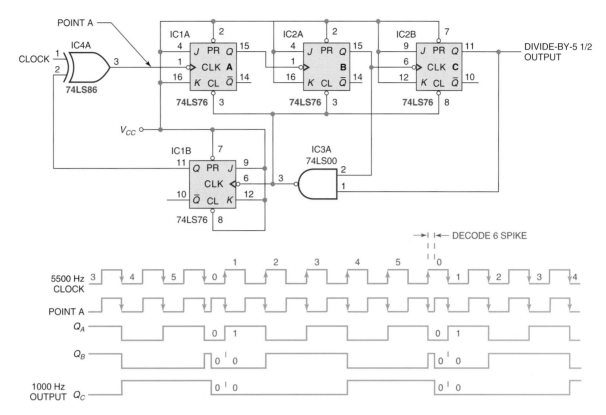

FIGURE 10-20 Decode-and-clear divide-by-5½

10.8 PROGRAMMING A CPLD (Optional)

Follow the procedure presented in the preceding chapters:

- Create a new folder within the Altera folder and name it **bincount**.

- Create a new .vhd file and name it **bincount**.

- Write a VHDL program that will describe a 4-bit binary up-counter. Have the counter increment one count for each HIGH-to-LOW transition of the clock. Invoke the ieee library by beginning your program with these three lines:

    LIBRARY ieee;

    USE ieee.std_logic_1164.all;

    USE ieee.std_logic_unsigned.all;

Line three of the ieee library statements enables us to use the addition sign in the ARCHITECTURE statement.

Declare **clk** to be a standard logic input and **Q** to be a 4-bit standard logic vector buffer. Q needs to be declared a buffer instead of just an output so that we can add '1' to its present value.

- Compile the program.

- Create a .vwf (vector waveform file) named **bincount.vwf** and use it to simulate the circuit. Figure 10-21 shows sample simulation waveforms for bincount. Note that the 4-bit counter counts from 0 to F_{16} (0000 to 1111_2) continuously.

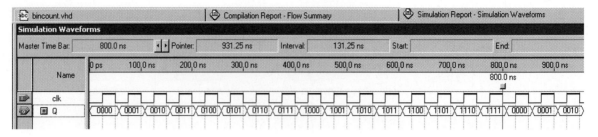

FIGURE 10-21 Simulation waveforms for bincount.vwf

- Assign pins on the CPLD to **clk** and each bit of **Q**. Recompile your program.

- Program your CPLD.

- Test the operation of the CPLD using input switches and LEDs.

bincount.vhd looks like this:

```
LIBRARY ieee;
USE ieee.std_logic_1164.all;
USE ieee.std_logic_unsigned.all;

ENTITY bincount IS
    PORT(clk: IN STD_LOGIC;
    Q:BUFFER STD_LOGIC_VECTOR(3 DOWNTO 0));
END bincount;

ARCHITECTURE a OF bincount IS
BEGIN
    PROCESS (clk)
    BEGIN
        IF (clk'EVENT AND clk = '0') THEN
                Q <= (Q+1);
        END IF;
    END PROCESS;
END a;
```

Figure 10-22 shows sample simulation waveforms for bincount with **Q** expanded into its 4 bits **Q(3) Q(2) Q(1) Q(0)**. **Q(0)** toggles on each HIGH-to-LOW transition of **clk**. **Q(1)** toggles on

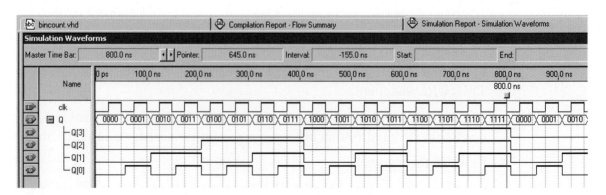

FIGURE 10-22 Simulation waveforms for bincount.vwf expanded

each HIGH-to-LOW transition of **Q(0)**. **Q(2)** toggles on each HIGH-to-LOW transition of **Q(1)**. **Q(3)** toggles on each HIGH-to-LOW transition of **Q(0)**. The frequency of **Q(0)** is half the frequency of **clk; Q(1)** is one-fourth the frequency of **clk; Q(2)** is one-eighth the frequency of **clk;** and **Q(3)** is one-sixteenth the frequency of **clk.** We have created a divide by 2, 4, 8, 16 circuit.

An alternate procedure is to declare Q as an output port and create a signal within the ARCHITECTURE statement and use it to keep track of the count. We will name this internal signal **count.** Then Q is equated to that signal just before the end of the ARCHITECTURE statement.

This alternate-procedure bincount_alt.vhd looks like this:

```
LIBRARY ieee;
USE ieee.std_logic_1164.all;
USE ieee.std_logic_unsigned.all;


ENTITY bincount_alt IS
        PORT(clk: IN STD_LOGIC;
            Q:OUT STD_LOGIC_VECTOR(3 DOWNTO 0));
     END bincount_alt;


ARCHITECTURE a OF bincount_alt IS
     SIGNAL count:STD_LOGIC_VECTOR(3 DOWNTO 0);
BEGIN
     Process (clk)
     BEGIN
         IF (clk'EVENT AND clk = '0') THEN
                count <= (count+1);
         END IF;
     END PROCESS;
     Q<=count;
END a;
```

Each of these solutions produces the same result. Each uses 4 macrocells and 9 pins of the CPLD.

No provision was made to reset the count to zero. Let's modify the program to include a reset signal, **res.** When **res** is HIGH, the output is locked at zero. When **res** returns LOW, the count begins from zero.

bin_with_reset.vhd looks like this:

```
Library ieee;
USE ieee.std_logic_1164.all;
USE ieee.std_logic_unsigned.all;

ENTITY bin_with_res IS
     PORT(clk,res: IN STD_LOGIC;
          Q:BUFFER STD_LOGIC_VECTOR(3 DOWNTO 0));
END bin_with_res;
```

```
ARCHITECTURE a OF bin_with_res IS
BEGIN
    Process (clk,res)
    BEGIN
        IF res ='1' THEN
                Q<="0000";
        ELSE
            IF (clk'EVENT AND clk = '1') THEN
                    Q<=(Q + 1);
            END IF;
        END IF;
    END PROCESS;
END a;
```

Notice these features of bin_with_res:

- Q is a 4-bit vector. When Q is reset to zero, four bits are shown. Multiple bits are included in quotes: Q<="0000";

- Two sets of IF statements are used. The first tests **res**, the second checks for a LOW-to-HIGH transition of **clk**.

Figure 10-23 shows sample simulation waveforms for bin_with_res with **Q** as a composite signal.

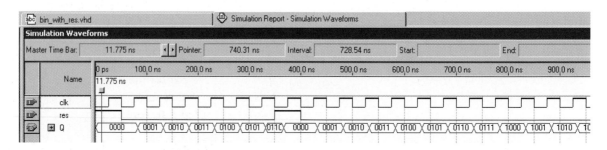

FIGURE 10-23 Simulation waveforms for bin_with_res.vwf

EXAMPLE 10-5

Create a program that describes the operation of a decade counter. Use it to divide an input clock by ten (use the most significant bit of the decade counter as the output). Call the original clock **clk** and call the divided-down clock **clk2**. Include an active-LOW reset signal that resets the output count to zero (LOW on **reset** causes the output to reset to zero).

Solution

```
Library ieee;
USE ieee.std_logic_1164.all;
USE ieee.std_logic_unsigned.all;
```

```
ENTITY decade IS
    PORT(clk,res: IN STD_LOGIC;
        clk2:OUT STD_LOGIC);
END decade;

ARCHITECTURE a OF decade IS
    SIGNAL Q:STD_LOGIC_VECTOR(3 DOWNTO 0);
BEGIN
    PROCESS(clk,res,Q)
    BEGIN
        IF (res ='0') THEN
            Q<="0000";
        ELSIF (clk'EVENT AND clk = '1') THEN
            IF Q = "1001" THEN
                Q<="0000";
            ELSE Q<=(Q + 1);
            END IF;
        END IF;
    clk2<=Q(3);
    END PROCESS;
END a;
```

Notice these features of decade.vhd:

- Within the PROCESS, first **res** is checked to see whether the output should be reset.

- To divide by ten, the counter should count up to nine and reset on the next clock transition. So, if **res** is HIGH and a LOW-to-HIGH transition of the clock has occurred, **Q** is checked to see whether it has reached nine. If so, **Q** resets to zero. Otherwise the count increments.

- At the end of the PROCESS, the output **clk2** is equated to the most significant bit of the counter, **Q(3)**. We only need the most significant digit. Its frequency is one-tenth of the input clock frequency.

Figure 10-24 shows sample simulation waveforms for decade. **clk2** completes one cycle for every ten cycles of **clk**.

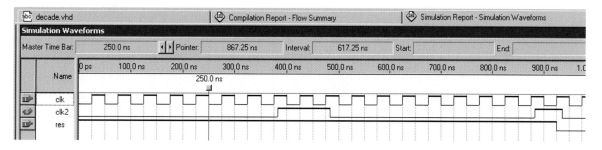

FIGURE 10-24 Simulation waveforms for decade.vwf

10.9 TROUBLESHOOTING COUNTERS

Digital circuits with counters in them are usually used to reduce a clock speed or to sequence the operation of other parts of the digital system. Because they seldom run at slow clock speed, an oscilloscope is one of the best choices to troubleshoot these types of circuits. Another good troubleshooting device for this type of circuit is a logic analyzer. A logic analyzer will store the logic value of several inputs (8 to 64 or more) at the transitions of a clock signal. It can also be set to start storing the data when it sees a preset bit pattern on the input lines. When this preset pattern appears on the inputs, the logic analyzer will start recording the logic values of the inputs at each clock transition. This data can be displayed to the user either as 1s and 0s or as waveforms in the time domain. If the technician knows the expected waveforms for a particular circuit, he can use the logic analyzer to see the actual waveforms being generated by the circuit, compare them to what he expects to see, and determine the fault in the circuit.

Of course, a good digital storage oscilloscope is also a valuable tool. The technician can set it to store a single trace and grab a piece of the counter outputs at the transitions of the clock signal. A good logic analyzer used to be quite expensive but today you can get boards to put into a regular personal computer for no more than the cost of a good oscilloscope. The personal computer gives the user the computing power to display the latched information in many formats, such as waveforms or binary data. As you will see in later chapters, the computer can also display the data as computer instructions if it is told which type of processor the data is coming from.

The circuit shown in Figure 10-25 is the display count circuit for the number of photos that have been exposed and printed in an automatic film processing machine. The count never goes above nine so the technician has been called in to repair it.

While the machine is running, the technician uses his oscilloscope to look at the incoming clock signal on pin 14 of U2. He finds a nice clean signal that seems to be running at the same rate as the light flash which exposes each print as the machine runs. He knows where to look for the signal because he studied the schematic for the counter circuit before he arrived at the photo lab to work on the machine.

He moves the oscilloscope probe to the clock input of the next counter in the chain, which is pin 14 of U1. This is the counter for the display which does not work, so the technician suspects that the clock is not getting to the IC or that the U1 is bad. The clock is present on the clock input (pin 14 of U1), so the technician checks to see if there are signals on the counter outputs. After checking all the outputs he finds that they are all LOW and not changing; the display reads zero all the time. But why? Looking a bit farther, he finds pins 3 and 2 of U1 are HIGH. This puzzles the technician because, if he has read the schematic correctly, they should be LOW. He is not exactly sure how the 74LS90 counter works, so he goes back to the data book and looks up the IC. What he finds is a description of the chip and schematic of the IC similar to that in Figure 10-17. After examining this, he realizes that pins 2 and 3 are the clear inputs for the counter and must be low if the counter is to run.

Going back to the running film processor, he looks at the voltage on both sides of resistor R_{20} and finds it to be ground potential. This means that the two pins are not connected to this

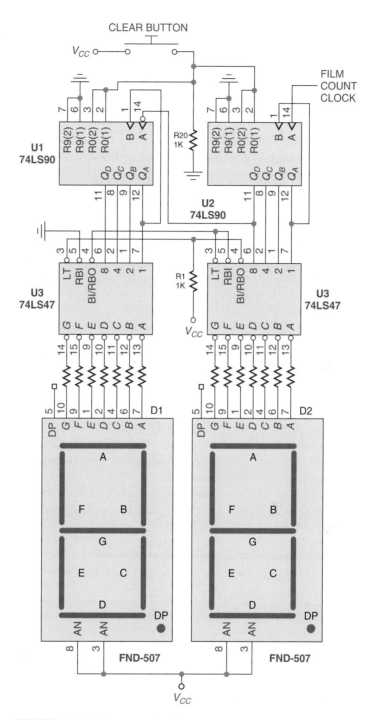

FIGURE 10-25 Counter circuit

resistor, as the schematic shows. Looking even closer at the board, he notices that there is a small 30-gauge wire which has been torn loose. The wire was a repair for a bad trace on the board—the very trace which connects the two pins to R_{20}. The technician takes a small pair of needlenose pliers, touches the loose wire to the end of resistor R_{20}, and the counter's seven-segment display starts to count. Fixing the wire was more work than finding it. The technician had to stop the running machine and disassemble a major part of it to reach the bottom of the board to solder the wire.

DIGITAL
APPLICATION Programmable Counter

The portion of the circuit shown is from a board which was designed to measure vibration levels at selected frequencies. It is used to monitor the operation of a large motor generator system and is controlled by a 68HC11 microcontroller. The 8254 (U31) is a programmable, presettable down counter. This chip has three of these counters in it. Counter 0 is used to set the frequency to look at for the motor end of the system and counter 1 is used for the generator side. This chip is designed to be placed on a computer bus, but in this application the data to program it is sent by a serial format. Notice the 74LS164 8-bit shift register. It is used to convert the serial data to a parallel format so it can be loaded into the 8254. (Schematic courtesy of Precise Power Inc.)

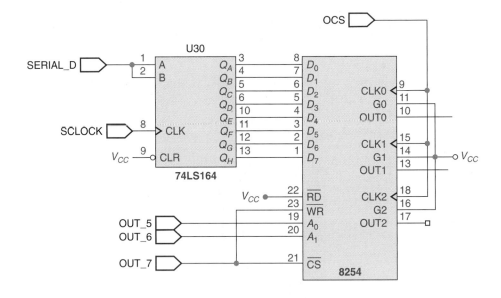

SUMMARY

- A ripple counter has the flip-flops clocked by the Q output of the previous flip-flop.

  The ripple counter can be constructed with JK flip-flops or edge-triggered D flip-flops. To make the ripple counter count down, the clock is clocked from the $\overline{Q}$ of the previous flip-flop. Because of the ripple down of one flip-flop clocking the next, the total time it takes for the ripple counter to change one count is the sum of the propagation delays for all the flip-flops in the counter. This means that the output of the counter is not stable during the time the count is rippling down the chain of flip-flops.

- The synchronous counter uses steering logic to cause the JK flip-flop to change to the correct state on the next clock.

  This allows all the flip-flops to be clocked at the same time and from the same clock. The synchronous counter is as fast as the propagation delay of one flip-flop and has no unstable state during the change from one count to the next. Synchronous down counters can also be made by using the $\overline{Q}$ outputs to drive the steering logic for the JK inputs.

- Divide-by-*N* counters can be made by the decode-and-clear method or by the synchronous method.

  The decode-and-clear method often will produce a small spike in the waveforms of the counter during the decode of the count which clears the counter. This can be a problem in some circuits which use the count to clock other flip-flops. The synchronous divide-by-*N* counter uses steering logic to set up the *JK* inputs of the flip-flops to clock to the correct state on the next clock. This prevents the decode spike from forming. The synchronous divide-by-*N* counter can be configured to count through any set of count values and in any order. This means it can be used as a device sequencer in a digital circuit.

- The presettable counter can be loaded with a count and then allowed to count either up or down from that point.

  Presettable counters can be used to make a programmable divide-by-*N* counter.

QUESTIONS AND PROBLEMS

1. Draw the logic diagram for a 5-bit ripple counter using negative edge-triggered *JK* flip-flops.

2. Draw the logic diagram of a divide-by-9 ripple counter using the decode-and-clear method. Use negative edge *JK* flip-flops.

3. Design a synchronous counter which will count in the following order: 000, 100, 001. Use negative edge-triggered *JK* flip-flops.

4. Make a list of negative-edge TTL and CMOS flip-flops in the data manuals.

5. Repeat number 4 for positive-edge flip-flops.

6. Draw the logic diagram for a synchronous counter which will count from 0 to 12. Use 7476 ICs and show pin numbers.

7. Repeat number 3 using positive edge-triggered flip-flops.

8. Draw the logic diagram for a divide-by-6 counter. Use the decode-and-clear method and one 7490 IC.

9. Draw the waveforms for the counter in number 8.

10. Draw the logic diagram for a ripple down counter which will count from 15 to 0 and start over again at 15. Use 7476 ICs and show pin numbers.

11. Draw the waveforms for the down counter in number 10.

12. Draw the logic diagram for a presettable counter which will count from 3 to 10 and start over again at 3. Use 7476 ICs and show pinouts.

13. Draw the waveforms for the counter in number 12.

14. Draw the logic diagram for a synchronous up-down counter which will count up from 0 to 15 or down from 15 to 0. Use 7476, 7432, 7408, and 7404 ICs and show pinouts.

15. Draw the waveforms for the up-down counter in number 14 with the following count: 0, 1, 2, 3, 4, 5, 4, 3, 2, 3, 4.

16. Design a divide-by-15 and ½ ripple counter.

17. Design a divide-by-7 and ½ synchronous counter.

18. Use 74LS93 ICs and 74LS90 ICs to divide a 1.8432-MHz clock to a frequency of 60 Hz.

19. Give two problems with a ripple decode-and-clear counter.

20. Draw the waveforms for the counter in question 17.

21. What would be the effect on the counter circuit in Figure 10-2 if the Q_A input to the NAND gate which clears the counter became disconnected?

22. Add another bit to the counter in Figure 10-4 and draw the logic diagram for the counter.

23. What would be the effect on the counter in Figure 10-10 if pin 5 of the 74LS08 AND gate became disconnected?

24. What would be the effect on the presettable counter in Figure 10-13 if pin 11 of IC1 went HIGH and did not ever go LOW?

25. Add another bit to the counter in Figure 10-16 and draw the logic diagram for the counter.

26. Explain the operation of the 74LS190 counter IC.

27. What would be the lowest frequency that could be obtained from a 74HC4020 counter IC if it had an input frequency of 983.04 kHz?

28. What would be the effect on the counter in Figure 10-20 if pin 2 of IC4A became disconnected?

29. Can a counter be made with transparent D flip-flops?

30. What would happen to the clock circuit in Figure 10-19B if pin 1 of IC22A became disconnected?

31. Write a program using VHDL that will divide the incoming clock frequency by 6.

32. Write a program using VHDL that will divide the incoming clock frequency by 13.

33. Write a program using VHDL that will divide the incoming clock frequency by 2, 4, 8, 16, and 32.

34. Write a program using VHDL that will cause the 8-bit output to count by 2s.

35. Write a program using VHDL that will implement an 8-bit binary down counter.

36. Write a program using VHDL that will implement a 4-bit counter that will count down from 9. Have an input control signal named **res9** that will cause the counter to reset to 9.

37. Write a program using VHDL that will implement an 8-bit binary up/down counter. If an input control signal is HIGH, have the counter count up; if LOW, have the counter count down.

38. Find and correct four errors in this VHDL program.

```
LIBRARY ieee;
USE ieee.std_logic_1164.all;
ENTITY countup IS
    PORT(clk: IN STD_LOGIC;
    Q:BUFFER STD_LOGIC_VECTOR(3 DOWNTO 0));
END countup;
ARCHITECTURE a OF countup IS
BEGIN
    PROCESS
    BEGIN
        IF (clk'EVENT AND clk = '1')
                Q <= (Q+1);
        END IF;
    END PROCESS;
END a;
```

39. Find and correct four errors in this VHDL program.

```
Library ieee;
USE ieee.std_logic_1164.all;
USE ieee.std_logic_unsigned.all;

ENTITY decade IS
    PORT(clk,res: IN STD_LOGIC;
        clk2:OUT STD_LOGIC);
END decade;

ARCHITECTURE a OF decade IS
    Q:STD_LOGIC_VECTOR(3 DOWNTO 0);
BEGIN
    PROCESS(clk,res,Q)
    BEGIN
        IF (res ='0') THEN
                Q ="0000";
        ELSIF (clk'EVENT AND clk = '1') THEN
                IF Q = 1001 THEN
                        Q<=0000;
                ELSE Q<=(Q + 1);
                END IF;
        END IF;
    clk2<=Q;
    END PROCESS;
END a;
```

OBJECTIVES

After completing this lab, you should be able to:

- Design a divide-by-N synchronous counter.
- Use typical TTL counter ICs.

COMPONENTS NEEDED

2	7476 dual JK flip-flop ICs
1	quad AND gate IC

PROCEDURE

PART I

Using negative edge-triggered JK flip-flops, design and construct a divide-by-10 synchronous counter. Find the missing values in the tables shown.

1. Define the operation of the flip-flop to be used.

Before Clock	After Clock		Before Clock	
Q	Q	$\rightarrow$	J	K
0	0	$\rightarrow$		
0	1	$\rightarrow$		
1	0	$\rightarrow$		
1	1	$\rightarrow$		

2. Define the JK for each flip-flop in the counter for each count required.

Before Clock				After Clock				JK State Before Clock							
D	**C**	**B**	**A**	**D**	**C**	**B**	**A**	J_D	K_D	J_C	K_C	J_B	K_B	J_A	K_A
0	0	0	0	0	0	0	1								
0	0	0	1	0	0	1	0								
0	0	1	0	0	0	1	1								
0	0	1	1	0	1	0	0								
0	1	0	0	0	1	0	1								
0	1	0	1	0	1	1	0								
0	1	1	0	0	1	1	1								
0	1	1	1	1	0	0	0								
1	0	0	0	1	0	0	1								
1	0	0	1	0	0	0	0								

3. Write the Boolean statement for each *J* and *K* input in the counter.

$$J_A = \qquad J_B = \qquad J_C = \qquad J_D =$$
$$K_A = \qquad K_B = \qquad K_C = \qquad K_D =$$

4. Convert the Boolean statements for the *JK* inputs to a logic diagram.

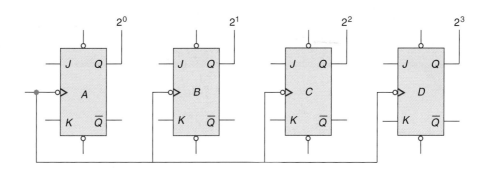

5. Construct the circuit you designed and use it to draw the following waveforms. Have your instructor check the operation.

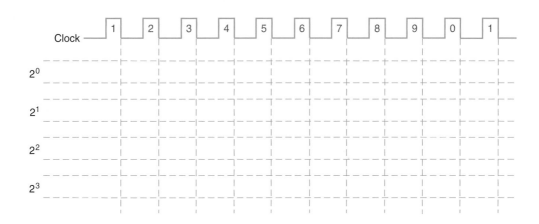

PART 2

Use a 7490 BCD counter to make the following counters. Have your instructor check the operation.

1. A divide-by-5 counter.
2. A divide-by-2 counter.
3. A divide-by-10 counter.

PART 3

Use a 7493 to make a divide-by-10 counter using the decode-and-clear method.

If your counter does not work properly, consider the following points:

1. Check all the power supply connections.
2. Check for any unconnected input on the *JK* flip-flop.
3. Use the oscilloscope to trace the signal from the input clock to the last flip-flop. They should match the expected waveform for a BCD counter.
4. If the waveforms do not match the expected BCD waveform, either your wiring is at fault or your design is incorrect.

LAB 10B   COUNTERS

OBJECTIVES

After completing this lab, you should be able to use Multisim to:

- Construct a counter.
- Troubleshoot a synchronous counter and a ripple counter.

PROCEDURES

1. Using Multisim, construct a 4-bit ripple counter from negative edge-triggered flip-flops. Use the digital clock to drive the counter and display the output waveforms on the logic analyzer.
2. After completing the 4-bit ripple counter, construct a 4-bit synchronous counter and use the logic analyzer as in problem 1 above.
3. Open file 10B-1.ms9. This is a 4-bit synchronous counter with two faults. Find the faults and fix the counter so it will run correctly.

Schmitt-Trigger Inputs and Clocks

OUTLINE

KEY TERMS

astable multivibrator   Schmitt trigger

hysteresis   voltage comparator

OBJECTIVES

After completing this chapter, you should be able to:

■ Explain the operation of a Schmitt-trigger input.

■ Use a Schmitt-trigger input to square up a sine wave.

■ Use the Schmitt trigger in the construction of a clock.

■ Describe how the 555 timer works and how it is used as a clock.

■ Use a 4001 CMOS IC to construct a crystal oscillator.

11.1 THE SCHMITT-TRIGGER INPUT

Figure 11-1 shows the graph of the input voltage versus the output voltage of a typical TTL **Schmitt-trigger** input. As the input voltage rises, the output stays at a LOW or 0 value until the input voltage reaches about 1.8 volts. At this upper threshold, the output snaps to a logic 1 value. When the input voltage drops, the output does not return to logic 0 until the input voltage drops below the lower threshold of about 0.8 volts. The difference in the upper and lower thresholds is called the **hysteresis** of the Schmitt trigger and is typically 1 volt for a TTL Schmitt-trigger input. The symbol for a Schmitt-trigger input is the graph in Figure 11-1, as shown in the noninverting gate on the right side of the figure.

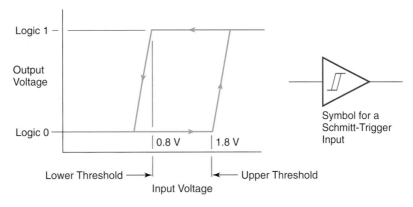

FIGURE 11-1 Output voltage versus input voltage of a Schmitt trigger

11.2 USING A SCHMITT TRIGGER TO SQUARE UP AN IRREGULAR WAVE

The fact that the Schmitt trigger has hysteresis is the reason it can be used to square up a wave such as a sine wave. As the input voltage rises and passes the upper threshold, the output voltage changes states. It will not change states again until the input voltage drops below the lower threshold. This is shown in Figure 11-2, which uses a 7414 Schmitt-trigger input to an inverter.

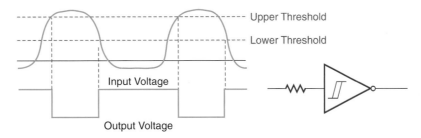

FIGURE 11-2 Using a Schmitt-trigger input to square up a sine wave

Notice the input voltage only goes 0.7 volts below ground. This is because the lower part of the sine wave has been clipped off by clamping diodes in the input of the 7414 IC. To protect this clamping diode, a current-limiting resistor is used to feed the sine wave.

The ability of the Schmitt trigger to square up a sine wave can be used to generate very accurate 60 and 120 Hz clock signals from the ac power grid. A power company must keep the frequency of the power grid at a very precise 60-Hz frequency to maintain the grid. This frequency can easily be used in digital applications. Figure 11-3 shows three methods of changing the ac sine wave to a digital TTL-level square wave.

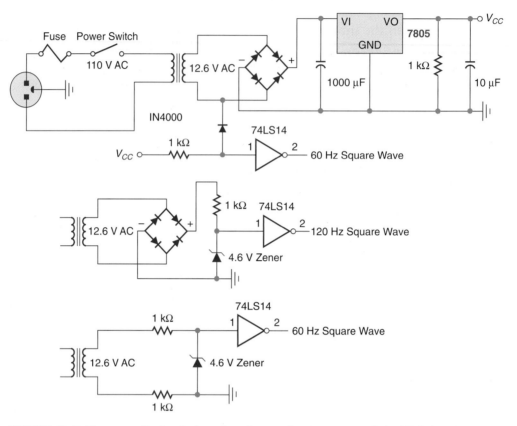

FIGURE 11-3 Three methods of changing the ac sine wave to a digital TTL-level square wave

11.3 A SCHMITT-TRIGGER CLOCK

A *clock* is an oscillator or, as it is sometimes called, an **astable multivibrator** and is used in a digital circuit. Figure 11-4 shows a simple clock made from a 7414 Schmitt trigger. When point A (which is the inverter output) is HIGH or at logic 1, the capacitor will charge through the 1-kΩ resistor and the TTL input as shown in Figure 11-5A. When the capacitor voltage reaches the upper threshold of the Schmitt trigger, the output of the inverter drops to 0 voltage or logic 0. This causes the capacitor to discharge through the 1-kΩ resistor, as shown in Figure 11-5B. When the capacitor voltage drops to the lower threshold, the inverter output changes back to a logic 1, thus completing one cycle of the clock as shown in Figure 11-5. Notice that the capacitor charges much faster than it discharges. This is because it can charge through the 1-kΩ resistor and the TTL input of the inverter, but can only discharge through the 1-kΩ resistor, which takes longer.

The frequency at which the clock will run is dependent on the *RC* time constant of charge and discharge. Because the TTL inverter input is a source load and helps to charge the capacitor,

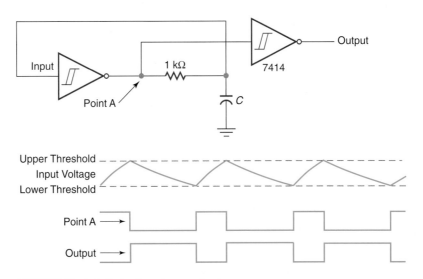

FIGURE 11-4 A Schmitt-trigger clock

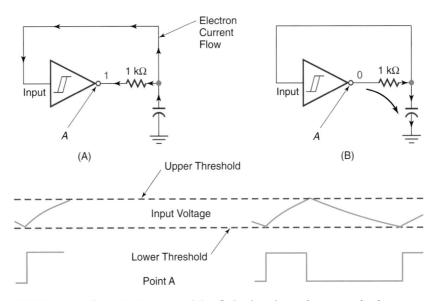

FIGURE 11-5 One clock cycle of the Schmitt-trigger inverter clock

the resistor cannot be much larger than 1 kΩ or the discharge voltage will never fall below the lower threshold of the Schmitt-trigger input. Therefore, we must keep the resistor, R, at about 1 kΩ, but we can change the capacitor, which will change the RC time constant for charge and discharge. The formula for the frequency of the clock is shown as a function of the capacitors, assuming that $R = 1$ kΩ.

$$f \approx \frac{6.69 \times 10^{-4}}{C}$$

The second inverter is used as a current buffer to drive other circuits without affecting the operation of the clock circuit.

The Schmitt-trigger oscillator shown in Figure 11-3 can be constructed from a 74C14 CMOS Schmitt-trigger inverter. When using a CMOS inverter, the input impedance is very high (around

10 MΩ), which means the input will not act as a source load as in the TTL inverter. This means that the resistor which is used in the *RC* circuit can be almost any value. A typical CMOS Schmitt-trigger input has about 2 volts hysteresis when the supply voltage is at 5 volts. The hysteresis voltage gets larger as the supply voltage rises. The formula for the approximate frequency of the CMOS Schmitt-trigger oscillator is shown below.

$$f \approx \frac{5.88 \times 10^{-4}}{RC}$$

(11-1)

EXAMPLE 11-1

Calculate the capacitor value for a 2-kHz oscillator made from a Schmitt-trigger circuit like the one in Figure 11-3.

Solution   This can be done using the formula given in the text.

$$f \approx \frac{6.69 \times 10^{-4}}{C}$$

$$C \approx \frac{6.69 \times 10^{-4}}{f}$$

$$C \approx \frac{6.69 \times 10^{-4}}{2 \text{ kHz}}$$

$$C \approx 0.33 \text{ μF}$$

The resistor must be 1 kΩ, and the final frequency will be approximately 2 kHz because of the differences in input impedances of the Schmitt-trigger inverter.

SELF-CHECK 1

1. What would be the approximate frequency of the Schmitt-trigger oscillator in Figure 11-3 if the capacitor were .15 microfarad.

2. Draw the waveforms for the input and output of a Schmitt-trigger inverter used to square up a sine wave.

3. What is the typical hysteresis for a TTL Schmitt-trigger input?

11.4 THE 555 TIMER USED AS A CLOCK

The 555 timer is a general-purpose timer IC which can be used for many applications. To understand its operation, we must first understand the operation of a **voltage comparator** made from an operational amplifier IC. Figure 11-6 shows an LM339 voltage comparator IC. On

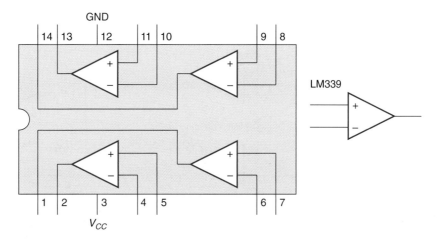

FIGURE 11-6 The LM339 voltage comparator

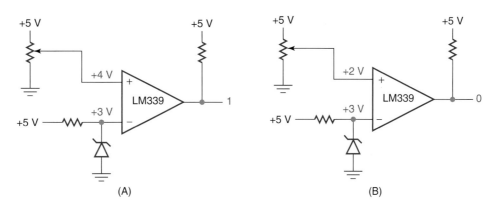

FIGURE 11-7 Operation of the LM339 voltage comparator

one IC there are four comparators which have two inputs each, one marked + and the other marked −. Each comparator also has one open-collector output, which is not typical of most operational amplifiers. This is because this operational amplifier is designed to be used as a voltage comparator in a digital circuit where the output will be ground or V_{CC} only.

The supply voltage for the IC can range from 3 V to 15 V, and the inputs have very high impedance. This means it can be used with CMOS circuits and that the inputs will not have any effect on a circuit to which they are connected.

If a reference voltage is placed on the negative input as shown in Figure 11-7A and the positive input is raised to a voltage greater than the negative input, the output will go to HiZ, which will produce a logic 1 because of the external pull-up resistor. If the positive input is now lowered to a voltage below the negative input voltage, the output will go to ground, which produces a logic 0, as shown in Figure 11-7B. In short, if the positive input is greater than the negative input, the output will be HiZ. If the positive input is less than the negative input, the output will be 0 or ground.

The 555 uses two of this type of comparator to set and reset a flip-flop. Figure 11-8 shows a 555 IC set up to make a clock. The reference voltage for the comparators is set by a voltage divider made up of three 5-kΩ resistors, which is where the chip got the name 555. This voltage divider places ⅓ of the supply voltage on the positive input of the lower comparator and

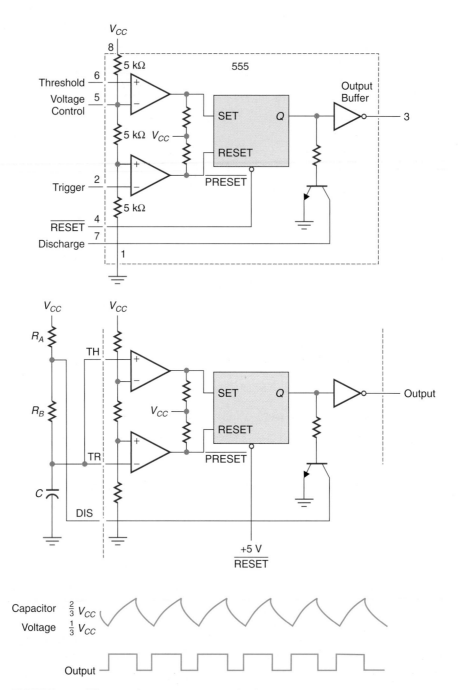

FIGURE 11-8 The 555 timer set up as a clock

⅔ of the supply voltage on the negative input of the upper comparator. The negative input of the lower comparator is called the trigger, and the positive input of the upper comparator is called the threshold.

If the threshold and trigger are tied together, they can be used to set and reset the flip-flop inside the 555 by raising them above ⅔ V_{CC} and lowering them below ⅓ V_{CC}.

When the trigger and threshold are above ⅔ V_{CC}, the top comparator is on or at logic 1 because the positive input is above the negative input, which is at ⅔ V_{CC} at all times. The bottom comparator is at logic 0 because the positive input is set at $\frac{1}{3}$ V_{CC} by the three 5-kΩ resistor voltage dividers, and the negative input is greater than that voltage. This sets the flip-flop to a 1 state, and the IC's output goes LOW because of the inverting buffer.

As the voltage on the trigger and threshold decreases below $\frac{2}{3}$ V_{CC} but greater than $\frac{1}{3}$ V_{CC}, the top comparator goes to 0 and the lower comparator stays at 0. This is the no-change state of the internal flip-flop. Therefore, it stays set to logic 1. When the input voltage on the inputs drops further to below $\frac{1}{3}$ V_{CC}, the bottom comparator turns on or goes to logic 1. This causes the flip-flop to be reset. Therefore, we can set or reset the internal flip-flop of the 555 by raising the two inputs, threshold, and trigger above $\frac{2}{3}$ V_{CC} and below $\frac{1}{3}$ V_{CC}.

The output of the internal flip-flop is also connected to the base of the discharge transistor inside the IC. When the Q output is at logic 1, it turns on the transistor connecting the discharge pin to ground. When the Q output is 0, the transistor is off and the discharge pin is at HiZ or not connected to anything.

Connecting a resistor capacitor circuit to the threshold, trigger, and discharge inputs as shown in Figure 11-8, will cause the 555 to oscillate or become a clock. When the flip-flop is reset to a 0 state, the discharge is in the HiZ state, which allows the capacitor to charge through R_A and R_B. When the capacitor voltage reaches a voltage just above $\frac{2}{3}$ V_{CC}, the flip-flop will change states to a logic 1. This causes the discharge transistor to be turned on, and the capacitor starts to discharge through R_B and the discharge transistor until the voltage drops to a little below $\frac{1}{3}$ V_{CC}. At this time the flip-flop is reset, and the whole cycle starts over. Figure 11-8 also shows the waveform for the 555 clock. Remember that the output of the 555 is the complement of the Q output of the flip-flop.

A formula for the frequency of the output can be computed by using the formula for the voltage charge of a capacitor as a function of the RC time constant and the time of charge.

$$\boxed{V_C = V_S(1 - e^{\frac{-T}{RC}})} \tag{11.2}$$

where V_C = capacitor voltage

V_S = supply voltage

T = time of charge

RC = resistance × capacitor, i.e., RC time constant

First, we need to solve the equation for the time of charge:

$$V_C = V_S(1 - e^{\frac{-T}{RC}})$$

$$\frac{V_C}{V_S} = 1 - e^{\frac{-T}{RC}}$$

$$\frac{V_C}{V_S} - 1 = -e^{\frac{-T}{RC}}$$

$$\frac{-V_C}{V_S} + 1 = e^{\frac{-T}{RC}}$$

$$\ln\left(\frac{-V_C}{V_S} + 1\right) = \frac{-T}{RC}$$

$$-T = (RC)\ln\left(\frac{-V_C}{V_S} + 1\right)$$

$$T = -(RC)\ln\left(\frac{-V_C}{V_S} + 1\right)$$

The time needed to change from $\frac{1}{3}\,V_S$ to $\frac{2}{3}\,V_S$ is equal to:

$$T = \left[-(RC)\ln\left(\frac{-\frac{2}{3}V_S}{V_S} + 1\right)\right] - \left[-(RC)\ln\left(\frac{-\frac{1}{3}V_S}{V_S} + 1\right)\right]$$

$$T = -RC\left[\ln\left(-\tfrac{2}{3} + 1\right) - \ln\left(-\tfrac{1}{3} + 1\right)\right]$$

$$T = -RC\left[\ln\left(\tfrac{1}{3}\right) - \ln\left(\tfrac{2}{3}\right)\right]$$

$$T = -RC\left[-1.10 - (-0.41)\right]$$

$$T = 0.69(RC)$$

Now we have an equation which will give us the time it will take to charge and discharge the capacitor in the middle $\frac{1}{3}$ of the voltage supply as a function of the RC time constant.

One cycle consists of one charge time and one discharge time. The RC time constant for the charge is $C(R_B + R_A)$ because the capacitor charges through both R_A and R_B; but the discharge RC time constant is CR_B because the capacitor discharges through R_B only. Therefore, the charge time will be longer than the discharge time because of the difference in the RC time constants. With this knowledge in hand, we can construct an equation for the total time of one cycle of the clock.

$$T_C = 0.69(R_B + R_A)C$$

and

$$T_{DC} = 0.69CR_B$$

where

$$T_C = \text{time to charge the middle } \tfrac{1}{3} \text{ of } V_{CC}$$

$$T_{DC} = \text{time to discharge the middle } \tfrac{1}{3} \text{ of } V_{CC}$$

Therefore the time for one cycle (P) is

$$P = T_C + T_{DC}$$

$$P = 0.69(R_B + R_A)C + 0.69CR_B$$

$$P = 0.69CR_B + 0.69CR_A + 0.69CR_B$$

$$P = 0.69C(R_B + R_A + R_B)$$

$$P = 0.69C(2R_B + R_A)$$

The frequency of a clock is equal to the reciprocal of the period P.

$$f = \frac{1}{P}$$

Therefore:

$$f = \frac{1}{0.69C(2R_B + R_A)}$$

$$f = \frac{1.44}{C(2R_B + R_A)}$$

This is the frequency formula for the 555 timer clock in Figure 11-8.

The 555 will produce a fairly stable output frequency from very long periods to about 0.5 MHz. It can run on 5 V to 18 V for a supply voltage and draws about 3 mA to 10 mA of current with no output load. One nice feature is its current output ability. The 555 can sink or source up to 200 mA of current, which means you can drive a heavy load with it.

EXAMPLE 11-2

What would be the resistor value for an oscillator made from a 555 timer using the circuit shown in Figure 11-8? The desired frequency is 1 kHz and the capacitor is to be 0.1 μF. Resistor A is to equal resistor B.

Solution   Solve the frequency equation for R.

$$f = \frac{1.44}{C(2R + R)}$$

$$f = \frac{1.44}{C(3R)}$$

$$3R = \frac{1.44}{Cf}$$

$$R = \frac{1.44}{3Cf}$$

$$R = \frac{1.44}{(3)(0.1 \times 10^{-6})(1 \times 10^{3})}$$

$$R = 4.8 \text{ k}\Omega$$

EXAMPLE 11-3

Use a 555 timer to produce a 5-second power on delay. Use the output of the 555 timer to turn on a relay.

Solution   Use the circuit in Figure 11-8 but disconnect the *DIS* pin from the RC timing circuit. This will keep the 555 timer from oscillating. When the power is off, the capacitor will be discharged through the resistance of the resistor in the *RC* timing circuit. After the power is applied, the capacitor will start to charge, and the relay will not be energized because the 555 timer output will be HIGH. When the capacitor reaches ⅔ of power supply, the output will go LOW and energize the relay. This is a typical circuit used to delay the starting of large motors until the power system has had time to come fully on-line.

To calculate the value of the capacitor, choose a value for the resistance in the *RC* circuit such as 100 kΩ, then calculate the value of the capacitor for the desired time.

$$T = -(RC) \ln\left[\frac{-V_C}{V_S} + 1\right]$$

$$T = -(RC) \ln\left[\frac{-\frac{2}{3}V_S}{V_S} + 1\right]$$

$$T = 1.1(RC)$$

$$C = \frac{T}{1.1R}$$

$$C = \frac{5}{1.1 * 100 \text{ k}\Omega}$$

$$C = 46 \text{ μF}$$

11.5 CRYSTAL OSCILLATORS

When very precise and stable clocks are needed, a quartz crystal is used to generate the frequency. Figure 11-9 shows a crystal oscillator made from a 4001 CMOS NOR gate. This circuit works well up to the limits placed on it due to the propagation delays of the CMOS NOR gate. The frequency of oscillation is determined by the vibrating frequency of the quartz crystal.

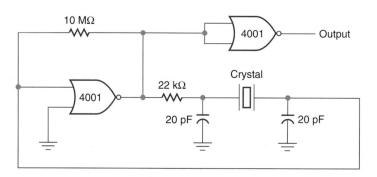

FIGURE 11-9 CMOS crystal oscillator

By placing a 10-MΩ resistor from the output to the input of the NOR gate, the input is biased at $\frac{1}{2}V_{DD}$. This basically makes a high-gain amplifier out of the gate. By placing a crystal and a PI network of capacitors from output to input, we can cause the amplifier to ring or oscillate at the frequency of the crystal. The buffer gate is used to keep the circuit being connected to the clock from affecting the operation of the clock.

Today the crystal oscillator used in most digital circuits is a hybrid circuit in a metal can. They have all the circuitry in the metal can to produce the desired frequency and a buffer that will sink and source approximately 20 to 30 mA of current. These types of oscillators are used in microcomputers and many other computer-controlled devices.

✔ SELF-CHECK 2

1. If R_A and R_B were 1 kΩ, what would be the value of the capacitor needed to produce a frequency of 1500 Hz in the 555 clock circuit shown in Figure 11-7?

2. What would be the upper threshold of the 555 timer if a 5-kΩ resistor were placed between the voltage control pin and V_{CC}?

11.6 TROUBLESHOOTING CLOCK CIRCUITS

Clock circuits are used in digital designs wherever an on/off digital signal is needed to drive the digital circuit. The type of clock used depends on the precision needed and the available gates and inverters in the design. A microcontroller design which is to run a precisely timed process will have a good crystal clock to drive the whole digital design because of the need for accurate timing. On the other hand, the same microcontroller may be used as a sequencer to turn on water valves one after the other. In this case, the precision of the clock driving the circuit is not critical and it can be made from available gates, such as a 74LS14, or even driven from the power line. Many microcontrollers have the ability to use a simple resistor and capacitor to set up a clock internal to the chip.

When troubleshooting clock circuits, the problem is usually quite evident because the whole system is down and there is no clock to run it. In most microcontroller and microprocessor designs, the clock is a hybrid circuit in a metal can and can easily be replaced. The need to replace the clock is not usually the reason for the loss of the clock. Most of the these clock modules are quite reliable and sturdy. Loss of the clock is usually caused by loss of power to the clock or the output being pulled down by a shorted load. The shorted load can often be seen with an oscilloscope. The clock signal will be present but the upper voltage level will be too low to be used as a good logic 1 in the digital circuit. This is a dead giveaway that there is an IC with a bad input or a capacitor has shorted to ground somewhere.

Clocks that are made from capacitive-discharge type circuits most often have problems with leaking capacitors. This can cause frequency changes or loss of the clock altogether. Heavy loads on the clock output can sometimes result in failure over time, but this is not usually the case. A voltmeter, an oscilloscope, and a good working knowledge of how the clock circuit works are what is needed to troubleshoot clock circuits.

The circuit in Figure 11-10 is the clock circuit that provides the fast and slow speeds for the film advance in a film processor. The operator of the film processor has noticed that the film advance has speeded up and once in a while this will cause the film to jam. The technician is called in to solve the problem.

The first thing the technician notices is that the film does seem to be moving faster than the last time he worked on the processor. To make sure this is true, he uses his oscilloscope to measure the frequency of the fast clock on pin 5 of U4A, which produces the clock signal. He

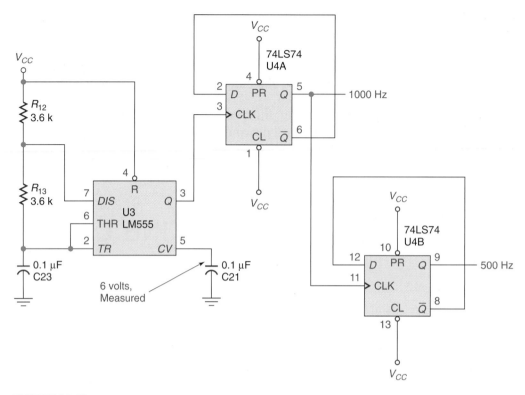

FIGURE 11-10

knows where to find this signal because he has a set of schematics for the film processor and has studied them. He finds just what he expected, the clock measures 1.35 kHz, about one-third too high. This should not interfere with the operation of the machine, but it could cause the film to jam because it was being advanced too fast. Looking at the clock output of the LM555 (pin 3 of U3), he measures a frequency of 2.7 kHz. This is just what it should be if the *D* flip-flop is working correctly.

After studying the LM555 clock circuit, he concludes that the capacitor, C23, must be leaking and thereby raising the output frequency. He puts his oscilloscope probe on pin 2 of U3 to see the charge/discharge voltage of the capacitor. He sees a typical charge/discharge voltage like that in Figure 11-8, but he notices that the upper and lower thresholds are not at one-third and two-thirds of the supply voltage. This should not be. The technician knows that an LM555 timer has an internal voltage divider of 5-kΩ resistors to set the upper and lower thresholds and the only way that can change is to put a resistor on the VC (voltage control) input. He then looks at the schematic and sees that there is a capacitor connected to this input (VC). He believes he has found the problem. This capacitor is leaking and pulling down the upper and lower thresholds. Using a pair of wire cutters he clips the cap (C21) out of the circuit and then measures the frequency at LM555 (pin 3 of U3). This time it measures a good solid 2 kHz.

After replacing the capacitor, the technician checks the operation and speeds of the stepper motors that drive the film advance and finds them working correctly. The problem is solved and the film will not jam because the film advance is now operating at the correct speed.

DIGITAL APPLICATION # A Schmitt-Trigger Clock

The circuit shown below creates the LOW-going load pulse to load a shift resister from the parallel outputs of a character generator. This is done on the rising edge of the 1-MHz character clock, which passes a one from the D input to the Q output of U3A (edge-triggered D flip-flop). This causes Q complement (pin 6) to go low also. This LOW signal from Q complement is delayed for a fraction of a microsecond by the one-shot made from U1B, U2C, C_2, and R_2. This delay circuit will be studied in depth in the next chapter. After the short delay, pin 1 (Clear) of the flip-flop goes LOW and clears the flip-flop, causing the Q output to go LOW again.

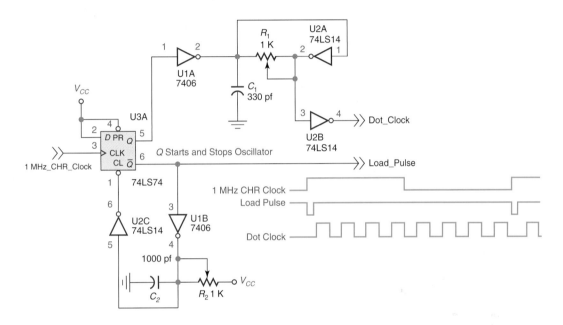

When the flip-flop's Q output went HIGH, the 8-MHz dot clock stopped because the output of the 7406 inverter went LOW. (U1A) When the Q output went back LOW, the output of the 7406 (U1A) went to HiZ, thus allowing the dot clock to run. The dot clock is a free-running TTL-level clock made from a 74LS14 Schmitt-trigger inverter. Notice the 1-kΩ pot (R_1), which is used to adjust the frequency of the clock. This clock produces the dots which makes up one row of a character on the screen of a standard TV. By increasing the dot clock frequency, the character is made thinner and there is more space between the characters on the screen.

This circuit was designed by the author as part of a system to display information on a large screen television.

SUMMARY

- A Schmitt-trigger input has hysteresis, which makes it useful in cleaning up poor digital input and producing clean square waves from slow-rising waveforms, such as the voltage across a capacitor as it charges.

The hysteresis also makes the Schmitt trigger a good element to use in the construction of a simple relaxation oscillator. The input impedance of the standard TTL Schmitt-trigger input can be a problem in design of a simple clock or oscillator. The CMOS Schmitt trigger has a very high input impedance, which eliminates this problem. The problem with the CMOS Schmitt trigger is the low output current drive of a CMOS device and the limiting speed of the CMOS device.

- The 555 timer is a very versatile IC. It can be used to produce clocks from very fast to very long time periods.

  The output can sink or source up to 200 mA, which makes it capable of driving a small relay or light if needed. The output frequency of the 555 timer can be accurately calculated and will remain quite stable. This IC has many uses other than the construction of an oscillator.

- Crystal oscillators are by far the most common type of clock used to run digital devices.

  The stability and accuracy of the quartz crystal are very good and make for an extremely stable clock. Today, most computers and other digital devices use a crystal module which has the crystal and all the needed parts to make the oscillator inside. These modules can usually drive up to 20 mA and come in a wide variety of frequencies.

QUESTIONS AND PROBLEMS

1. Using the Data Manual, find the upper and lower thresholds for a 74C14 when V_{DD} is 10 V.

2. Find the operating frequency of a 555 timer wired as in Figure 11-8. Use the following values for the components.

	R_B	R_A	C
a.	1 kΩ	1 kΩ	0.01 μF
b.	3 kΩ	1 kΩ	0.1 μF
c.	1 kΩ	5 kΩ	10 μF

3. Draw the waveforms for the circuit in Figure 11-11.

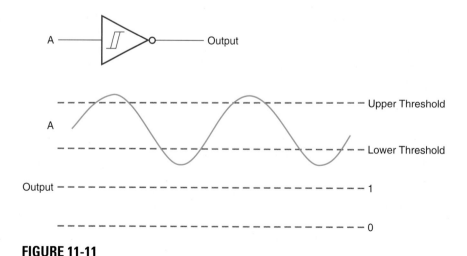

FIGURE 11-11

4. How much time will it take for the capacitor shown in Figure 11-12 to charge to $\frac{1}{4} V_S$, $\frac{1}{2} V_S$, and $\frac{3}{4} V_S$ from no charge when the switch is closed?

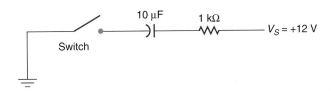

FIGURE 11-12

5. Using the approach shown in section 11.4 to compute the formula for the frequency of the 555 timer clock, compute the formula for the CMOS Schmitt clock shown in Figure 11-13. Remember the CMOS input is very HiZ.

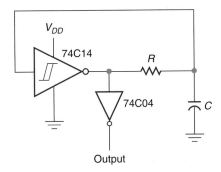

FIGURE 11-13

6. Draw the logic diagram for a Schmitt-trigger clock, like the one shown in Figure 11-4, which will oscillate at a frequency of 5 kHz. Use a 7414 IC and show pinouts and component values.

7. Why does it take longer to discharge the capacitor in Figure 11-4 than to charge it?

8. What is the output value of an LM339 voltage comparator if the positive input is greater in voltage than the negative input?

9. Draw the logic diagram for a clock which will produce a 2-kHz even-duty cycle TTL square wave. Use a 555 timer and a 7476 IC and show pinouts.

10. Draw the waveforms for the clock in number 9. Show the capacitor voltage of the 555 timer, the output of the 555 timer, and the output of the *JK* flip-flop.

11. Compute the formula for the frequency of the 555 time clock in Figure 11-8 if a 5-kΩ resistor is placed from the voltage control to ground.

12. Repeat number 11 but place the resistor from the voltage control at V_{CC}.

13. Draw the logic diagram for a Schmitt-trigger circuit to square up a 5 V peak-to-peak ac sine wave and produce a 10 V HIGH-to-LOW dc wave. Use a 7414 and a 7407.

14. Draw the waveforms for the circuit in number 13.

15. Compute the formula for the frequency of the clock in Figure 11-4 if a 74C14 IC is used, and the V_{DD} is 5 V.

16. What would be the lower threshold of a 555 timer with a 10-kΩ resistor connected between the voltage control pin and ground?

17. Why is the resistor in the TTL Schmitt-trigger clock kept no larger than 1 kΩ?

18. What is the period of a 1500-Hz clock?

19. What are the approximate upper and lower threshold voltages for a 7414 Schmitt trigger?

20. Draw the symbol to indicate a Schmitt-trigger input.

21. What are the upper and lower thresholds for the 555 timer in Figure 11-8 if a 5-kΩ resistor is placed from the voltage control (pin 5) to ground?

22. What happens to the clock in Figure 11-4 if a 10-kΩ resistor is used instead of the 1-k shown?

23. What type of output does the LM339 have?

24. Give three criteria needed for a circuit to oscillate.

25. Why is a 74C04 used in Figure 11-13?

26. How much time will it take for the capacitor voltage in Figure 11-12 to reach 5 volts?

27. Draw the waveform for the input to the Schmitt-trigger inverter (pin 1) of the top circuit in Figure 11-3.

28. What can be done to lower the zero voltage level of the Schmitt-trigger input in question 27?

29. What happens to the output of the circuit in Figure 11-11 if pin 12 of U4B is disconnected?

30. What happens to the output of the circuit in Figure 11-11 if resistor R_{12} becomes 5.6 kΩ?

OBJECTIVES

After completing this lab, you should be able to:

- Explain the operation of a Schmitt trigger and measure the upper and lower thresholds.
- Explain the operation of an oscillator made from a Schmitt-trigger inverter.
- Explain the operation of a 555 timer used as an oscillator.

COMPONENTS NEEDED

1	7414 hex Schmitt-trigger inverter IC
2	1-kΩ ¼-W resistors
2	capacitors (value to be computed)
1	555 timer IC
1	0.01-µF capacitor
2	20-pF capacitors
2	crystals of different frequency below 1 MHz
1	10-MΩ, ¼-W resistor
1	22-kΩ, ¼-W resistor
1	4001 CMOS quad NOR gate IC

PROCEDURE

PART 1

1. Construct the circuit for the figure shown and supply the input with a 4-V ac to 5-V peak ac input waveform at about 1 kHz frequency.

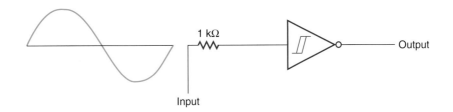

2. Use the scope to measure the upper and lower thresholds. Using graph paper, draw the input and output waveforms.

PART 2

1. Construct the oscillator shown.

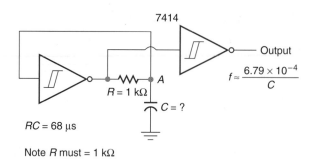

2. Compute the expected frequency for the oscillator.

3. Display the output wave on the oscilloscope and measure the frequency.

4. Display the output wave and point A on the oscilloscope. Using graph paper, draw the two waveforms.

PART 3

1. Construct the 555 astable multivibrator shown.

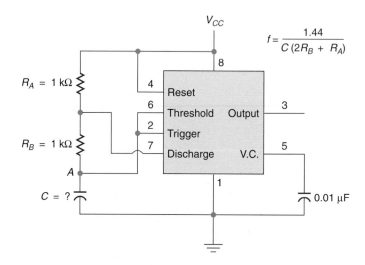

2. Compute the value of C that will produce a frequency of 9.6 kHz.

3. Display the output and point A on the oscilloscope and measure the frequency. Using graph paper, draw the waveforms.

PART 4

1. Construct the crystal oscillator shown using the crystal provided.

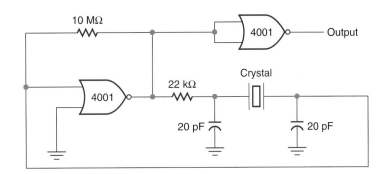

2. Measure the output frequency of the oscillator on the oscilloscope. Does it match the frequency of the crystal?

3. Change the crystal to a different crystal with a different frequency and measure the output frequency again. Does the output frequency match the crystal?

LAB 11B

CLOCKS

OBJECTIVES

After completing this lab, you should be able to use Multisim to:

- Construct a Schmitt-trigger oscillator.
- Measure and design typical oscillators.
- Troubleshoot an oscillator circuit.

PROCEDURES

1. Using Multisim, construct a Schmitt-trigger oscillator that will run at 1 kHz.
2. Open file 11B-1.ms9 and complete the instructions contained in the circuit description box.
3. Open file 11B-2.ms9 and complete the instructions contained in the circuit description box.
4. Open file 11B-3.ms9 and complete the instructions contained in the circuit description box.

One-Shots

OUTLINE

KEY TERMS

data separator

nonretriggerable one-shot

pulse stretcher

retriggerable one-shot

OBJECTIVES

After completing this chapter, you should be able to:

- Describe how to use an *RC* network to debounce a switch.
- Describe how to make a pulse stretcher.
- Describe how to condition the input of the pulse stretcher to make a retriggerable one-shot.
- Use the 555 timer as a one-shot.
- Use the 74121 and 74122 one-shots.
- Make a data separator from one-shots.

12.1 A ONE-SHOT DEBOUNCE SWITCH

The circuit shown in Figure 12-1 uses an *RC* time constant and a Schmitt trigger to debounce a momentary switch or button. When the button is pushed, the capacitor is discharged very rapidly. When the button is released, the bounce of the metal contacts makes and breaks the circuit in a random manner. The open switch allows the capacitor to start to charge through the resistor connected to V_{CC}. The time for the voltage to reach the upper threshold of the Schmitt-trigger input is dependent on the *RC* time constant. Therefore, the switch must remain open for a certain period of time before the output will change states.

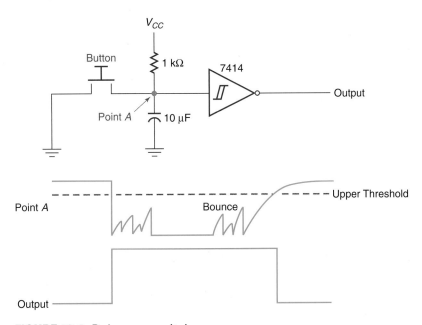

FIGURE 12-1 Debounce switch

This is a **retriggerable one-shot**, because each time the switch closes, the capacitor is discharged and the timing cycle begins again. The input may bounce, but the output will not change until the switch has stayed open for a period of time determined by the *RC* time constant.

 # 12.2 THE PULSE STRETCHER

By adding an open-collector gate such as a 7406 to the input of the debounce switch in Figure 12-1 and changing from a 7414 to a 74C14, a **pulse stretcher** can be made, as shown in Figure 12-2. The input to the 74C14 is held LOW by the output of the 7406 as long as the 7406 input is HIGH. When the input goes back LOW, the capacitor starts to charge through the resistor. When the capacitor voltage reaches the upper threshold of the 74C14 Schmitt trigger, the output will snap LOW. This will stretch the positive-going pulse by the amount of time it takes the capacitor to charge to the upper threshold of the 74C14 Schmitt trigger.

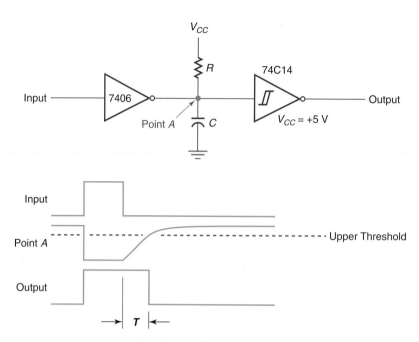

FIGURE 12-2 Pulse stretcher

If we use a CMOS Schmitt trigger, we can ignore the input impedance because a CMOS input has very high impedance; therefore, the only things which will affect the charge time are the resistor and capacitor used. By using a little algebra, we can devise a formula for the time which this circuit will stretch the input pulse.

$$T = -(RC) \ln \left(\frac{-V_C}{V_S} + 1 \right)$$
(12-1)

where V_C = capacitor voltage
  V_S = supply voltage
  T = time of charge
  RC = resistance × capacitor, i.e., RC time constant

Using the CMOS data specification for a 74C14, we find that the upper threshold voltage is 3.6 V at a V_{DD} of 5 V. Substituting this in the equation, we get a simpler formula for the time stretch for the circuit in Figure 12-2.

$$T = -(RC) \ln \left(\frac{-3.6 \, V}{5 \, V} + 1 \right)$$

$$T = 1.27 \, (RC)$$

This formula does not take into account the propagation delays for the two gates used; however, this is of no concern except in very short total time pulses.

EXAMPLE 12-1

What would be the value of the resistor and capacitor needed to stretch a pulse 1.5 msec using the circuit in Figure 12-2?

Solution

$$T = 1.27 \, (RC)$$

Solve for R and then choose a capacitor of reasonable value, such as 0.1 μF.

$$R = \frac{T}{1.27C}$$

$$R = \frac{1.5 \text{ msec}}{1.27 \times .1 \text{ μF}}$$

$$R = 11.8 \text{ kΩ}$$

12.3 THE RETRIGGERABLE ONE-SHOT

In the circuit just discussed, the output could not time out or change states until the input returns LOW. This can be changed by conditioning the input of the 7406 with a capacitor, resistor, diode, or edge-triggered circuit as shown in Figure 12-3.

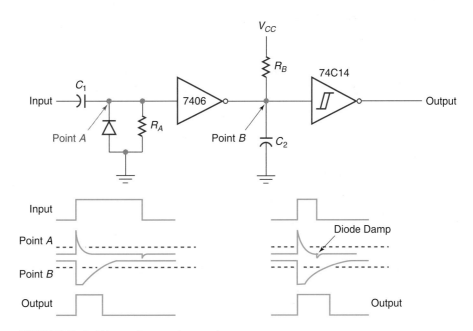

FIGURE 12-3 Edge-triggered one-shot

When the input goes HIGH, capacitor C_1 has no voltage drop, and all the voltage drop is across the resistor R_A. This causes the 7406 output to go LOW. After the capacitor charges to a large enough voltage, the input of the 7406 will be LOW, causing the 7406 output to go to HiZ.

This means that on the positive edge of the input, the 7406 output will put out a very short negative-going pulse. This negative-going pulse discharges capacitor C_2 and starts the timing cycle of the 74C14 Schmitt trigger.

The *RC* time constant for the 7406 conditioned input should be very short so that output of the 7406 will be a very short negative-going pulse. If this pulse is very short compared to the time of the 74C14, it can be ignored in the calculations of the time out for the whole circuit. Therefore, the formula for the time out of the one-shot is the same as for the pulse stretcher.

The advantage of this circuit is that the output pulse width is independent of the length of the input pulse width. If the input is retriggered by a positive edge before the 74C14 times out, capacitor C_2 is discharged and the timing cycle starts all over again. This is shown in the waveform in Figure 12-4.

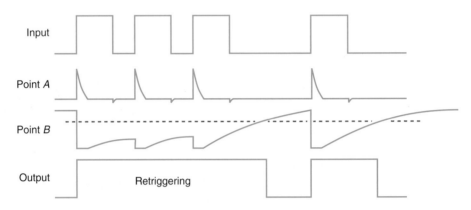

FIGURE 12-4 Triggering the one-shot in Figure 12-3

Notice that if the one-shot is retriggered quickly enough, the capacitor C_2 never reaches the upper threshold voltage of the 74C14 and the output stays HIGH. This is what is meant whenever we say a one-shot is retriggerable.

The diode in parallel with R_A prevents the discharging capacitor from pulling the input voltage to the 7406 any lower than the forward bias voltage of the diode, which is 0.7 V. This diode does the same thing as the clamping diodes inside the 7406. The diode can be left out in many cases when the size of the capacitor is small and the discharge current is also small.

 12.4 THE NONRETRIGGERABLE ONE-SHOT

If it is not desired to have the one-shot retriggerable, an OR gate can be used to inhibit the input for the period of the one-shot's timing period. This is shown in Figure 12-5.

Notice that the OR gate is a 74ALS32. This is because the output of the CMOS Schmitt-trigger inverter will only supply .36 mA. The input of the 74ALS32 will require .1 mA. This will not load down the Schmitt-trigger inverter's output and will leave enough current drive to connect the output of the one-shot to another gate in a circuit.

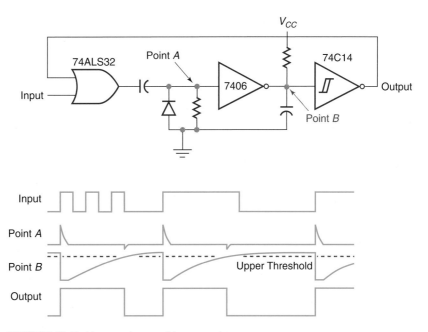

FIGURE 12-5 Nonretriggerable one-shot

This interface problem can be eliminated by using a standard 7414 TTL IC which has more drive ability, but we have to figure in the input impedance of its input to the *RC* timing formula used to compute the time out of the one-shot.

 SELF-CHECK 1

1. What would be the value of the capacitor needed to stretch a pulse 1 msec using the circuit in Figure 12-2? The resistor is to be 10 kΩ.

2. What is a retriggerable one-shot?

 12.5 THE 555 AS A ONE-SHOT

Using some of the techniques we just discussed, the 555 timer can be made into a stable one-shot. The timing period can be long or short. Figure 12-6 shows a 555 set up as a one-shot. Notice the input conditioning of the *RC* and the diode to produce edge triggering. This is a nonretriggerable one-shot whose timing period is dependent on the *RC* time constant of R_A and C_A.

When a LOW-going edge trigger is input, the flip-flop is reset and the discharge pin goes to HiZ. This allows capacitor C_A to start to charge. When it reaches $\frac{2}{3}$ V_{CC}, the flip-flop will be set, causing the discharge transistor to turn on discharging capacitor C_A, thus ending the timing cycle until another LOW-going edge is input.

The time of pulse width is the time it takes C_A to change to $\frac{2}{3}$ V_C. From previous calculations we know the equation for the time of charge for a resistor and capacitor, and with a little algebra we can compute a formula for the timing period of the 555 one-shot in Figure 12-6.

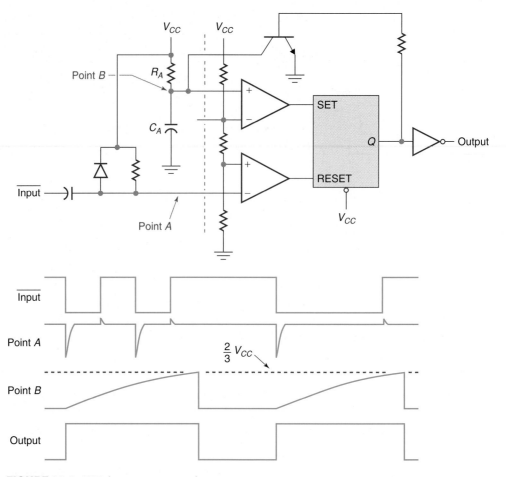

FIGURE 12-6 555 timer as a one-shot

$$T = -(RC) \ln \left(\frac{-V_C}{V_S} + 1 \right)$$

$$T = -(RC) \ln \left(\frac{-\frac{2}{3} V_C}{V_S} + 1 \right)$$

$$T = -(RC) \ln (\frac{1}{3})$$

$$T = 1.1 \, (RC)$$

EXAMPLE 12-2

Design a 555 one-shot with a time out of 5 seconds using the circuit shown in Figure 12-6.

Solution   The first step is to calculate the resistor and capacitor of the RC timing part of the circuit.

$$T = 1.1 \ (RC)$$

$$R = \frac{T}{1.1C}$$

Choose an appropriate capacitor such as 10 μF.

$$R = \frac{5 \text{ sec}}{1.1 \times 10 \text{ μF}}$$

$$R = 455 \text{ k}\Omega$$

The value of the edge-triggering capacitor and resistor is not critical as long as it produces a LOW-going pulse to trigger the one-shot. A .1 μF and a 10-kΩ resistor will do.

12.6 THE 74121 AND 74LS122

The 74121 is a nonretriggerable one-shot which has three inputs to the timing circuits. The 74LS122 is a retriggerable one-shot with an active LOW clear. Figure 12-7 shows the pin outputs and truth tables. The pulse width can be controlled by an external resistor and capacitor or by an external capacitor and the internal resistor. These are handy ICs for many applications. The 74123 is a dual retriggerable one-shot with clear.

SELF-CHECK 2

1. Use a 555 timer to design a one-shot like the one in Figure 12-6. Make the time-out period 1 second.

2. Use a 74122 one-shot to design a retriggerable one-shot that will trigger on the rising edge of the input and have a time-out period of 10 msec.

12.7 THE DATA SEPARATOR

A **data separator** is a digital circuit which separates the data from the system clock of stored serial data. This data can be stored on magnetic tape or a magnetic disk, and it can be stored by several different methods.

The data separator shown in Figure 12-8 is designed to separate the clock from the data for the input to a UART. This type of serial data transmission was discussed in the chapter on shift registers. The data separator in Figure 12-8 is designed to work with the asynchronous receiver shown in the chapter on shift registers. The data separator produces a clock 12 times the baud rate of the data. This is the clock frequency the receiver needs to input the serial data and transfer it to the parallel register.

To understand the operation of this data separator, we first need to look at the data as it comes off the tape. Data is stored on the tape using two sine wave frequencies. A logic 1 is indicated by a 3.6 kHz frequency and a 0 by one-half of the 3.6 kHz frequency, 1.8 kHz. The baud rate or bit rate is one-twelfth of the one frequency or 3.6 kHz divided by 12, which equals 300 bits per second. This is a commonly used baud rate for serial data stored on tape recorders.

121 One-Shots

Truth Table

Inputs			Outputs	
A_1	A_2	B	Q	$\overline{Q}$
L	X	H	L	H
X	L	H	L	H
X	X	L	L	H
H	H	X	L	H
H	↓	H	⊓	⊔
↓	H	H	⊓	⊔
↓	↓	H	⊓	⊔
L	X	↑	⊓	⊔
X	L	↑	⊓	⊔

Pulse Width = 0.7(RC)

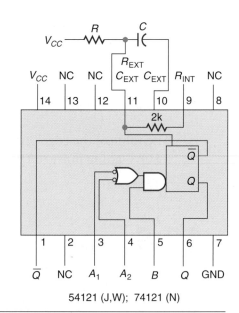

54121 (J,W); 74121 (N)

122 Retriggerable One-Shots with Clear

Truth Table

Inputs					Outputs	
Clear	A_1	A_2	B_1	B_2	Q	$\overline{Q}$
L	X	X	X	X	L	H
X	H	H	X	X	L	H
X	X	X	L	X	L	H
X	X	X	X	L	L	H
X	L	X	H	H	L	H
H	L	X	↑	H	⊓	⊔
H	L	X	H	↑	⊓	⊔
H	X	L	H	H	L	H
H	X	L	↑	H	⊓	⊔
H	X	L	H	↑	⊓	⊔
H	H	↓	H	H	⊓	⊔
H	↓	↓	H	H	⊓	⊔
H	↓	H	H	H	⊓	⊔
↑	L	X	H	H	⊓	⊔
↑	X	L	H	H	⊓	⊔

Pulse Width = 0.45(RC)

54LS122 (J,W); 74LS122 (N)

Notes: ⊓ = one high-level pulse, ⊔ = low-level pulse.
To use the internal timing resistor of 54121/74121, connect R_{INT} to V_{CC}
An external timing capacitor may be connected between C_{EXT} and R_{EXT}/C_{EXT} (positive)
For accurate repeatable pulse widths, connect an external resistor between R_{EXT}/C_{EXT} and V_{CC} with R_{INT} open-circuited
To obtain variable pulse widths, connect external variable resistance between R_{INT} or R_{EXT}/C_{EXT} and V_{CC}.

FIGURE 12-7 One-shot truth tables

The first 741 op amp is used to amplify the input signal from the tape recorder's output. The second 741 op amp is a zero crossing detector to make a square wave which changes state on the zero transition of the sine wave. This zero detector has a hysteresis of a little over 1 V to help eliminate any noise problem at the transition point. A diode is used to prevent the input signal to the 339 op amp from going below ground. The 339 op amp converts the +12 V square wave to a standard TTL square wave of 0 V to +5 V, as shown in Figure 12-8 at point B.

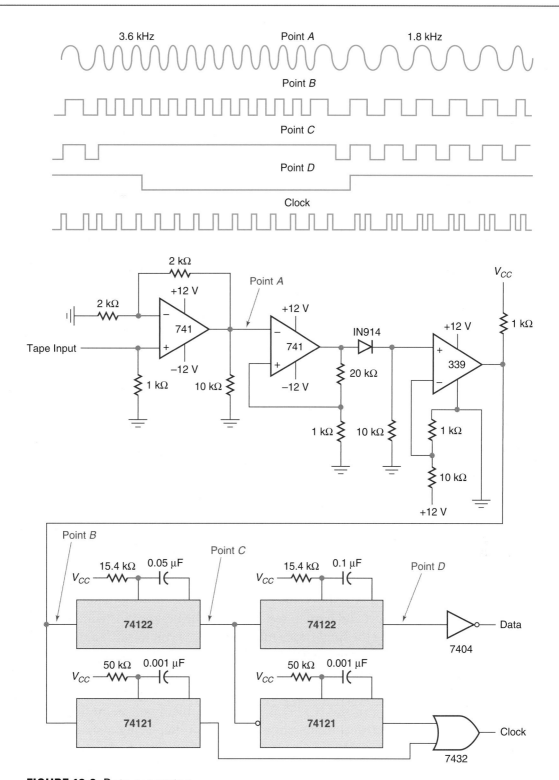

FIGURE 12-8 Data separator

After the signal is recovered and converted to a TTL-level input, it is fed to a 74122 retrigger-able one-shot whose pulse width is set to 1.25 times the period of the 1 frequency. This means that the 74122 one-shot will be retriggered when the frequencies are 3.6 kHz or 1 and the output will not time out and return to a 0 level. When the input frequency to the 74122 one-shot is 1.8 kHz or a 0, the output will time out, producing a square wave whose frequency is that of the 0 frequency or 1.8 kHz. This is shown on the timing diagram in Figure 12-8 at point C.

The output of the first 74122 is fed to the input of the second 74122. The pulse width of this one-shot is 1.25 of one cycle of the 0 frequency or 1.8 kHz. When the first one-shot is HIGH because it is being retriggered, the second one-shot will time out and go LOW for the period of time a 1 is on the incoming signal to the first one-shot. When the incoming frequency drops to the 0 frequency, the first one-shot passes on the 1.8 kHz frequency to the second one-shot. This causes the second one-shot to be retriggered, and the output will go HIGH for the time that a 0 is being fed to the first one-shot. This is shown in Figure 12-8 at point *D*. Notice the second one-shot produces a 0 when a 1 frequency is fed to the first one-shot and a 1 when a 0 frequency is fed to the first one-shot's input. This is corrected by the inverter on the output of the second one-shot.

The clock is recovered by two nonretriggerable 74121 one-shots which have a short pulse width. Their outputs are ORed together to produce the clock, which is 12 times the baud rate. One of the 74121 one-shots is fed from the positive edge of the incoming TTL signal and produces the proper clock when the incoming frequency is at 3.6 kHz or a 1 frequency, but will only produce half that when the incoming frequency drops to a 0 value. At this point the second 74121 supplies the missing clock pulse because it is fed from the falling edge of the output of the first one-shot, which will only be there when a 0 frequency is present. This is shown in Figure 12-8. Notice that the duty cycle of the clock wave is not even, but it will still drive the asynchronous receiver quite well.

This method of recording the receiver data and clock on the tape at the same time removes the problem of uneven baud rates due to the uneven mechanical speed of the tape recorder. It also allows us to use a device which was really designed for storing analog speech to store digital data.

12.8 TROUBLESHOOTING ONE-SHOTS

Circuits containing one-shots can often be difficult to troubleshoot because the operation of the one-shot is short and hard to see. Probably the best test instrument to use is a good storage oscilloscope. Using the single trace mode will allow you to catch a single pulse from the output of the one-shot and compare it to the input signal which triggered the one-shot. One-shots are often affected by heavy loads on their outputs and by bad capacitors. The chips which are used to make the one-shot are often a problem because of upper and lower threshold variations in the manufacturing of the ICs. This type of problem will be noticed if the precision of the one shot time is critical to the circuit operation. Replacing a standard TTL Schmitt-trigger device with an HC or HCT device will affect the one-shot timing because of the differences in the thresholds.

The circuit shown in Figure 12-9 is a one-shot created out of a standard 74LS74 *D* flip-flop, 7406 inverter, 74HC14 Schmitt-trigger inverter, resistors, and a capacitor. The LM555 timer will generate a clock signal of about 20 Hz. When the rising edge of this clock is seen at the clock input of the flip-flop, the 1 value on the *D* input is passed to the *Q* output and the *Q* complement is driven LOW. The LOW on the *Q* complement is delayed by the pulse-delay circuit made from the 7406 and 74HC14. When the delayed LOW signal reaches the flip-flop's "Clear" input, the flip-flop is cleared, the *Q* output goes LOW, and the *Q* complement goes

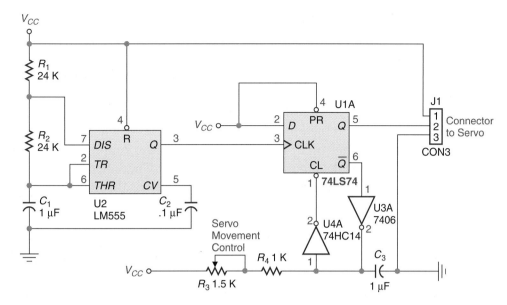

FIGURE 12-9

HIGH. This ends the pulse started with the rising edge of the clock from the LM555. The pot will vary the length of the pulse from about 1 ms to 2 ms in length. This circuit will produce a 1- to 2-ms pulse every one-twentieth of a second.

The circuit is designed to drive a standard model servo such as those used on radio-controlled airplanes. The technician designed and built this circuit to test the servos used by the shop which makes models for the film studio where he works. The device worked well for a month or more and then quit altogether.

After opening up the aluminum box which the technician had made to hold the printed circuit board and the other parts of the device, he noticed that one of the wires coming from the servo connector was burned off. It was the middle one, which feeds to the flip-flop's Q output. He surmised that one of the model builders had gotten the connectors mixed up and placed a battery connector on the device's output connector, thus placing a HIGH voltage on the flip-flop's Q output. So he quickly replaced the wire and put in a new 74LS74 flip-flop and 7406 open-collector inverter. He was able to do this easily because he had placed IC sockets on the PC board when he had made it, in anticipation of this type of problem. This fixed the circuit and the technician was able to control the movement of a servo with the control pot.

Unfortunately, the servo's, movement still was not smooth. It jerked and did not move with the pot movement well at all. The technician got out his storage oscilloscope to look at the pulses being generated by the circuit. He set his oscilloscope on auto-triggering and storage. This allowed him to see the pulse coming from the circuit at the slow rate of 20 Hz. He saw that the pulse width did not follow the movement of the pot. He had expected to see this, so he removed the wire going to the pot and placed it on a new pot which he just held in his hand. Moving the new pot, he could very precisely control the servo's movement. When he took a good look at the old pot, he noticed that there was some cyanoacrylate glue on the outside of the pot. This is the type of glue which is fast drying and is used by the model builders. The glue had gotten into the pot and onto the resistive surface which the wiper rubs. The technician put the device back together with the new pot and took it back to the model shop.

DIGITAL APPLICATION Pulse Stretcher

The circuit shown below is used to supply the driving signals to an IGBT (insulated gate bipolar transistor) driver. The IGBTs are set up in an "H" pattern to turn on, off, and reverse the current in the exciter coil in a large written-pole generator. When changing the current from one direction to the other, one set of the IGBTs in the H bridge must be off before the other set can be turned on or they will short the power supply and burn up. This is the reason for the two one-shot delay circuits in the schematic. They are made from two 7406 and 74LS14 ICs (U9A, U11A, U9D, U11B). (Schematic Courtesy of Precise Power Inc.)

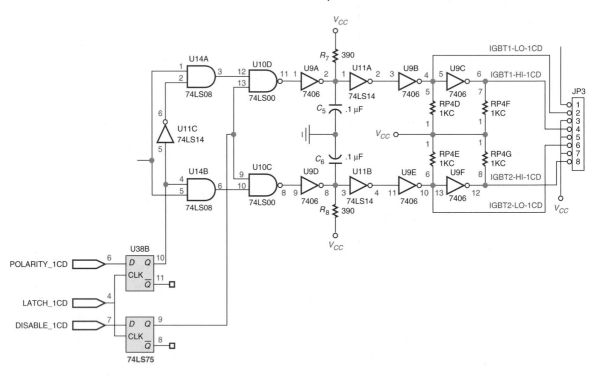

SUMMARY

- The pulse stretcher is a circuit that will add time to a pulse.

  The Schmitt-trigger input combined with an *RC* timing circuit is a good method of doing this. If a CMOS Schmitt-trigger input is used, the timing calculations are fairly simple and accurate. This pulse stretcher is often used to control multiplexed addresses to dynamic RAM and to lengthen pulses to devices which require longer length pulses.

- A one-shot is a device which will produce a predetermined pulse length when triggered.

  The length of the pulse can be accurately set. A retriggerable one-shot will reset the timing period each time the one-shot is triggered. A nonretriggerable one-shot will time out after the trigger and not reset the timing period during the output pulse time if triggered during that output pulse.

- The 555 timer can be configured to make an accurate one-shot which can have very long time-out periods.

This IC is quite often used as a one-shot because of its output drive current capabilities. The 555 could be used for driving the display light of a radio display that would come on when the radio was turned on or tuned, then go out after a set period of time. The 555 would have enough drive current to run the small lamp for this application.

- For most digital applications of one-shots, the IC versions of one-shots are used.

  Such ICs as the 74LS121, 74LS122, and 74LS123 are used instead of constructing a one-shot from many ICs.

QUESTIONS AND PROBLEMS

1. What is the total pulse width for the circuit shown?

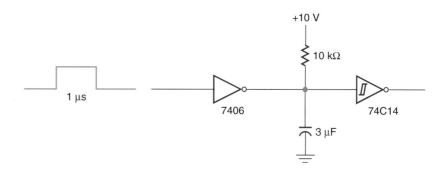

2. What would be the capacitor's value to produce a 1-μs pulse using a 74121 if the resistor used is 10 kΩ?

3. At what frequency will the circuit in Figure 12-3 stop timing out and have a constant 1 for the output? ($R = 1$ kΩ and $C = 0.01$ μF)

4. Complete the waveform for the one-shot described in number 3 for the given input waveform.

5. Draw the logic diagram for a pulse stretcher which will stretch the incoming pulse 5 microseconds. Use a 74C14 IC, a capacitor and resistor, and a 7406 IC.

6. Draw the waveform for the input signal, capacitor voltage, and the output waveform of the circuit in number 5.

7. Draw the logic diagram for an edge-triggered one-shot which has a 15-millisecond time-out pulse. Use a 74C14 IC, a 7406 IC, and two capacitors and resistors.

8. Using the nonretriggerable one-shot circuit in Figure 12-5, draw the logic diagram for a one-shot with a pulse width of 30 microseconds.

9. Draw the logic diagram for a one-shot using a 555 timer with a time-out period of 30 seconds.

10. Use a 74121 to produce a one-shot with a time-out period of 2 microseconds. Show pin numbers.

11. At what input frequency will a retriggerable one-shot, made from a 74122, stop producing a pulse on its output if the capacitor used is 0.1 μF and the resistor is 1 kΩ?

12. Design a retriggerable one-shot which will stop producing a pulse on its output when the input frequency is 1.5 kHz.

13. Make a list of CMOS one-shots and draw the pinouts.

14. If the capacitor is 0.5 microfarad and the resistor is 3.3 kilohms, what is the time-out period for the one-shot in Figure 12-2?

15. Use a 74122 to design a one-shot that will stop producing a pulse at an input frequency greater than 2 kHz.

16. Design a pulse stretcher that will stretch a pulse 20 msec.

17. How long will it take for a 0.1-μF capacitor to charge to 5 volts through a 100-kΩ resistor if 20 volts are applied to the *RC* circuit?

18. What is meant by the term nonretriggerable one-shot?

19. What would be the resistor value needed to produce a 4-μsec. pulse using a 74121 one-shot and a 0.001-μF capacitor?

20. Design a 555 timer one-shot like the one in Figure 12-6 to time out in three seconds.

21. What would be the effect on the operation of the circuit in Figure 12-2 if the resistor raised in value?

22. What would be the effect on the operation of the circuit in Figure 12-2 if the capacitor was shorted?

23. What happens to the operation of the circuit in Figure 12-9 if the 74HC14 is replaced with a 7414 IC?

24. What is the purpose of the diode in the circuit of Figure 12-3?

25. What could happen to the circuit in Figure 12-3 if the diode was missing or open?

26. What would be the effect on the operation of the circuit in Figure 12-6 if a 10-kΩ resistor is placed from the negative input of the top comparator to ground? This is the voltage control input on the LM555.

27. Why is the diode in the circuit of Figure 12-8?

28. What happens if a 10-kΩ resistor is placed from the negative input of the top comparators in Figure 12-6 to V_{CC}?

29. Use an LM339 comparator to create a delay one-shot.

30. Draw the logic diagram for a one-shot using a 74LS74 *D* flip-flop, 7406 inverter, and 74LS14 IC.

ONE-SHOTS

OBJECTIVES

After completing this lab, you should be able to:

- Construct a one-shot from a 74C14 and an *RC* circuit.

- Use a 74121 to shorten a positive pulse.

- Use the oscilloscope to observe the waveforms for the circuits in this lab.

COMPONENTS NEEDED

1   7414 TTL hex Schmitt-trigger input inverter IC

1   74C14 CMOS hex Schmitt-trigger input inverter IC

1   7406 hex open-collector output inverter IC

2   1-kΩ, $\frac{1}{4}$-W resistor

1   0.01-μF capacitor

1   1N914 diode or equivalent

PROCEDURE

1. Construct the circuit shown and use the ac signal generator to produce a 20-kHz input signal.

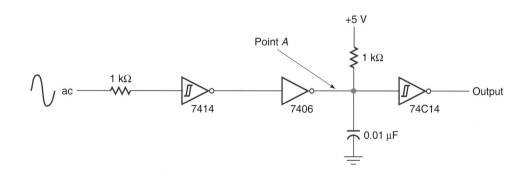

2. Using the figure shown in number 1, draw the expected waveforms and the actual waveform seen on the scope for the output and point *A*. Use graph paper.

3. Make the one-shot edge-triggered by adding the diode, capacitor, and resistor as shown.

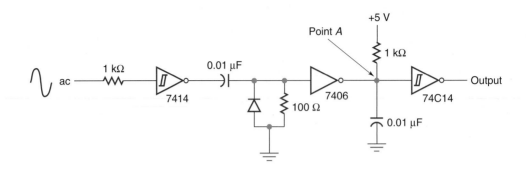

4. Using the figure shown in number 3, draw the expected waveforms and the actual waveform seen on the oscilloscope for the output and point *A*. Use graph paper.

5. Draw the logic diagram for a circuit which will take an ac 10-kHz even-duty cycle wave and produce a 10-kHz wave which is HIGH 25 percent of the time and LOW 75 percent of the time. Use a 7414 and a 74121 to construct the circuit.

LAB 12B   ONE-SHOTS

OBJECTIVES

After completing this lab, you should be able to use Multisim to:

- Understand the operation of a retriggerable one-shot.
- Troubleshoot a retriggerable one-shot.

PROCEDURES

1. Open file 12B-1.ms9 and run the simulation. This is a retriggerable one-shot made from an open-collector inverter and a Schmitt-trigger inverter. Change the value of the capacitors and observe the changes in the waveforms. When you are comfortable with its operation, move to Step 2 of this lab.

2. Open file 12B-2.ms9 and run the simulation. This is the same circuit as in Step 1, but there is one fault in the circuit. Find the fault and repair it.

CHAPTER 13

Digital-to-Analog and Analog-to-Digital Conversions

OUTLINE

KEY TERMS

analog-to-digital
binary ladder
digital-to-analog

flash converter
successive approximation
$2R$ ladder

OBJECTIVES

After completing this chapter, you should be able to:

- Use resistor networks for digital-to-analog conversion.

- Explain the operation of a TTL digital-to-analog converter.

- Use voltage comparators to produce an analog-to-digital converter.

- Describe the count-up and compare method of analog-to-digital conversion.

- Describe the successive approximation method of analog-to-digital conversion.

13.1 RESISTOR NETWORKS FOR DIGITAL-TO-ANALOG CONVERSION

We will look at two resistor networks to do the job of converting a binary number to a proportional analog voltage. The first is the binary ladder. Figure 13-1 shows the binary ladder made with a switch for each binary bit instead of TTL outputs. This will help to simplify the explanation.

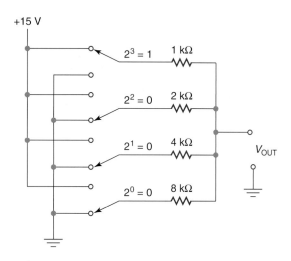

FIGURE 13-1 The binary ladder D-to-A converter

The binary number which is set by the switches is 1000 or decimal number 8. The largest number that the four switches can express is 1111 or decimal number 15. In this case a 1 is +15 V, and a 0 is ground. Therefore, if the binary number 1111 or decimal number 15 is put on the switches, the output of the binary ladder is tied to the +15-V supply through all the resistors in parallel, as shown in Figure 13-2. This produces a 15-V output voltage. If all the switches are switched to the 0 position, the output is 0 V or ground, as shown in Figure 13-3.

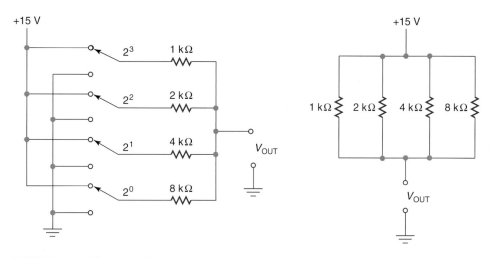

FIGURE 13-2 Binary ladder with all ones on the input

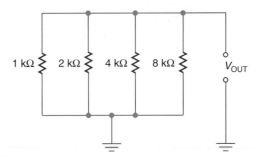

FIGURE 13-3 Binary ladder with all zeros on the input

Now let us analyze the switch configuration in Figure 13-1. The equivalent circuit is shown in Figure 13-4. If we reduce the circuit to two series-equivalent resistors, the voltage output will be the voltage drop across R_B. Using the voltage divider formula, we can find the output voltage for the binary number 1000 or decimal number 8 to be 8 V.

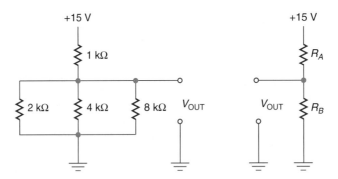

FIGURE 13-4 Equivalent circuit for the binary ladder with 1000 on the input

$$R_B = \cfrac{1}{\cfrac{1}{2\text{ k}\Omega} + \cfrac{1}{4\text{ k}\Omega} + \cfrac{1}{8\text{ k}\Omega}} = 1.1429\text{ k}\Omega \qquad (13\text{-}1)$$

$$V_{OUT} = V_S\left(\frac{R_B}{R_A + R_B}\right)$$

$$V_{OUT} = 15\text{ V}\left(\frac{1.1429\text{ k}\Omega}{1\text{ k}\Omega + 1.1429\text{ k}\Omega}\right)$$

$$V_{OUT} = 8\text{ V}$$

The output voltage for the other possible binary number inputs can be computed in a similar manner. You will find the voltage increments to be 1 V for this **binary ladder**. The binary number equivalent to 10 produces a voltage of 10 volts and the binary number equivalent to 7 produces a voltage of 7 volts. In other words, the supply voltage is in increments equal to the supply voltage divided by the largest binary number which can be input to the resistor network. Therefore, the voltage increment for a binary ladder is found by using the following formula.

$$\text{Binary ladder voltage increment} = \frac{V_S}{2^N - 1} \tag{13-2}$$

where V_S = voltage supply

N = number of bits in the binary number input

When N equals the number of bits in the binary number input to the binary ladder, $2_N - 1$ is the largest number that can be expressed for a binary number of N bits. If this formula is the amount of each increment, the final output voltage must be equal to the binary number input times the binary ladder voltage increment. The following formula is used for the voltage output of the circuit shown in Figure 13-1.

$$\text{Binary ladder: } V_{\text{OUT}} = \text{binary number input} \left(\frac{V_S}{2^N - 1} \right) \tag{13-3}$$

The values of the resistors in the binary ladder are divided by 2 for each binary power increase, that is, a 2^0 resistor is 8 kΩ, a 2^1 resistor is 4 kΩ, a 2^2 resistor is 2 kΩ, and a 2^3 resistor is 1 kΩ. If a fifth bit is added to the binary ladder, the resistor value is one-half the 2^3 resistor or 500 Ω. You can see that the larger the binary number, the smaller the resistor. It is not easy to get resistors with the exact values that fit this pattern.

These two problems can be eliminated by using another type of resistor network to produce the proportional output voltage. This is the $2R$ network shown in Figure 13-5. Using the same method shown in Figure 13-4, we can determine the output voltage for the binary number switched into the $2R$ network in Figure 13-5. This is shown in Figure 13-6.

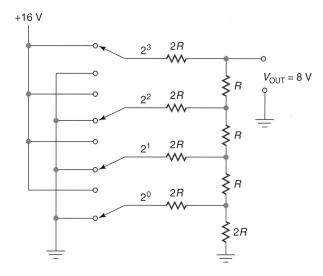

FIGURE 13-5 $2R$ D-to-A converter

The **$2R$ ladder** is similar to the binary ladder except that the voltage increments are equal to the voltage supply divided by the total number of combinations in the binary number input. The total number of combinations in a binary number of N bits is 2^N. Therefore, the formula

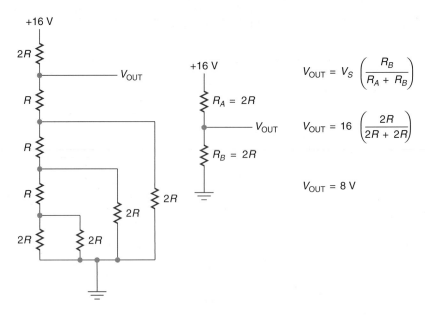

FIGURE 13-6 Equivalent circuit for $2R$ network

for the voltage output of the $2R$ network is equal to the binary number input times the voltage supply divided by 2^N.

$$2R \text{ network: } V_{OUT} = \text{binary number} \left(\frac{V_S}{2^N} \right)$$

(13-4)

Both of these networks give accurate output voltages as long as the output load impedance is very high with respect to the network's impedance. If the load resistance is lowered, it reduces the linearity of the output. To minimize this problem, a HiZ buffer, such as an operational amplifier, is usually used to drive the analog load.

EXAMPLE 13-1

What would be the voltage increment, ΔV, of an 8-bit $2R$ network with a voltage supply of 25.6 volts? What would be the output voltage if the binary number 78 were placed on the digital input?

Solution   First, find the voltage increment ΔV.

$$\Delta V = \frac{V_S}{2^N}$$

$$\Delta V = \frac{25.6}{2^8}$$

$$\Delta V = 0.1 \text{ volts}$$

Next, use the voltage increment ΔV to obtain the output voltage.

$$V_{OUT} = \text{Binary Number} \times \frac{V_S}{2^8}$$

$$V_{OUT} = 78 \times 0.1$$

$$V_{OUT} = 7.8 \text{ volts}$$

EXAMPLE 13-2

What would be the voltage increment ΔV of a 5-bit binary ladder with a voltage supply of 15.5 volts?

Solution

$$\Delta V = \frac{V_S}{2^N - 1}$$

$$\Delta V = \frac{15.5 \text{ V}}{2^5 - 1}$$

$$\Delta V = \frac{15.5 \text{ V}}{31}$$

$$\Delta V = .5 \text{ volts}$$

13.2 THE TTL DIGITAL-TO-ANALOG CONVERTER

The binary number in the two previous networks was not a TTL-level input. To make the network work, the 1 voltage must be the supply voltage, and the 0 voltage must be ground or 0 volts. The TTL output voltage will give a good 0 voltage, or, in the worst case, 0.4 volts, but the 1 voltage is typically about 3.5 volts. By using an open-collector output, such as the 7406 inverter or 7407 has, and a pull-up resistor to V_S, we can convert the TTL-level voltages to the voltage needed by the D-to-A network. The output voltage of the 7406 is not absolute ground or absolute V_S, but it is very close. This is shown in Figure 13-7.

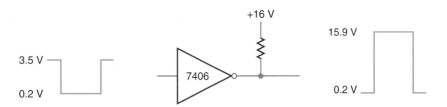

FIGURE 13-7 Converting TTL voltage levels to D-to-A levels

Figure 13-8 shows this type of buffer used to make a TTL D-to-A converter where the top output voltage is 15 volts. When the output of the 7406 goes LOW, it will go to 0.1 V or 0.2 volts above ground at best, which introduces some error into the D-to-A converter; and when the 7406 output goes to HiZ, the 1-kΩ resistor pulls the 20-kΩ resistor up to $+16$ V. In doing so, it adds its 1-kΩ resistor to the 20-kΩ resistor. This means the $2R$ resistor is a little larger (5 percent in this case) when it is at $+16$ V and does not quite reach ground when it is brought LOW.

These errors can be eliminated by other methods; but for many uses of D-to-A converters, these errors are tolerable. Figure 13-9 shows a D-to-A converter used to control the speed of a small dc motor such as might be used on a robotic arm. Notice the operational amplifier used to buffer the D-to-A converter.

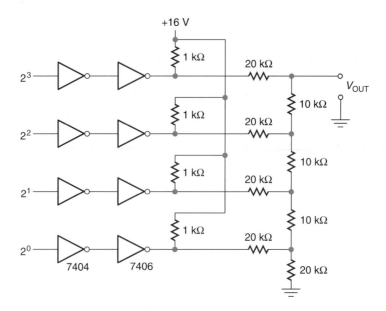

FIGURE 13-8 A TTL 2*R* D-to-A converter

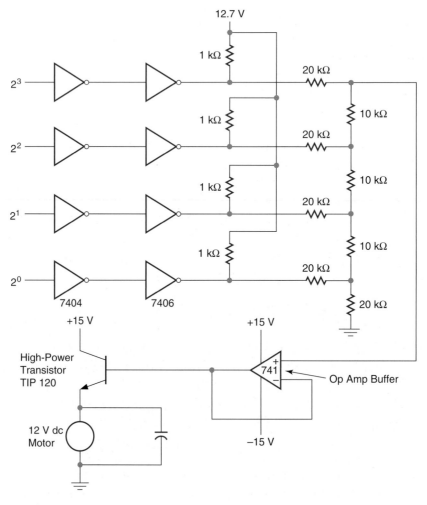

FIGURE 13-9 A TTL 2*R* D-to-A converter used to control a small motor

SELF-CHECK 1

1. What are the main disadvantages in the use of the binary ladder network?

2. What is the voltage increment ΔV for a digital-to-analog converter of seven bits that uses a $2R$ ladder and has a supply voltage of 24 volts?

3. What are the two main error-causing defects with the digital-to-analog circuit in Figure 13-8?

13.3 ANALOG-TO-DIGITAL CONVERSION USING VOLTAGE COMPARATORS

The voltage comparator, which was discussed in the chapter on clocks, can be used to make a very fast analog-to-digital converter. An **analog-to-digital** converter produces a binary number which is in direct proportion to an analog voltage input.

Figure 13-10 shows a 3-bit A-to-D converter made from seven LM339 voltage comparators. The negative input of each comparator is tied to a resistor voltage divider which divides the 8-volt supply into 1-volt increments. Each voltage comparator has a reference voltage of 1 volt greater than the previous comparator. All of the positive inputs to the comparators are tied together so the input voltage will increase on all comparators at the same time.

If the input voltage increases to 2.5 volts, the output of the first two comparators will be $+5$ V or logic 1 because the positive input will be greater than the negative input; but the rest of the comparators' outputs will be at ground or logic 0. The output of the LM339 is an open-collector output; so by using a pull-up resistor to $+5$ V, the output will be standard TTL levels even though the input may increase to 8 volts. As the voltage increases to 3.5 volts, the third comparator's output will change to a logic 1. If the analog voltage increases to above 7 volts, all the comparators will be at logic 1. The comparators will go to logic 0 when the input voltage goes below the reference voltages set by the voltage divider.

The output of all the comparators is decoded to form a 3-bit binary number by the logic gates shown in Figure 13-10. When all the comparators' outputs are at logic 0, the outputs of their corresponding NAND gates are 1, because any 0 into a NAND gate produces a 1 output. If the analog input voltage rises to 1.5 volts, the first comparator's output goes to logic 1, which is fed to the two-input NAND gate. The other input to this NAND gate is from the inverter which comes off the output of the second comparator. The output of the second comparator is still 0 because the analog voltage is still 1.5 volts and is not yet large enough to change the state of the second comparator. This 0 is inverted to a 1 and fed to the input of the NAND gate for the first comparator. At this point, the first NAND gate has 1 and 1 on its inputs and 0 on its output, which is fed to the 2^0 NAND gate. This produces a logic 1 or binary number 1 on the output of the 2^0 NAND gate. A binary number 1 on the output means that the analog input voltage lies between 1 and 2 volts.

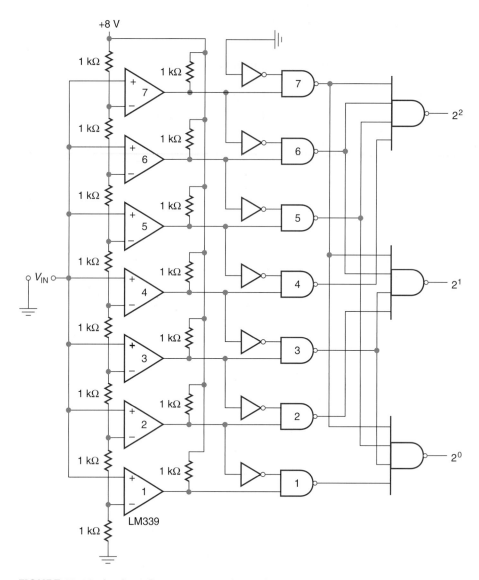

FIGURE 13-10 An A-to-D converter using voltage comparators

When the input voltage increases to 2.5 volts, the second comparator's output will also be logic 1. This produces a 0 on the inverter's output, which inhibits the first NAND gate, thus removing the 1 on the 2^0 NAND gate's output. The logic 1 on the second comparator's output enables the second NAND gate, producing a 0 on its output which produces a logic 1 or binary number 10 on the output of the 2^1 NAND gate. This means that the analog input lies between 2 and 3 volts. It can be seen that as the voltage increases, the binary output of the A-to-D converter will change to reflect the value of the analog input.

In order to increase the accuracy of the comparator in Figure 13-10, you need more comparators and NAND gates. This is a major drawback for this type of A-to-D converter; but it operates very fast. The only thing to slow it down is the propagation delay of the comparator and NAND gates, which is on the order of 50 ns to 75 ns. Because of its speed, this type of A-to-D converter is often called a **flash converter**.

13.4 THE COUNT-UP AND COMPARE ANALOG-TO-DIGITAL CONVERTER

This type of A-to-D converter uses one voltage comparator and a D-to-A converter. Figure 13-11 shows a count-up and compare A-to-D converter using an LM339 voltage comparator and a 2R resistor network.

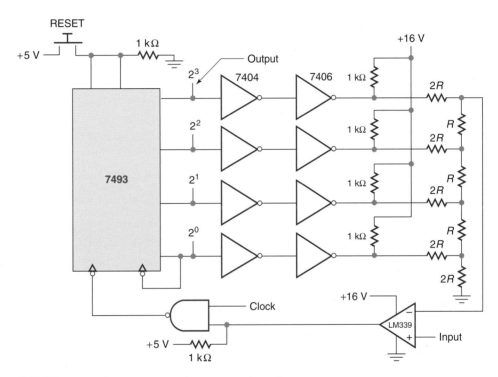

FIGURE 13-11 A count-up and compare A-to-D converter

The D-to-A converter is driven by a 7493 binary counter which can count from 0000 to 1111 in binary or 0 to 15 in decimal. The output of the D-to-A converter is fed to the negative input of a voltage comparator, and the positive input comes from the analog voltage which we want to measure. The output of the LM339 voltage comparator is used to enable or inhibit a NAND gate which feeds the clock to the 7493 counter. If the NAND gate is inhibited, the counter will not receive a clock pulse and will stop counting.

When the RESET button is pushed, the counter is cleared, and all its outputs go to logic 0. This puts a 0 or ground voltage on the negative input of the voltage comparator. Let us say that the analog input to the voltage comparator has 7.5 volts on it. This means that the positive input is greater than the negative input of the voltage comparator and its output will be at logic 1. A logic 1 on the input of the NAND gate which is fed from the comparator enables the NAND gate, and the clock is passed on to the 7493 counter. As the counter counts, its display gets larger and the analog voltage output of the $2R$ resistor network also gets larger.

When the voltage at the negative input of the voltage comparator gets larger than the analog voltage input to the positive input of the comparator, the output of the comparator goes LOW. This inhibits the NAND gate and stops the clock and the 7493 counter. The stopped counter will now contain the binary number which produces a voltage one increment of the D-to-A converter larger than the analog input voltage.

The A-to-D converter stays at this point until the RESET button is pushed or until the analog input voltage increases, at which time the counter simply counts back up to the new voltage. If the analog input voltage drops after the counter stops, the counter does not change and has to be reset to count back up to the lower voltage. Figure 13-12 shows the waveform for the analog output of the D-to-A converter as the analog input voltage changes.

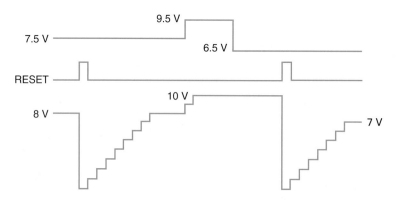

FIGURE 13-12 D-to-A converter waveforms

To increase the accuracy of the count up and compare A to D converter, increase the size of the counter and the D-to-A converter. In the case of the A-to-D converter in Figure 13-11, if you added one more 7493 to the converter, the +16 V would be divided into 256 increments, as compared to 16 increments with one 7493 counter. The main drawback to this type of A-to-D converter is speed because of the time it takes to count-up and compare.

EXAMPLE 13-3

Draw the logic schematic for an analog-to-digital converter like the one in Figure 13-11 that has 8-bit resolution.

Solution   Refer to Figure 13-13.

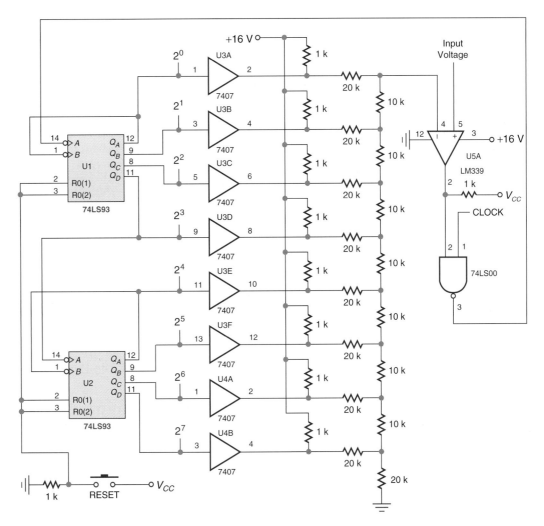

FIGURE 13-13

13.5 THE SUCCESSIVE APPROXIMATION ANALOG-TO-DIGITAL CONVERTER

This type of A-to-D converter also uses a D-to-A converter and a voltage comparator, but it uses a different technique to determine the binary number to feed into the D-to-A converter.

To study this technique, let us use the D-to-A converter and voltage comparator in Figure 13-14. To determine the correct binary number using the **successive approximation** method, you first set the most significant bit of the D-to-A converter to a logic 1. Then test the output of the D-to-A converter against the input analog voltage which is to be measured to see if the D-to-A output is larger or smaller. If the voltage generated by the D-to-A converter is smaller, leave the most significant bit at a logic 1. If it is larger, bring it LOW, to a logic 0. In the example in Figure 13-14, the LED is off, indicating that the binary number 1000 is too small. Therefore, we keep the 1 in the most significant place and bring the next place to a logic 1. Now we have the binary number 1100 or decimal number 12 input to the D-to-A converter. The LED is still off because the analog voltage being measured is greater than 12 volts; therefore, leave

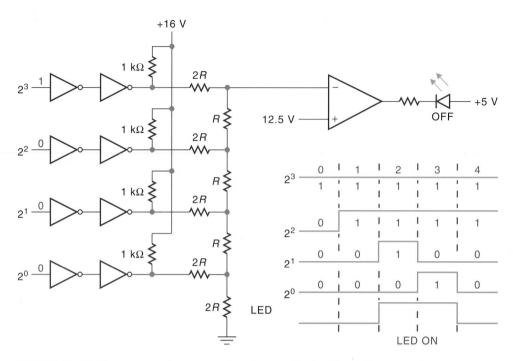

FIGURE 13-14 The successive approximation method of A-to-D conversion

the second bit at logic 1 also. Next, we bring the third bit HIGH, and we have binary number 1110 or decimal number 14 on the input to the D-to-A converter. The voltage output of the D-to-A converter is now larger than the 12.5 volts of the analog input voltage and the LED is on. In this case we remove the 1 on the third bit and put a 1 on the last and least significant bit. This gives us the binary number 1101 input to the D-to-A converter. The output voltage is now 13 volts, which is still greater than the analog input voltage and the LED is still on. Therefore, we remove the 1 and make it a 0. This gives us the final binary number 1100 or decimal number 12, which is one increment of the D-to-A converter less than the actual analog input voltage being measured.

The successive approximation method just described takes only four cycles to determine the correct binary number for the given analog input voltage, and it takes the same time to determine a large number as a small one. Therefore, the successive approximation method is faster than the count-up method but not as fast as the voltage comparator method.

Figure 13-15 shows a successive approximation A-to-D converter which uses a CP and CP' nonoverlapping clock generator, shift register and storage registers. When the RESET button is pushed, the clock generator and A flip-flop are preset to 1 while all the rest of the flip-flops are reset to 0. The A-to-D converter will stay in this configuration until the RESET button is released, thus allowing the clock generator flip-flop to run. After the clock is allowed to run, Q_A is preset to 1 which places a 1 on the 2^3 or most significant bit of the D-to-A converter. The contents of the comparator are fed to an AND gate which is controlled by the CP clock of the nonoverlapping clock generator. When it goes HIGH, the content of the comparator is passed through the AND gate to the A_S storage flip-flop NAND gate. If the comparator is 1, the flip-flop will be set; if it is 0, the flip-flop will not be set. Next comes the CP' clock, which shifts the A, B, C, and D flip-flop shift register 1 place. This makes Q_B a 1 and the cycle is repeated. After the 1, which was generated in Q_A by the RESET pulse, is shifted out of the A, B, C, and D flip-flop shift register, the correct binary number is present on the 2^0 and 2^3 outputs.

Notice that it took only four CP' pulses to obtain the correct binary number. Figure 13-16 shows the waveforms for the operation of the A-to-D converter in Figure 13-15.

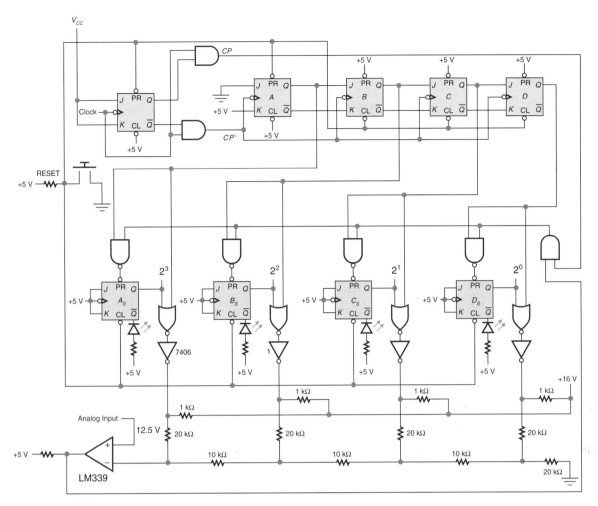

FIGURE 13-15 A successive approximation A-to-D converter

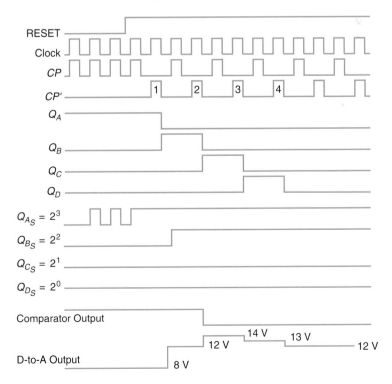

FIGURE 13-16 Waveforms for the successive approximation
A-to-D converter in Figure 13-15

SELF-CHECK 2

1. Which analog-to-digital conversion method will be the fastest for all values of voltages to be converted?

2. Draw the waveforms for the D-to-A output in Figure 13-16 as if the analog input voltage of the successive approximation counter were 7.5 volts.

3. Design a count-up and compare A-to-D converter like the one shown in Figure 13-11. Use one 7407, one 74LS93, one 74LS08, and one LM339 IC. Show pin numbers.

13.6 THE DAC0830 DIGITAL-TO-ANALOG CONVERTER INTEGRATED CIRCUIT

Today most digital-to-analog converters are made for use with microprocessors and have built-in logic to latch data from a data bus. National Semiconductor's DAC0830 D-to-A converter is a typical example of an IC made for microprocessor use. Even though the IC is designed for the microprocessor environment, the insides and the way it works are the same as the D-to-A converters we have just studied.

Figure 13-17A shows the internal circuitry of the DAC0830 and the connections to a small operational amplifier to buffer and amplify the output voltage. The output voltage is generated by the $2R$ ladder inside the IC. These resistors are thin film silicon-chromium (SiCr or Si-chrome) resistors and are switched to the digital inputs by SPDT current switches made from CMOS technology.

The control of these SPDT switches is from the Q outputs of a set of eight transparent D latches that are cascaded together. This method of using two sets of latches to store two values for D-to-A conversions is called double buffering. This will allow the computer to load the first set of D latches at any time. The number can then be transferred to the next buffer at a time determined by some other control signal when the conversion is needed. This is often done to time the change in output voltage of an external clock which would control the rate of signal change, such as would be used to reproduce speech.

Figure 13-17A shows the voltage reference on pin 11 coming from the voltage drop across a Zener diode. This provides a 2.5-volt voltage reference. The change in output voltage for one change in the binary input would be 2.5 volts divided by 256. This is the same as the $2R$ ladders we studied earlier. The output voltage is buffered by a 741 operational amplifier which amplified the voltage by 2, as can be seen in Figure 13-17A.

This IC is a CMOS chip and can have a V_{CC} supply up to a maximum of 17 volts; but all the inputs are TTL compatible even when the V_{CC} is greater than 5 volts. It is recommended that the applied reference voltage be less than 5 volts and the V_{CC} supply be at least 9 volts greater than the reference voltage. This is to ensure the correct operation of the CMOS switches to the $2R$ ladder network.

Figure 13-17B (page 480) shows a different way to use a $2R$ ladder to produce an output voltage. This method requires the use of an operational amplifier to convert the proportional

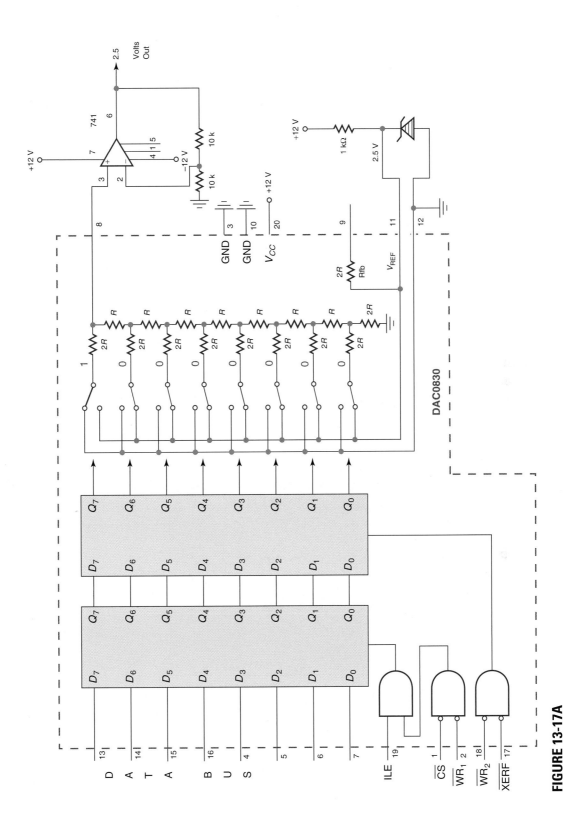

FIGURE 13-17A

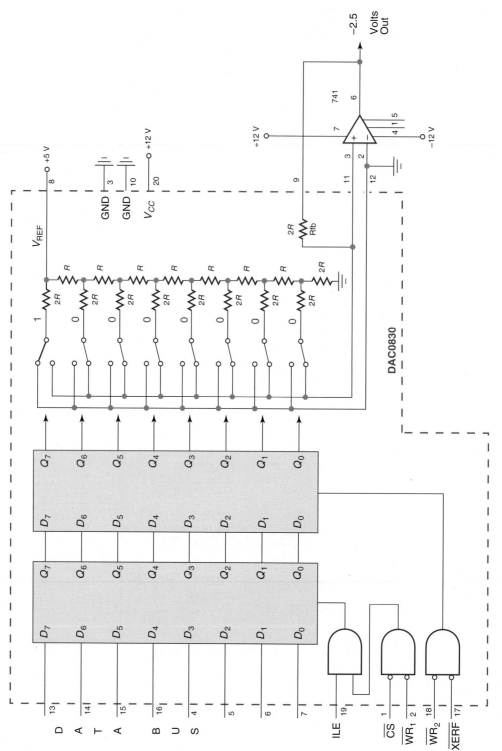

FIGURE 13-17B

current flow to a voltage. The reference voltage is connected to the normal output of the $2R$ ladder, and the switch poles are connected to the $+$ and $-$ inputs of the operational amplifier. A feedback loop is supplied to the operational amplifier's output through a built-in $2R$ resistor called Rfb. This method produces a negative voltage and requires a negative supply to run the operational amplifier.

13.7 MAKING THE LOGIC FOR A 3-BIT VOLTAGE COMPARATOR ANALOG-TO-DIGITAL CONVERTER

Section 13.3 of this chapter discusses the operation of an analog-to-digital converter made from a series of op amp voltage comparators. This type of A-to-D converter is very fast but requires a great deal of logic to construct if a large binary output is required. The example in Figure 13-10 has only a 3-bit binary output and requires seven op amps, seven two-input NAND gates, six inverters, and three four-input NAND gates. This would take five separate ICs if you used 74LSXX chips.

Here is a situation where a programmable logic device can be useful. We can construct all the logic needed to make the A-to-D converter in one chip, except for the op amps. Here is a VHDL program named **a_to_d.vhd** that configures a CPLD to provide the A_to_D hardware. The outputs from the comparators 1 through 7 in Figure 13-10 are used as the inputs to the CPLD, **input(1)** through **input(7)**. The 3-bit binary outputs are called **output(2)**, **output(1)**, and **output (0)**.

```
LIBRARY ieee;
USE ieee.std_logic_1164.all;
ENTITY a_to_d IS
        PORT(input:IN STD_LOGIC_VECTOR (7 DOWNTO 1);
             output:OUT STD_LOGIC_VECTOR (2 DOWNTO 0));
END a_to_d;
ARCHITECTURE a OF a_to_d IS
BEGIN
    output(0) <= NOT ((NOT(input(1) AND NOT input(2))) AND
    (NOT(input(3) AND NOT input(4))) AND (NOT (input(5) AND NOT
    input(6))) AND NOT input(7));
    output (1) <= NOT ((NOT(input(2) AND NOT input(3))) AND
    (NOT(input(3) AND NOT input(4))) AND (NOT(input(6) AND NOT
    input(7))) AND NOT input(7));
    output(2) <= NOT ((NOT(input(4) AND NOT input(5))) AND
    (NOT(input(5) AND NOT input(6))) AND (NOT(input(6) AND NOT
    input(7))) AND NOT input(7));
    END a;
```

13.8 TROUBLESHOOTING DIGITAL-TO-ANALOG CONVERTERS

Most digital-to-analog and analog-to-digital converters are single IC with a few external components to make them work. Therefore, a technician must first know how the chip works before he can determine if it is bad or not. Also, most D-to-A and A-to-D converters are designed to work with a microprocessor bus. To test the operation of the device, the microprocessor must run a program to put the proper information out to the chip. This often requires a logic analyzer to determine just what was output to the chip. Many times, the system manufacturer will have test software for the technician to use to test parts of the system including any A-to-D chips on the system.

The circuit shown in Figure 13-18 is a digital-to-analog converter which gets the digital number from a microprocessor output port. The voltage which is produced by the D-to-A converter is used to set the time-out time of a one-shot made from an LM339 (U2A). The LOW-going pulse coming from the one-shot is used to turn on a relay which will control the on/off state of a lightbulb. The light is used to expose a photo print in an automatic photo processor. The microprocessor calculates the correct exposure time based on the paper used and the light intensity falling on the paper. The light intensity is measured by a set of CDS cells mounted in the table where the print paper is exposed. The microprocessor moves the negative into place

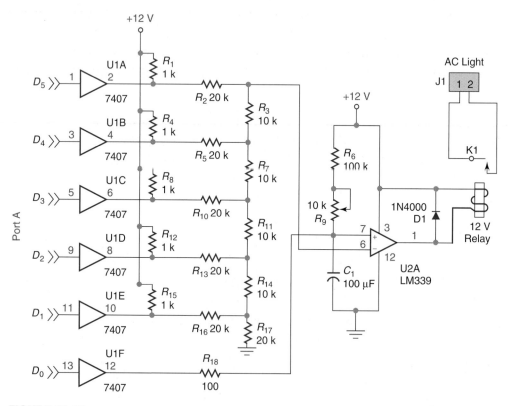

FIGURE 13-18

and turns on the light bulb before the print paper is moved into place. The microprocessor then uses an analog-to-digital converter to sample the voltage across each of the CDS cells and then determines the exposure length for the print.

This device had been working well, but lately the machine operator had complained that the length of exposure was not always correct and that short exposures were never quite right. Given this information, the technician was called in to get it working accurately again.

After studying the schematics for the device, the technician decided to place the machine in the manual mode and make it expose the print at a fixed time. By doing this, he could place a logic analyzer on the computer output port and see what binary number was there when the one-shot was reset. He also placed a storage oscilloscope on the D-to-A output (pin 6 of U2A). This way, he thought, I can see if the voltage being produced is the correct voltage for the binary number being placed on the input of the D-to-A circuit.

Hooking the logic analyzer clock to D_0 of the output port (pin 12 of U1F), the technician proceeded to start with the smallest and work to the largest exposure times. He found was that the voltage went to 0 on the four count, and then came back up on the five count as if the count had been one. This made sense to him. He reasoned that the 2^3 bit was shorted to ground. To test this, he placed another probe of the logic analyzer on the output of the 7407 used for the D_3 bit of the output port (pin 6 of U1C). Just as he thought, the output of the 7407 did not go HIGH when the input did which would explain the incorrect voltages he had seen.

He inspected the board for any apparent problems. Pin 6 of the 7407 had a large amount of solder on it and on the top of the board there was a small strand of wire shorting pin 7 to pin 6. How did that get there, he wondered. Then he remembered that he had replaced a wire two weeks earlier. He must have let a small strand of the wire fall onto the board when he stripped the insulation from the wire. Pin 7 is the IC ground pin and shorting that pin to pin 6 would never let pin 6 go HIGH.

After carefully removing the wire strand, the technician ran the numbers again and they worked correctly. Before he left, he ran the timing setup software to calibrate the one-shot time out. He did this by adjusting R_9 when the computer set a predefined number on the output port.

DIGITAL APPLICATION ## Analog-to-Digital Converters

The circuit shown on page 484 uses a 68HC11 microcontroller to monitor and regulate the output voltage of a large written-pole generator. The two ICs at the top of the schematic are analog-to-digital converters used to set the level of current in the exciter coil. This coil is used to write the magnets on the generator's rotor. The chips are serially fed by the micro-controller using output port pins PB0, PB1, PB2, PB3, and PC0. (Schematic courtesy of Precise Power Inc.)

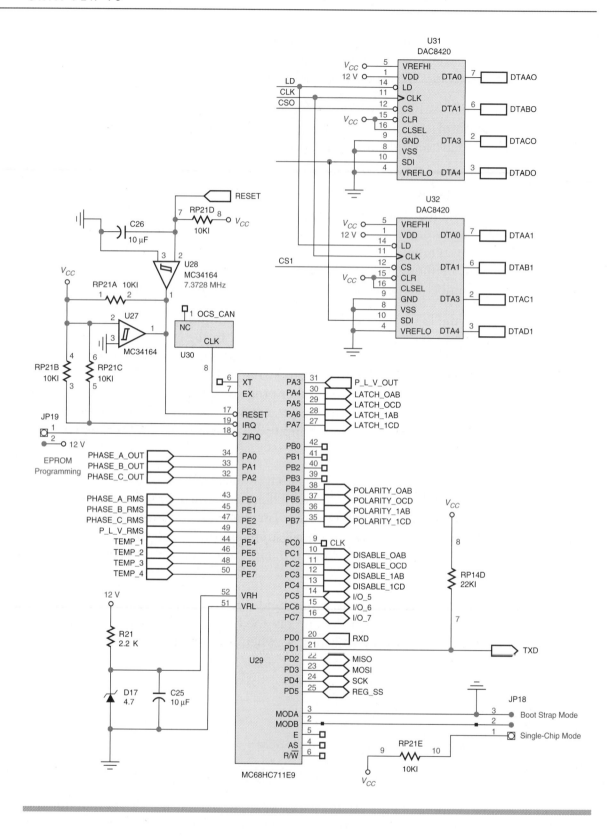

SUMMARY

- There are two main resistor networks used for digital-to-analog conversion (binary ladder and the $2R$ ladder).

  The binary ladder can be used for small and simple D-to-A converters. Because the binary ladder uses precise resistor values that are multiples of each other, it is impractical to make very large or accurate binary ladder networks. The $2R$ ladder needs only two values of resistors and can be as large as needed.

- The conversion from TTL voltage levels to the voltages needed by the resistor network can be done reasonably well using an open-collector output and a pull-up resistor.

  This method does introduce some error in the final output voltage due to the added resistance of the pull-up resistor and the fact that the open-collector will not pull to exact ground.

- Analog-to-digital conversion can be accomplished by three different methods—flash conversion, count-up and compare, and successive approximation.

  Flash conversion is by far the fastest of the three and is often used for such devices as frame grabbers, which capture TV pictures into a computer. The count-up and compare method is simple to design but requires the most time to do the conversion. The successive approximation method is faster than the count-up and compare but takes much more hardware to implement.

- Today, the two main A-to-D conversion methods used in ICs are the flash and successive approximation methods.

  Most D-to-A and A-to-D ICs are designed for use with computers and have extra interfacing logic added to them to interface with the computer bus.

QUESTIONS AND PROBLEMS

1. Draw the logic diagram for a TTL $2R$ D-to-A converter with 256 increments and 10 volts of maximum output.

2. What is the output voltage if the binary number 0110 is put on the TTL inputs of the D-to-A converter in Figure 13-8?

3. What is the voltage increment of the D-to-A converter in Figure 13-8 if the supply voltage to the $2R$ network is changed to 5 volts, 10 volts, and 32 volts?

4. What is the binary number output of the A-to-D converter in Figure 13-10 if 6.3 volts are put on the analog input?

5. Draw the waveform for the analog output of the successive approximation A-to-D converter in Figure 13-14 with 5.5 volts applied.

6. Why is the 7406 open-collector inverter used in Figure 13-8?

7. What is the purpose of the op amp-buffer in Figure 13-9?

8. Draw the logic diagram for an A-to-D converter like the one in Figure 13-10 but make the output four bits wide.

9. Draw the waveform for the A-to-D converter in Figure 13-11 if the converter is first reset and then the input voltage follows this pattern.

    0 V to 3.3 V to 6.4 V to 5.2 V to 7.3 V

10. Repeat number 9 with the A-to-D converter in Figure 13-15.

11. Find the value of the resistor if the 2^4 bit is added to Figure 13-1.

12. Find the supply voltage of the circuit in Figure 13-5 if the output voltage changes by 3 volts for a change in the binary number equivalent to 2.

13. Repeat number 12 for the circuit in Figure 13-2.

14. Draw the logic diagram for a count-up A-to-D converter which has an 8-bit output. Use two 7493s, two 7407s, one 7400, one LM339, and resistors for a $2R$ ladder. Show pinouts and make the supply voltage to the $2R$ ladder 12 volts.

15. Draw the waveforms for the analog output of the A-to-D converter in number 14 for an input voltage of 1.2 volts.

16. List two problems with using a binary ladder.

17. Design a 6-bit count-up and compare A-to-D converter using a binary ladder. Show pin numbers of the ICs used.

18. What is the voltage increment V for an 8-bit $2R$ ladder with 15 volts as the supply voltage?

19. What is the voltage increment V for an 8-bit binary ladder with 15 volts as the supply voltage?

20. What would be the output voltage for the $2R$ ladder in question 18 if the binary number 10111000 were placed on the inputs?

21. What is the effect on the output voltage of the analog-to-digital converter shown in Figure 13-9 if the 2^0 bit was shorted to ground?

22. What is the effect on the circuit in Figure 13-10 if the negative input to the number 4 LM339 comparator becomes grounded?

23. What happens to the operation of the circuit in Figure 13-11 if the output of the LM339 comparator becomes disconnected?

24. What is the effect on the count-up and compare analog-to-digital converter in Figure 13-13 if pin 1 of U2 becomes disconnected?

25. What happens to the output of the successive approximation A-to-D converter in Figure 13-15 if the *CP* line becomes grounded? Why?

26. Why was Q complement used to drive the LEDs in Figure 13-15?

27. Draw the logic circuit for an added extra bit to the circuit in Figure 13-15.

28. Which type of A-to-D converter studied in this chapter is the fastest?

29. What is the exposure time for the light in Figure 13-18 if the A-to-D is putting out 6 volts?

30. What voltage on the output of the A-to-D converter in Figure 13-18 produces an exposure time of ten seconds?

OBJECTIVES

After completing this lab, you should be able to:

- Construct a D-to-A converter.
- Use an LM339 voltage comparator to construct an A-to-D converter.
- Use the oscilloscope to observe the stair-step waveform.

COMPONENTS NEEDED

1	7493 4-bit ripple counter IC
1	7404 hex inverter IC
1	7406 hex open-collector output inverter IC
1	7400 quad NAND gate IC
1	LM339 quad op amp comparator
4	1-kΩ, $\frac{1}{4}$-W resistors
1	1 k or larger pot
4	10-kΩ, $\frac{1}{4}$-W resistors
5	20-kΩ, $\frac{1}{4}$-W resistors
4	red LEDs
4	330-Ω, $\frac{1}{4}$-W resistors

PROCEDURE

1. Construct the A-to-D converter shown in the following order.

   a. Wire up the 7493 counter and get it to run.

   b. Wire up the 7404 and 7406 D-to-A converter and display the stair-step pattern on the oscilloscope.

   c. Wire up the op-amp comparator and the 7400 NAND gate.

2. Set the clock at the 1 Hz rate, put 5.5 V on the analog input, and reset the counter. At which binary number does the counter stop and why?

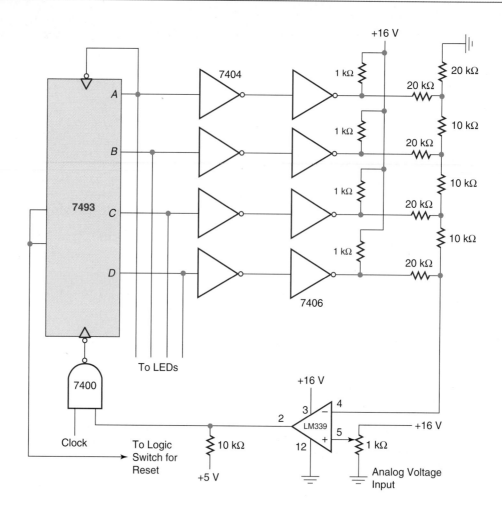

3. Now raise the voltage to 10.5 V. What is the number now?

4. Now lower the voltage to 6.5 V. What is the number now and why?

If your circuit does not work properly, consider these points:

1. Check the power supply connections to all the components in the circuit.

2. Disconnect pin 2 of the LM339 from the NAND gate input. This will allow the counter to count from 0 to 15.

3. Put channel one of the oscilloscope on pin 4 of the LM339. This is the output of the D-to-A converter. You should see the characteristic stair-step waveform from about 0.2 V to 15 V. If the waveform is not in even 1-V increments, check the resistor network for wiring errors.

4. If the stair-step waveform is okay, put the oscilloscope probe on the output of the LM339 comparator. You should be able to vary the duty cycle of the output waveform by varying the input voltage on pin 5 of the LM339. If you cannot do this, something is wrong with the comparator part of the circuit.

OBJECTIVES

After completing this lab, you should be able to use Multisim to:

- Construct a binary and 2R ladder.
- Troubleshoot a count-up and compare analog-to-digital converter.

PROCEDURES

1. Using Multisim, construct a 4-bit binary ladder that will divide 16 volts into 1-volt increments. Use switches and pull-up resistors as shown in Figure 13-2.

2. Construct a 4-bit 2R ladder that will divide 16 volts into 1-volt increments. Use switches and pull-up resistors as shown in Figure 13-5.

3. Open file 13B-1.ms9. This is a count-up and compare 4-bit analog-to-digital converter. Run the simulation until you are comfortable with its operation.

4. Open file 13B-2.ms9. This is a count-up and compare 4-bit analog-to-digital converter with two faults. Run the simulation, find the faults, and repair them.

CHAPTER 14

Decoders, Multiplexers, Demultiplexers, and Displays

OUTLINE

KEY TERMS

demultiplexer   multiplexer
full decoder   partial decoder
LED   seven-segment display
liquid crystal display

OBJECTIVES

After completing this chapter, you should be able to:

- Explain the operation and use of decoders.

- Explain the operation and use of demultiplexers.

- Explain the operation and use of multiplexers.

- Use a multiplexer to reproduce a desired truth table.

- Use typical multiplexer and demultiplexer ICs.

- Use multiplexers in a typical application of an 8-trace oscilloscope.

- Explain the operation of LEDs.

- Explain the operation of seven-segment LEDs and their decoders.

- Explain the operation of LCDs and how to drive them.

14.1 DECODERS

Figure 14-1 shows a full 2-bit decoder which will enable one and only one of the four AND gates for each possible binary number input to the 2^0 and 2^1 inputs of the decoder. Figure 14-2 is the truth table for the full 2-bit decoder shown in Figure 14-1.

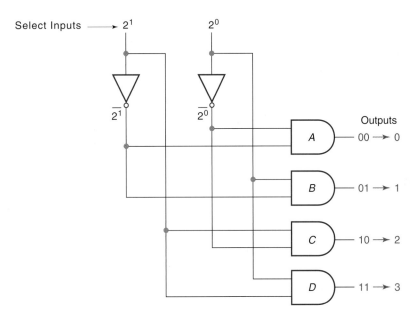

FIGURE 14-1 A full 2-bit decoder

2^1	2^0	$\overline{2^1}$	$\overline{2^0}$	A	B	C	D
0	0	1	1	1	0	0	0
0	1	1	0	0	1	0	0
1	0	0	1	0	0	1	0
1	1	0	0	0	0	0	1

FIGURE 14-2 The truth table for a full 2-bit decoder

The two inverters in the decoder in Figure 14-1 provide $\overline{2^0}$ and $\overline{2^1}$, which are fed along with 2^0 and 2^1 to the AND gates in the proper order to enable each gate when the proper binary number is input. This is called a **full decoder** because it has an active output line for each possible binary number which can be input to the decoder.

If you were to increase the size of the binary number input to the decoder by 1 bit, to 3 bits, the number of outputs would be 2^3 or 8 if the decoder was to be a full decoder, as shown in Figure 14-3. Notice that as the number of bits in the input increases so does the number of inputs of the AND gates used. If the binary number gets very large, a full decoder can become very large. Consider a full 8-bit decoder; it would need 256 AND gates, each with eight inputs.

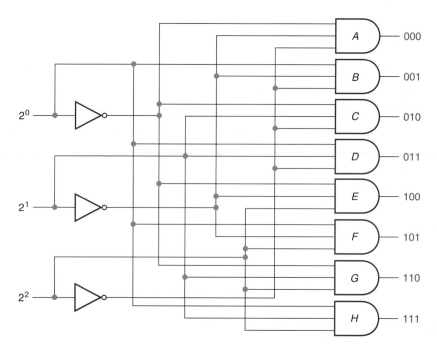

FIGURE 14-3 A full 3-bit decoder

In most cases we do not need to decode every bit of a long binary number. Therefore, a full decoder is not really necessary. An address decoder for a typical Z80 computer output port system may only need unique outputs for 0, 1, 2, and 3 from an 8-bit binary address. All other possible outputs are not needed. Therefore, the decoder would be constructed as shown in Figure 14-4.

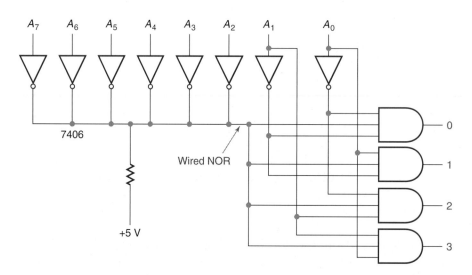

FIGURE 14-4 A partial 8-bit decoder

By using a wired NOR gate made from open-collector inverters, if any 1 is put on the upper address bits (bit A_2 to A_7), the output of the wired NOR gate will go LOW, causing the four AND gates to be inhibited. The only numbers which can enable any of the AND gates are 0, 1, 2, and 3 because they do not have a 1 in bits A_2 to A_7. This is called a **partial decoder** and is frequently used in computers.

14.2 DEMULTIPLEXERS

A **demultiplexer** is a digital switch which allows us to switch one input to one of many possible output lines. The line which we want the input to be connected to is determined by a binary number which is input to the demultiplexer. Figure 14-5 shows a 1-to-4 demultiplexer. It looks very much like a decoder; in fact, the two are almost the same. The only difference is in the use of the enable line of the decoder. A demultiplexer uses the enable line as a data input. Notice that the data appears on the select output when the corresponding binary number is input to the select inputs.

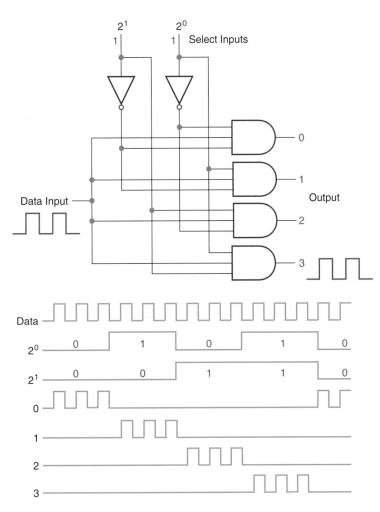

FIGURE 14-5 A 1-to-4 demultiplexer

 14.3 MULTIPLEXERS

The **multiplexer** is the opposite of the demultiplexer. It selects one channel as an input and connects it to a signal output. Figure 14-6 shows a full 4-to-1 multiplexer. The outputs of the AND gates are ORed together to produce a common output. The AND gate, which will control the output, is selected by the binary number input to the select inputs.

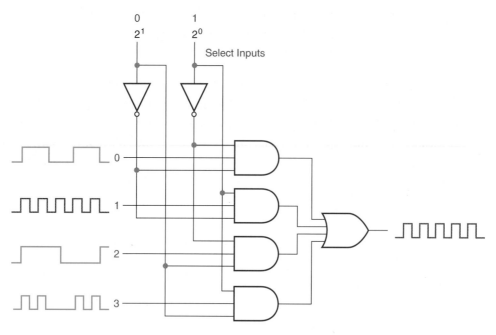

FIGURE 14-6 A 4-to-1 multiplexer

14.4 USING A MULTIPLEXER TO REPRODUCE A DESIRED TRUTH TABLE

To build a digital circuit which will conform to the truth table in Figure 14-7, use a 4-to-1 multiplexer, as shown in Figure 14-7. The select inputs become the truth table input variables, and the channel inputs are made LOW or HIGH to reflect the desired output for a given

B	A	Y
0	0	1
0	1	0
1	0	1
1	1	0

FIGURE 14-7 Using a multiplexer to reproduce a truth table

combination of A and B inputs. When the A and B select inputs are sequenced through the values of the truth table, the output of the multiplexer will go LOW or HIGH according to the values which are on the channel inputs, thus reproducing the truth table.

At first glance, this method of using a multiplexer to reproduce a truth table seems to require a multiplexer with at least the same number of select inputs as inputs in the truth table. With a little ingenuity, we can make our 2-select input multiplexer work like a 3-select input multiplexer. Consider the truth table in Figure 14-8 (A). This is the standard method of writing a truth table, starting with 000 and counting in binary to 111, which is the largest binary number that the 3-bit number can express. This gives all the possible combinations for the given three inputs A, B, and C.

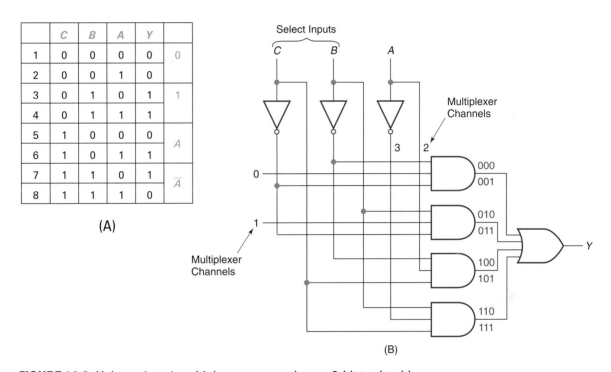

(A)

(B)

FIGURE 14-8 Using a 4-to-1 multiplexer to reproduce a 3-bit truth table

Notice the two most significant bits in the truth table, C and B, change in value every other line. Therefore, you can group the lines of the truth table into four groups of 2 each in which C and B are the same. In the first group, C and B are both 0 and A is 0, then 1; but the outputs for both lines 1 and 2 are 0. If C and B are placed on the select input of a 4-to-1 multiplexer and then 0 on the channel 0 of the multiplexer, the output is 0 when the C and B input is 0 no matter what A is. This is shown in Figure 14-8 (B). Lines 3 and 4 of the truth table are similar; but in this case, the output Y is 1 in both cases. Therefore, we make the 1 channel HIGH, or 1, which gives us a 1 on the multiplexer output no matter what A is. The next two lines of the truth table (lines 5 and 6) do not have the same output value. When A is 1, the output Y is 1; and when A is 0, the output Y is 0. Therefore, we simply tie input A to the input for channel 2. This will cause the output to follow the A input when C and B are 1 and 0, respectively, thus fulfilling the truth table. The last two lines (lines 7 and 8) do not have the same output either. When A is 0, the Y output is 1; and when the A input is 1, the Y output is 0, or opposite the value of A. Therefore, we use $\overline{A}$ to feed the third input channel of the multiplexer, thus completing the circuit for the truth table. This method of using a multiplexer to produce a pattern of pulses is quite handy for sequencing the operation of a digital machine.

EXAMPLE 14-1

Implement the truth table in Figure 14-9 using four 74LS22 dual four-input NAND gates with open-collector outputs and one 74LS04 Hex inverter.

	2^3	2^2	2^1	2^0	Y		2^3	2^2	2^1	2^0	Y
U2A	0	0	0	0	1		0	0	0	1	1
U2B	0	0	1	0	0		0	0	1	1	0
U3A	0	1	0	0	0		0	1	0	1	1
U3B	0	1	1	0	1		0	1	1	1	0
U4A	1	0	0	0	0		1	0	0	1	1
U4B	1	0	1	0	1		1	0	1	1	1
U5A	1	1	0	0	0		1	1	0	1	0
U5B	1	1	1	0	0		1	1	1	1	0

FIGURE 14-9

Solution   Refer to Figure 14-10.

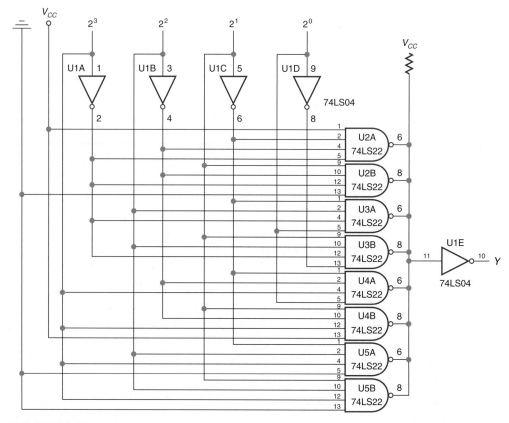

FIGURE 14-10

14.5 MULTIPLEXER AND DEMULTIPLEXER ICS

There are many different types of multiplexers and demultiplexers which are put on ICs. Figure 14-11 shows the logic drawing for three typical demultiplexer/decoders which are TTL in construction. Notice the 74138 has three enable inputs which can be used as data inputs or enables. The 74154 is a full 4-bit decoder with 16 output lines and two enable lines. Notice all three of these ICs have active LOW outputs. Figure 14-12 shows the 74150 and 74151 multiplexer ICs.

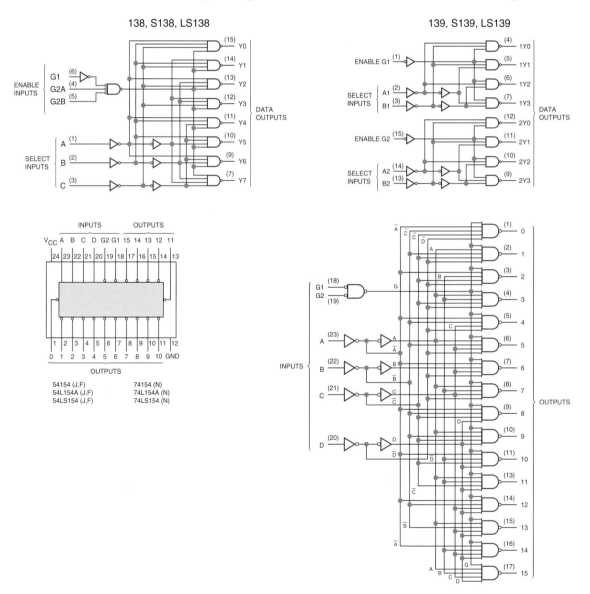

FIGURE 14-11 Decoders and demuliplexers

In the CMOS family there are several analog multiplexers and demultiplexers, such as the 4051, 4052, and 4053 ICs. An analog multiplexer can pass an analog signal from the channel input to the output. These types of ICs can be used to multiplex the inputs to a multitrace oscilloscope or multiplex analog phone lines.

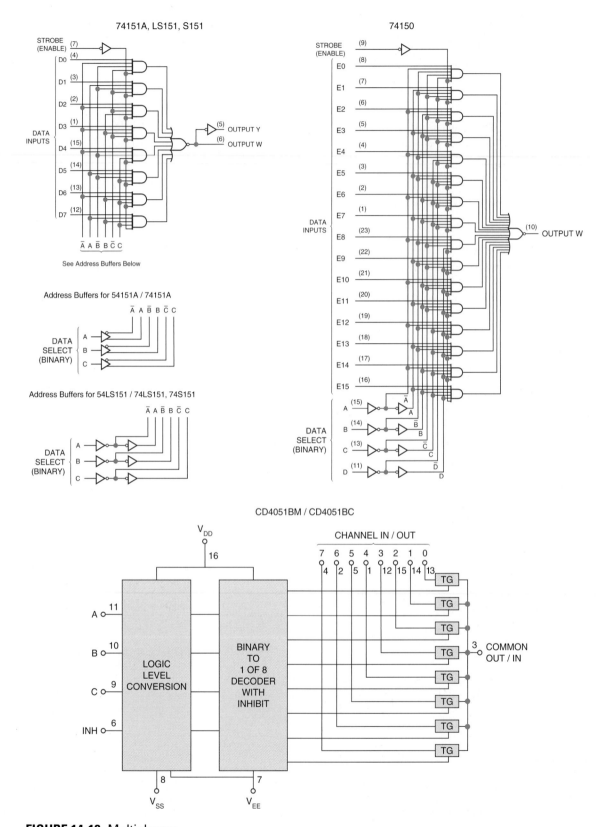

FIGURE 14-12 Multiplexers

The 4051 CMOS IC will multiplex eight analog inputs to one analog output. The peak-to-peak analog voltage is set by the positive V_{DD} voltage and the negative V_{EE} voltage. A typical voltage set up for this IC would be a V_{DD} of $+5$ volts, a V_{SS} of 0 volts, and a V_{EE} of -5 volts. This would allow an analog voltage of 10 volts peak-to-peak to be controlled by a CMOS 0- to

5-volt digital signal. The ON resistance for a selected channel is typically 120 ohms and the OFF leakage current is typically 0.001 nanoamps. This IC is often used to select the analog input to an A-to-D converter.

14.6 THE 8-TRACE OSCILLOSCOPE MULTIPLEXER

Figure 14-13 shows the circuit for an 8-trace oscilloscope multiplexer. This is a very handy gadget when working with digital circuits having several signals which are to be observed at the same time and in relation to each other.

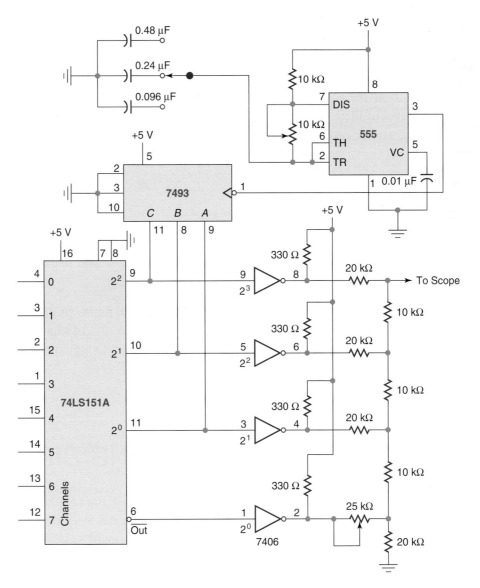

FIGURE 14-13 Digital 8-trace oscilloscope multiplexer

The 7493 counter is used as a divide-by-8 binary counter. Its outputs are fed to the select inputs of an 8-to-1 multiplexer (the 74151A) and the upper three bits of a 4-bit D-to-A converter. The least significant bit of the D-to-A converter is driven by the output line of the multiplexer. As the 7493 counts, the analog voltage at the output of the D-to-A converter increases by increments of two. This is because the counter's least significant bit is connected to the 2^1 input of the D-to-A converter.

When the output of the multiplexer changes, the analog voltage changes by 1 increment because the 2^0 input of the D-to-A converter is controlled by the output of the multiplexer. Therefore, the data on channel 0 will be displayed at the 0 to 1 increment of the D-to-A converter. Channel 1 will be displayed at the 2 to 3 increment and channel 2 will be displayed at the 4 to 5 analog voltage level of the D-to-A converter. This will continue until the divide-by-8 counter starts over at 0.

When the counter is run at a speed of at least ten times less than the sweep frequency, all eight inputs will appear on the oscilloscope at their separate voltage levels. This is because the counter is running faster than the eye can detect. All the waves will appear on the oscilloscope at the same time.

A 555 timer is used to provide a clock for the counter. The rotary switch changes the frequency of the counter to produce a multiplex rate of at least ten times less than the sweep frequency. An external sync probe connected to the slowest input frequency to the multiplexer is the best way to sync the oscilloscope to the wave patterns. The 25-kΩ pot is used to adjust the 0 to 1 logic level of the eight waves on the oscilloscope. Similar schemes can be used in conjunction with an analog multiplexer to produce an analog multitrace oscilloscope.

 SELF-CHECK 1

1. Draw the logic diagram for a 4-bit full decoder.

2. Draw the logic diagram for an 8-bit partial decoder that will decode the numbers 2, 4, 8, and 16.

3. Design a multiplexing circuit that will multiplex eight channels of digital data to one channel and then demultiplex the one channel to one of eight channels at the receiving end. Use one 74LS138, one 74LS151, and any other ICs you may need.

 # 14.7 THE LIGHT-EMITTING DIODE

When electrons fill a positive hole at the junction of *PN* material, the electrons lose some energy. This energy is given off as heat and light. All *PN* junctions do this, but the *PN* junctions made of gallium emit sufficient amounts of light to be used as a visible light source. Depending on the type and amount of crystal doping, the light emitted can be red, green, or yellow.

Because the **LED** is a *PN* junction, it exhibits all the properties of a typical diode. The LED will produce light when the diode is forward biased but not when it is reverse biased. This is shown in Figure 14-14.

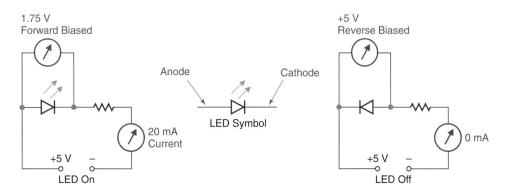

FIGURE 14-14 LED biasing

When the LED is forward biased, the voltage drop is about 1.75 volts for a typical red LED. The forward-biased voltage is higher for yellow and green LEDs. To be easily seen, a typical LED needs between 5 mA to 20 mA. The resistor in Figure 14-14 is used to limit the current through the diode. If the resistor were not there, the diode would burn out due to high current.

The intensity of light produced by the LED is directly proportional to current flowing through the diode and can be used as a modulated light source. The on/off speed is also quite high, about ten nanoseconds for a typical red LED. The speed of the LED lends it to such uses as a high-speed optocoupler. LEDs come in several packages, as shown in Figure 14-15. The cathode of the LED can be found by looking for the flag lead inside the plastic case.

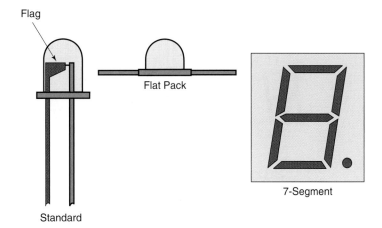

FIGURE 14-15 LED packages

When driving an LED with a TTL output, it is best to design the circuit so the LED will be forward biased (or on) when the TTL output is LOW or at ground potential. This is because a typical TTL output can produce up to 16 mA when at the LOW state without raising the 0 voltage above 0.4 volts. This is shown in Figure 14-16.

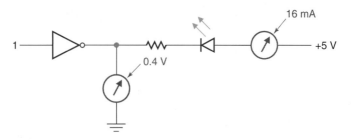

FIGURE 14-16 Driving an LED with a TTL output

Forward biasing an LED with a TTL 1 voltage will produce enough current to light a typical red LED; but the output voltage may be below the 2.0-volt limit for a logic 1. Two good ICs to use when driving LEDs which need higher currents are the 7406 or 7407. These two ICs are open-collector and can sink up to 40 mA of current. They can also have up to 30 volts applied to the output pull-up resistor. This makes them very useful when driving a display that must use +12 volts or higher.

The 74ALS1005 is an open-collector hex inverter IC which can sink up to 24 mA. This IC can be driven by CMOS ICs that are running with V_{DD} at +5 volts.

 # 14.8 THE SEVEN-SEGMENT DISPLAY

As can be seen in Figure 14-17, the **seven-segment display** is actually eight separate LEDs (seven segments and one decimal point). The seven-segment display format is used in other types of displays and can display any number from 0 to 9. Figure 14-17 shows the typical segments used to display 0 through 9.

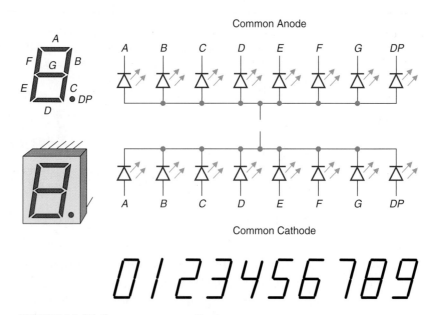

FIGURE 14-17 Seven-segment displays

There are two types of LED seven-segment displays: common cathode and common anode. As can be seen in Figure 14-17, the common cathode has all the cathodes of the seven segments connected together, and the common anode is the same except that the anodes are all tied together. Also, notice the way the segments are labeled. This is a de facto standard for seven-segment displays and MSI ICs designed to work with seven-segment displays.

Figure 14-18 shows a logic diagram for a 7447 and 7448 TTL decoder driver. These ICs will decode a 4-bit BCD number into the proper output to display the BCD number on the seven-segment display.

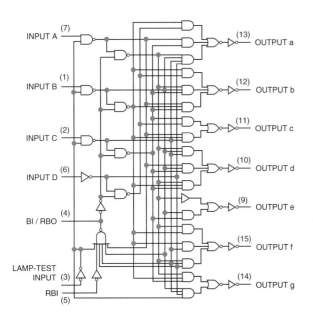 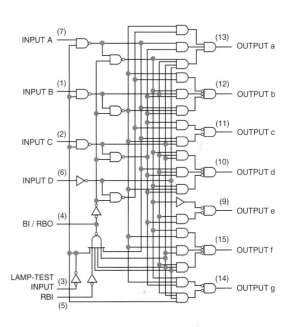

General Description

The 46A, 47A, and LS47 feature active-low outputs designed for driving common-anode LEDs or incandescent indicators directly; and the 48, LS48, and LS49 feature active-high outputs for driving lamp buffers or common-cathode LEDs. All of the circuits except the LS49 have full ripple-blanking input/output controls and a lamp-test input. The LS49 features a direct blanking input. Display patterns for BCD input counts above nine are unique symbols for authenticating input conditions. All of the circuits except the LS49 incorporate automatic leading and/or trailing-edge, zero-blanking control (RBI and RBO). Lamp test (LT) of these devices may be performed at any time when the BI/RBO node is at a high logic level. All types (including LS49) contain an overriding blanking input (BI) which can be used to control the lamp intensity (by pulsing), or to inhibit the outputs.

FIGURE 14-18 Seven-segment decoder drives

Note 1: BI/RBO is a wire-AND logic serving as blanking input (BI) and/or ripple-blanking output (RBO).

Note 2: The blanking input (BI) must be open or held at a high logic level when output functions 0 through 15 are desired. The ripple-blanking input (RBI) must be open or high if blanking of a decimal zero is not desired.

Note 3: When a low logic level is applied directly to the blanking input (BI), all segment outputs are H (46, 47); L (48) regardless of the level of any other input.

Note 4: When rippleblanking input (RBI) and inputs A, B, C, and D are at a low level with the lamp-test input high, all segment outputs go H and the ripple-blanking output (RBO) goes to a low level (response condition).

Note 5: When the blanking output (BI/RBO) is open or held high and a low is applied to the lamp-test input, all segment outputs are L.

H = High Level,
L = Low Level,
X = Don't Care

46A, 47A														
Decimal or Function	Inputs						BI / RBO(1)	Outputs						Note
	LT	RBI	D	C	B	A		a	b	c	d	e	f	g	
0	H	H	L	L	L	L	H	L	L	L	L	L	L	H	
1	H	X	L	L	L	H	H	H	L	L	H	H	H	H	
2	H	X	L	L	H	L	H	L	L	H	L	L	H	L	(2)
3	H	X	L	L	H	H	H	L	L	L	L	H	H	L	
4	H	X	L	H	L	L	H	H	L	L	H	H	L	L	
5	H	X	L	H	L	H	H	L	H	L	L	H	L	L	
6	H	X	L	H	H	L	H	H	H	L	L	L	L	L	
7	H	X	L	H	H	H	H	L	L	L	H	H	H	H	
8	H	X	H	L	L	L	H	L	L	L	L	L	L	L	
9	H	X	H	L	L	H	H	L	L	L	H	H	L	L	
10	H	X	H	L	H	L	H	H	H	H	L	L	H	L	
11	H	X	H	L	H	H	H	H	H	L	L	H	H	L	
12	H	X	H	H	L	L	H	H	L	H	H	H	L	L	
13	H	X	H	H	L	H	H	L	H	H	L	H	L	L	
14	H	X	H	H	H	L	H	H	H	H	L	L	L	L	
15	H	X	H	H	H	H	H	H	H	H	H	H	H	H	
BI	X	X	X	X	X	X	L	H	H	H	H	H	H	H	(3)
RBI	H	L	L	L	L	L	L	H	H	H	H	H	H	H	(4)
LT	L	X	X	X	X	X	H	L	L	L	H	L	L	L	(5)

48														
Decimal or Function	Inputs						BI / RBO(1)	Outputs						Note
	LT	RBI	D	C	B	A		a	b	c	d	e	f	g	
0	H	H	L	L	L	L	H	H	H	H	H	H	H	L	
1	H	X	L	L	L	H	H	L	H	H	L	L	L	L	
2	H	X	L	L	H	L	H	H	H	L	H	H	L	H	(2)
3	H	X	L	L	H	H	H	H	H	H	H	L	L	H	
4	H	X	L	H	L	L	H	L	H	H	L	L	H	H	
5	H	X	L	H	L	H	H	H	L	H	H	L	H	H	
6	H	X	L	H	H	L	H	L	L	H	H	H	H	H	
7	H	X	L	H	H	H	H	H	H	H	L	L	L	L	
8	H	X	H	L	L	L	H	H	H	H	H	H	H	H	
9	H	X	H	L	L	H	H	H	H	H	L	L	H	H	
10	H	X	H	L	H	L	H	L	L	L	H	H	L	H	
11	H	X	H	L	H	H	H	L	L	H	H	L	L	H	
12	H	X	H	H	L	L	H	L	H	L	L	L	H	H	
13	H	X	H	H	L	H	H	H	L	L	H	L	H	H	
14	H	X	H	H	H	L	H	L	L	L	H	H	H	H	
15	H	X	H	H	H	H	H	L	L	L	L	L	L	L	
BI	X	X	X	X	X	X	L	L	L	L	L	L	L	L	(3)
RBI	H	L	L	L	L	L	L	L	L	L	L	L	L	L	(4)
LT	L	X	X	X	X	X	H	H	H	H	H	H	H	H	(5)

FIGURE 14-18 (*continued*)

EXAMPLE 14-2

Design a BCD counter that will count from 0 to 999 and display the output on three FND-507 seven-segment LED displays. Use three 74LS90 BCD counters and three 74LS47 BCD-to-seven-segment decoder ICs. Implement the lamp test and the leading zero blanking.

Solution   Refer to Figure 14-19.

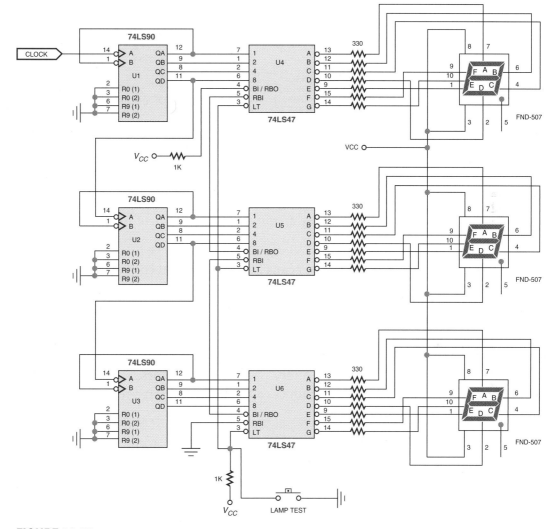

FIGURE 14-19

 # 14.9 THE LIQUID CRYSTAL DISPLAY

There are two types of LCDs in use today: dynamic and field-effect LCDs. The two types use different materials for the **liquid crystal** and work differently. Neither type emits any light and each must have an external light source to be seen.

The dynamic LCD has the liquid crystal material sandwiched between clear pieces of glass. The seven-segment pattern which is etched on the front glass plate is made of a clear electrically conductive material such as indium oxide. The back glass is coated with this clear conductor, which corresponds to the seven segments, as shown in Figure 14-20. In this way, only the digit segments will be seen when an electric current is applied to the LCD.

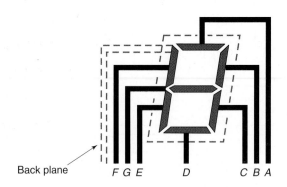

Back plane   F G E   D   C B A

FIGURE 14-20 Clear LCD segment conductors

When a voltage is applied between the segment pattern and the back conductor on the back sheet of glass, the liquid crystal diffuses the light. This happens because the index of refraction changes randomly, causing the light to be refracted randomly as it passes through the liquid crystal material. The scattering action causes the segment to appear milky white in color.

A dc voltage will produce this effect on a dynamic LCD; however, an ac voltage is used. This is because even a small dc current can cause the segment conductor to be electrically plated with material from the liquid crystal. An ac current will prevent this. The current for a typical dynamic seven-segment LCD is very small, about 25 μA at 30 V_{pp}, 60 Hz. This is the main reason for using LCD displays.

The field-effect or twisted nematic LCD is the most commonly used LCD display. This is the type of LCD used by most battery-operated calculators, watches, and computers. The most common LCD of this type produces a black segment on a reflective background.

To understand how this LCD display operates, you must first understand the operation of a polarized sheet of glass. Figure 14-21 shows a vertically-polarized sheet of glass. Notice only light rays that are vertically polarized will pass through the glass. The light which passes through the glass is all vertically polarized and, of course, less in intensity because some of the light rays cannot pass.

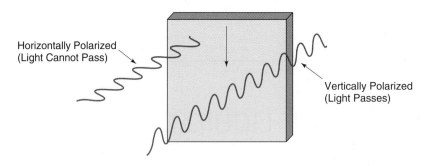

Horizontally Polarized
(Light Cannot Pass)

Vertically Polarized
(Light Passes)

FIGURE 14-21 Vertical polarized glass

If two polarized sheets of glass are placed at right angles, no light will pass through them. This is because the first polarized glass will stop all light rays which are not vertically polarized and the second polarized glass will only pass horizontally-polarized light. Therefore, no light can pass through when using both sheets. This is shown in Figure 14-22.

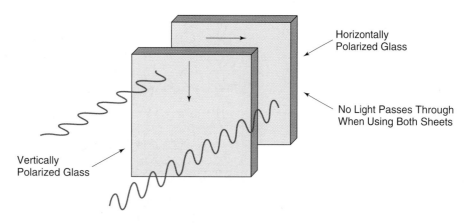

FIGURE 14-22 Filtering all lights

If it were possible to twist the vertical light rays passing the first vertically polarized glass 90°, they would pass through the second horizontally polarized glass. This is exactly what the liquid crystal material can do, that is, twist light 90°. By placing the liquid crystal material between the two polarized glass sheets, the vertically polarized light is twisted 90° and passed through the rear horizontally polarized glass, thus passing the light through the two polarized sheets of glass. The twisting of the vertical light rays will continue until an electrical current is passed through the liquid crystal material. When a current passes through the liquid crystal, it stops twisting the light and passes it unaltered to the horizontally polarized glass. This blocks the passage of the light because the light is vertically polarized. This is shown in Figure 14-23. Notice that only the area under the segment conductor will be affected, thus producing a black segment.

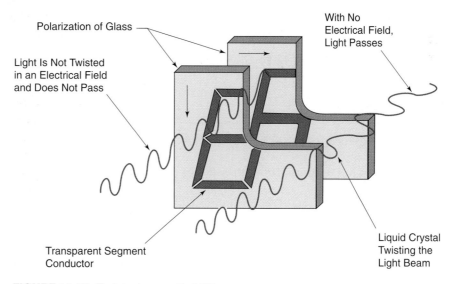

FIGURE 14-23 Twisted nematic LCD

The field-effect or twisted nematic LCD typically runs on a 60-Hz ac, 8-V_{pp} voltage with about 300 μA of current. Again, the main advantage of LCDs of this type is low current. The ac voltage used to drive an LCD display can be obtained with some CMOS XOR gates and a 50 percent duty-cycle clock. As can be seen in Figure 14-24 (page 510), when the input is 0, the voltage differential between the segment conductor and the back plane is 0. When the input is 1, the voltage potential is an ac voltage.

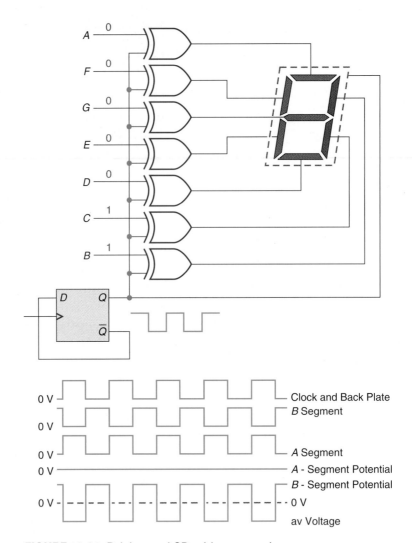

FIGURE 14-24 Driving an LCD with an ac voltage

 SELF-CHECK 2

1. Draw a diagram of a seven-segment display and label the segments with the proper letters.

2. Why is it not a good procedure to use a TTL logic 1 to light an LED?

3. What is the typical forward-biased voltage for a red LED?

Today, the twisted LCD display is produced in large panels that have very small individual pixels set in a matrix that allows the pixels to be turned on and off individually. This type of display is used to make TV screens and computer displays.

By placing a backlight behind the display and by using color filters, a color LCD display can be created. By placing a red, green, and blue filter over a set of pixels—and varying the voltage on each pixel in order to produce different light intensity—any color can be created.

This is the basic method for producing color used by the old cathode ray tube in most TVs. However, with the cathode ray tube, a stream of electrons is focused on the front screen of the tube, where they collide with a material that gives off light when hit by electrons. There are three different materials used: one gives off red light, the second, green, and the third, blue. Sometime take a magnifying glass and look at your color TV screen. You will see the three very small color pixels that make up the final color that you see on the screen.

The material that produces the light on the old color TV tube has what is called persistence. This means that once the material has been hit by the electrons, the light it produces will persist for a short time after the electron beam leaves. To achieve this effect and to speed up the color LCD, small, thin-film transistors able to charge a small capacitor are etched in the matrix for each color pixel. This causes the pixel to stay lit longer. These types of displays are called active displays and are used today on most color computer displays.

14.10 PROGRAMMING A CPLD (Optional)

In this section, we will develop a sequence of Quartus II projects that leads to a 2-digit seven-segment display that displays a count of 00–99.

We will follow this sequence:

Project 1: Design a 4-bit binary counter that counts from 0000 to 1111 and displays the count on LEDs.

Project 2: Add the code needed to display each count on a common-cathode seven-segment display as one hexadecimal digit, 0 to F.

Project 3: Design a 4-bit decade counter that counts from 0000 to 1001 continuously and displays the count on LEDs.

Project 4: Add the code needed to display each count on a common-cathode seven-segment display as one decimal digit, 0 to 9.

Project 5: Design a two-digit decade counter that counts from 00 to 99 continuously and displays the count as two BCD digits and two decimal digits on two common-anode seven-segment displays.

PROJECT 1: bincount

Follow the procedure presented in previous chapters:

- Create a new folder within the Altera folder and name it **bincount**.

- Create a new .vhd file and name it **bincount.vhd**.

- Write a VHDL program that will implement a 4-bit binary counter that counts the full count 0000–1111. Have each bit appear as an output. (In Chapter 10, **decade.vhd** brought out only the most significant digit to function as a divide-by-10 circuit.)

Invoke the ieee library by beginning your program with these three lines:

LIBRARY ieee;

USE ieee.std_logic_1164.all;

USE ieee.std_logic_unsigned.all;

The third line is needed so that we can use binary addition in the ARCHITECTURE statement.

Declare **clk** to be standard logic and have the circuit count on the negative edge of **clk** (HIGH-to-LOW transition).

Declare **Q** to be a 4-bit buffer of the standard logic vector type, so that all four bits will become outputs and the value of **Q** can be used within the ARCHITECTURE statement, Q <= Q + 1.

- Compile the program.

- Create a .vwf (vector waveform file) named **bincount.vwf** and use it to simulate the circuits.

Assign pins to the four bits of **Q**. On the PLDT-2 trainer from RSR Electronics, use pins 49, 50, 51, and 52 since they are jumpered to LEDs. Assign the **clk** input to one of the pins that is physically close to the stitches, such as pin 75.

- Program your CPLD. Don't forget to remove all jumpers from **HD1** first.

- Test the operation of the CPLD using input switches and LEDs. Your outputs should count from 0000 to 1111 and repeat.

bincount.vhd will look something like this:

```
LIBRARY ieee;
USE ieee.std_logic_1164.all;
USE ieee.std_logic_unsigned.all;
ENTITY bincount IS
     PORT
     (clk: IN STD_LOGIC;
     Q:BUFFER STD_LOGIC_VECTOR (3 DOWNTO 0));
END bincount;
ARCHITECTURE a OF bincount IS
BEGIN
     PROCESS (clk)
     BEGIN
          IF (clk'EVENT AND clk = '0') THEN
               Q <= Q + 1;
          END IF;
     END PROCESS;
END a;
```

PROJECT 2: sevenseg

Now that we have a binary counter, let's add programming to drive the common-cathode LEDs on the trainer. As in **bincount.vhd**, we will use a PROCESS to check for a HIGH-to-LOW transition on the input **clk**. When the transition occurs, the 4-bit signal **Q** will increment one count. The PROCESS will then be ENDed so that **Q** will update to the new count. A second PROCESS with **Q** as its sensitivity list will begin. In the second process, the value of **Q** will be checked with a series of IF and ELSIF statements. For each count, a 7-bit value is assigned to the signal **sevseg**. The 7-bit code will drive the common-cathode seven-segment display. A HIGH will cause the corresponding segment to light and a LOW will turn it off. The outputs of this circuit will cause four LEDs to display the count in binary—and the equivalent hexadecimal digit will light on the seven-segment display.

```vhdl
Library ieee;
USE ieee.std_logic_1164.all;
USE ieee.std_logic_unsigned.all;
ENTITY sevenseg IS
    PORT
        (clk:IN STD_LOGIC;
        Q:BUFFER STD_LOGIC_VECTOR (3 DOWNTO 0);
        sevseg:OUT STD_LOGIC_VECTOR (6 DOWNTO 0));
END sevenseg;
ARCHITECTURE a OF sevenseg IS
BEGIN
    PROCESS (clk)
    BEGIN
        IF (clk'EVENT AND clk = '0') THEN
            Q <= Q+1;
        END IF;
    END PROCESS;
    PROCESS (Q)
    BEGIN
        IF Q = "0000" THEN
            sevseg <= "1111110";
        ELSIF Q = "0001" THEN
            sevseg <= "0110000";
        ELSIF Q = "0010" THEN
            sevseg <= "1101101";
        ELSIF Q = "0011" THEN
            sevseg <= "1111001";
        ELSIF Q = "0100" THEN
            sevseg <= "0110011";
        ELSIF Q = "0101" THEN
            sevseg <= "1011011";
        ELSIF Q = "0110" THEN
            sevseg <= "1011111";
        ELSIF Q = "0111" THEN
            sevseg <= "1110000";
        ELSIF Q = "1000" THEN
            sevseg <= "1111111";
        ELSIF Q = "1001" THEN
            sevseg <= "1110011";
        ELSIF Q = "1010" THEN
            sevseg <= "1110111";
        ELSIF Q = "1011" THEN
            sevseg <= "0011111";
        ELSIF Q = "1100" THEN
            sevseg <= "0001101";
        ELSIF Q = "1101" THEN
            sevseg <= "0111101";
        ELSIF Q = "1110" THEN
            sevseg <= "1001111";
```

```
            ELSIF Q = "1111" THEN
                    sevseg <= "1000111";
            End IF;
        END PROCESS;
    END a;
```

Before you program and run the CPLD, study the values of **sevseg** for digits A through F in order to determine how those digits will appear on the seven-segment display.

The most significant bit of **sevseg** controls segment A (top) of the seven-segment display. B through G follow in order, with the least significant bit controlling segment G (middle). On the PLDT-2 trainer from RSR Electronics, assign pins to **sevseg** as follows:

Segment A = sevseg(6) > pin 58

Segment B = sevseg(5) > pin 60

Segment C = sevseg(4) > pin 61

Segment D = sevseg(3) > pin 63

Segment E = sevseg(2) > pin 64

Segment F = sevseg(1) > pin 65

Segment G = sevseg(0) > pin 67

PROJECT 3: tencount

Now modify the binary counter in **bincount.vhd** to reset after a count of 9 (1001). Name the program **tencount.vhd** (0–9 makes 10 counts).

We need all three ieee statements so we can use the binary addition function in the ARCHITECTURE statement. **Q** is declared a four-bit buffer so we can use its output value back in the program. The first IF statement checks for the negative edge of **clk**. The second set of IFs checks to see if the count is nine when the clock arrives. If it is, it has been at nine for a full clock cycle and needs to be reset. Remember to use quotes around multiple bits, "1001" and "0000".

tencount.vhd looks something like this:

```
Library ieee;
USE ieee.std_logic_1164.all;
USE ieee.std_logic_unsigned.all;
ENTITY tencount IS
    PORT(clk:IN STD_LOGIC;
        Q:BUFFER STD_LOGIC_VECTOR (3 DOWNTO 0));
END tencount;
ARCHITECTURE a OF tencount IS
BEGIN
    PROCESS(clk,Q)
    BEGIN
        IF(clk'EVENT AND clk = '0') THEN
            IF Q = "1001" THEN
                Q<= "0000";
```

```
                    ELSE Q<= (Q + 1);
                    END IF;
              END IF;
         END PROCESS;
    END a;
```

PROJECT 4: SevSegDec (SEVEN-SEGMENT DECIMAL)

Now that we have a decimal or decade counter with all four bits of **Q** appearing as outputs, let's assign a seven-bit code to **sevseg** at each count to drive the seven-segment display. As before, this will be done in a second PROCESS. The first PROCESS sets the value of **Q**, and the second PROCESS assigns the 7-bit code to **sevseg**.

```
Library ieee;
USE ieee.std_logic_1164.all;
USE ieee.std_logic_unsigned.all;
ENTITY SevSegDec IS
    PORT(clk: IN STD_LOGIC;
         Q:BUFFER STD_LOGIC_VECTOR (3 DOWNTO 0);
         sevseg:OUT STD_LOGIC_VECTOR (6 DOWNTO 0));
END SevSegDec;
ARCHITECTURE a OF SevSegDec IS
BEGIN
    PROCESS(clk,Q)
    BEGIN
        IF(clk'EVENT AND clk = '0') THEN
            IF Q = "1001" THEN
                    Q <= "0000";
            ELSE Q<=(Q + 1);
            END IF;
        END IF;
    END PROCESS;
    PROCESS (Q)
    BEGIN
        IF Q = "0000" THEN
            sevseg <= "1111110";
        ELSIF Q = "0001" THEN
            sevseg <= "0110000";
        ELSIF Q = "0010" THEN
            sevseg <= "1101101";
        ELSIF Q = "0011" THEN
            sevseg <= "1111001";
        ELSIF Q = "0100" THEN
            sevseg <= "0110011";
        ELSIF Q = "0101" THEN
            sevseg <= "1011011";
        ELSIF Q = "0110" THEN
            sevseg <= "1011111";
        ELSIF Q = "0111" THEN
            sevseg <= "1110000";
```

```
            ELSIF Q = "1000" THEN
                    sevseg <= "1111111";
            ELSIF Q = "1001" THEN
                    sevseg <= "1110011";
            END IF;
        END PROCESS;
    END a;
```

PROJECT 5: SevSegDec2 (SEVEN-SEGMENT DECIMAL 2)

Now for the big step. We will modify **SevSegDec.vhd** into a 2-digit decimal counter, 00 to 99, and a 2-digit seven-segment display driver. Name the new project **SevSegDec2.qpf** (Quartus Project File).

Library ieee;

USE ieee.std_logic_1164.all;

USE ieee.std_logic_unsigned.all;

The most significant digit will be kept by **P**, a 4-bit signal, and the least significant digit will be kept by **Q** as before.

```
ENTITY SevSegDec2 IS
    PORT(clk: IN STD_LOGIC;
        Q:BUFFER STD_LOGIC_VECTOR (3 DOWNTO 0);
        P:BUFFER STD_LOGIC_VECTOR (3 DOWNTO 0);
        sevsegQ:OUT STD_LOGIC_VECTOR (6 DOWNTO 0);
        sevsegP:OUT STD_LOGIC_VECTOR (6 DOWNTO 0));
    END SevSegDec2;
```

The following PROCESS uses IF statements to check for 99.

```
ARCHITECTURE a OF SevSegDec2 IS
BEGIN
   PROCESS(clk,Q,P)
   BEGIN
       IF(clk'EVENT AND clk = '0') THEN
            IF (P = "1001" AND Q = "1001") THEN    --PQ = 99?
                        P <= "0000";
                        Q <= "0000";
            ELSIF (P = "1001" AND Q < "1001") THEN --count in the 90s?
                        P <= "1001";
                        Q <= Q + 1;
            ELSIF (P < "1001" AND Q = "1001") THEN --count 89 or 79 or ...?
                        P <= P + 1;
                        Q <= "0000";
            ELSE Q <= (Q + 1);
            END IF;
       END IF;
   END PROCESS;
```

Values are assigned to **P** and **Q** at the end of the PROCESS. A second PROCESS uses IF statements to check the count of **Q** and assigns the appropriate 7-bit code to **sevsegQ**.

```
PROCESS (Q)
BEGIN
    IF Q = "0000" THEN
            sevsegQ <= "1111110";
    ELSIF Q = "0001" THEN
            sevsegQ <= "0110000";
    ELSIF Q = "0010" THEN
            sevsegQ <= "1101101";
    ELSIF Q = "0011" THEN
            sevsegQ <= "1111001";
    ELSIF Q = "0100" THEN
            sevsegQ <= "0110011";
    ELSIF Q = "0101" THEN
            sevsegQ <= "1011011";
    ELSIF Q = "0110" THEN
            sevsegQ <= "1011111";
    ELSIF Q = "0111" THEN
            sevsegQ <= "1110000";
    ELSIF Q = "1000" THEN
            sevsegQ <= "1111111";
    ELSIF Q = "1001" THEN
            sevsegQ <= "1110011";
    END IF;
END PROCESS;
```

A third PROCESS uses IF statements to check the count of **P** and assigns the appropriate 7-bit code to **sevsegP**.

```
PROCESS (P)
BEGIN
    IF P = "0000" THEN
            sevsegP <= "1111110";
    ELSIF P = "0001" THEN
            sevsegP <= "0110000";
    ELSIF P = "0010" THEN
            sevsegP <= "1101101";
    ELSIF P = "0011" THEN
            sevsegP <= "1111001";
    ELSIF P = "0100" THEN
            sevsegP <= "0110011";
    ELSIF P = "0101" THEN
            sevsegP <= "1011011";
    ELSIF P = "0110" THEN
            sevsegP <= "1011111";
    ELSIF P = "0111" THEN
            sevsegP <= "1110000";
```

```
        ELSIF P = "1000" THEN
                sevsegP <= "1111111";
        ELSIF P = "1001" THEN
                sevsegP <= "1110011";
        END IF;
    END PROCESS;
END a;
```

Q is the least significant digit and will appear on the right.

On the PLDT-2 trainer from RSR Electronics, assign pins to **sevsegQ** as follows:

Segment A = sevsegQ(6) > pin 58

Segment B = sevsegQ(5) > pin 60

Segment C = sevsegQ(4) > pin 61

Segment D = sevsegQ(3) > pin 63

Segment E = sevsegQ(2) > pin 64

Segment F = sevsegQ(1) > pin 65

Segment G = sevsegQ(0) > pin 67

P is the most significant digit and will appear on the left. Assign pins to **sevsegP** as follows:

Segment A = sevsegP(6) > pin 69

Segment B = sevsegP(5) > pin 70

Segment C = sevsegP(4) > pin 73

Segment D = sevsegP(3) > pin 74

Segment E = sevsegP(2) > pin 76

Segment F = sevsegP(1) > pin 75

Segment G = sevsegP(0) > pin 77

P(3)P(2)P(1)P(0) represents the most significant digit. Assign **P** to pins 44, 45, 46, and 48 to light LEDs 1 through 4. **Q(3)Q(2)Q(1)Q(0)** represents the least significant digit. Assign **Q** to pins 49, 50, 51, and 52 to light LEDs 5 through 8.

Assign **clk** to pin 80. The seven-segment displays should show the count in decimal, and the LEDs should show the corresponding count in BCD.

Let's use the CPLD as a 2-to-1 multiplexer. If the select signal (**sel**) is LOW, then input 0 (**in0**) will pass to the output unaltered. If **sel** is HIGH, then input 1 (**in1**) will pass to the output unaltered.

```
LIBRARY ieee;
USE ieee.std_logic_1164.all;
ENTITY mux_2to1 IS
    PORT(sel,in0,in1: IN STD_LOGIC;
    Q:OUT STD_LOGIC);
END mux_2to1;
```

```
ARCHITECTURE a OF mux_2to1 IS
BEGIN
      Process (sel,in0,in1)
      BEGIN
          IF (sel = '0') THEN
                Q <= in0;
          ELSIF (sel = '1') THEN
                Q <= in1;
          END IF;
      END PROCESS;
END a;
```

Simulation waveforms for **mux_2to1.vwf** are shown in Figure 14-25. When **sel** is LOW, **in0** passes to the ouput Q; when **sel** is HIGH, **in1** passes to the output.

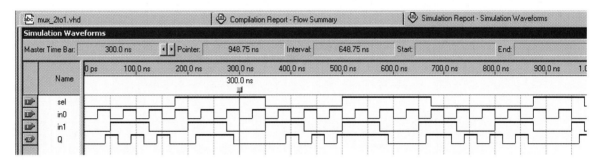

FIGURE 14-25 Simulation waveforms for mux_2to1.vwf

 # 14.11 TROUBLESHOOTING DECODERS

Decoders and multiplexers are often used in microprocessor control circuits. Almost every microcomputer has some type of decoder to decode the address bus to select RAM, ROM, and I/O ports. If the system being worked on has the ability to be programmed, then the technician can write a short loop program which will read or write to whichever memory or I/O port the decoder selects. This should give a continuous select pulse on the decoder's output. In this way, the technician can test the operation of the decoder. A good oscilloscope and even a logic analyzer are useful for debugging decoders.

The circuit shown in Figure 14-26 is the address decoder for an input/output port on a clone computer. It is designed so it can be set to any standard MSDOS I/O address. MSDOS and even Windows only decode the first ten address lines (A_0 to A_9). The decoder shown in Figure 14-26 was designed for use only in the old IBM XT I/O bus. This input/output bus was designed to interface with the old Intel 8088 CPU. This I/O bus is still a standard on all clones, even if it has a 400-MHz Pentium CPU. The address was decoded in blocks of 32 bytes, so only the last five address lines needed to be decoded (A_5 to A_9).

The circuit shown in Figure 14-26 is used to interface a clone computer to the old film processor which keeps going down. This time the computer quit receiving data from the film

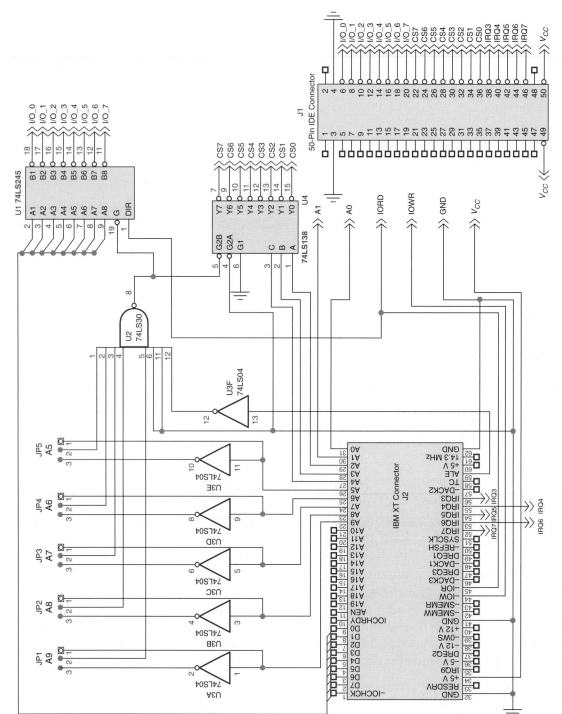

FIGURE 14-26 Address decoder for IBM clone computer

processor and the technician was called in to repair it. The technician suspected the address decoder board because the clone computer could still send commands to the film processor. This meant that the output port was working. The technician remembered that he had an extension card for the old IBM bus slot in his toolbox; so he got it out and placed the interface card on it. He placed the card above the computer so he could easily get oscilloscope probes on it. When he removed the card, he noticed that the edge connector was very dirty, so before he put it in the extension card he used a pencil eraser to clean the edge connector. This made the edge connector shine but did not remove any metal, as sandpaper would.

Our technician is excellent at working on hardware, but he also has a good knowledge of 8088 real-mode programming. He placed his Windows computer system into MSDOS mode and went into DEBUG. DEBUG is a standard program which has been packaged with IBM clones ever since the first IBM PC appeared on the market back in 1980, and it is still there in Windows. With this program, our technician used the "Assemble" command to place a very short program in memory and ran it. This program is shown in Program 14-1. It simply reads and writes to I/O port 2A0 Hex continuously. This is the I/O port to which the decoder should be set if the board's documentation could be trusted.

```
-A CS:0100                     ;enter the assemble command at the DEBUG prompt
1DDD:0100 MOV DX,2A0           ;place the I/O port address (2A0 Hex) in the DX register
1DDD:0103 MOV AL,AA            ;place AA Hex (10101010 Binary) in the AL register
1DDD:105 OUT DX,AL             ;output register AL to I/O port addressed in register DX
1DDD:106 IN AL,DX              ;input into register AL the content of I/O port
1DDD:0107 JMP 1DDD:0100        ;jump to the beginning and do it again
```

PROGRAM 14-1

After starting the program the technician took his oscilloscope and started looking for the decoded select pulse which should be present. First he looked at pin 15 of U4 and found nothing. This indicated that the input port which the computer used to get information from the film processor was not working. He needed to find out why.

Next the technician looked at pin 5 of U4. This is the select input which selects the decoder. There was no pulse on it either. Then he noticed that the 74LS30 was hanging part of the way out of its socket. Looking down, he saw the cause. Another technician had added a modem to the computer yesterday and when he had put it in the bus he had hit the 74LS30, dislodging it. Our technician quickly reseated the chip and went back to looking for the select pulse. It was there this time. To ensure that the card was working correctly, he changed his program to test all the I/O addresses which the 74LS138 U4 could address (2A0 Hex to 2BF Hex). After taking the interface card off the extension card and reseating it in the computer, he started the Windows software which normally runs the film processor and it worked quite well.

Program 14-1 is a short program that will output and input from the I/O port. The address of the I/O port is placed in the 16-bit register DX inside the CPU. This program is shown as it would appear if it were loaded using the DEBUG program. The comments to the right of the program would not be entered in DEBUG and are there only for the reader's use.

DIGITAL APPLICATION

Decoder for 68HC11 Microcontroller Serial Communicator

The circuit shown below is the slave select decoder for a high-speed serial communication system used to communicate with up to seven 68HC11 computer systems. The data is carried on a ribbon cable which is connected to a JP21 connector. This connector carries the MOSI (master-out slave-in), MISO (master-in slave-out), and SCK (serial clock) for the synchronous high-speed communication. This system runs at just below 1 M bit per second of serial data transfer.

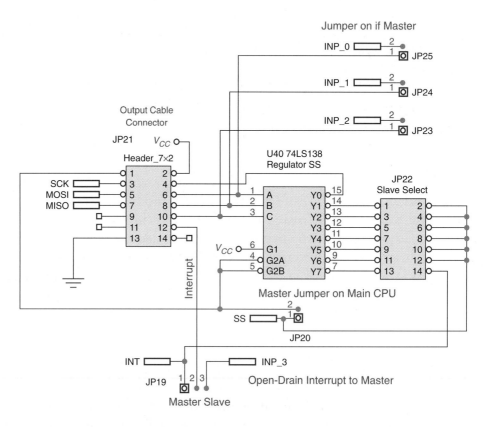

Note: (1) Bit 3 of Port C (PC3) must be set to output open drain.
(2) Slave select 7 will cause an interrupt of all slave CPUs. The master CPU should not have slave select jumper connected.

If the computer system is a master controller, then the jumpers on JP23, JP24, and JP25 are placed on the pins and the master 68HC11 computer can place the 3-bit binary code for the slave it wishes to communicate with on the ribbon cable. The slave computer does not have these jumpers shorted. The slave computer has a jumper placed on JP22; the position of this jumper determines the address of the slave. The decoder chip used is a 74LS138 decoder chip. The output bit, INP_3, is an open-drain output used by the slave to interrupt the master computer to request communication. (Schematic courtesy of Precise Power Inc.)

SUMMARY

- A decoder is a digital circuit that will produce a unique active output for each of a given set of binary numbers input to the decoder.

  A decoder is a full decoder if each possible binary number is decoded. A partial decoder only decodes a subset of binary numbers. Decoders are widely used in computer applications to decode the address bus of the computer to select memory banks or input/output ports.

- The decoder can be used as a demultiplexer if a common enable line is added to each of the AND gates in the decoder.

  This will allow the data on the enable to be switched to the channel that is selected on the demultiplexer's select inputs.

- A multiplexer is a decoder which has a channel input to each of the AND gates used.

  The output of each of the AND gates is ORed together to produce one output from the multiplexer. The data coming out of the one output is selected by the binary number placed on the select inputs of the multiplexer.

- The *PN* junction will produce light when forward biased.

  The *PN* junction made from properly doped gallium will emit light in sufficient amounts to be used as a light source. This is called an LED or light-emitting diode. The LED works just like a regular diode except that the forward bias is about 1.75 volts for a red LED and somewhat higher for green and yellow LEDs. Seven-segment LEDs use 7 or 8 LEDs configured in patterns that can be used to reproduce numbers. They have either the cathodes or the anodes of all the diodes tied together to produce a common anode or cathode. The seven-segment LED has a de facto segment labeling system that places the segments on the face of the LED.

- The LCD (liquid crystal display) does not produce any light but blocks light to make a display.

  The LCD is used in applications where low power is needed. The field-effect or twisted nematic LCD uses polarized glass and is the most widely used LCD. LCDs are usually driven by an ac voltage to reduce plating on the surface of the display.

QUESTIONS AND PROBLEMS

1. Draw the logic diagram for a 3-bit full decoder.

2. Draw the logic diagram for a partial decoder which will give active-LOW output for FB, FA, FC, and FF.

3. Use a 74150 multiplexer to reproduce the following truth table. Draw the logic diagram.

Inputs					Output		Inputs					Output
E	D	C	B	A	Y		E	D	C	B	A	Y
0	0	0	0	0	0		1	0	0	0	0	0
0	0	0	0	1	1		1	0	0	0	1	0
0	0	0	1	0	1		1	0	0	1	0	0
0	0	0	1	1	1		1	0	0	1	1	0
0	0	1	0	0	0		1	0	1	0	0	1
0	0	1	0	1	0		1	0	1	0	1	1
0	0	1	1	0	1		1	0	1	1	0	0
0	0	1	1	1	0		1	0	1	1	1	1
0	1	0	0	0	1		1	1	0	0	0	1
0	1	0	0	1	0		1	1	0	0	1	0
0	1	0	1	0	0		1	1	0	1	0	0
0	1	0	1	1	1		1	1	0	1	1	1
0	1	1	0	0	1		1	1	1	0	0	1
0	1	1	0	1	1		1	1	1	0	1	1
0	1	1	1	0	0		1	1	1	1	0	1
0	1	1	1	1	0		1	1	1	1	1	0

4. What is the typical forward-biased voltage for a red LED?

5. Draw and label the segments of a typical seven-segment LED.

6. Using a data book, draw the logic diagram for a three-digit, seven-segment LED display using 7447 decoders and implement the leading 0 suppression.

7. What are the two types of LCD displays?

8. Why are LCD displays driven with an ac voltage?

9. Which type of display is faster, the LCD or the LED?

10. Could a twisted nematic LCD display be made to have a black background and white segments?

11. Draw the logic diagram for a partial decoder which will decode the first eight binary numbers of an 8-bit number. Use 74138 and 7406 ICs. Show pinouts.

12. Redesign the 8-trace oscilloscope multiplexer in Figure 14-13 to be a 16-trace oscilloscope multiplexer.

13. Make a list of CMOS analog multiplexers and show the pinouts.

14. Look up the 4511 CMOS IC and describe its operation.

15. Look up the 74C945 CMOS IC and describe its operation.

16. Use a 74LS47 IC and an FND-507 seven-segment LED to display the count of a 74LS90 counter IC. Show pin numbers.

17. List four multiplexer ICs.

18. Use one 7406, one 74LS30, and one 74LS138 IC to make a partial decoder for address 0ff0 to 0ff7 hexadecimal. Show pin numbers.

19. What is polarized glass?

20. What does LCD stand for?

21. What will happen to the operation of the circuit in Figure 14-4 if the output of inverter A_6 shorts to ground?

22. What kind of outputs must the inverters in Figure 14-4 have for the circuit to work correctly?

23. In Figure 14-11 and Figure 14-12, which chips can be used as multiplexers?

24. What will happen to the operation of the digital 8-trace oscilloscope multiplexer shown in Figure 14-13 if pin 11 of the 74LS151A IC becomes disconnected?

25. What will happen to the display in Figure 14-19 if pin 2 of U6 becomes grounded?

26. Which pins on the jumper JP1 in Figure 14-26 would be shorted to make the decoder see the address line (A9) as a one?

27. What size address block does each CSx produce in Figure 14-26?

28. What is the total address range of the decoder in Figure 14-26?

29. Which pins would be shorted to decode the address 380 Hex in Figure 14-26?

30. What pins will be used as the data input for a demultiplexer if the 74LS138 is used as the chip?

31. Write a program using VHDL that will implement a 4-to-1 multiplexer. A 2-bit select input will determine which one of the 4 inputs will pass to the output.

32. Write a program using VHDL that will implement an 8-to-1 multiplexer. A 3-bit select input will determine which one of the 8 inputs will pass to the output.

33. Write a program using VHDL that will implement a 1-to-4 demultiplexer. A 2-bit select input will determine which one of the 4 outputs will receive the input waveform. The other outputs should be LOW.

34. Write a program using VHDL that will implement a 1-to-8 demultiplexer. A 3-bit select input will determine which one of the 8 outputs will receive the input waveform. The other outputs should be LOW.

35. Write a program using VHDL that will implement a 2-to-4 decoder. A 2-bit select input will determine which one of the 4 outputs will go LOW. The other 3 outputs should be HIGH.

36. Write a program using VHDL that will implement a 3-to-8 decoder. A 3-bit select input will determine which one of the 8 outputs will go LOW. The other 7 outputs should be HIGH.

MULTIPLEXERS, LEDs, AND SEVEN-SEGMENT DISPLAYS

OBJECTIVES

After completing this lab, you should be able to:

- Use a 74150 to construct a circuit to reproduce a 5-bit input truth table.
- Construct a one-digit seven-segment LED display.
- Test the operation of a red LED.

COMPONENTS NEEDED

1	FND-510 seven-segment LED common anode
1	7447 BCD-to-common-anode seven-segment LED driver
8	330-Ω, ¼-W resistors
1	red LED
1	50-Ω, 1-W resistor
1	1-kΩ pot, 1 W
1	74150 multiplexer

PROCEDURE

1. Determine the pinout of the seven-segment LED by the following method:

   a. Connect pin 1 to +5 V.

   b. Use a 330-Ω resistor connected to ground as a probe and test all the other pins to see if you can light a segment.

   c. If no segment lights, move the 5-V connector to the next pin and repeat the probing with the 330-Ω resistor connected to ground. When a segment lights, you have found the common anode.

   d. After the first segment lights, leave the +5-V connection on that pin and use the 330-Ω resistor to determine which pins are the *A*, *B*, *C*, *D*, *E*, *F*, *G*, and decimal point segments.

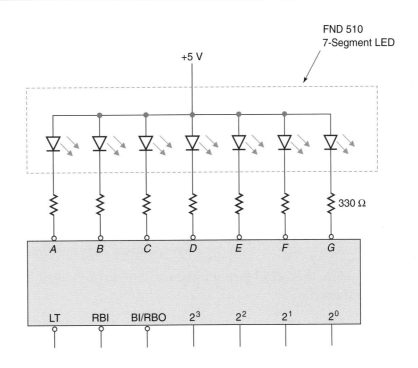

2. Use your data book to determine the pinout for the 7447 decoder driver and connect the seven-segment LED to it as shown. Have your instructor check its operation.

3. Construct the simple circuit shown. Complete the table and draw the graph for current versus voltage.

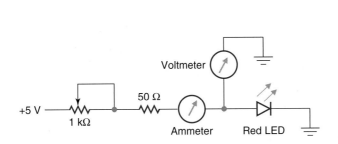

Current (mA)	Forward-Biased Voltage
0	
0.5	
1	
2	
3	
4	
5	
10	
20	
30	
40	
50	
60	

4. Implement a four-input expression using a 74150 multiplexer as follows:

   a. Write a four-input truth table that you wish to design.

Inputs				Output
D	C	B	A	Y
0	0	0	0	1
0	0	0	1	0

Etc.

   b. Place each value of Y on the corresponding data input pin.

   c. Sequence the 74150 through the truth table by using a 7493 and the pulser switch on the trainer.

   d. Observe the outputs on pin 10. (You may want to invert the output.)

5. Implement a five-input as follows:

   a. Write a five-input truth table that you wish to design.

Inputs					Output
E	D	C	B	A	Y
0	0	0	0	0	1
0	0	0	0	1	0
0	0	0	1	0	1
0	0	0	1	1	1

Etc.

   b. For the first two lines of the truth table place the appropriate value on data Channel 0.

Inputs					Output
E	D	C	B	A	Y
0	0	0	0	0	0
0	0	0	0	1	0
0	0	0	1	0	1
0	0	0	1	1	1
0	0	1	0	0	0
0	0	1	0	1	1
0	0	1	1	0	1
0	0	1	1	1	0

Etc.

Place 0 V on Channel 0

Place 5 V on Channel 1

Place E on Channel 2

Place E on Channel 3

LEDs

OBJECTIVES

After completing this lab, you should be able to use Multisim to:

- Construct a red LED test circuit.
- Troubleshoot a 16-bit partial decoder.

PROCEDURES

1. Open file 14B-1.ms9 and run the simulation. When you have become comfortable with the operation of the decoder, continue on to the next part of this lab.

2. Open file 14B-2.ms9 and run the simulation. There is one fault in this circuit. Find it.

CHAPTER 15

Tri-State Gates and Interfacing to High Current

KEY TERMS

bidirectional bus driver   bus

buffer   tri-state

OBJECTIVES

After completing this chapter, you should be able to:

■ Explain the operation of a tri-state gate.

■ Interface logic gates to transistors for higher current control.

■ Construct a circuit using tri-state gates that will multiplex two or more signals onto two or more displays.

■ Describe the use of tri-state gates with computer buses.

■ Use relays and optocouplers to isolate circuits.

15.1 TRI-STATE GATES

In the gates with totem-pole outputs studied earlier in this text, either the top transistor was on, or the bottom transistor was on, and the gate output was a 1 or a 0. In tri-state gates, both transistors can be turned off, and the output is not pulled up to V_{CC} or pulled down to ground. The gate assumes a third state or high-impedance state. In this HiZ state, the gate has no effect on the gates to which it is connected. If the outputs of several gates are connected together, only one of the ICs can be active at a time. The remaining gates must be in their HiZ state.

Figure 15-1 shows two **tri-state buffers**. In the first buffer, the control is not bubbled. A 1 on the control enables the IC, and the output is either HIGH or LOW as determined by the input. A 0 on the control inhibits the gate, and the gate enters the high-impedance state. The second buffer in Figure 15-1 has a bubble on the control line. This implies that a 0 enables the gate and the output assumes a 1 or 0 state. When the control line goes HIGH, the output enters the HiZ state, and another tri-state gate can control the output.

FIGURE 15-1 Tri-state gates

Figure 15-2 shows three tri-state buffers connected together with the common output driving an OR gate. Since the control input is active HIGH (no bubble), only one control input can be HIGH at a time. Table 15-1 shows the wide variety of tri-state devices that are available.

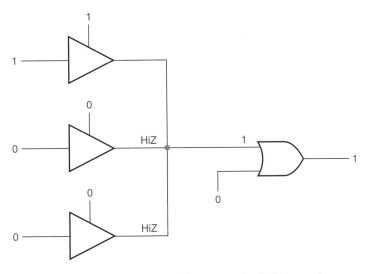

FIGURE 15-2 Only one control input can be HIGH at a time

TABLE 15-1   Available tri-state gates

Device Number		Description
54175/74125	54LS125/75LS125	Tri-state quad buffer
54126/74126	54LS126/74LS126	Tri-state quad buffer
54S134/74S134		Tri-state 12-input NAND
54LS240/74LS240	54S240/74S240	Tri-state inverting octal buffer
54LS241/75LS241	54S241/75S241	Tri-state octal buffer
54LS242/74LS242	54S242/75S242	Tri-state octal buffer
54LS243/75LS243	54S243/75S243	Tri-state octal buffer
54LS244/75LS244	54S244/75S244	Tri-state octal buffer
54S244/74S244		Tri-state octal buffer
54LS245/75LS245		Tri-state octal buffer
54365/74365	54LS365/75LS365	Tri-state hex buffer
54366/74366	54LS366/74LS366	Tri-state hex buffer
54367/74367	54LS367/74LS367	Tri-state hex buffer
54368/74368	54LS368/75LS368	Tri-state hex buffer
545940/745940		Tri-state octal buffer
545941/745941		Tri-state octal buffer
54173/74173	54LS173/74LS173	Tri-state quad-D registers
54251/74251 545251/745251	54LS251/74LS251	Tri-state data selector/multiplexer
54LS254/74LS253	54S253/74S253	Tri-state data selector/multiplexer
54LS257/74LS257	54S257/74S257	Tri-state quad 2-data selector/multiplexer
54LS258/74LS258	54S258/74S258	Tri-state quad 2-data selector/multiplexer
54S299/74S299		Tri-state 8-bit universal shift/storage registers
54LS353/74LS353		Tri-state data selector/multiplexer
54LS373/74LS373	54S373/74S373	Tri-state octal latch
54LS374/74LS374	54S374/74S374	Tri-state octal latch
54LS670/74LS670		Tri-state 4 × 4 register file
54C173/74C173		Tri-state quad-D flip-flop
54C373/74C373		Octal latch with tri-state outputs
54C374/74C374		Octal D-type flip-flop with tri-state outputs
4034BM/4034BC		8-stage tri-state bidirectional parallel/serial input/output bus register
4043BM/4043BC		Quad tri-state NOR rs latches
4048BM/4048BC		Tri-state expandable 8-function 8-input gate
4076BM/4076BC		Tri-state quad-D flip-flop
4503BM/4503BC		Hex noninverting tri-state buffer

15.2 TRI-STATE INVERTERS AND BUFFERS

Buffers are single-input circuits that do not alter the signal; 1 in, 1 out. Several of the ICs listed in Table 15-1 are octal buffers, such as the 74LS240, 241, 242, 243, and 244. They vary in their configurations. For example, Figure 15-3 shows the pinout for a 74LS240. There are two groups of four tri-state inverters. One group is controlled by $2\overline{G}$ on pin 19. When $2\overline{G}$ goes LOW, the top set of inverters, 2A1, 2A2, 2A3, and 2A4 are enabled, and data can pass inverted from 2A1 to 2Y1, 2A2 to 2Y2, 2A3 to 2Y3, and 2A4 to 2Y4. When $2\overline{G}$ goes HIGH, 2Y1, 2Y2, 2Y3, and 2Y4 enter a high-impedance state. Other tri-state gates could control those output lines. $1\overline{G}$ on pin 1 controls the other set of inverters. Each group is convenient for handling four bits in parallel.

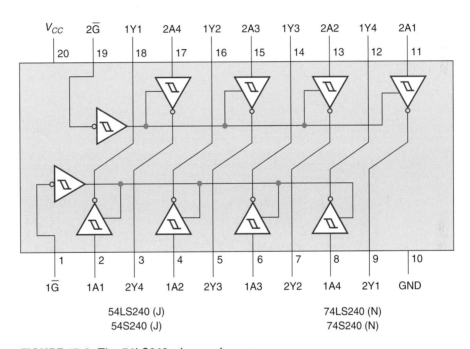

FIGURE 15-3 The 74LS240 tri-state inverter

The 74LS241, Figure 15-4, is designed in a similar manner except that the gates do not invert, and the top group of four gates is enabled when 2G goes HIGH. Even though the gates do not alter the signal, they do buffer the output from the input. Buffers are used to provide increased current drive capabilities. Often a source cannot provide the current drive required by the following circuitry. In those cases, the source drives a buffer, and the buffer drives the following circuitry. Figure 15-5 shows the 74LS242 quad-bus transceiver.

The 74LS242 shown in Figure 15-5 has eight buffers, but they are arranged in pairs. One pair is shown in Figure 15-6. When GBA is HIGH, $\overline{GAB}$ must be HIGH also. Data can pass inverted from 1B to 1A through inverter 1. The output of inverter 2 is in the high-impedance state and does not compete with the signal at 1B. Data passes from line or bus 1B to bus 1A. When $\overline{GAB}$ is LOW, inverter 2 is enabled. GBA must be LOW to put inverter 1 into its

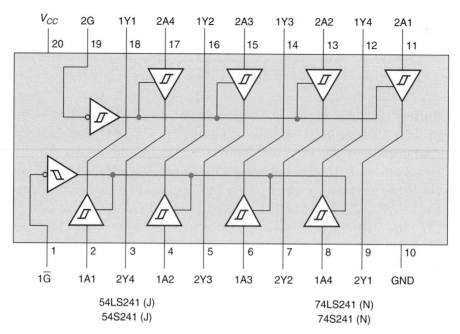

FIGURE 15-4 The 74LS241

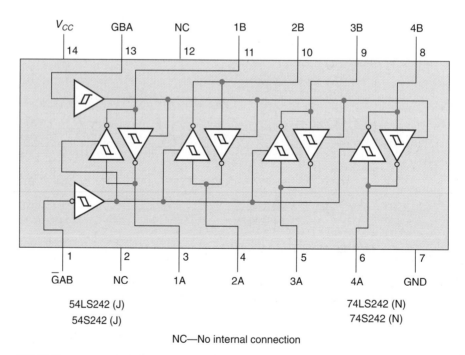

NC—No internal connection

FIGURE 15-5 The 74LS242 quad-bus transceiver

high-impedance state so that its output does not interfere with the signal on 1A. By controlling these two tri-state inverters, data can flow in either direction between 1A and 1B. This combination is called a **bidirectional bus driver** or a bus transceiver, since it can transmit or receive data. The 74LS242 is a quad-bus transceiver.

Digital systems are often configured in parallel, with each bit having its own data line. Microprocessor-controlled systems often work with 8 or 16 data lines called a bus. ICs like the 74LS245 shown in Figure 15-7 are ideal for controlling the bus of microprocessors. When $\overline{G}$ is LOW, gates 1 and 2 are both enabled. When direction control DIR goes HIGH, gate 2 produces a 1 out, which enables all eight buffers that pass data from A1 through A8 to B1 through B8. The output of gate 1 is 0, and the buffers that pass data from B1 through B8 to A1 through A8 enter the HiZ state. When direction control goes LOW, gate 1, a NOR gate, produces a 1 out, and the eight buffers that pass data from A1 through A8 to B1 through B8 enter the HiZ state. Only one set is active at a time and no conflict occurs.

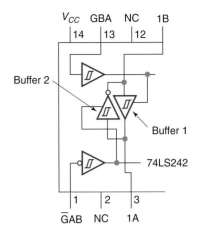

FIGURE 15-6 Only one of the buffers can be enabled at a time

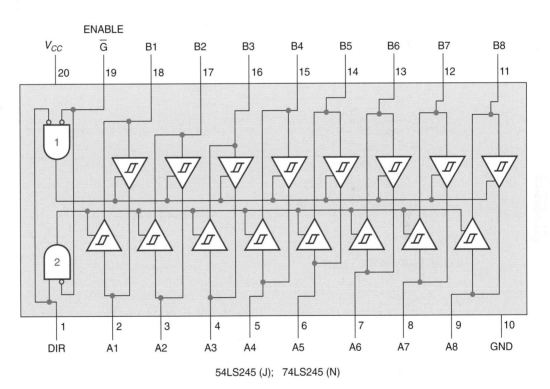

FIGURE 15-7 The 74LS245

54LS245 (J); 74LS245 (N)

15.3 COMPUTER BUSES AND THE TRI-STATE GATE

A **bus** is a group of conductors used to transfer digital binary numbers from one device to another. Microcomputers use three types of buses (address, control, and data buses). As shown in Figure 15-8, the address bus is used by the central processor to select the memory location or input/output device to be accessed. This bus is unidirectional and transfers the address or

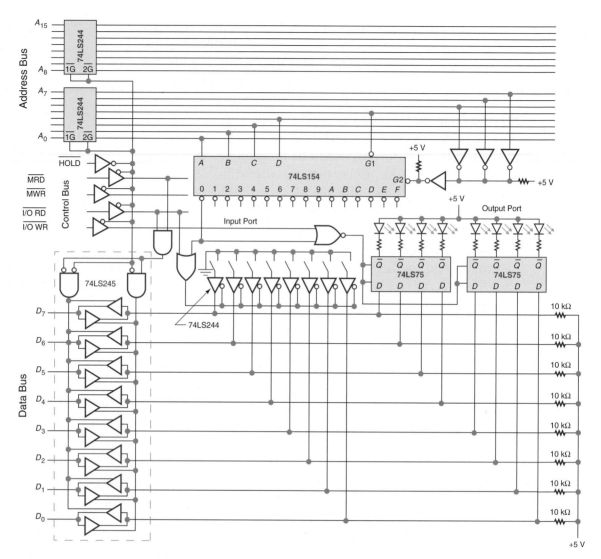

FIGURE 15-8 An 8085 bus

data in the direction from the CPU to the devices connected to the bus. Figure 15-8 shows a typical input/output port wired to the bus of an 8085 CPU. The address bus is buffered with 74LS244 tri-state gates. This gives it good current driving capability, plus it can be put into HiZ by a control signal called $\overline{\text{HOLD}}$. This ability to disconnect the CPU from its bus allows other devices, such as a direct memory access device, to gain control of the bus and use any of the memory or input/output devices on the bus.

The data bus is used to transfer data in the form of binary numbers to and from the CPU and memory or input/output devices. This bus is bidirectional because data flows both ways. To buffer this bus, a bidirectional bus driver such as the 74LS245 must be used. The 74LS245 IC gives good current driving ability and can be switched to control the direction of data flow. The data bus can have many devices connected to it, each of which draws current from it. This means that a typical NMOS IC, such as a CPU, cannot supply the current needed to produce a good logic 0 or 1, therefore the need for current buffering.

Tri-state gates are also used to input data to a data bus. As shown in Figure 15-8, the input device is a 74LS244 which is controlled by the combination of an address decoder and the control signal $\overline{\text{I/O RD}}$. When the CPU wishes to receive data from this input port, it places 00000000 on the lower eight bits of the address bus, which enables the 0 select line of the address decoder 74LS154. The CPU then makes the $\overline{\text{I/O RD}}$ signal LOW, which causes the tri-state gates of the input port to turn on and control the bus. When the CPU has latched the data into a register inside the CPU, it raises the $\overline{\text{I/O RD}}$ to a 1, thus inhibiting the tri-state gates and putting them in HiZ state. In this way, the CPU can cause one and only one device to have control of the bus at a time.

SELF-CHECK 1

1. What two types of TTL outputs can be tied together?

2. Which direction will data flow in a 74LS2445 bidirectional bus driver if the DIR pin is HIGH and the ENABLE G pin is LOW?

3. Why are tri-state gates used in microprocessor circuits?

 # 15.4 BUFFERING TO HIGH CURRENT AND HIGH VOLTAGE

The digital engineer and technician would like to turn the world on and off with a 5-V signal, but the world runs on such voltages as 120-V ac, 440-V 3-phase ac, and currents of milliamperes to megamperes. This has not deterred them, though; instead, they design ways to control these large voltages and currents with their 5-V digital signal.

Control of dc voltages and currents, in excess of that which a digital IC can supply, can be accomplished by use of a buffer IC or transistor. The ac voltages can be controlled by triacs if they are not extremely large or too high in current. If the control does not need to be extremely fast, a relay can be used to buffer the digital signal to extremely high currents and voltages. Relays offer high-current capability and complete circuit isolation from the high current and high voltage. Complete circuit isolation can also be obtained with optocouplers. Each of these options will be examined more closely later in this chapter.

Figure 15-9 shows the use of a transistor to supply enough current to energize a relay which will turn on a pump motor. The 7406 is an open-collector inverter which is designed to be used for buffering to high voltages (up to 30 V) and high currents (up to 40 mA). When a TTL logic 1 is input to the 7406 input, the inverter's output goes LOW, causing a current of about 23 mA to flow through the emitter-base junction of the transistor. This saturates the transistor and turns it on hard, producing a large collector current to energize the relay. The current through the transistor's collector is limited only by the resistance of the relay's coil, 50 Ω, and therefore can be quite large. The collector current in Figure 15-9 is 12 V divided by 50 Ω or 240 mA.

When the TTL logic signal goes to 0, the inverter's output goes to HiZ and no base current can flow, thus shutting off the transistor and stopping the current flow in the relay coil. This causes the magnetic field of the coil to fall in on its own coil windings, inducing a back emf or voltage

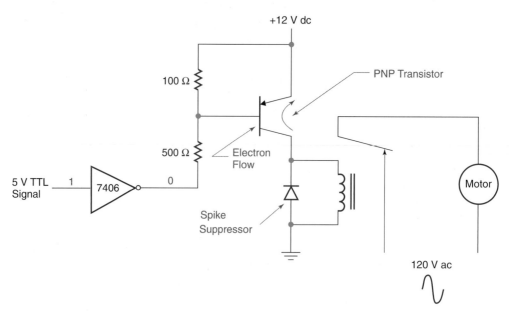

FIGURE 15-9 A buffering to high current

opposite of the voltage which produced the magnetic field. This reverse voltage spike can be quite large and damage the transistor or other components in the circuit. To prevent this spike from being too large, a diode is placed across the relay's coil in a reverse-bias mode to the circuit's voltage. The diode will be forward biased by the back emf generated by the relay's coil. This diode prevents the back voltage from going any larger than −0.7 V, which is the forward-biased voltage of the diode.

There are two things to consider when designing a circuit using a transistor to turn on or off a higher current. The first is to use a power transistor which has an I_C or collector current large enough to handle the current of the device being controlled. In Figure 15-9, the transistor must be able to handle at least 0.25 A or it will burn up. The second is to supply enough base current to turn on the transistor hard so there will be little voltage drop across the transistor. If the base current is too little, the transistor will decrease the current flow from emitter to collector, thus producing a voltage drop across the transistor. This voltage drop causes more power to be dissipated by the transistor, and it may burn up from heat. Remember, power equals current times voltage. When the voltage drop across the transistor is almost 0, the power dissipated by the transistor decreases.

An IC designed to interface CMOS or TTL logic levels to a relay is the 74C908. This IC can supply a current of 250 mA at 30 V. This is sufficient to energize most small relays.

The 75XXX series of ICs primarily consists of interfacing ICs with high current and voltage driving ability. The 75491 and 75492 are good examples of this series of ICs. They are high-current drivers, as shown in Figure 15-10. The 75492 is a set of darling pair transistors with the emitter of the output transistor tied to ground and the collector as the output. The 75491 has both the emitter and collector as outputs, so the designer can use it to either sink or source current. Both of these driver ICs can handle up to 250 mA of output current and up to 20 volts on the collector. The 75XXX series of ICs is used to drive such things as relays, stepping motors, servo motors, printer heads, and displays.

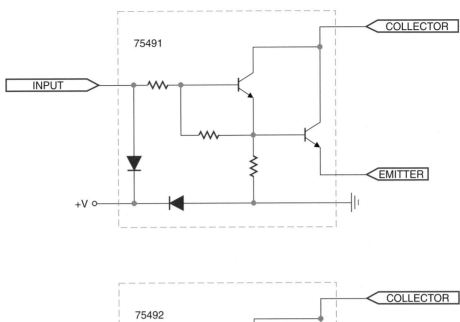

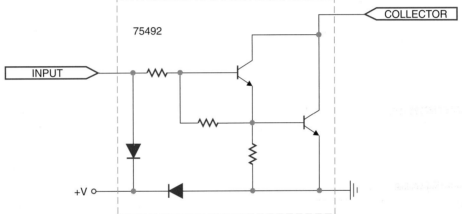

FIGURE 15-10

EXAMPLE 15-1

Design a current-driving circuit to drive a 2.5-amp, 24-volt relay. Use a 7407 IC and a TIP-125 PNP transistor. The TIP-125 transistor typically has an Hfe of 2500 and a maximum collector current of 8 amps. This makes it a good choice for this type of circuit.

Solution   Refer to Figure 15-11. The collector current will determine the required base current needed to place the transistor in saturation. It is a good idea to use two to four times the needed current in the base circuit of the transistor to ensure that the transistor will always be in saturation. If the transistor slips out of saturation and starts to produce a voltage drop across the emitter-to-collector, the power dissipated by the transistor will increase, and the transistor may burn up. The total current needed for the above design is only 1 mA, so we will use 4 mA in the design. The calculated resistance is 5825 ohms, but that is not a standard resistance for resistors, so we will use a 5-kΩ resistor. That will increase the base current to a little over 5 mA. The current through resistor Rb is calculated by dividing the 1.4 voltage drop of the forward-biased emitter base junction by the 1 k resistance of Rb. This gives 1.4 mA, which, when added to the 5-mA current through Ra, produces a 6.4-mA total current. This is well within the current limits of the 7407 IC.

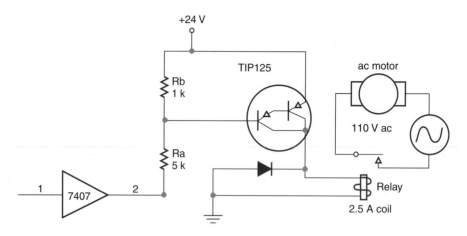

FIGURE 15-11

$$\text{Hfe} = \frac{\text{Ic}}{\text{Ib}} = 2500 \text{ typ} \quad \text{Ib} = \frac{2.5 \text{ A}}{2500} = 1 \text{ mA}$$

Use an Ib 2 to 4 times larger.

$$\text{Ra} = \frac{24 - 0.7}{4 \text{ mA}} \quad \text{Ra} = 5825\Omega$$

15.5 MULTIPLEXING SEVEN-SEGMENT LED DISPLAYS

The circuit in Figure 15-12 uses a 74LS241 octal buffer to multiplex two digits onto two seven-segment displays. When the Q output of the 7476 is LOW, the BCD digit 1 passes through the tri-state buffers into the 7447, where it is decoded to drive a seven-segment display. The displays are common anode. The common anode must be connected to a positive supply for the displays to be lit. The $\overline{Q}$ output is inverted by inverter number 1 of the 7406. The LOW level turns on Q1, and seven-segment display number 1 is connected to +12 V through the turned on transistor.

During this time outputs 2Y1, 2Y2, 2Y3, and 2Y4 of the 74241 are in their high-impedance state and do not interfere with BCD digit number 1. The 7406 is open collector. When Q goes LOW and gets inverted by the 7406 inverter number 2, the output is pulled HIGH by the 470-Ω resistor. The emitter-base junction of Q2 is not forward biased and Q2 turns off. Only seven-segment display number 1 is on to display the BCD digit number 1.

When the 7476 changes states, Q is HIGH, the BCD digit number 2 passes through to the 7447 and is decoded. The 7406 inverter number 2 goes LOW and Q2 turns on. The BCD digit number 2 is displayed on the seven-segment display number 2. Display number 1 is off.

This process alternates with the output of the 7476. The 7476 is wired to toggle. Its output has an even-duty cycle so that each set of tri-state buffers is enabled for an equal amount of time,

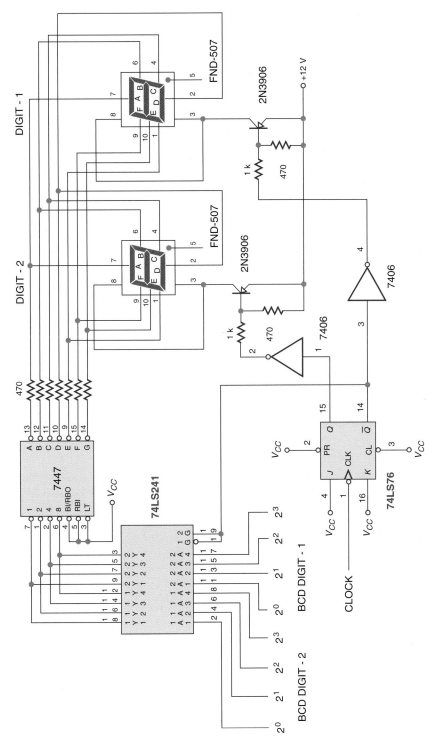

FIGURE 15-12 Multiplexing two seven-segment displays

and the brightness of each seven-segment display should be about the same. Each display is on for one-half of each clock cycle. They appear to be about one-half as intense as they would be if they were on continuously. The current-limiting resistors could be reduced in value to allow more current to flow through the LEDs during the time they are on. If this is done and the clock stops, the LEDs can be destroyed from excessive current.

SELF-CHECK 2

1. Using the procedure outlined in section 15.4, design a high-current interface to a TTL-level control signal to drive a 12-volt 500-mA relay. Use a TIP-120 transistor.

2. Use a 74492 IC to drive a 12-volt 100-mA relay.

3. What is the main advantage and disadvantage of using a relay to control a high-current device?

15.6 ISOLATING ONE CIRCUIT FROM ANOTHER WITH OPTOCOUPLERS

The circuit in Figure 15-13 uses an optocoupler to transfer the 60-Hz line frequency to a standard TTL logic signal. The optocoupler also isolates the ac voltage from the digital signal. The optocoupler has a regular LED which, when forward biased, will cause a photodiode to conduct a current. This current then turns on an NPN transistor to supply a TTL signal to a Schmitt-trigger input inverter.

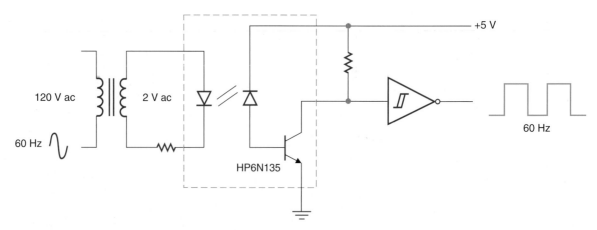

FIGURE 15-13 Optocoupler to isolate the ac voltage

The only thing connecting the two circuits is the light given off by the LED. Optocouplers completely isolate one circuit from another, just as relays do, but much faster. The speed of the optocoupler makes it very useful for such applications as serial interface isolation.

15.7 INSULATED GATE BIPOLAR TRANSISTOR (IGBT)

The IGBT is designed to control very high voltages and currents with a small voltage of about 0 to 15 volts at low mA currents. Regular transistors that can stand voltages of 300 to 1200 volts and currents of 50 to 600 amps require heavy base currents, and MOSFETs are typically limited to voltage under 500 volts because the resistance of the drain region becomes too high. The IGBT does this by using a MOSFET transistor and bipolar transistor together. This effectively overcomes the high-resistance drain problem but does produce a parasitic transistor that can cause the device to latch up like an SCR if the current is too high. The latch current has been greatly increased through modifications in chip construction.

Figure 15-14 shows the equivalent circuit for a typical IGBT. Notice that the device has an NPN transistor, PNP transistor, and a small resistance called the shunting resistance. This resistance is made very small, which will keep the NPN transistor off under normal operation.

If the current gets large enough to forward bias the NPN transistor, the device will latch up and not turn off until the current drops to 0. Nevertheless, IGBTs are made that can handle up to 1200 volts at 600 amps and turn on and off at 20 kHz. A typical IGBT made for this range is the CM800HA-24H manufactured by POWERX. Most of the IGBTs made for high currents are in a modular package with large screw-type terminals and heat sink backplates. The currents and voltages are much too large for printed circuit boards.

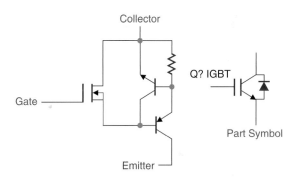

FIGURE 15-14 IGBT equivalent circuit

These parts are quite often used to turn off and on inductive loads such as motors or large coils. Therefore, most of the devices have a kick-back diode built into them to carry the back current produced by an inductive device when it is turned off, as was discussed in Section 15.4. The part symbol shown in Figure 15-14 shows this diode in the symbol. The CM800HA-24H only requires 6 to 20 volts to turn it on and will operate at 20 to 25 kHz. The gate current is very low because it is the gate to a MOSFET transistor. The author has found that a gate voltage of 15 volts works best in most designs.

Because of the high voltages and currents involved, any stray inductance in leads which carry the high current can produce high back emf voltage spikes and radiated noise which can damage the IC and other electronic parts. For this reason, the IGBT is usually driven with an optocoupler. In this way the control electronics is isolated from the high voltage and currents of the IGBT circuit. The optocoupler and gate drive circuit can be designed from standard off-the-shelf parts, but most companies that make IGBTs also make a driver module designed to control their products. These modules have all the optocouplers and other parts necessary to drive the IGBT.

Figure 15-15 shows the M57959L made by POWERX. This device has an optocoupler in it with a built-in current-limiting resistor. The device is supplied with an isolated power source of 15 volts and will turn on and off the gate to the IGBT. It also has a short circuit detection

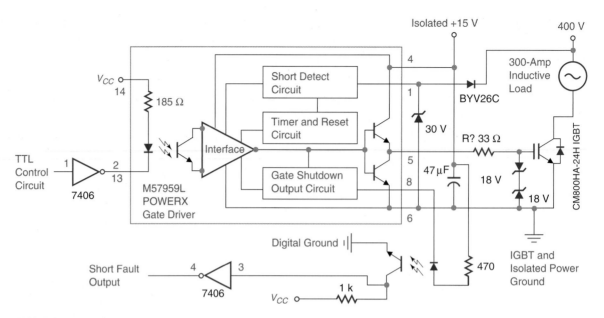

FIGURE 15-15 IGBT control circuit

capability that will shut down the IGBT if a high voltage is detected on the collector for more than a short time after the gate is brought HIGH. This condition will produce a FAULT output on pin 8 which can be fed back to the controlling electronics through an optocoupler.

The power to run the IGBT driver is supplied by an isolated power source such as a small ac-to-dc converter. The design of the printed circuit board that contains the IGBT drive and the isolated power supply can be critical to the proper operation of the device. Most often, the reason for IGBT failure is a gate short, which usually results in a large bang and smoke. For this reason, the length of the gate track and the stray inductance can be very important. Notice the two back-to-back Zener diodes in the circuit. This protects the gate from voltage spikes.

IGBTs are used in such applications as large power inverters, uninterruptable power supplies, variable-speed motor controls, and written-pole motors and generators. They are used as the interface of the 5-volt digital signal to the high-voltage large current world.

15.8 TROUBLESHOOTING HIGH-CURRENT DIGITAL CIRCUITS

High-current interfaces to digital control circuits are not hard to troubleshoot. When they fail, the smoke and flame is quite evident. The reason for the failure may not be as easy to see. High-current devices such as motors and coils have inductive loads and inductive kick voltages when turned on and off quickly. The small stray inductance of lead wire to the motors or coils can produce spike voltages large enough to damage digital devices. This is why optocouplers are often used to isolate the high-current world from the digital circuit. Today, the high-current interface which uses the transistor or IGBT is becoming more and more prevalent. Devices such as switching power supplies, variable-frequency motor speed controls, and uninterruptable power supplies use high-current interfaces to digital circuits.

The circuit shown in Figure 15-16 is the high-current interface to the dc motor that drives the film advance in a film processor. The operator smelled a burning odor and called in the technician to find out what was wrong. Just by using his nose, he soon found the source of the odor. It was a transistor on a motor driver board which had burned up. This will be easy to fix, thought the technician, but before I replace the transistor I had better figure out why it burned up in the first place. He got out the schematic and found the transistor number. Using a data manual he soon found the basic data such as Hfe and collector current he needed to analyze the circuit.

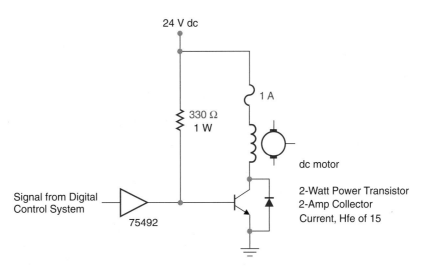

FIGURE 15-16 Motor control circuit

The first thing he noticed was a 1-amp fuse in line with the motor. This must be blown, he thought, but when he checked it with a volt-ohm meter he found it was good. This puzzled him. How could the transistor draw enough current to burn up and not blow the fuse? Looking at the driver board, he noticed that the 330-Ω resistor was dark and discolored. Removing it and measuring it he found it to be a little over 500 ohms. This made sense because he knew that the way resistors fail is to get larger, and this one had gotten hot also. But why had it gotten hot?

Using the data on the transistor he determined the base current required to saturate the transistor at a collector current of one amp to be about 66 mA. He used a collector current of one amp because the fuse was a one-amp fuse and that was the maximum the collector current could ever be or the fuse would blow. He reasoned that the motor's actual current draw was something less. This meant that the resistor would dissipate a little over 1½ watts of power. The resistor he removed and measured was a one-watt resistor. Now things made sense. The resistor had gotten hot because it was working on the edge of its power rating and the resistance had increased. This caused the transistor to come or of saturation, which produced a voltage drop form emitter to collector. The transistor started to dissipate power, got hot, and finally burned out.

After this analysis of the problem, the technician replaced the 330-Ω resistor with a 2-watt resistor and also replaced the transistor. The circuit worked properly and the 330-Ω resistor and the transistor were cool to the touch. The technician also filled a service report with the manufacturer of the machine so the problem could be corrected on future machines.

Typical Use of an Optocoupler

The schematic shown is an optocoupled RS-232 communication line. The circuit uses an optocoupler to electrically isolate one RS-232 device from another on the other end of the four wires. This is done where differences in ground potential or other electrical problems can cause damage to the RS-232 devices. The author has used this circuit for a number of years to connect several buildings to a VAX computer system. Before the use of the optocouplers, the RS-232 driver ICs were being burned quite frequently due to differences in ground potential between buildings during electrical storms.

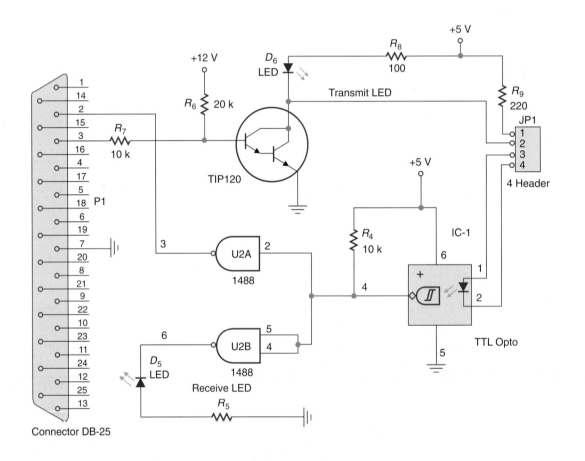

SUMMARY

- The tri-state gate has an extra input called the control input.

  When the control input to a tri-state gate is active, the output of the gate will be HiZ. The output can be either HIGH or LOW when the control input is inactive. This gives the tri-state gate three possible outputs. It also makes it possible to connect the outputs of the tri-state gates together and then control which output can be on at a time.

- Tri-state gates are used for buffering and usually have higher current-driving capabilities than normal TTL gates.

The bidirectional bus driver is a set of tri-state gates set back-to-back to allow data to be transferred in both directions through the IC. The tri-state gate is used extensively in computers to control the many bus systems of the computer.

- Devices that require high currents or voltages to operate can be interfaced to TTL and CMOS logic by using power transistors and special interfacing ICs.

  If the power requirements of a device are greater than can be handled by transistors or the device is an ac device, relays or optocouplers can be used.

- Relays are slow but offer very good circuit isolation and very high voltage and current capabilities.

  Optocouplers use the light of an LED to control a photo transistor or triac. This provides very good circuit isolation and speed, but the current and voltage capabilities, although good, are not as high as a relay.

- The 75XXX series of ICs primarily consists of interfacing ICs designed for many different types of interfacing problems.

  The 7406 and 7407 are open-collector TTL ICs that have relatively high output voltage and current levels. This makes them good choices for moderate voltage and current interfacing.

- The TIP-125 and TIP-120 transistors are PNP and NPN darling pairs which can handle up to eight amps on the collector and have an Hfe of over 2000. This makes them good choices for interfacing designs.

QUESTIONS AND PROBLEMS

1. Make a list of CMOS tri-state gates. (Use your CMOS data book.)

2. Use Figure 15-9 as the circuit and compute the values for the base resistor for a collector current of 300 mA. (The transistor Hfe is 15.)

3. Expand the multiplexing circuit in Figure 15-10 to multiplex four displays.

4. Expand the bus in Figure 15-8 to include an input port at address 000E hexa decimal.

5. Expand the bus in Figure 15-8 to include an output port at address 00F0 hexa decimal.

6. Which direction will data flow in a 74LS245 IC if pin 1 is LOW and pin 19 is LOW?

7. What is the address bus in a typical computer used for?

8. What two types of TTL outputs can be tied together?

9. Why is the diode used in the circuit of Figure 15-9?

10. Give two reasons why relays are good devices for interfacing high-current devices with a TTL output.

11. Draw the circuit to drive a 250-mA 12-volt lightbulb using a 75492 IC.

12. What is the main difference between the 75491 IC and the 75492 IC?

13. Use a 74LS06 IC and a TIP-120 transistor to drive a one-amp lightbulb.

14. Add a third digit to the multiplexed display in Figure 15-12.

15. Use a full wave rectifier to produce a 120-Hz clock from the circuit shown in Figure 15-12.

16. What are the advantages of using an optocoupler to drive a device?

17. List two IC D latches which have tri-state outputs.

18. Draw the pinout for the 74LS244 IC.

19. What type of seven-segment LED is used in Figure 15-12?

20. What will happen if the clock of the circuit in Figure 15-12 is lowered to 15 Hz?

21. What will happen to the input of the OR gate in Figure 15-2 if the bottom tri-state gate has a 1 on the control input?

22. What will happen to the circuit in Figure 15-9 if the 500-Ω resistor increases to 2 kΩ?

23. Why are PNP transistors used to multiplex the two seven-segment LEDs in Figure 15-12?

24. What happens to the circuit in Figure 15-12 if three of the 74LS76s become grounded?

25. What other type of digital output, besides a tri-state gate, can be wired together without damaging the device?

26. Draw the symbol for an IGBT.

27. What happens to the operations of the circuit in Figure 15-13 if the resistor connected to the input of the Schmitt trigger opens up to infinite resistance?

28. What type(s) of digital output do the three inverters in the upper right-hand corner of Figure 15-8 have?

29. A diode is shown across the emitter to collector of the IGBT in Figure 15-15. What is it for?

30. What is the reason for the optocoupler which is used to supply the shot-fault output in Figure 15-15?

TRI-STATE GATES

OBJECTIVES

After completing this lab, you should be able to:

- Use a transistor to control higher current.
- Multiplex seven-segment displays.
- Use tri-state gates to control a bus.

COMPONENTS NEEDED

1	7447 BCD to common-anode seven-segment LED driver IC
1	74LS241 tri-state gate buffer IC
1	7476 dual *JK* flip-flop IC
1	7406 hex open-collector output inverter IC
2	FND-507 common-anode seven-segment LEDs
2	small-signal PNP transistors
9	470-Ω, ¼-W resistors
2	1-kΩ, ¼-W resistors

PROCEDURE

1. Construct the circuit in Figure 15-12.

2. Pick the transistor to do the current buffering.

3. Have the instructor check the operation.

4. Run the displays at frequencies of 1000 Hz and 1 Hz. At which speed does the display appear brightest and why?

OBJECTIVES

After completing this lab, you should be able to use Multisim to:

- Construct a high-current interface for a TTL output.
- Troubleshoot a transistor high-current TTL interface.

PROCEDURES

1. Open the file 15B-1.ms9. This is a TTL high-current interface designed to control a relay.

2. What is the wattage of the lamps which are turned on and off by the relay?

3. What is the resistance of the relay coil?

4. Open the file 15B-2.ms9. This is the same circuit as before but an extra-heavy load has been placed on the output transistor. This was done by placing a 5-ohm resistor across the relay coil. The circuit will work with this increasd load.

   Increase the resistance of the resistor connected to the base of the PNP output transistor. What happens to the voltage across the emitter to collector of the PNP transistor? What will this cause? The secret to keeping transisotrs from burning up is to either keep the current through them low or keep the voltage drop across them low. Remember $P = V * I$, and dissipated power is what makes things get hot and burn up.

Memories and an Introduction to Microcomputers

OUTLINE

KEY TERMS

Central Processing Unit (CPU)
dynamic RAM
EEPROM
EPROM
flash memory

PROM
random-access memory (RAM)
read-only memory (ROM)
static RAM

OBJECTIVES

After completing this chapter, you should be able to:

■ Use and understand semiconductor memories.

■ Understand the basic microcomputer structure and operation.

■ Have some basic knowledge of the Z-80 CPU and its operation.

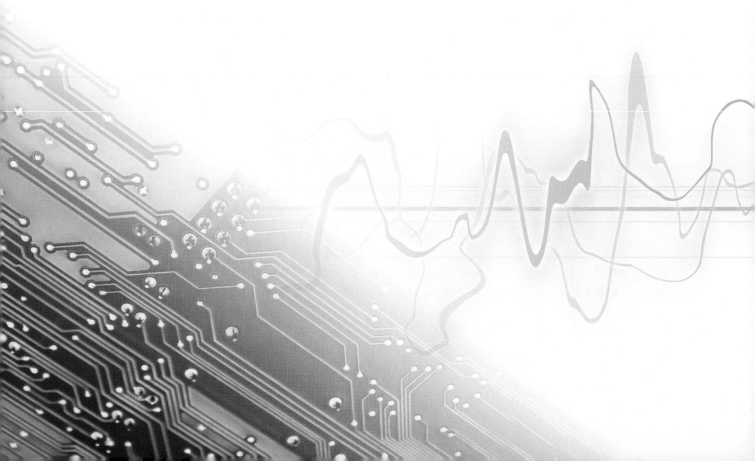

16.1 THE MICROCOMPUTER AND ITS PARTS

Much of what you have learned in this textbook is used in the world of the computer and its hardware. The computer is a digital machine with four main parts: central processor, memory, input/output, and program. The central processor is the digital brain behind the computer. The CPU (**Central Processing Unit**) controls all other parts of the computer and is responsible for interpreting the program stored as binary numbers in the memory.

Memory is a large set of storage registers that can be accessed by the CPU. It is used to store the program and data. We will discuss several types of memory later in this chapter.

The program is stored in the memory as a set of binary numbers. These sequential binary numbers are instructions used by the CPU to accomplish a task that the programmer wishes done by the computer. This part of the computer is not a set of gates or semiconductors but is a set of instructions created by the mind of the computer programmer. For this reason the program is called *software*, in contrast to hardware, which is the electronics of the computer. The hardware of the computer would be of little use if the program were not there to run it. You might say the computer program is the fuel that the computer hardware uses to accomplish a job. As many computer hardware manufacturers have learned the hard way, you can have a very well-designed, efficient piece of hardware, but if no one has developed good software for it, it just will not sell.

The last basic part of a computer is the input/output to the computer. This is the human interface to the computer. The computer can do many different things very fast, but the human cannot pry off the lid to the CPU and look at what is going on inside. The computer must output its answer to some device that the human can see and understand. Also, the computer must be able to get input from the outside world. This input/output is not always from a human; it can come from other machinery or devices controlled by the computer.

We will look at the four main parts of a computer and how they work together.

 # 16.2 THE CENTRAL PROCESSING UNIT

Before the advent of single-chip CPUs, central processors were made from many chips and placed on printed circuit boards. They were large and power-hungry. Large-scale integration of the CPU has produced a great number of single-chip central processors, such as the 6800, 6502, 8080A, 8085, and Z-80. Because all of these CPUs are in a single 40-pin package, they are called *microprocessors.* The term "micro" seems to infer that the computing power is somewhat less than a minicomputer or mainframe CPU, and that is true for most of the 8-bit CPUs. An 8-bit CPU has an 8-bit data bus. Today's new 16-bit and 32-bit single-chip CPUs, such as the Z382, 8086, 68000, 68040, 80386, 80486, and others, are quite powerful and can no longer be thought of as being small. Because the Z-80 central processor is still used and is a very good example of an 8 bit processor, we will use it as an example. This chapter is not designed to be a complete documentation on microprocessors, so we will not go into detail on many topics. The subject of microprocessors would take a whole textbook.

Figure 16-1 shows a Z-80–based CPU that is part of a simple computer trainer. The Z-80 CPU has 16 address lines labeled A_0 to A_{15}. It uses these lines to supply the binary address of the memory or I/O with which the CPU wishes to communicate. This is a unidirectional bus and is buffered for more current drive by a set of 74LS245 ICs. This buffering is required because the Z-80 can only supply a little over 1 TTL load on any of its outputs. A 1-TTL output load is typical of most NMOS LSI CPUs.

The data bus is eight bits wide, bidirectional, and used for data transfer to and from the CPU. The direction of data flow through the 74LS245 is controlled by a Z-80 control signal called $\overline{RD}$. This control signal will go LOW any time the CPU wishes to bring data into the CPU from the outside. When pin 1 of the data bus buffer (74LS245) goes LOW, data will be transferred from pin 18 to pin 2, from pin 17 to pin 3, and so on.

The Z-80 generates several control signals to control the operation of the data bus and other parts of the computer. These control signals are buffered with a 74LS245 also and make up the third Z-80 bus called the control bus. There are four basic control signals used to control the memory and input/output of the computer: $\overline{RD}$, $\overline{WR}$, $\overline{MREQ}$, and $\overline{IORQ}$.

$\overline{MREQ}$ will go LOW when a valid address is placed on the address bus by the Z-80 for a memory read or write. If the memory operation is to be read, then the $\overline{RD}$ signal will also go LOW. If the memory operation is to be a write to memory, then the $\overline{WR}$ signal will go LOW. The $\overline{MREQ}$ and the $\overline{RD}$ are ORed together to produce a control signal called $\overline{MRD}$, which will be used to select the memory for a read. The $\overline{MREQ}$ and the $\overline{WR}$ are ORed together to make the $\overline{MWR}$ signal, which will be LOW for a write to memory.

When the CPU reads or writes to an input or output device on the data bus, the $\overline{IORQ}$ will go LOW but not the $\overline{MREQ}$. By ORing the $\overline{RD}$ and $\overline{WR}$ to $\overline{IORQ}$, we can produce two new input/output control signals called $\overline{IORD}$ and $\overline{IOWR}$. IC6 is a 74LS32 which does this in the schematic of Figure 16-1.

One more control signal on the control bus of this computer is called $\overline{M1}$. $\overline{M1}$ will go LOW when the CPU is fetching a byte from memory that will be used as the next instruction in the program the computer is executing. When this signal is ORed with the $\overline{IORQ}$, a new signal is produced, called $\overline{INTACK}$, or interrupt acknowledge. $\overline{INTACK}$ is used to read a byte from a special input port called an interrupt vector port. This byte is used to tell the computer where to get the next instruction to execute. Interrupts are a method of breaking the execution of a program and forcing the CPU to do something else. Because of their complexity, they will not be discussed in this chapter to any great extent.

You will probably notice that the chip select (CS) pins of all the 74LS245 buffers are connected to a signal through an inverter called $\overline{BUSAK}$, or bus acknowledge. This signal is made LOW, causing all the bus drivers to be placed in their HiZ state when the $\overline{BUSRQ}$ is made LOW by some outside device. When this happens, the CPU will finish the current instruction and then place all the bus buffers in HiZ. This is a method for an outside device to gain control of the computer bus system. This is commonly called *direct memory access* or DMA.

The $\overline{RESET}$ pin is controlled by a power-on one-shot made from an *RC* timing constant and a Schmitt trigger, much like the one-shots studied in this text. The $\overline{RESET}$ input is used to start the Z-80 fetching instructions from address 0000 hex. When the $\overline{RESET}$ input is brought LOW for at least six clock cycles and then returned HIGH, the CPU will start fetching the next

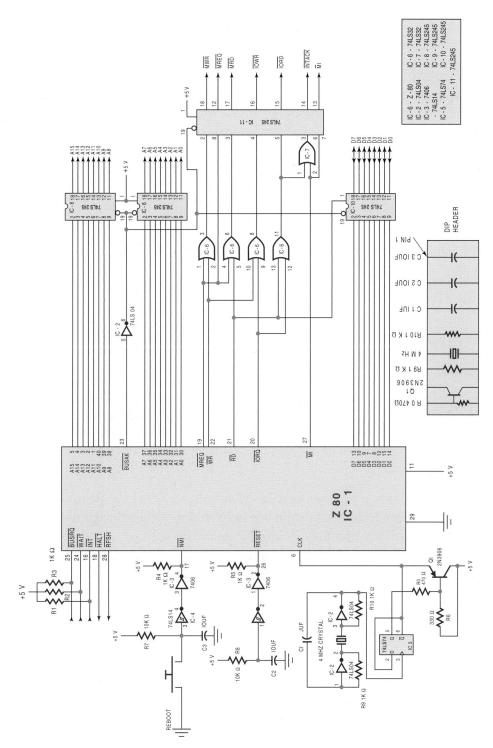

FIGURE 16-1 Z-80–based CPU in a simple computer trainer

instruction at address 0000 hex. This is how the computer is started at the right place in the program each time the power is applied.

The $\overline{\text{NMI}}$ or nonmaskable interrupt is used to reboot the Z-80 after the computer has had power applied. This input causes the program to begin from 0066 hex when brought LOW. The *RC* one-shot is the same as the $\overline{\text{RESET}}$ except that a button has been placed in the circuit to retrigger the one-shot.

The $\overline{\text{RFSH}}$ signal is an output used by the Z-80 to refresh dynamic memory. This signal will go LOW when the address contains the address for refreshing the dynamic memory. The Z-80 has a hidden refresh cycle for dynamic memory, which makes it very easy to use with dynamic memory. We will take a look at dynamic memory later.

The $\overline{\text{HALT}}$ output indicates that the CPU has executed the halt instruction and is halted. Only an $\overline{\text{NMI}}$ or $\overline{\text{RESET}}$ can make the CPU continue after the halt instruction has been executed.

The CPU may be interrupted by bringing the $\overline{\text{INT}}$ pin LOW. This interrupt can be masked by the CPU through programming, and there are three different modes for interrupts in the Z-80.

The $\overline{\text{WAIT}}$ input is used to make the CPU stop and wait for slow memory, which takes longer to get the data on the data bus than the normal bus cycle.

16.3 COMPUTER MEMORY

A semiconductor memory is an integrated circuit capable of storing a binary number and recalling it when addressed or selected by a computer or other digital device. A simple latch made from *D* flip-flops, such as in Figure 7-27, can be considered memory because it can store a binary number.

There are two major types of memory, ROM and RAM. ROM stands for **read-only memory**. This type of memory has a preset value set in its memory cells that cannot be easily changed. RAM is **random-access memory**. RAM is read/write memory. This means that the value stored in its memory cells can be changed to a new value easily and quickly. The name random-access memory is not a good name because it implies that memory cells can be accessed randomly or in any order, not that they are read/write memory cells. The fact is, most ROM and RAM can be accessed randomly. Nevertheless, the acronym RAM has come to mean read/write semiconductor memory.

16.4 ROM

Figure 16-2 shows how a ROM can be constructed by using a decoder, four tri-state gates, and some diodes. This ROM can store eight 4-bit numbers or 32 bits of information. Each 4-bit word, or *nibble*, can be read or placed on the output line by supplying the correct address to the inputs of the decoder and enabling the tri-state gates with a logic LOW on the IC chip select ($\overline{\text{CS}}$) input. If address 111 is placed on the address inputs, output number 7 of the decoder will go LOW, or 0 volts. This will forward-bias the diode between the output

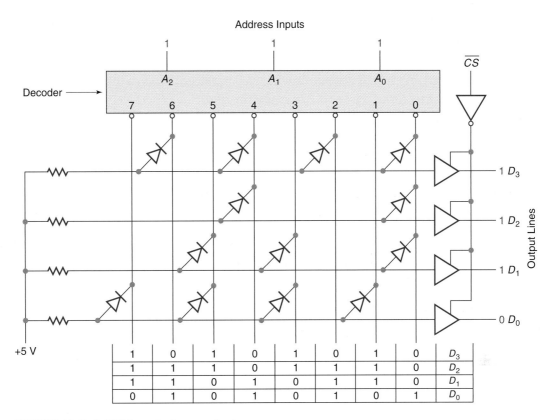

Address Inputs

FIGURE 16-2 A ROM made from a diode array

number 7 and D_0 output, thus pulling the D_0 output to a logic LOW. All of the other diodes connected to the D_0 output are reverse biased because the other decoder outputs are 1, or positive voltage. Because D_1, D_2, and D_3 do not have any diodes to pull them LOW, their output value is 1. Each time the address to the ROM is changed, the D_0 to D_3 outputs reflect the value stored in the ROM. That value is determined by the placement of diodes between the decoder output and the D_0 to D_3 outputs.

Memories are measured by the number of memory addresses and the number of bits each address can store. The memory in Figure 16-2 is an 8-by-4 ROM. This means that it has eight address locations of four bits each, or 32 bits of total storage. ROMs of the type in Figure 16-2 are manufactured as single-IC semiconductors with predefined bit patterns stored in them. They are used for computer memories, character generators, and code converters.

16.5 PROM

The problem with ROM is that once the IC is made, the bit pattern in it cannot change, and to have a new IC manufactured is quite expensive. The PROM was introduced to help alleviate this problem. **PROM** stands for *programmable read-only memory*; its bit pattern can be set by the user. The programming is done by blowing a small semiconductor fuse in the memory cell in which you wish to make a 1. This is shown in Figure 16-3.

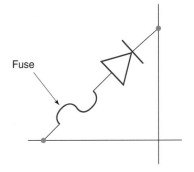

FIGURE 16-3 A PROM memory cell

To program a typical PROM, place the address of the memory location on the address lines of the PROM, place the data to be stored on the data output lines, hold $\overline{CS}$ HIGH, and pulse the program input PM HIGH for a few milliseconds. This will blow the fuse on each diode where a 1 is placed, but not on the diode where a 0 is placed. When a fuse is blown, the memory cell will, from that time on, read as a 1, because the diode has effectively been disconnected. Figure 16-4 shows the circuit for a 4-by-4 PROM that is programmed as just described.

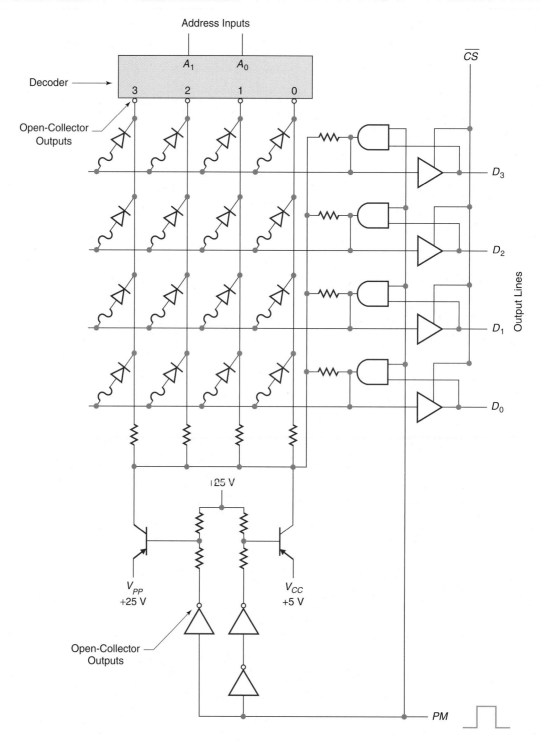

FIGURE 16-4 A 4-by-4 PROM

16.6 EPROM

Once a PROM has been programmed by blowing the fuses at the bit locations where 1s were needed, it cannot be reprogrammed. Once a fuse is blown, it cannot be put back together again. The EPROM solved the problem by allowing the IC to be cleared and then programmed to a new bit pattern. **EPROM** stands for *erasable programmable read-only memory.*

EPROMs use a light-sensitive memory cell that, when exposed to ultraviolet light, will return to a 1 value. Therefore, most EPROMs have all 1s in their memory cells after they have been cleared by exposing them to ultraviolet light for about 20 minutes.

As shown in Figure 16-5, the memory cell of an EPROM has a floating gate to the field-effect transistor that can be charged by placing a high voltage of about 12.5 to 25 volts on it. The exact programming voltage depends on the type of EPROM being programmed. You should check the IC specifications for the exact programming voltage to be used. Charging this floating gate causes the memory cell to become a 0. Because the electrons are forced across a very thin barrier of silicon dioxide (an insulator) to charge the floating gate, they will not cross back unless their energy levels are increased artificially. The ultraviolet light will cause the floating gate to lose its charge, thus making the memory cell regain its 1 value. Most EPROMs will be completely erased by exposure to ultraviolet light with a wavelength of 2500 Å or less and an intensity of 15 W-s/cm^2 for 15 to 20 minutes. A standard 15-watt fluorescent germicidal lamp (barbers put their cutting instruments under such lamps) works quite well for erasing EPROMs. The EPROMs should be placed about one inch below the light. The light should be enclosed so that your eye is not exposed to it for long periods. Actual erasing begins at a wavelength of about 4000 Å. This means that a standard fluorescent light used for visual lighting can erase an EPROM in about three to four years. Direct sunlight can do the job in as little as one week. Therefore, the EPROM should have the optical window in the chip covered to prevent external light from entering the chip.

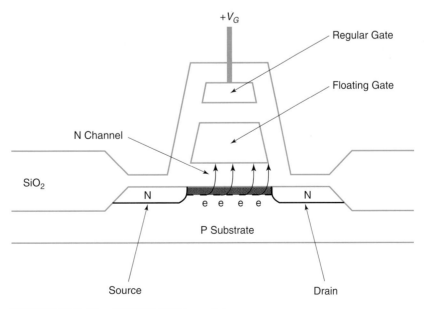

FIGURE 16-5 The EPROM FET transistor

The 2716 is a 2K-by-8 EPROM, which is a typical EPROM. We will study its operation and programming methods because it is typical. It should be noted that the 2708, which was a very early EPROM, is somewhat different in its programming methods. It is still found in older equipment.

Figure 16-6 shows the pinout of the 2716 EPROM. There are 11 address lines (A_0 to A_{10}) that select the memory location to read or program and eight data lines (D_0 to D_7) which are used to output data or place data into memory. The IC has one +5-V power supply and +25 V on the V_{PP} input for programming of memory. The $\overline{OE}$ (output enable) controls the internal tri-state gates on the D_0 to D_7 output pins. The $\overline{CE}$ (chip enable) also controls the output tri-state gates. The difference in the two is that when $\overline{CE}$ returns to the inactive state (HIGH) the 2716 goes into its standby mode, which causes it to draw about 75 percent less power.

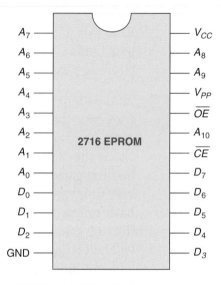

To program the 2716 EPROM, the V_{PP} pin is brought up to +25 V, the $\overline{OE}$ is made HIGH, and the $\overline{CE}$ is used to control programming. The byte to store in the memory is placed on the output lines (D_0 to D_7), and the $\overline{CE}$ pin is pulsed HIGH from a LOW state for 50 ms. You can program the memory locations randomly or sequentially. Once programmed, a 0 will remain a 0 until it is erased by exposure to ultraviolet light.

FIGURE 16-6 2716 EPROM

EPROMs have quickly become the major ICs used for storage of boot programs and operating systems in computers today. Some typical chips used today are listed in Figure 16-7. The 2716 and the 2732 are 24-pin chips, and the rest are 28-pin chips. They have a similar pin configuration, which allows the designer to make provisions for EPROM upgrade in the design with some simple jumpers on the board.

16.7 EEPROM

EEPROM stands for *electrically erasable programmable read-only memory*. This type of memory will retain its stored bit pattern when power is removed. It can have the bit pattern programmed, and it can be changed by applying an electric field to the memory cell. The chief advantage to this type of memory is the ease with which it can be changed.

EPROMs cannot be selectively cleared, nor can they be cleared very fast. The EEPROM is an improvement over the basic EPROM technology. Figure 16-8 shows the basic EEPROM memory transistor and the floating gate. In the EPROM, the electrons are forced across the silicon dioxide insulation by placing a high voltage across the P substrate and the normal gate. The electrons then collect on the floating gate and are trapped, thus charging the gate. When the gate is charged, the field-effect transistor will not conduct.

In the EEPROM, the floating gate and normal gate have a protruding portion that comes very close to the drain of the transistor. Electrons are forced onto the floating gate by producing a

27256	2764	2732A	2716
V_{PP}	V_{PP}		
A_{12}	A_{12}		
A_7	A_7	A_7	A_7
A_6	A_6	A_6	A_6
A_5	A_5	A_5	A_5
A_4	A_4	A_4	A_4
A_3	A_3	A_3	A_3
A_2	A_2	A_2	A_2
A_1	A_1	A_1	A_1
A_0	A_0	A_0	A_0
O_0	O_0	O_0	O_0
O_1	O_1	O_1	O_1
O_2	O_2	O_2	O_2
GND	GND	GND	GND

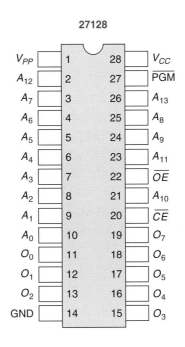

27128

2176	2732A	2764	27256
		V_{CC}	V_{CC}
		$\overline{PGM}$	A_{14}
V_{CC}	V_{CC}	N.C.	A_{13}
A_8	A_8	A_8	A_8
A_9	A_9	A_9	A_9
V_{PP}	A_{11}	A_{11}	A_{11}
$\overline{OE}$	$\overline{OE}/V_{PP}$	$\overline{OE}$	$\overline{OE}$
A_{10}	A_{10}	A_{10}	A_{10}
$\overline{CE}$	$\overline{CE}$	$\overline{CE}$	$\overline{CE}$
O_7	O_7	O_7	O_7
O_6	O_6	O_6	O_6
O_5	O_5	O_5	O_5
O_4	O_4	O_4	O_4
O_3	O_3	O_3	O_3

Pin Names

$A_0 - A_{13}$	Addresses
$\overline{CE}$	Chip Enable
$\overline{OE}$	Output Enable
$O_0 - O_7$	Outputs
$\overline{PGM}$	Program
N.C.	No Connect

2716 — (2K by 8)
2732 — (4K by 8)
2764 — (8K by 8)
27128 — (16K by 8)
27256 — (32K by 8)

FIGURE 16-7 Typical EPROMs

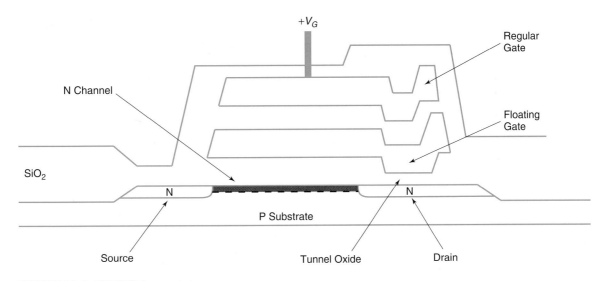

FIGURE 16-8 EEPROM transistor

high voltage of − to + from drain to normal gate. Then, just as in the EPROM transistor, the electrons collect on the floating gate, thus charging it negatively. Reversing the voltage polarity removes electrons and reverses the charge. This gives the EEPROM the ability to be cleared and then reprogrammed quickly, with a voltage of about 21 volts.

EEPROM technology has not yet produced the ultimate read/write memory. The number of storages is not unlimited. It stands at about 100,000 at present, and it takes much more time to write than typical RAM. Because of these limitations, the EEPROM will not be used as a RAM. The EEPROM is being used today to store the configuration information for such devices as computer terminals, printers, and modems. The operator of the equipment can clear the EEPROM and reprogram it with a new configuration for a piece of computer equipment without removing the chip or using any special light. The configuration can be easily changed and will remain unchanged even when the power is turned off.

 # 16.8 STATIC RAM

The **static RAM** IC uses a simple set/reset flip-flop made from cross-coupled transistors as a memory cell. This memory cell can be either set or reset and will hold the set or reset value until the power is turned off. Typical read/write speed for MOS (metal-oxide semiconductor) RAMs ranges from 55 to 450 ns. This is quite fast enough for almost all computer operations today. All static RAM ICs are volatile, which means they lose their memory patterns when electrical power is lost. The old magnetic core memory is nonvolatile and static. Magnetic core memory used an array of magnetic doughnut-shaped circles to store bits. This memory was the major type used in computers before the advent of good semiconductor memories. The term *core memory,* which is often used today to mean the central RAM of a computer, came into use because in the past, the RAM was made from magnetic core memory.

Static RAM is made with four basic technologies: MOS (metal-oxide semiconductor), CMOS (complementary metal-oxide semiconductor), TTL (transistor-transistor logic), and ECL (emitter-coupled logic). TTL RAM is much faster than MOS RAM but is not as dense. A typical TTL RAM, like Motorola's MCM93415, is a 1K-by-1 RAM with an access time of 45 ns. CMOS RAM is slower than TTL RAM, somewhat denser, and uses much less power. When a CMOS RAM is not being read from or written to, it usually uses no power at all. This means that a small long-life battery, such as a silver oxide cell, can be used as a backup power supply for the RAM when the main power is turned off. This makes the CMOS RAM appear to be nonvolatile, and it is used in many applications, such as portable computers. A typical CMOS RAM is Motorola's MCM61L16P20, which is a 2K-by-8 CMOS RAM with an access time of 200 ns.

ECL RAM is the fastest of the four basic technologies, but it uses more power. A typical ECL RAM, such as Motorola's MCM10474-15, is a 1K-by-4 RAM with a 15-ns access time. The access time for a memory IC is the time it takes for the number stored in the memory to become stable on the data lines after the address and CS have become active. Figure 16-9 shows some typical static RAMs.

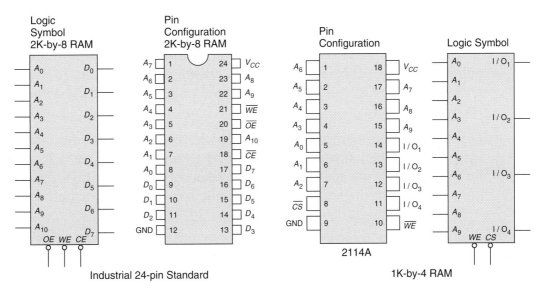

FIGURE 16-9 Typical static RAM

16.9 DYNAMIC RAM

Dynamic RAM uses a single transistor and a small capacitor for a memory storage cell. Because of the few components used per memory cell, the dynamic RAM is very dense. At present, the dynamic RAM can hold more memory cells per IC than any other type of IC memory. Because of the high density and relatively low power consumption, the dynamic RAM has quickly become the major computer memory used today. The only problem with dynamic RAM is that it must have every memory cell read or written to every 2 ms or the small capacitor in the memory cell will discharge, and the RAM will lose the stored bit pattern. This refresh operation takes computer time and other circuits to accomplish. The Z-80 CPU has a built-in transparent refreshing system for dynamic RAM that does not waste any CPU time. This makes the Z-80 CPU very attractive to designers.

The 4164 dynamic RAM is a 64K-by-1 NMOS (N-channel metal-oxide semiconductor) made by many companies. It has a single +5-V power supply, and its outputs can handle two TTL loads. Figure 16-10 shows the pinout of the RAM.

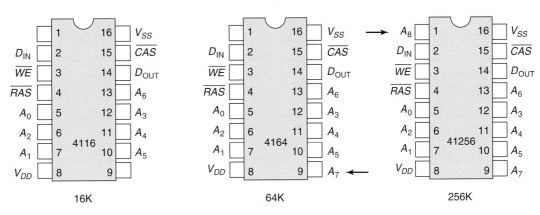

FIGURE 16-10 The industrial standard 16-pin package pinout for dynamic RAMs

The RAM comes in a 16-pin package. To address 64K of memory, you need 16 address pins (A_0 to A_{15}). To get all 16 address inputs, one data input, V_{CC}, ground, and the control inputs into one 16-pin package, the address inputs were cut in half and multiplexed into the chip. This means that only eight pins are used to supply the RAM IC with 16 address inputs. To do this, two new control lines were added: $\overline{RAS}$ (row address strobe) and $\overline{CAS}$ (column address strobe). These two strobes latch address information into the row and column of the memory cell matrix used by the RAM.

Figure 16-11 shows the memory cell array and the row and column address latches. As can be seen in Figure 16-10, the industrial standard pin package for the three dynamic RAMs is the same except for the addition of one address input for each larger IC memory. This means that a computer design could be made that would allow an upgrade in memory size by simply changing memory ICs.

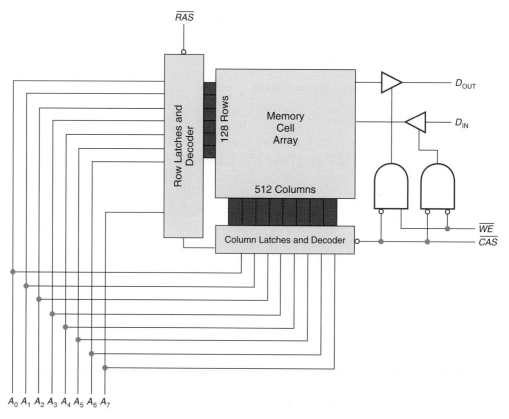

FIGURE 16-11 Typical DRAM configuration

To read or write to the RAM, the least significant byte (LSB) of the address is placed on the A_0 to A_7 address inputs, and then the $\overline{RAS}$ (row address strobe) is brought LOW. This latches the LSB of the address into the row latches of the memory array. Then the most significant byte (MSB) of the address is placed on the A_0 to A_7 address inputs and latched into the column latches of the memory array when $\overline{CAS}$ (column address strobe) goes LOW. Shortly after the LOW-going CAS signal, the data is put on the D_{OUT} pin, or the data on the D_{IN} pin is stored in the RAM. Reading or writing is controlled by the $\overline{WE}$ input to the IC.

To refresh the memory cells in the array, you only need to latch in a new row address, because each time a new row is selected, all the memory cells on that row are refreshed. Notice that the row address is actually only seven bits wide, and the column address is nine bits wide. Therefore, only a 7-bit refresh address is needed to refresh the entire dynamic memory.

During the M1 machine cycle or instruction fetch machine cycle of the Z-80 CPU, the 7-bit internal refresh register is incremented and placed on the lower seven bits of the address bus.

Then the $\overline{\text{MREQ}}$ and $\overline{\text{RFSH}}$ go LOW. By logically ORing these two signals, you can produce a dynamic refresh signal.

The schematic shown in Figure 16-12 is the memory decoders and ROM/RAM for the computer trainer shown in Figure 16-1. The ROM is a 2716, 2K-by-8 EPROM, and the RAM is two 2114, 1K-by-4 static RAM chips. The address decoder is a 74LS138 octal decoder. This decoder was studied in this text, so its operation should be clear. The decoder will allow the ROM to be accessed from address 0000 hex to 07FF hex and the RAM from 0800 hex to 0BFF hex. The decoder supplies a LOW output for each of the first eight 1K blocks of memory of the possible 64K the CPU can address.

The 74LS245 is used to add drive current to the data bus, because the ROM and RAM ICs can only supply about one TTL load to an output. The direction of the data flow is determined by the $\overline{\text{MRD}}$ control signal from the CPU, and the bus driver is turned on and off with the address decoder. The bus driver will only be on if an address that is in the memory area of the ROM and RAM is present on the address bus.

The 4116, 4164, and the 41256 are small dynamic RAM today. Dynamic RAM with the storage capacity of 8 and 16 M bytes are used in today's computers. DRAMs such as Toshiba's TC59S6408 can be auto- or self-refreshed by programming the mode register in the chip. The older computer board would have large portions of the board with rows of DRAM to make the central memory for the computer. Now the DRAM is placed on smaller printed circuit boards and plugged into the main board. These newer and much larger DRAMs still use the CAS and RAS strobes for addressing and they work very much the same way but faster.

16.10 THE INPUT/OUTPUT OF THE COMPUTER

The input and output of a computer can be a wide variety of things, from a typical terminal to the control and operation of a running motor generator on a power grid. Figure 16-13 shows the I/O for the computer trainer shown in Figure 16-1. It consists of three 8-bit output ports with LEDs and a hex keypad for input of machine code to program the computer. The address decoder is the same old 74LS138 as was used for the memory, but this time the address is only eight bits wide instead of 16. This is because the Z-80, just like the older 8080, has only an 8-bit address for input/output ports. The decoder will supply a LOW select line for the first eight possible addresses. These select lines are NORed together with $\overline{\text{IORD}}$ and $\overline{\text{IOWR}}$ to produce the latch enables for the output ports and the one input port. The three output ports are made of two 7475 quad D transparent latches. When the enable or clock input to these D flip-flops is brought HIGH, the data on the data bus is passed on to the Q and $\overline{Q}$ outputs of the latches. When the enable or clock returns LOW, the data is latched on the Q and $\overline{Q}$ outputs of the D latches. When a 1 is present on the data line, the $\overline{Q}$ is made LOW, which turns on the associated LED. In this way the CPU can display three bytes in binary to the computer user. In this computer, the first two ports (0 and 1) are used to display the current memory address being used and the third (2) is the data contained in that address.

The fourth output port (3) is a 4-bit 7475 latch used to hold the current key that the computer wishes to test to see if it has been pressed by the operator. This is done by decoding the 4-bit binary number with a 74150 multiplexer. The 74150 multiplexer selects the logic level of one of the 16 keys

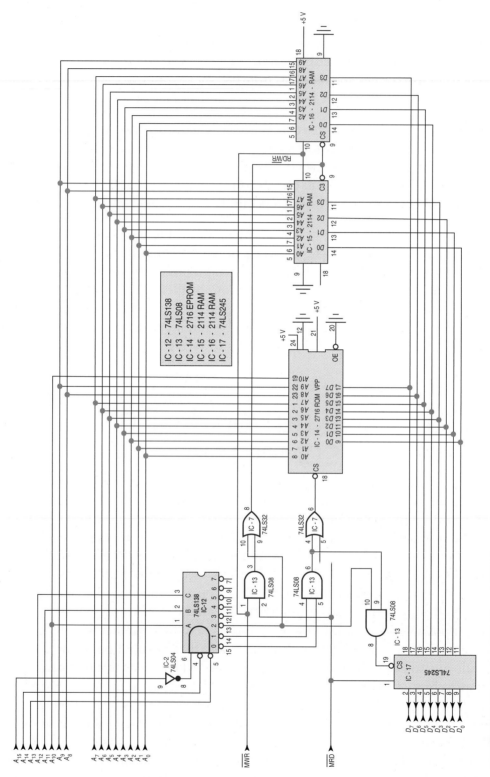

FIGURE 16-12 Trainer RAM and ROM

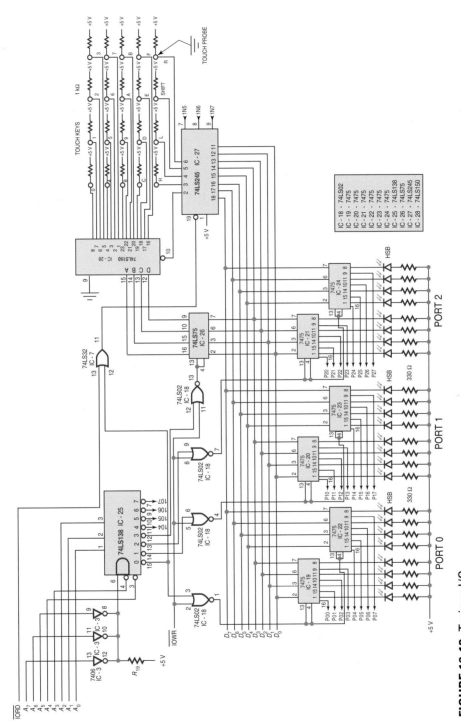

FIGURE 16-13 Trainer I/O

and places it on the most significant bit of the input port 0. The CPU can then read this port and, by examining the logic level of that bit, decide if the key was pressed. If a key is pressed, the input to the multiplexer (74150) will be LOW, which is inverted and passed on the computer input port.

EXAMPLE 16-1

Add an extra output port to the trainer I/O in Figure 16-13. The output port will drive two seven-segment LEDs. Use one 74LS273 IC, two 7447 ICs and two seven-segment LEDs.

Solution   Refer to Figure 16-14.

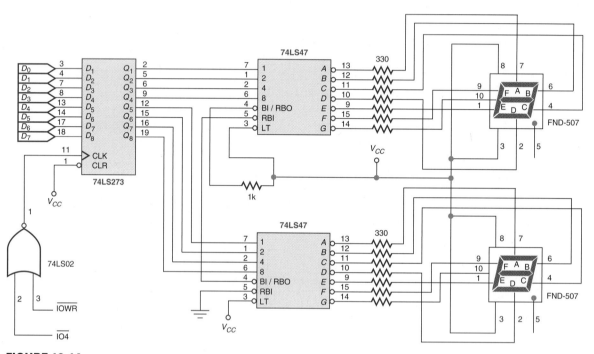

FIGURE 16-14

✔ **SELF-CHECK 1**

1. What are the four main parts of a computer?

2. What does EPROM stand for?

3. What is the difference between static RAM and dynamic RAM?

4. How many output ports are in the schematic shown in Figure 16-13?

16.11 THE PROGRAM

As you can see, the computer must have a program that continually changes the binary number on the output port that selects the key to be tested and then reads the input port to see if it has been pressed. This program is stored in the ROM so that it will be present each time the power is applied to the computer.

To create a program for a microprocessor, you must first understand the microprocessor's internal architecture and its instruction set. Each microprocessor has a set of internal registers that can be used in different ways to manipulate binary numbers. The internal register set for the Z-80 is shown in Figure 16-15. The A register is the accumulator, which is an 8-bit storage register where the answer to arithmetic instructions is stored. The F or flag register is a collection of 1-bit flip-flops that indicate the status of the last instruction that has just occurred. For example, suppose the binary number 70 hex that was stored in the B register was subtracted from the A register, which contains 70 hex also. The answer 00 hex would be stored in the A register. The zero flag in the flag register would be set to 1, indicating that the result of the previous subtraction was zero. The microprocessor has other instructions that can test the state of the zero flag and do one of two things based on its value. The Z-80 has the following flags in its flag register.

S — Sign flag, used in signed 2's complement arithmetic

Z — Zero flag, indicates a zero answer

H — Half-carry, indicates a carry from D_3 to D_4, used for binary-to-BCD conversion

P/V — Parity and overflow flag, used to indicate even or odd parity or an overflow in 2's complement

N — Negative flag, used to indicate a subtraction

C — Carry flag, used to indicate a carry-out or borrow from the MSB in an arithmetic instruction

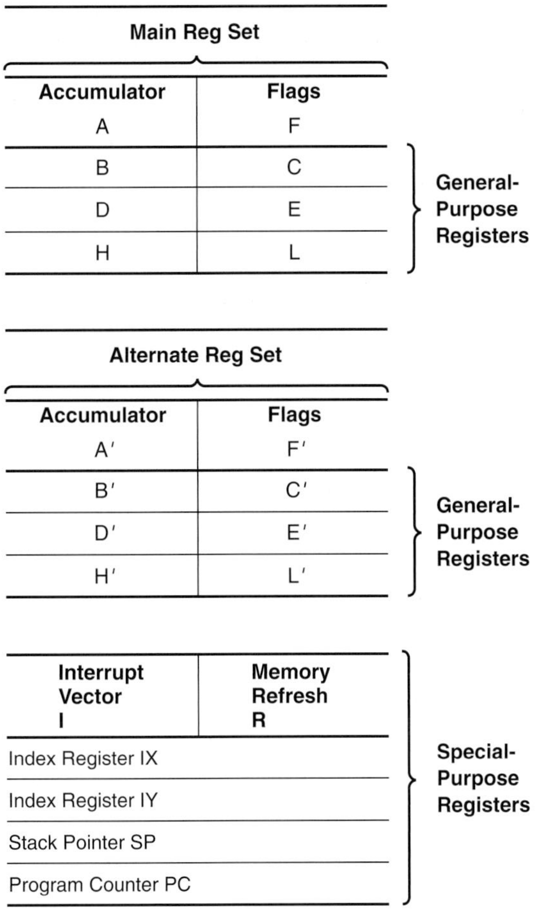

FIGURE 16-15 Internal register set for the Z-80

The BC, DE, and HL registers are used to store binary numbers or to hold addresses for memory locations where other binary numbers can be stored. They can be used as 8-bit registers or 16-bit registers when needed.

The IX and IY registers are called index registers. They are used to hold address locations for tables of binary numbers stored in memory.

The SP is the stack pointer. It holds the address of a last-in-first-out storage stack in memory for temporary storage of the CPU registers.

The PC is the program counter, which contains the address of the next instruction to be executed. This register keeps track of the location of the program in memory and the location of the next instruction.

The R or refresh register is used to produce an incremental address for the hidden refresh of dynamic memory if it is used for the computer's memory.

The I or interrupt register is used in the vectored interrupt mode of the Z-80. Interrupts are a method for an outside device to interrupt the flow of the Z-80 program and cause it to jump to a new program, which will survive the device that produced the interrupt. The Z-80 has three interrupt modes which are beyond the scope of this discussion.

The AF′, BC′, DE′, and HL′ registers are an alternative set of registers that can be exchanged with the regular set at any time with an instruction in the program.

The instructions for a microprocessor program are stored as binary numbers in the memory and called *op codes*. Op codes are read into the CPU and decoded to determine which instruction should be executed. Each op code will work on or affect another binary number, such as the number stored in the A register. The binary number operated on by the instruction is called the *operand*. The operand can be another register or binary number stored in the memory.

To help in writing programs, each of the major types of instructions have been given short alpha codes to help the programmer remember them. These alpha codes are called *mnemonics*. An instruction that will load the A register with the B register is

Op	Mnemonic	Operand	Comment
78	LD	A,B	;Load register A from register B

Programs can be written using the mnemonics of the instruction set alone and then processed by a program called an assembler to produce the actual op codes. This is a much easier way to produce a program than by looking up the op codes and placing them in memory by hand.

There is much more to programming a microprocessor than is discussed here, but you must learn programming if you wish to completely understand the operation of a microcomputer.

16.12 THE MICROCONTROLLER

The microcontroller is a computer on a chip. These ICs have the CPU, RAM, ROM, and many different kinds of IO devices on the same chip. They do not have large memories and do not run at the high rates of speed that the newer processors do, but they are very small and can do many of the things which a designer needs a computer to do. They are used mainly as control computers for machinery and other devices where the computer control is simple. Microwave ovens, sprinkler timers, automobiles, and even pacemakers for heart patients have microcontrollers in them. These are called embedded systems because the computer is programmed to do a single job and cannot be reprogrammed by the user to do something different.

Every company that makes CPUs has a line of microcontrollers. All of them can be ordered in high volume with a program built into the ROM for your application. This is good if you are mass-producing a product such as an automobile, but can be a problem in cost if you only want a few of the chips. Most microcontroller manufacturers make the chips with EPROM or EEPROM ROM also. These are used for low-volume production or product development.

Motorola's 68HC11 series of microcontrollers is a typical example of this type of computer chip. Figure 16-16 shows the logical layout of this computer. This microcontroller has an asynchronous serial port, which can be used to communicate with a terminal or modem, and a high-speed synchronous serial port for communication between microcontrollers. There are also an 8-bit A-to-D converter, four output compare timers, four input capture timers, a pulse

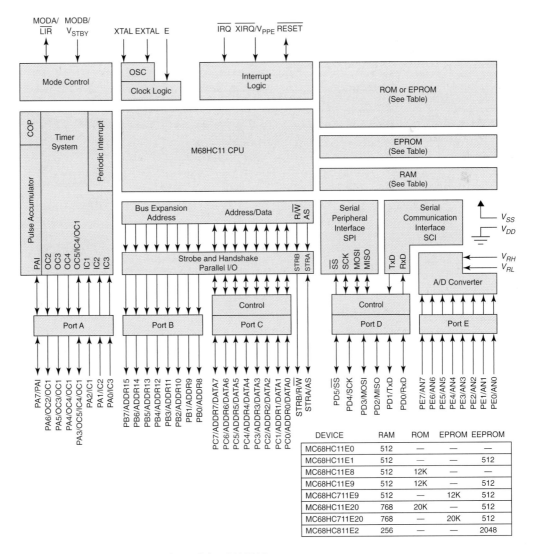

DEVICE	RAM	ROM	EPROM	EEPROM
MC68HC11E0	512	—	—	—
MC68HC11E1	512	—	—	512
MC68HC11E8	512	12K	—	—
MC68HC11E9	512	12K	—	512
MC68HC711E9	512	—	12K	512
MC68HC11E20	768	20K	—	512
MC68HC711E20	768	—	20K	512
MC68HC811E2	256	—	—	2048

FIGURE 16-16 Block drawing of the 68HC11

accumulator, periodic real-time interrupt, and up to 24 bits of general input/output. There are 512 bytes of RAM, 12-k bytes of ROM, and 512 bytes of EEPROM. The 68HC711E9 has the 12 k of ROM as EPROM with a "Clear" window on the chip so it can be erased. This chip is used for development work. It can be run in single-chip mode or external mode. In external mode the address and data bus are made available on ports B and C so the designer can add external RAM, ROM, or I/O devices to the system. This is a common feature in most microcontrollers.

Figure 16-17 shows the programming model for the 68HC11. This model is very similar to the 6800 microprocessor, which was one of Motorola's first processors. The complete operation and programming of this chip is beyond the scope of this textbook. If you are interested in this chip, the documentation by Motorola is very good. Motorola also makes an evaluation board (the M68HC11EVB board) which has an external ROM called the Buffalo ROM on it. The Buffalo program is a small monitor program which will allow you to communicate with the board from a standard computer via the RS-232C port.

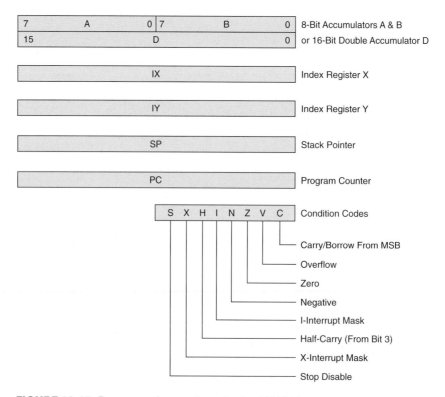

FIGURE 16-17 Programming register in the 68HC11

DIGITAL APPLICATION Schematic of the CPU Portion of a 68HC711E9 System

This schematic on page 575 is of the CPU portion of a 68HC711E9 microcontroller control system. The system is used to log data in nonvolatile RAM and to monitor machine vibration. Notice connector JP18, which is used to connect a logic analyzer for system diagnosis. (Schematic courtesy of Precise Power Inc.)

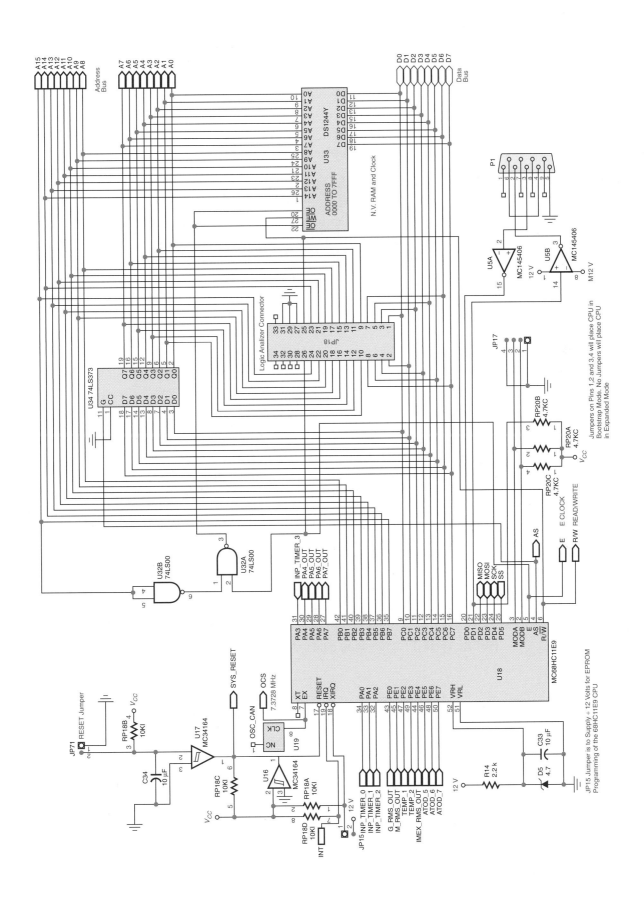

16.13 FLASH MEMORY

Flash Memory is a type of EEPROM (**E**lectrically **E**rasable **P**rogrammable **R**ead-Only **M**emory) that can be erased in blocks, written to in blocks, and read in blocks. Flash memory came out of the work of Dr. Masuoka in the early 1980s. It was designed to look like a mechanical disk drive, and the idea was to replace the disk drive with semiconductor memory.

Most flash memory is made from the NAND type of construction. This type of construction places the floating gate FET transistors in series instead of in parallel as in the NOR type of construction. This allows for a smaller bit size and a one transistor per bit cell. This also means that NAND type flash memory can be erased in large blocks. The erasing process is a bit different also, as is shown in Figure 16-18. The voltage to charge the floating gate is placed from the P substrate to the regular gate instead of the drain or source. This gives a larger surface to transfer the electrons to the floating gate. The voltage needed is less and the transistor will last longer.

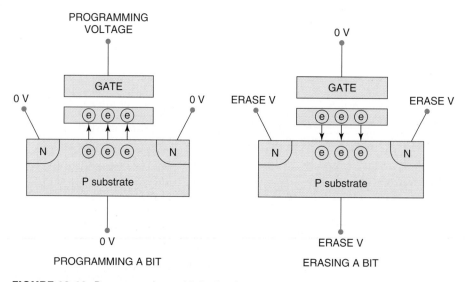

FIGURE 16-18 Programming a bit in flash memory

This type of memory is used in the SmartMedia technology. This is the small memory card that is commonly used in digital cameras and music recording devices. These cards range from 16 megabytes to 1 gigabyte. The size of this type of memory will no doubt get even larger in the future. Figure 16-19 shows the pinout of the Toshiba flash memory chip family, and Figure 16-20 shows the SmartMedia card and its connector configuration. All the control pins for the memory chip are brought out on the card. Because the chip uses 8 I/O lines to input and output both data and address information, the same control lines will work for different-sized memory chips.

The notch on the corner of the card determines the voltage used by the memory chip inside. For the card shown in Figure 16-20, the supply voltage would be 3.3 volts; if the notch were on the other corner, the supply voltage would be 5 volts.

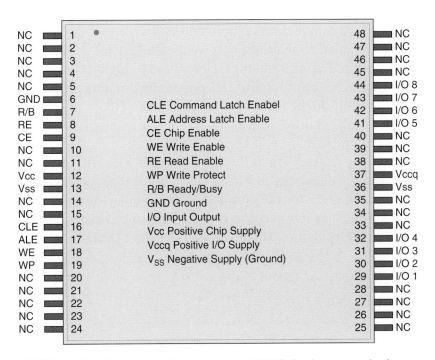

FIGURE 16-19 The pinout of the standard NAND flash memory in the TSOP package

To access the memory chip, the $\overline{CE}$ signal is brought LOW to enable the chip; then the ALE and CLE signals are used to select the register that is to be read from or written to. These are active HIGH signals, and when both are LOW the data register is selected. The data register is 528 bytes, which is the size of the page of data which can be written or read to the memory chip. The size of the memory page is 2^9 bytes or 512 bytes plus 16 more bytes for error codes such as a CRC code.

To **READ DATA** from the memory chip, you must first send the read command to the command register. This is done by making the CLE signal HIGH, then placing the read command code (00h) on the I/O

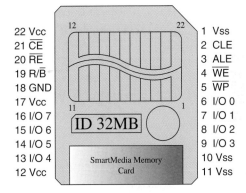

FIGURE 16-20 The pinout of a SmartMedia memory card

lines, and bringing the $\overline{WE}$ line LOW and then back HIGH. The address of the page to be read into the 528-byte data register is then written to the chip. This is done by bringing the ALE line HIGH, then placing the address of the page to be read on the I/O lines, and again bringing the $\overline{WE}$ line LOW, and then back HIGH. This must be done three times for most chips and four for the larger chips in order to get the complete address into the chip. The first address byte sets the data pointer that is used to start the output of data to the I/O lines when the data is read from the data register. The next two bytes are the address of the 528-byte page that is to be read from the memory. Because the page is 528 bytes long, in order to allow the data pointer to be set from 256 to 512, the read command (01h) is used. If it is desired to set the data pointer to the last 16 bytes, the read command (50h) is used. It should be noted that the complete 528-byte page is read from memory to the data register and only the data pointer used to start the data transfer to the I/O lines is changed.

After the read command and page address is written to the chip, the R/$\overline{\text{B}}$ line will go LOW for a period of about 25 microseconds. When the R/$\overline{\text{B}}$ line goes back HIGH, the data may be read from the chip. The ALE and the CLE are both brought LOW to select the data register, and the first byte can be read from the data register by bringing the $\overline{\text{RE}}$ LOW. This will place the data of the first byte on the I/O lines. Bringing the $\overline{\text{RE}}$ HIGH and LOW again will read out the next byte. This will continue until all the data has been read from the data register. If the $\overline{\text{RE}}$ is toggled again, the R/$\overline{\text{B}}$ line will again go LOW, indicating that the chip is reading the next page into the data register. When the R/$\overline{\text{B}}$ line again goes HIGH, the next page of memory can be read from the data register. This is called sequential reading. It should be noted that the $\overline{\text{CE}}$ must be held LOW for the duration of the reading operation or the data read operation will be ended and a new read command will have to be entered to read any more data. The waveforms for the read operation are shown in Figure 16-21.

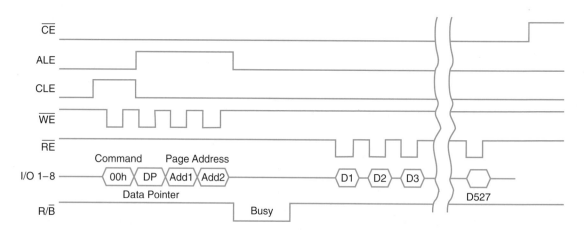

FIGURE 16-21 Waveforms for the read operation

Some of the newer chips have a $\overline{\text{CE}}$ "don't-care" feature which will allow the $\overline{\text{CE}}$ to return to a HIGH state between byte reads of the data register. In this case, the read operation is not terminated until a new command is entered into the command register.

To **WRITE** to the flash memory, the command byte (80h) is written to the command register. This sets the chip up to write data from the data register in the chip to a page in the memory. Next, the address of the page to be written to is sent to the address buffers by bringing the ALE line HIGH, and the CLE LOW. Just as in the read operation, the first byte of the address is the location of the data pointer; this indicates where in the data register the data is to be started when writing it to the memory. In most cases, this byte is (00h), because the data is to start being written from the beginning of the data register. The ALE and the CLE are brought LOW to set the chip to receive data into the data register, and the 528 bytes of data are sent to the chip one after the other. After the data register in the chip has been loaded with the data to be written to the memory, a command of (10h) is sent to the command register to start the writing process. At this time the R/$\overline{\text{B}}$ line will go LOW, indicating that the chip is busy writing the data into memory. This typically takes about 250 μsec. Figure 16-22 shows the waveforms for the write operation.

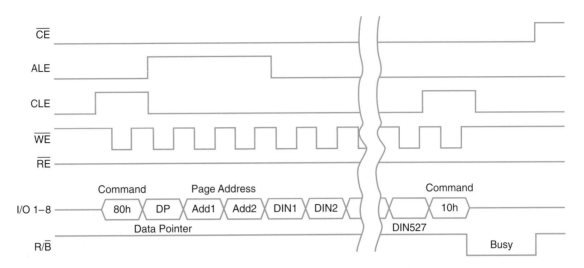

FIGURE 16-22 Waveforms for the write operation

BLOCK ERASE is done by first sending the command byte (60h) to the command register in the chip. This sets the chip up to erase a block of memory. Erasing a block will set all the bits in the block to 1. When writing to memory from the data register, only zeros in the data are written to the memory. Blocks are 32 pages in size, or 16896 bytes of memory. This amount is usually referred to as 16 k of memory. After the command word to erase memory is sent, the address is sent to the address buffers. The address for erasing is only 2 bytes long because the data pointer byte is not needed. The first 3 bits of the address are not used because the block being erased is 32 pages long. Once the chip has the address of the block to be erased, the command word (D0h) is written to the command register. This is done by bringing the CLE LOW, the ALE HIGH, placing the word (D0h) on the I/O lines, and toggling the $\overline{WE}$ signal. The chip will then start to erase the addressed block of memory. This will take about 2 msec. to do. During this time the R/$\overline{B}$ line will go LOW, indicating that the chip is busy. Figure 16-23 shows the waveforms for the erase operation.

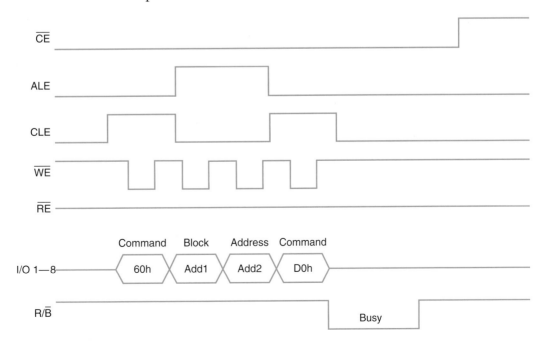

FIGURE 16-23 Waveforms for the erase operation

—	STATUS REGISTER
I/O PIN	**PIN DEFINITION**
I/O 0	Program/Erase 1=OK 0=Error
I/O 1	Reserved for Future Use
I/O 2	Reserved for Future Use
I/O 3	Reserved for Future Use
I/O 4	Reserved for Future Use
I/O 5	Reserved for Future Use
I/O 6	1=Busy 0=Ready
I/O 7	1=Not Write Protected 0=Protected

FIGURE 16-24 Bit definitions of the status register

READING THE STATUS REGISTER is done by sending the command (70h) to the command register, and then reading the data register. Figure 16-24 shows the meaning of the bits in the status register.

The status register should be read after the writing of a page or the erasing of a block to check for errors in the operation. Because EEPROM has a finite life, the memory will eventually wear out and errors will start to appear. When this happens, software should mark the page or block and not use it for data storage.

HARDWARE INTERFACING is not a problem with most computer systems or microprocessors. To interface the SmartMedia card with the Z-80 computer shown in this chapter would require 1 input bit from an input port, 3 bits of one output port, and the 8-bit data bus of the computer. The input bit would be used to read the R/$\overline{B}$ signal so that the computer could tell when the flash memory was busy. The three output bits would be used for the $\overline{CE}$, ALE, and CLE lines of the flash memory. The computer data bus would be used for the I/O lines of the flash memory that would be connected to the computer data bus D_0 through D_7. The $\overline{RE}$ and $\overline{WE}$ would use the $\overline{IOWR}$ and $\overline{IORD}$ control signals, ORed with an address decoder select line of the computer.

The R/$\overline{B}$ input bit could come from input port 0 of the computer. This input port is used to scan the keyboard, to read which key has been pressed. This input port has 3 extra bits which are not being used. One of these bits could be used. The output port for the $\overline{CE}$, ALE, and CLE would have to be constructed. This port could be a copy of one of the 7475 transparent latch output ports that are used for the LED display. The computer I/O address decoder has 4 extra select lines which are not being used. One of these could be used to select this output port. This could be output port 4. The next select line (5) from the address decoder could be ORed with $\overline{IORD}$ to make the $\overline{RE}$ control signal for the flash memory, and with $\overline{IOWR}$ to make the $\overline{WE}$ control signal for the flash memory.

With this extra hardware and a little programming to control the flash memory, a very large amount of permanent memory could be added to this simple Z-80 computer. This circuit interface is shown in Figure 16-25.

Figure 16-26 shows some of the chip numbers that are manufactured by Toshiba Co.

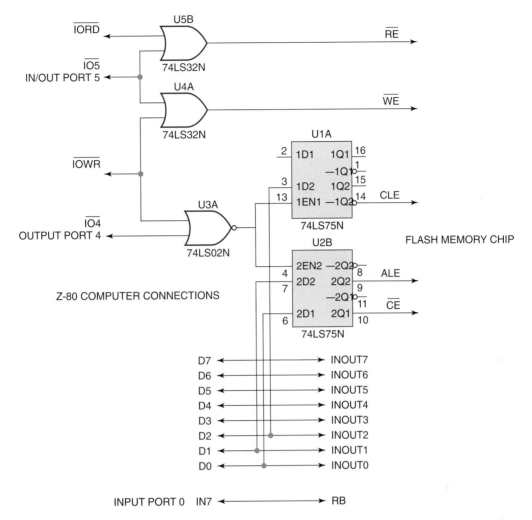

FIGURE 16-25 Interface for flash memory to the Z-80 computer

Density	0.16 micron Page = 528 B Block = 16 kB	0.13 micron Page = 528 B Block = 16 kB	0.13 micron Page = 2112 B Block = 128 kB
64 Mb	TC58V64BFT (ST)		
128 Mb	TC58128AFT (ST)	TC58DVM72A1FT00 (ST)	
128 Mb	TC581282AXB (CEDC)	TC58DVM72A1XB11 (CEDC)	
256 Mb	TC58256ATF (ST)	TC58DVM82A1FT00 (ST)	
256 Mb	TC58256AXB (CEDC)	TC58DVM82A1XB11 (CEDC)	
512 Mb	TC58512FT (ST)	TC58DVM92A1FT00 (ST)	
512 Mb		TH58DVM92A1XB11 (CEDC)	
1 Gb	TH58100FT (ST)	TC58DVG021FT00 (ST)	TC58NVG0S3AFT00 (CEDC)
2 Gb			TH58NVG1S3AFT00 (CEDC)

(ST) = Standard CE   (CEDC) = Chip Enable Don't Care

FIGURE 16-26 Flash memory manufactured by Toshiba Co.

SUMMARY

- The four main parts of a computer are the central processing unit (CPU), memory, input/output, and program.

  The CPU controls the rest of the computer's parts by executing the program stored in the memory. A program is a sequential list of binary numbers which are part of the CPU's instruction set.

- There are several types of read-only memory (ROM).

  PROM stands for programmable read-only memory, which can be programmed but, once programmed, cannot be changed. The EPROM can be erased by exposure to ultraviolet light. The EEPROM can be erased by using an electrical current. All ROMs are nonvolatile and are used to store programs which cannot be lost when power is lost.

- RAM stands for random-access memory.

  RAM is read/write memory and is used for the main memory in a computer. Static RAM can be written to and the memory of the RAM will not change until the power is lost. Dynamic RAM must be written to or read from every 2 msec. or it will lose the bit pattern programmed into it. The reading of a dynamic memory to keep it from losing its memory is called refreshing the memory. Dynamic memory is the most dense of all the memory types and it is used for the main memory of most computers today.

- The microcontroller is a computer on a chip containing the CPU, RAM, ROM, and various I/O devices.

QUESTIONS AND PROBLEMS

1. What is the frequency of the clock fed to pin 6 of the Z-80 in Figure 16-1?

2. What are the names of the three buses in the Z-80 computer in Figure 16-1?

3. What will happen if $\overline{BUSRQ}$ is brought LOW?

4. What is the CPU doing if the $\overline{M1}$ control signal is LOW in Figure 16-1?

5. What would be the address of the 2716 ROM in Figure 16-8 if the inverter on address line 15, which drives pin 6 of the 74LS138, were not there?

6. Why is the 74LS245 buffer IC used in the computer in Figure 16-1?

7. How many output ports can be added to the computer in Figure 16-9 if the address decoder is not changed?

8. Why were 74LS02 NOR gates used in Figure 16-9 instead of 74LS32 OR gates?

9. Draw the standard pinout for a 16-pin 64-k dynamic RAM.

10. Redraw the ROM in Figure 16-2 to be 8-by-8.

11. What are the four main parts of a computer?

12. List three 16- or 32-bit CPU ICs.

13. What does ROM stand for?

14. What does EPROM stand for?

15. What is the difference between static RAM and dynamic RAM?

16. Draw the pinout for a standard 64-k dynamic RAM IC.

17. How many address lines does it take to refresh a 4164 RAM?

18. Which register in the Z-80 CPU will always point to the next instruction to be read from memory?

19. What is mnemonic code?

20. Which register in the Z-80 is used to store the result of arithmetic operations?

21. What happens if the input to the inverter (IC-2, pin 9) in Figure 16-12 becomes grounded?

22. What is the address range and block size of the decoder in Figure 16-12?

23. What happens to the operation of the circuit in Figure 16-14 if pin 3 of the 74LS02 is disconnected?

24. Which port of the 68HC11 is used for the A-to-D inputs?

25. What is the difference between a microcontroller and a microprocessor?

26. In the 68HC11 circuit shown in the Digital Application section, how many bytes is the nonvolatile RAM?

27. In the 68HC11 circuit shown in the Digital Application section, what is the purpose of the MC14506 IC?

28. In the 68HC11 circuit shown in the Digital Application section, what is the purpose of the 74LS373 IC?

29. Which type of memory uses ultraviolet light to clear the memory?

30. What will happen to the circuit in Figure 16-1 if pin one of IC 6 becomes disconnected?

OBJECTIVES

After completing this lab, you should be able to:

- Understand the read and write operation of a static RAM.
- Understand the use of a memory as a code translator.

COMPONENTS NEEDED

2	8-pin dip switches
14	10-kΩ, ¼-W resistors
8	330-Ω, ¼-W resistors
1	FND-510 common anode seven-segment LED
2	2114 1K-by-four static RAM
2	7406 quad open-collector inverter ICs

PROCEDURE

This circuit uses a pair of 2114 RAM ICs to convert a standard hex digit expressed in four bits of binary to the 8-bit binary code necessary to display the hex digit on a seven-segment LED. Code conversion is quite often done with ROM instead of RAM. A typical example of this is the character generator ROM used in a CRT controller circuit. This ROM generates the correct code to be shifted out across the computer monitor screen from the ASCII code input to the address pins of the ROM.

Follow the procedure below to program the RAM with the correct code for each hex number. After you have programmed the RAM with the correct code for each hex number, the seven-segment LED will display the corresponding 4-bit binary input.

1. Wire the circuit shown and open all the switches before connecting the power. This will place the RAM outputs in the HiZ state, and SW1 cannot pull down an output that might be 1. If this happens, the RAM could be damaged.

2. Now set the four RAM address inputs A_0, A_1, A_2, and A_3 to ground or to the 0 value.

3. Use SW1 to display a 0 on the seven-segment display.

4. Store the 0 code in the RAM by first making $\overline{WE}$ LOW, and then bringing $\overline{CS}$ LOW and then back HIGH. This procedure will write the binary value on the RAM outputs into memory location 0.

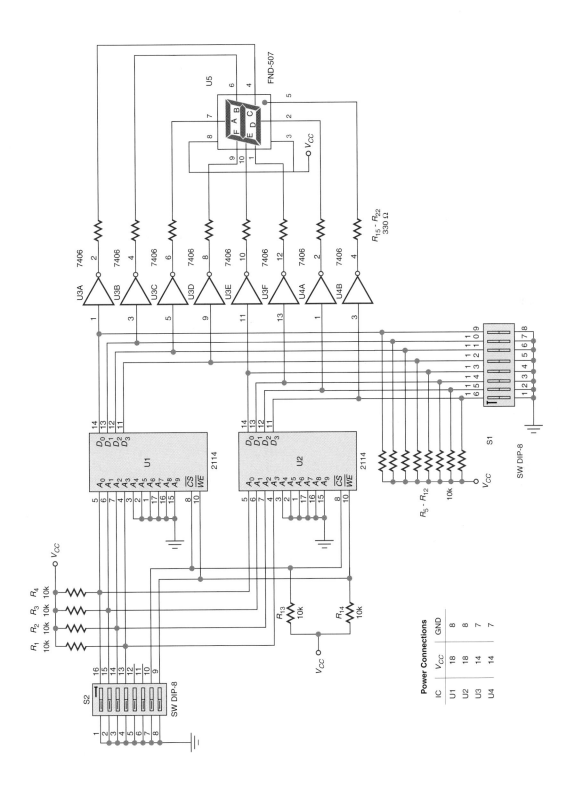

5. Next change the 4-bit memory address to binary 1 and repeat steps 3 and 4 to program the next RAM location with the correct code for binary number 1.

6. Repeat this procedure for all 16 numbers of the hex number system. Use lowercase b and d for binary 1011 and 1101.

7. After you have programmed the RAM, place all the switches in SW1 to the open position and bring $\overline{WE}$ HIGH.

8. Now place a binary number on the 4-bit address of the RAMs. The corresponding hex number should be displayed on the seven-segment LED.

9. Connect A_4 of the RAMs to SW2 pin 12 and make it HIGH. Now reprogram the seven-segment LED with the same codes as before, but make the hex point light each time. This will make the A_4 input the hex point of one of the next hex numbers in a two-number hex value.

10. Use the remaining switch on SW2 for A_5 of the RAMs and program two more sets of different codes for the seven-segment LED.

APPENDIXES

Lab Trainer Plans

LAB TRAINER PLANS

The labs in this text are designed to be constructed on a solderless breadboard or protoboard and require an external power supply, clock, or frequency generator and some buffered LEDs for logic indicators. These can be purchased separately or as a complete trainer which has everything in one package.

Another and much more desirable option is to build the equipment needed yourself. Schematics are shown in the following figures: Figure A-1 shows the debounce switch; Figure A-2 shows 8 buffered LED logic indicators; Figure A-3 shows 8 logic switches; Figure A-4 shows the clock generator; and Figure A-5 shows a power supply to operate these components and your lab circuit.

These parts are easily obtained and the trainer can be constructed in many different ways. The parts list for the digital trainer follows.

Quantity	Part Description
2	IC 1-2, 7406 open-collector inverters
1	IC 3, 7408 quad NAND gate
2	IC 4-5, 4050 CMOS to TTL buffers
1	IC 6, 555 timer
1	7805 +5 V voltage regulator
1	LM317 positive variable voltage regulator
8	1-kΩ, 0.5-W resistors
6	1-kΩ, 0.25-W resistors
1	240-Ω, 0.25-W resistor
1	5-kΩ potentiometer
1	20-kΩ potentiometer
8	Red LEDs
10	SPDT switches
1	5-position rotary switch
2	0.01-μF capacitors
1	0.1-μF capacitor

Quantity	Part Description
1	1-μF capacitor
3	10-μF capacitors
1	100-μF capacitor
2	4000-μF capacitors, 25 V dc
1	4 A bridge rectifier
1	2 A, 18 V center tap transformer
8	10-kΩ, 0.25-W resistors

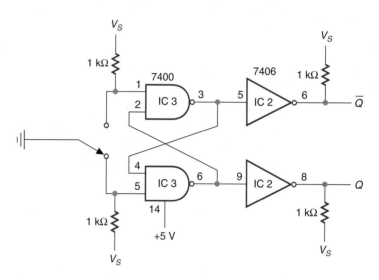

FIGURE A-1 Debounce switch

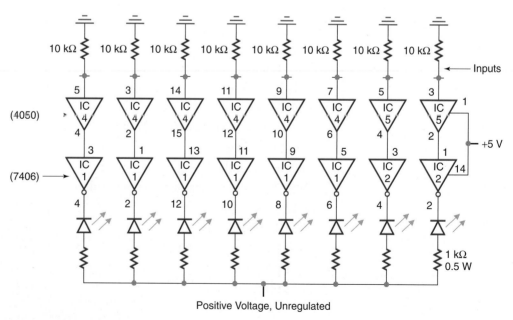

Note: 4050 V_{DD} is pin 1.
       4050 GND is pin 8.
       7406 V_{CC} is pin 14.
       7406 GND is pin 7.

FIGURE A-2 8 buffered LEDs

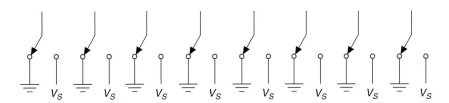

FIGURE A-3 8 logic switches

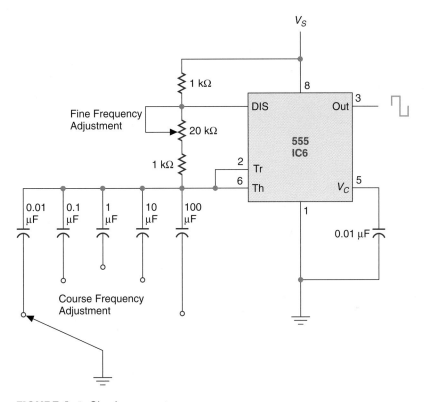

FIGURE A-4 Clock generator

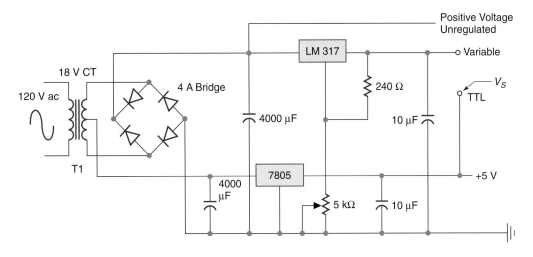

FIGURE A-5 Power supply

Equipment Needed

EQUIPMENT NEEDED FOR THE LABS

Most electronics labs in schools have all the needed equipment shown in the following list and more, but some items such as oscilloscopes may be few in number or nonexistent. In these cases, the instructor can omit portions of a lab or use other methods to explain the lab in a class demonstration.

The labs were designed to be constructed on protoboards such as explained in Lab 1. This protoboard can be a separate board all by itself or part of a complete trainer which has its own power supply, clock, debounce switch, etc. Such a trainer is very useful in the completion of these labs. There are several companies which produce a trainer or you can construct your own from the plan in Appendix A.

The equipment list needed to do the labs follows.

Quantity	Description	Labs Used In
1	Digital or analog multimeter	all labs
1	10-MHz dual-trace oscilloscope	8, 9, 10, 11, 12, 13
1	0- to 20-V power supply	13
1	An ac signal generator	9, 11
1	Digital trainer with protoboard or 1 protoboard, 5-V power supply, TTL signal generator and 1 debounce switch	all labs
3	7400 ICs	2, 3, 7, 8, 9, 13
1	7402 IC	2, 3, 8, 9
1	7404 IC	2, 3, 6, 8, 13, 14
2	7406 ICs	12, 13, 14, 15, 16
1	7408 IC	2, 3, 5, 7, 8, 9, 10, 13
1	7410 IC	9
1	7411 IC	2, 3, 8
1	7414 IC	9, 11, 12
1	7420 IC	9
3	7432 ICs	2, 3, 5, 8

Quantity	Description	Labs Used In
1	7447 IC	14, 15
1	74LS74	9
2	7475 ICs	9
2	7476 ICs	8, 9, 10, 15
2	7483 ICs	1, 5
2	7486 ICs	4, 5
1	7490 IC	10, 15
1	7493 IC	10, 13, 14
3	7495 ICs	9
1	74121 IC	12
1	74150 IC	14
1	74LS241	15
1	74C14 IC	6, 12
1	74LS164	9
1	4069	2
1	4071	2
1	4081	2
1	4011	2
1	4070	4
1	4012	4
1	4009	4
2	74180 ICs	4
1	555 timer	11
1	LM339 IC	13
1	4001 IC	2, 6, 11
1	1-$k\Omega$ resistor array	6, 9, 11, 12, 13, 15
2	10-$k\Omega$ resistor arrays	13, 16
1	20-$k\Omega$ resistor array	13
1	330-Ω resistor array	all labs
1	10-$M\Omega$ resistor	11
1	22-$k\Omega$ resistor	11
2	100-Ω resistors	5, 6, 12
8	Red LEDs	all labs
2	FND-507 7-segment displays	14, 15, 16
2	0.01-μF capacitors	11, 12
1	0.68-μF capacitor	11
2	20-pF capacitors	11
2	crystals of various frequencies	11
2	PNP power transistors	15
1	1-$k\Omega$ pot	6, 13, 14
1	914 diode	12
10	470-Ω resistors	15
2	8-pin DIP switches	16
2	2114 1K-by-4 static RAMs	16

Pinouts

PINOUTS OF THE ICs USED IN THE LABS (TTL)

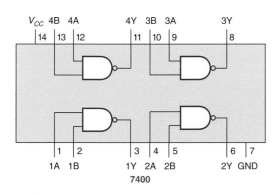

7400

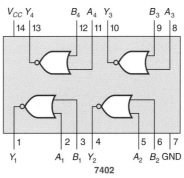

7402

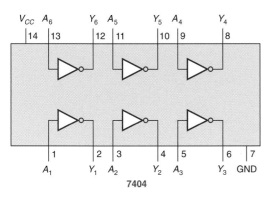

7404

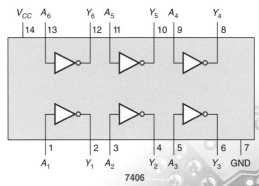

7406

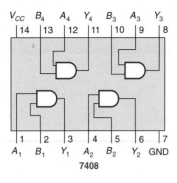

7408

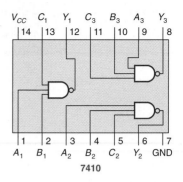

7410

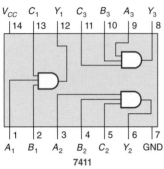

7411

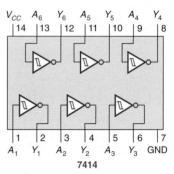

7414

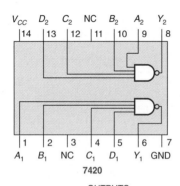

7420

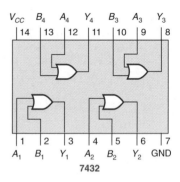

7432

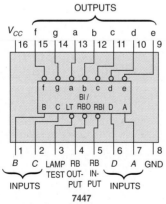

7447

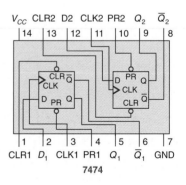

7474

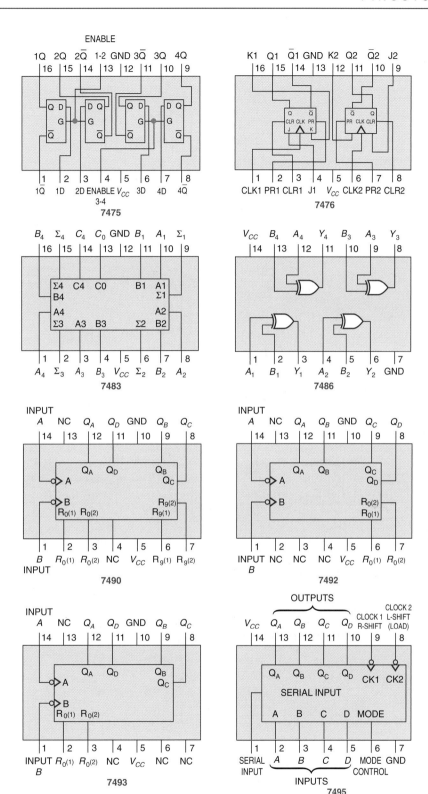

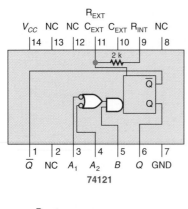

R_{EXT}
V_{CC} NC NC C_{EXT} C_{EXT} R_{INT} NC
14 13 12 11 10 9 8

2 k

$\overline{Q}$
Q

1 2 3 4 5 6 7
$\overline{Q}$ NC A_1 A_2 B Q GND
74121

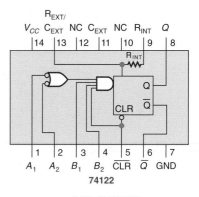

$R_{EXT/}$
V_{CC} C_{EXT} NC C_{EXT} NC R_{INT} Q
14 13 12 11 10 9 8

R_{INT}

Q
$\overline{Q}$

CLR

1 2 3 4 5 6 7
A_1 A_2 B_1 B_2 $\overline{CLR}$ $\overline{Q}$ GND
74122

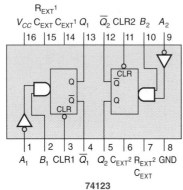

R_{EXT}^1
V_{CC} C_{EXT} C_{EXT}^1 Q_1 $\overline{Q_2}$ CLR2 B_2 A_2
16 15 14 13 12 11 10 9

CLR
Q
$\overline{Q}$
Q
$\overline{Q}$
CLR
Q

1 2 3 4 5 6 7 8
A_1 B_1 CLR1 $\overline{Q_1}$ Q_2 C_{EXT}^2 R_{EXT}^2 GND
C_{EXT}
74123

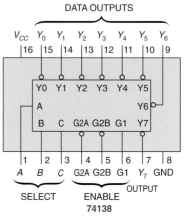

DATA OUTPUTS
V_{CC} Y_0 Y_1 Y_2 Y_3 Y_4 Y_5 Y_6
16 15 14 13 12 11 10 9

Y0 Y1 Y2 Y3 Y4 Y5
A Y6
B C G2A G2B G1 Y7

1 2 3 4 5 6 7 8
A B C G2A G2B G1 Y_7 GND
SELECT ENABLE OUTPUT
74138

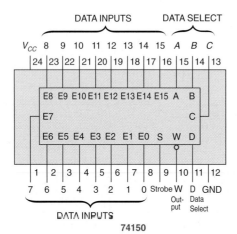

DATA INPUTS   DATA SELECT
V_{CC} 8 9 10 11 12 13 14 15 A B C
24 23 22 21 20 19 18 17 16 15 14 13

E8 E9 E10 E11 E12 E13 E14 E15 A B
E7 C
E6 E5 E4 E3 E2 E1 E0 S W D

1 2 3 4 5 6 7 8 9 10 11 12
7 6 5 4 3 2 1 0 Strobe W D GND
Output Data Select
DATA INPUTS
74150

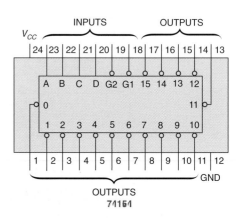

INPUTS OUTPUTS
V_{CC}
24 23 22 21 20 19 18 17 16 15 14 13

A B C D G2 G1 15 14 13 12
0 11
1 2 3 4 5 6 7 8 9 10

1 2 3 4 5 6 7 8 9 10 11 12
GND
OUTPUTS
74154

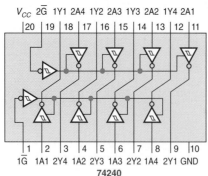

V_{CC} $2\overline{G}$ 1Y1 2A4 1Y2 2A3 1Y3 2A2 1Y4 2A1
20 19 18 17 16 15 14 13 12 11

1 2 3 4 5 6 7 8 9 10
$1\overline{G}$ 1A1 2Y4 1A2 2Y3 1A3 2Y2 1A4 2Y1 GND
74240

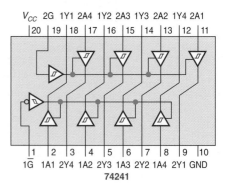

V_{CC} 2G 1Y1 2A4 1Y2 2A3 1Y3 2A2 1Y4 2A1
20 19 18 17 16 15 14 13 12 11

1 2 3 4 5 6 7 8 9 10
$1\overline{G}$ 1A1 2Y4 1A2 2Y3 1A3 2Y2 1A4 2Y1 GND
74241

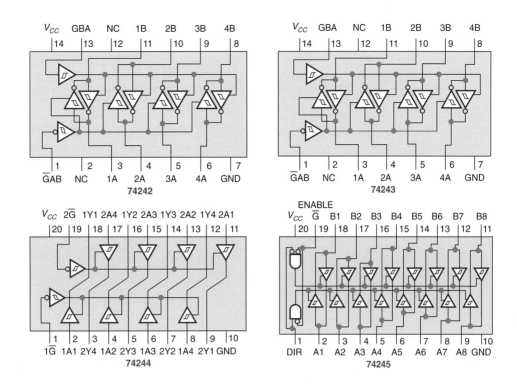

74242

74243

74244

74245

PINOUTS OF THE ICs USED IN THE LABS (CMOS)

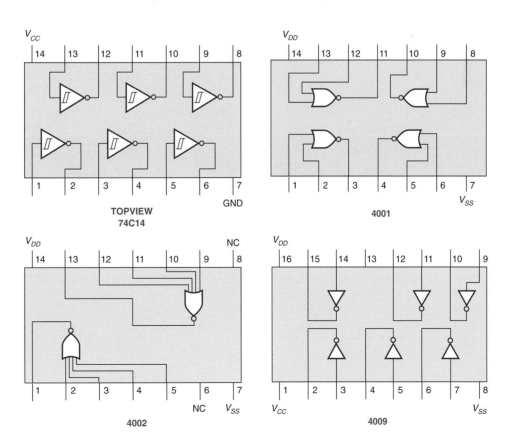

TOPVIEW
74C14

4001

4002

4009

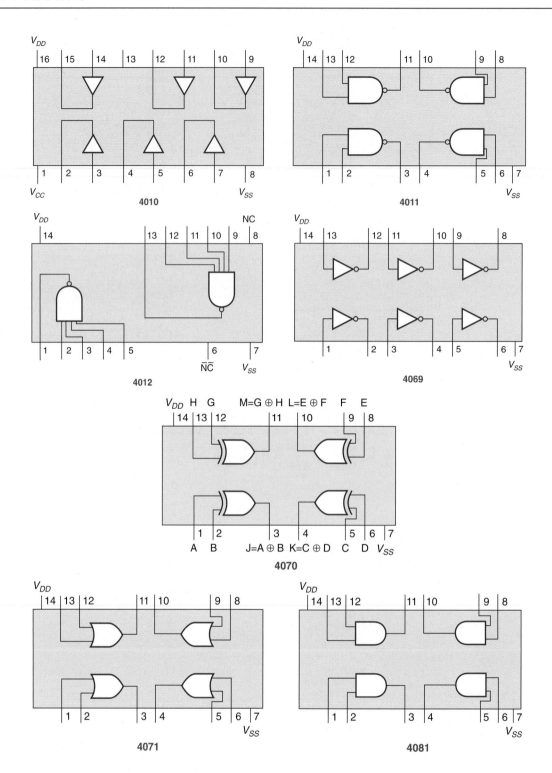

PINOUTS OF THE ICs USED IN THE LABS (ANALOG)

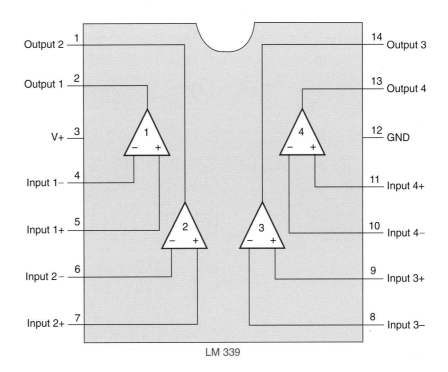

LM 339

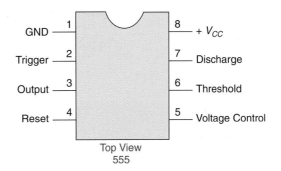

Top View
555

Metal Can Package

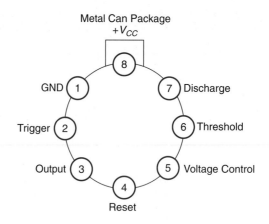

FND-507

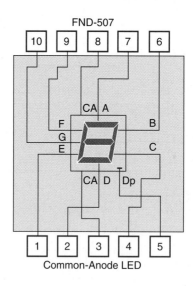

Common-Anode LED

NAND Gate, MOS, and CMOS

TTL NAND GATE

Figure D-1 shows the internal circuit of a TTL NAND gate. Although the gate can be used without knowledge of the internal circuitry, the TTL characteristics can be better understood by investigating the circuitry. All of the transistors in Figure D-1 are N-P-N silicon. Remember that in an N-P-N silicon transistor, the voltage on the base with respect to the emitter must be about $+0.7$ V to forward-bias the transistor and turn it on. When the transistor is turned on hard and saturated, the voltage on the collector with respect to the emitter is less than $+0.4$ V. Also, when a transistor is turned on, collector current flows and a drop occurs across the collector resistor. For example, when Q_2 in Figure D-1 is on, current flows through R_3, and most of the supply voltage is dropped across R_3. The voltage at the collector of Q_2 goes LOW. When Q_2 turns off, no collector current flows and the collector voltage rises.

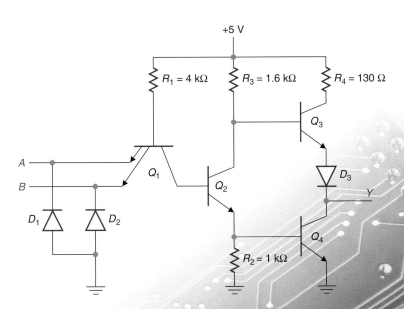

FIGURE D-1 TTL two-input NAND gate

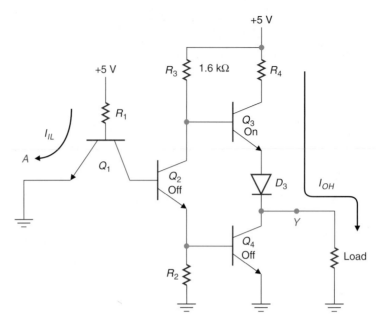

FIGURE D-2 NAND gate: Any 0 in, 1 out

Q_1 is a multiemitter transistor, one emitter for each input, connected in a collector-follower circuit. Any 0 into a NAND should yield a 1 out. Suppose A is a 0, as shown in Figure D-2. The base-emitter junction of Q_1 is forward-biased and conventional current flows through R_1. This current is I_{IL} and is -1.6 mA maximum. The negative indicates flow out of the gate, as shown in Figure D-2.

With the emitter grounded and the emitter-base junction forward-biased, the voltage on the base drops to about 0.7 V. Since 0.7 V on the base is not sufficient to forward-bias the base-collector junction of Q_1 and the base-emitter junction of Q_2, Q_2 turns off. With Q_2 off, there is no emitter current through R_2. With no voltage developed across R_2, the base-emitter junction of Q_4 will be turned off. With Q_4 off, the output Y will not be connected to ground.

Since Q_2 is off, there is no collector current flowing through R_3, and the collector voltage at Q_2 is HIGH. The emitter-base junction of Q_3 and diode D_3 are forward-biased. Q_3 is turned on and Y is connected to $+5$ V through a saturated transistor and a forward-biased diode. The output Y is HIGH, or 1. A 0 in at A causes Y to go HIGH. The current flowing out of the gate at Y is I_{OH} and is 400 μA maximum.

If both A and B are HIGH, as shown in Figure D-3, then the base-emitter junction of Q_1 is not forward-biased. Q_2 can now be forward-biased by $+5$ V dropped across R_1, the collector-base junction of Q_1, the base-emitter junction of Q_2 and R_2 to ground. The arrow in Figure D-3 shows this path. Q_2 turns on and saturates. With Q_2 on, emitter current through R_2 causes a voltage drop across R_2 that forward-biases the base-emitter junction of Q_4 and saturates it. With Q_4 on, output Y has a path to ground through Q_4. Q_4 is capable of sinking 16 mA, I_{OL}, and still maintaining a 0 level of 0.4 V or less. The collector voltage at Q_2 is equal to the collector-emitter drop across Q_2, approximately 0.3 V, plus the base-emitter drop across Q_4, approximately 0.7 V. The voltage at the collector of Q_2 is about 1.0 V. One volt is not sufficient to forward-bias both the base-emitter junction of Q_3 and the diode D_3, Q_3 turns off and Y does not have a path to $+5$ V through Q_3. D_3 ensures that Q_3 cannot turn on while Q_4 is on. With A and B inputs both 1, the output Y is 0.

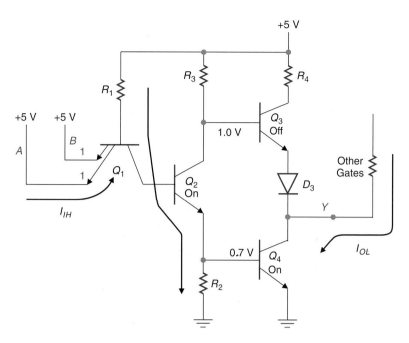

FIGURE D-3 NAND gate: All 1s in, 0 out

The current into the gate at A and B, I_{IH}, is leakage current with a maximum of 40 μA. If inputs A and B to the NAND gates are left floating (not connected to a signal) then the base-emitter junction of Q_1 is not forward-biased. Q_1 behaves as if the inputs were tied HIGH. In TTL, unused inputs are usually interpreted as one level by the ICs. Unused inputs are often tied HIGH through a 1-kΩ to 10-kΩ resistor. The diodes D_1 and D_2 on the inputs are normally reverse-biased. When switching transients cause the inputs to drop below ground, D_1 and/or D_2 turn on to clamp the input voltage. V_I, input clamp voltage, is a parameter that specifies the amount of negative swing that can occur. For the NAND gates and inverters, V_I is listed as -1.5 V maximum when the input current, I_I, is -12 mA. The manufacturer guarantees that when the input swings LOW enough to draw -12 mA, the input voltage will not fall below -1.5 V.

MOS TRANSISTORS

The symbols for the N-channel and P-channel enhancement mode transistors are shown in Figure D-4. Channel refers to the path through the transistor from drain to source. The symbol shows the channel broken into three parts.

The channel has to be completed or "enhanced" for conduction through the transistor to take place. For an N-channel device, the drain and source are N-type material. The substrate is P-type. Note that the arrow points from the P-type substrate toward the N-type channel. The gate is isolated from the channel by a thin layer of silicon dioxide insulator. The gate, channel, and insulator form a small capacitor. This capacitive input determines many of the characteristics of CMOS ICs. If the substrate and source are connected to ground and the drain to a positive voltage as shown in Figure D-5, the gate can control the amount of current that flows through the channel.

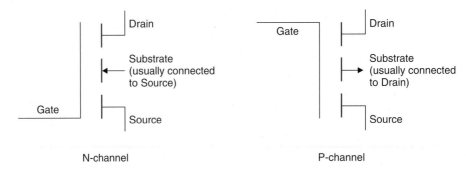

N-channel   P-channel

FIGURE D-4 Enhancement mode MOS schematic symbols

If the gate is held at a voltage near ground, the channel remains incomplete and only leakage current flows through the channel. If a positive voltage is applied to the gate (Figure D-5), free electrons in the P-type substrate are attracted to the channel. The N-channel is completed or enhanced and conventional current can flow from V_{DD} through the channel to ground. As the voltage on the gate becomes more positive, the number of free electrons that are attracted into the channel region and the amount of drain current that can flow through the transistor increases.

For digital applications, the gate inputs are either near V_{DD} for a 1 or near V_{SS} for a 0. The transistor is either completely enhanced (saturated) or off.

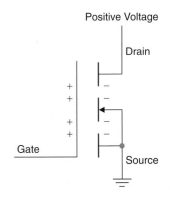

FIGURE D-5 N-channel MOS

The symbol for a P-channel enhancement mode transistor (Figure D-4) differs from the N-channel in two respects. The arrow on the substrate lead points away from the P-type channel toward the N-type substrate, and the gate is drawn from top to bottom. The drain and source are P-type material.

If the substrate and drain are connected to a positive voltage, as shown in Figure D-6, and the source is grounded, the gate can control the amount of current that flows through the transistor. To enhance the channel, P-type carriers in the substrate must be attracted into the channel region. This happens when a low voltage is applied to the gate as shown in Figure D-6.

If a positive voltage is applied to the gate, the P-type carriers are driven away from the channel. The channel is no longer complete, and the transistor turns off.

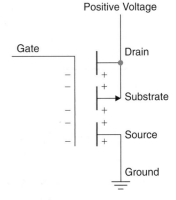

FIGURE D-6 P-channel MOS

CMOS

CMOS stands for complementary metal-oxide semiconductor. Complementary means that a P-channel transistor and an N-channel transistor work together in a totem-pole arrangement as shown in Figure D-7. Metal-oxide refers to the silicon dioxide layer between the gate and channel.

In Figure D-7, when *A* is HIGH, the N-channel on the bottom is enhanced and the output *Y* is connected to ground through a completed channel. The P-channel MOS on the top is turned off. When *A* is LOW, the P-channel is enhanced, and the N-channel is turned off. *Y* is connected to V_{DD} through the P-channel. These two transistors produce an inverter.

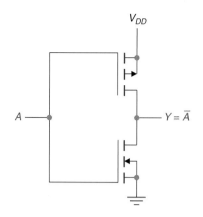

FIGURE D-7 CMOS inverter

Figure D-8 shows the simplicity of a four-input CMOS NAND gate. Each input controls a P-channel transistor and an N-channel transistor. The four N-channel transistors are connected in series. All four have to be enhanced by a 1 for *Y* to be connected to ground. All 1s in, 0 out. The four P-channel transistors are connected in parallel. If any one of the four is enhanced by a low level input, *Y* is pulled up to V_{DD} through the enhanced channel. Any 0 in, 1 out.

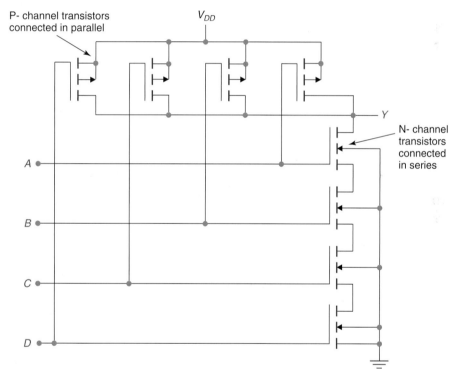

FIGURE D-8 Four-input CMOS NAND gate

10K One series of emitter coupled logic integrated circuits.

100K One series of emitter coupled logic integrated circuits.

A

AC (Advanced CMOS) Subfamily of CMOS. 74ACXX, 54ACXX.

ACT (Advanced CMOS-TTL Compatible) Subfamily of CMOS. 74ACTXX, 54ACTXX.

Active HIGH input A circuit input that is "looking for" or "waiting for" a 1 to cause the circuit to activate or function.

Active HIGH output A circuit output that is normally 0 and switches to a 1 when activated by the circuit.

Active LOW input A circuit input that is "looking for" or "waiting for" a 0 to cause the circuit to activate or function.

Active LOW output A circuit output that is normally 1 and switches to a 0 when activated by the circuit.

AHC (Advanced High-Speed CMOS) Subfamily of CMOS. 74AHCXX, 54AHCXX.

AHCT (Advanced High-Speed CMOS-TTL Compatible) Subfamily of CMOS. 74AHCTXX, 54AHCTXX.

ALS (Advanced Low-Power Schottky) Subfamily of TTL. 74ALSXX, 54ALSXX.

ALU (Arithmetic Logic Unit) An integrated circuit that performs arithmetic operations or logical operations on the inputs.

ALVT (Advanced Low-Voltage TTL Compatible) Subfamily of CMOS. 74ALVXX, 54ALVXX. Operates at 2.5 V or 3.3 V. Inputs can rise to TTL levels.

Analog Pertaining to data that is continuously variable and not broken into discrete units. An example of an analog device is an ordinary car speedometer.

Analog-to-digital Converting a continuous or analog quantity to a signal with a proportional value, usually a binary number.

AND gate A circuit that combines ones and zeros according to the rule, "all 1s in, 1 out or any 0 in, 0 out."

AND-OR-INVERT An integrated circuit that combines inputs through two layers of gates, first AND and then NOR.

Anode One lead of a diode or LED which is connected toward the positive supply terminal to forward-bias the diode.

ANSI American National Standard Institute.

ARCHITECTURE Segment of a VHDL program that describes the relationships between input and output signals.

AS (Advanced Schottky) Subfamily of TTL. 74ASXX, 54ASXX.

ASCII (American Standard Code for Information Interchange) A 7-bit code that represents the decimal digits, letters of the alphabet, symbols, and control characters.

Astable multivibrator A free-running clock or oscillator.

Asynchronous serial data transmission A system whereby data is transmitted one bit at a time on a single data line at a predetermined baud rate. Asynchronous implies that there is no specific time between the start of one word and the start of the next.

B

Baud rate The number of signal transitions per second that are transmitted or received, usually bits per second.

BCD (Binary-Coded Decimal) The code in which each decimal digit is represented by four bits.

BCD counter A 4-bit binary counter that counts from 0000 to 1001 and then resets to 0000. It advances one number for each pulse received.

Binary Base-2 number system which uses two digits, 0 and 1.

Binary ladder A resistive network based on resistor values which increase in value by a power of 2. The resistor network will produce a proportional voltage for a given binary number input to the network.

Bit A contraction for binary digit. Each place in a binary number is a bit; for example, 1011 is a 4-bit number.

Bit-vector A multi-bit signal in VHDL programming.

Boolean algebra An algebra used to express the output of a base-2 digital circuit in terms of its inputs and used to reduce the output to lowest terms.

Boolean expression Terms written in Boolean algebra that express the output of a circuit in terms of the input.

Broadside load A parallel load in which all bits of a data word are loaded into a register on a single clock pulse.

Bubble The small circle used on inputs and outputs of logic symbols to indicate the complement operation.

Byte A binary number eight bits long.

C

Carry-in The carry into the first stage of an adder from a previous addition. It is sometimes called C_0.

Carry-out The carry from the last stage of an adder.

Cathode One lead of a diode or LED which is connected toward the negative supply to forward-bias the diode.

Cell A position in a Karnaugh map.

Channel The path of the current flow in an MOS transistor.

Clear To reset or turn off a flip-flop and cause the Q output of a flip-flop to assume a 0 level.

Clock A continuous rectangular waveform used for timing.

CMOS (Complementary Metal-Oxide Semiconductor) One family of digital integrated circuits.

Collector-follower stage Configuration in which the input signal is on the emitter and output is on the collector. It is the input stage of TTL circuits.

Combinational logic Use of more than one gate to produce the required output.

Comparator Digital circuit that compares two binary numbers and outputs whether they are equal.

Complement (1) Invert; (2) A number which when added to a given number yields a constant. For example, the 9's complement of 7 is 2.

Conventional current Indicates the current flow is from positive to negative.

CPLD (Complex Programmable Logic Device) An integrated circuit containing multiple SPLD's that can be configured by the user.

CPU Central processing unit. The part of a computer that interprets the instruction fetched from the memory and executes it.

D

Data bus Circuits that generate, store, use, input, or output data are connected to the data bus. One line is provided for each bit of data.

Data separator A circuit that can separate multiplexed data into its component parts.

Decimal A base-10 number system which uses the digits 0 through 9.

Decoder A circuit that converts a number from another number system or code back into the decimal system.

Delayed clock A nonoverlapping clock or a dual clock system. The two rectangular waves are offset so that only one is HIGH at a time.

Delay/Power Product A parameter that gives an indication of a logic family's propagation delay and power dissipation, measured in picoJoules.

DeMorgan's theorems Two theorems of Boolean algebra where $\overline{A \cdot B} = \overline{A} + \overline{B}$ and $\overline{A + B} = \overline{A} \cdot \overline{B}$.

Demultiplexer A logic circuit which will switch the digital or analog data coming in on one input to one of many possible output lines. The output line selected to receive the input data is chosen by a binary number input to the logic circuit.

Digital Pertaining to data that is noncontinuous in nature. It changes in discrete steps. The data is represented by zeros and ones.

Digital-to-analog Converting a number (usually binary) to a proportional continuous analog quantity.

Diode A semiconductor device that conducts in one direction but not in the other. It has one anode lead and one cathode lead.

DIP (Dual In-Line Package) One style of integrated circuit package which has two rows of leads. Pin pitch = 0.1".

Drain An element of an MOS transistor. Analogous to the collector of a bipolar transistor.

Dynamic RAM Read-write memory that must have all the memory locations read every two ms to maintain the bit pattern stored in the memory.

E

EAC (End-Around Carry) The process in which the overflow in a 1's complement subtraction problem is added to the least significant column (rightmost).

ECL (Emitter-Coupled Logic) A family of digital logic integrated circuits noted for speed.

Edge-triggered flip-flop A flip-flop in which the data on the inputs is clocked into the flip-flop and appears at the output all on the same edge of the clock, unlike the master-slave flip-flop.

EEPROM Electrically erasable programmable read-only memory. Nonvolatile memory that can be programmed and erased by electrical means.

EIA (Electronic Industries Association) A trade association that helps set standards.

Electron current flow The actual flow of the electrons in the conductor from negative to positive.

Enable To supply a control signal for one of the basic gates that will allow data to pass through the gate.

Encoder A circuit that converts a number from decimal into another number system or code.

ENTITY Segment of a VHDL program which declares the input and output signals.

EPROM Erasable programmable read-only memory. Nonvolatile memory that can be programmed and erased with an ultraviolet light.

ESD (Electrostatic Discharge) Current caused by high static voltage build-up.

Even parity A system used to detect binary data transmission errors that uses a parity bit to make the total number of ones in a word even.

Exclusive-NOR gate A two-input gate that produces a 1 out when its inputs are alike. Also called nonexclusive OR.

Exclusive-OR gate A two-input gate that produces a 1 out when its inputs are different.

Expand To use additional gates to increase the number of inputs to a gate.

F

Fan-out A measure of how many loads a circuit can drive.

FAST (Fairchild Advanced Schottky TTL) Subfamily of TTL. 74FXX, 54FXX.

Fast carry A carry signal is developed at the same time that the other signals are being developed. Carry does not have to ripple through other gates. A look-ahead carry.

Flip-flop A bistable multivibrator. A circuit that is capable of assuming one of two conditions, on or off, in response to input signals, and maintaining that condition until signaled to change by the input.

FPGA (Field Programmable Gate Array) An integrated circuit that produces the required output by storing truth tables called look-up tables.

Frequency The number of cycles that a waveform completes in one second. It is measured in hertz.

Full adder A circuit that adds three inputs and outputs a sum and a carry.

Full decoder A circuit which will make one output line active for a given binary number input to the circuit. Each possible binary number has its corresponding output line.

Functional logic symbol An alternate symbol used to represent the operation of one of the basic gates.

G

Gates A circuit used to combine ones and zeros in specific ways. The basic gates are AND, NAND, OR, and NOR.

Generic Array Logic (GAL) One family of programmable logic devices.

Gullwing Surface-mount integrated circuit package leads that bend down and out.

H

H (High Speed) Subfamily of TTL. 74HXX, 54HXX.

Half adder A circuit that adds two inputs and outputs a sum and a carry.

HC (High-Speed CMOS) Subfamily of CMOS. 74HCXX.

HCT (High-Speed CMOS-TTL Compatible) Subfamily of CMOS. 74HCTXX.

Hexadecimal A base-16 number system which uses the 16 digits, 0 through 9 and A through F.

High-Speed, Low-Power, Low-Voltage (HLL) CMOS Subfamily of CMOS.

HiZ (High Impedance) The term used to indicate very high impedance in the order of 10 MΩ to 20 MΩ or higher.

I

IC Integrated circuit.

Inhibit To supply a control signal for one of the basic gates that will keep data from passing through the gate.

Inverted logic symbol An alternate symbol used to represent the operation of one of the basic gates. A functional logic symbol.

Inverter A circuit with one input and one output which functions according to the rule, "1 in, 0 out," or "0 in, 1 out."

J

JK flip-flop A flip-flop that can be turned on, turned off, toggled, or left the same according to the control signals on the J and K inputs.

J-lead Leads on a surface-mount integrated circuit that bend back under the package in a J shape.

K

Karnaugh map Systematic, graphical method of reducing Boolean expressions.

L

L (Low Power) Subfamily of TTL. 74LXX, 54LXX.

Latch A data flip-flop. A circuit that assumes an on state or an off state according to the input signal.

LCC (Leadless Chip Carrier) Surface-mount package that has no external leads. Solder connects the integrated circuit to the printed circuit board.

LCD (Liquid Crystal Display) A non-light-emitting method of displaying information.

Leading edge The first transition of pulse, be it LOW-to-HIGH or HIGH-to-LOW.

Leading zeros Zeros to the left of the most significant nonzero digit.

LED (Light-Emitting Diode) Emits light when forward biased.

LED display A light-emitting diode used to make a letter or number.

Logic diagram A schematic showing the gates, flip-flops, and other modules used in the circuit.

Look-ahead carry A carry signal is developed at the same time the other outputs are being generated. Carry does not have to ripple through other stages. A fast carry.

Low-Voltage (LV) Subfamily of CMOS. 74LVXX.

Low-voltage CMOS (LVC) Subfamily of CMOS. 74LVCXX.

Low-voltage Technology CMOS Subfamily of CMOS. 74LVTXX. Operates on 3.0–3.6 volts. Inputs can rise to TTL levels.

LS (Low-Power Schottky) Subfamily of TTL. 74LSXX, 54LSXX.

LSB (Least Significant Bit) The rightmost bit in a binary number.

LSI (Large-Scale Integration) An IC that contains circuitry equivalent to 100 gates or more.

M

Master-slave *D* flip-flop A flip-flop in which the input data is latched into the master section of the flip-flop on the leading edge of the clock and into the slave on the trailing edge.

Microminiaturization Shrinking the size and power requirements of electronic circuits.

Military specifications Military standard for IC construction.

Minuend A first or top number in a subtraction problem.

Mnemonic The alpha code for machine level instructions used in assembly language.

MSB (Most Significant Bit) The leftmost bit in a binary number.

MSI (Medium-Scale Integration) An IC that contains circuitry equivalent to more than 11 and less than 100 gates.

Multiplexer A logic circuit which will switch one of the many inputs to one output. The input which is connected to the output is selected by a binary number input to the logic circuit.

N

NAND gate A circuit that combines ones and zeros according to the rule, "all 1s in, 0 out or any 0 in, 1 out."

Negative edge The transition of a signal from HIGH to LOW.

Nine's complement The decimal number that is formed by subtracting a decimal number from all nines.

Nine's complement subtraction Method in which 9's complement of the subtrahend is added to the minuend.

Noise immunity One method of expressing the tolerance of a family of ICs to noise. It measures the range of acceptable input levels from the positive supply or ground.

Noise margin One method of expressing the tolerance of a family of ICs to noise. Measures the voltage difference between an acceptable input level and the corresponding acceptable output level.

Nonexclusive-OR gate A two-input gate that produces a 1 out when its inputs are alike. Also called an Exclusive-NOR.

Nonoverlapping clock Delayed clock or a dual clock system. The two rectangular waves are offset so that only one is HIGH at a time.

Nonretriggerable one-shot A logic device which, when triggered by the rising or falling edge of a digital signal, will produce an output pulse of a predetermined length. It cannot be triggered again until the output pulse has timed out.

NOR gates A circuit that combines ones and zeros according to the rule, "any 1 in, 0 out or all 0s in, 1 out."

O

Octal Base-8 number system which uses the 8 digits, 0 through 7.

Odd parity A system used to detect binary data transmission errors. It uses a parity bit to make the total number of ones in a word odd.

One's complement The binary number formed by inverting each bit of a binary number.

One's complement subtraction Method of subtraction in which the 1's complement of the subtrahend is added to the minuend.

One-shot A monostable multivibrator. A logic device which, when triggered by the rising or falling edge of the input pulse, will produce a pulse of a predetermined output pulse width.

Open collector TTL device in which the output has no internal path to the power supply. An external pull-up resistor is usually added.

Open drain CMOS device in which the output has no internal connection to the power supply. An external pull-up resistor is usually added.

Operational amplifier A high-gain amplifier with an inverting and noninverting input.

OR gate A circuit that combines ones and zeros according to the rule, "any 1 in, 1 out or all 0s in, 0 out."

Oscillator A circuit that switches its output from 0 to 1 and back according to its internal circuitry. The three types include the astable or free running (generates a clock), the monostable or one-shot (generates a single pulse), and the bistable or flip-flop.

Output port A register that latches data to be moved from the system to the outside world.

Overflow The carry from the most significant column (leftmost) in an addition problem.

P

Parallel data Each bit has its own data line. The entire word is transmitted on the same clock pulse.

Parity A system used to detect errors in binary data transmission.

Parity bit An extra bit used with data bits to make the total number of ones even or odd.

Parity checker Circuit that can determine whether the total number of ones in a binary word is even or odd.

Parity generator Circuit that can generate the proper parity bit for an even- or odd-parity system.

Partial decoder A logic circuit which will give an active signal on one output line for a given binary number input to the circuit. The circuit does not have a unique output line for each possible binary number input to it as a full decoder has.

PCB (Printed Circuit Board) A board that has copper tracks or traces to interconnect components.

Period Time required for a signal to complete one cycle.

Pin pitch Distance between centers of adjacent pins on an IC package.

PLCC (Plastic Leaded Chip Carrier) Surface-mount integrated circuit package with leads that bend back under the package.

PLD (Programmable Logic Device) IC containing gates that can be configured by the user.

Positive edge The transition of a signal from LOW to HIGH.

Preset To set or turn on a flip-flop and cause the output, Q, of a flip-flop to assume a 1 level.

Presettable counter A counter that can be loaded with the starting number. It then advances one count for each pulse received.

PROCESS A segment of a VHDL program containing conditional statements such as IF . . . THEN.

Programmable Array Logic (PAL) One family of programmable logic devices.

Programmable Logic Device (PLD) An integrated circuit containing AND-OR combinations that can be configured by the user.

PROM Programmable read-only memory. Nonvolatile memory that can have the bit pattern programmed once.

Propagation delay The measure of time between change in input and corresponding change in output.

Pull-up resistor A resistor that provides a connection to the power supply external to the IC. It is used with open-collector devices.

Q ————————————————

QuartusII Development system by Altera Corp. for programming programmable logic devices.

R ————————————————

RAM Random access memory. The read-write memory in a computer.

RC **time constant** Resistance in ohms times capacitance in farads yields seconds. The time for the capacitor to charge to 63.2% of applied voltage.

RESET To clear or turn off a flip-flop and cause the output, *Q*, of a flip-flop to assume a 0 level.

Retriggerable one-shot A logic device which, when triggered, will produce an output pulse of predetermined pulse width. The time width of the pulse or time-out period will start again each time the circuit is triggered even if the output has not yet timed out.

Ring counter A shift register in which the output of the last flip-flop is fed back to the inputs of the first flip-flop.

Ripple counter A counter designed so that each flip-flop generates the clock pulse for the flip-flop that follows. A ripple delay results.

ROM Read-only memory. The nonvolatile memory in a computer.

RS-232 (Recommended Standard-232) A voltage and format standard for serial data transmission.

S ————————————————

S (Schottky) Subfamily of TTL. 74SXX, 54SXX.

Schmitt-trigger input An input to a digital device which will change the output logic level at a fixed upper and lower threshold voltage of the input.

Serial data A single data line exists and data is transmitted or received one bit at a time.

SET To preset or turn on a flip-flop and cause the output, *Q*, of a flip-flop to assume a 1 level.

SET-RESET flip-flop A flip-flop that can be turned on by a signal on the SET input and turned off by a signal on the RESET input.

Seven-segment display An alphanumeric display made up of seven segments.

Shift counter Also called a Johnson Counter. It produces output waveforms that are offset timewise and is used for producing control waveforms.

Signed two's complement System in which a sign bit signifies whether a number is positive or negative and the remaining bits specify the magnitude. Negative numbers are represented in 2's complement form.

Sinks To provide a path for conventional current to flow to ground.

SMT (Surface-Mount Technology) Type of IC that is soldered to pads on the printed circuit board without passing through the board.

SO (Small Outline) Surface mount dual-inline-style integrated circuit package.

SOIC (Small Outline Integrated Circuit) Dual-inline-style surface-mount IC package. Pin pitch = 0.05."

SOP (Small Outline Package) Dual-inline-style surface-mount IC package. Pin pitch = 0.05".

Source Element of a field-effect transistor. Analogous to the emitter of a bipolar transistor.

Sources To provide a path for conventional current to flow from the power supply.

SPGA (Staggered Pin Grid Array) IC package in which pins are staggered for closer packing.

SPLD (Simple Programmable Logic Device) An integrated circuit containing AND-OR combinations that can be configured by the user.

SSI (Small-Scale Integration) An IC that contains circuitry equivalent to less than 12 gates.

SSOP (Shrink Small-Outline Package) Surface-mount IC package. Pin pitch = 0.026" or 0.02".

Standard logic (std_logic) Signal type used in VHDL programming.

Standard logic vector (std_logic_vector) Multi-bit signal type used in VHDL programming.

Static RAM Read-write memory that does not need to be read or refreshed to maintain the memory's bit pattern.

Substrate The silicon materials, P type or N type, upon which a transistor is fabricated.

Subtrahend The second or bottom number in a subtraction problem.

Successive approximation A method of using a digital-to-analog converter and a voltage comparator to produce a binary number which is proportional to the given analog input voltage.

Successive division Method for converting a decimal number to a binary number.

Switch bounce The making and breaking of switch contacts during a single switch closure.

Synchronous counter A counter designed so that each flip-flop receives the clock pulse at the same time.

Synchronous serial data transmission Data transmitted on a single line without framing bits for each byte.

T

Ten's complement The decimal number that is found by subtracting a decimal number from all nines and then adding one.

Ten's complement subtraction Method in which the 10's complement of the subtrahend is added to the minuend.

Through-hole construction Component leads pass through holes in the printed circuit board and are soldered to pads on the opposite side of the board from the component.

Time-out period Time required for a one-shot multivibrator to turn itself back off after being turned on.

Toggle To change state. To switch from 1 to 0 or 0 to 1.

Totem pole Circuit in which output has internal paths to the power supply and ground.

Trailing edge The second transition of a pulse as it returns to its original level, be it HIGH-to-LOW or LOW-to-HIGH.

Transparent *D* flip-flop A flip-flop that allows the input data to pass through to the output unaltered during one phase of the clock and latches the input data as the clock changes levels.

True magnitude Actual value as opposed to a complemented value.

Truth table A table listing all possible inputs to a circuit and the corresponding outputs.

TSSOP (Thin-Shrink Small-Outline Package) Surface-mount IC package. Pin pitch = 0.026" or 0.02".

TTL (Transistor-Transistor Logic) One of the popular families of digital integrated circuits.

2*R* ladder Sometimes called a 2*R* binary ladder. A resistive network based on two resistor values which will produce an output voltage proportional to the binary number input to the network.

Two's complement The binary number formed by inverting each bit of a binary number and adding 1.

Two's complement subtraction Method of subtraction in which the 2's complement of the subtrahend is added to the minuend.

U

Unique state The output state of one of the basic gates that occurs for only one combination of inputs.

Up-down counter Counter that can increment or decrement according to a control signal.

USB (Universal Serial Bus) A synchronous high-speed serial port.

V

V$_{CC}$ Positive supply voltage in a TTL IC (5 V). Sometimes used to designate the power supply voltage for a CMOS IC.

V_{DD} Positive supply voltage in a CMOS IC (+3 V to +18 V).

VHDL (Very High-Speed Hardware Description Language) Language used to configure the gates inside a programmable logic device.

Voltage comparator A circuit that compares the relative amplitudes of two input signals. The output is HIGH when the voltage on the noninverting input exceeds that on the inverting input.

Waveforms A graphical representation of a signal. It is the plot of the amplitude as a function of time.

Z

Zener diode A diode that conducts in the reverse direction at a definite voltage level.

Zero Insertion Force Socket (ZIF) IC socket with a lever that allows ICs to be inserted without force.

Answers to Self-Checks and Odd-Numbered Questions and Problems

CHAPTER 1

Self-Check 1

1.

11111	101011	110111
100000	101100	111000
100001	101101	111001
100010	101110	111010
100011	101111	111011
100100	110000	111100
100101	110001	111101
100110	110010	111110
100111	110011	111111
101000	110100	1000000
101001	110101	
101010	110110	

2. 63

3. 64

4. 22

5. 49

6. 110010010

7. 1001111

8. 1001010110

9. 1111110

Self-Check 2

1. 760
761
762
763
764
765
766
767
770
771
772
773
774
775
776
777
1000

2. 762_8

3. 111110101100000_2

4. F0F
F10
F11
F12
F13
F14
F15
F16
F17
F18
F19
F1A
F1B
F1C
F1D
F1E
F1F
F20

5. $E18_{16}$

6. 0100110010110000_2

7. 001001011000_{BCD}

8. 904_{10}

9. $F8_{16}$

10. 2784_{10}

11. 60_{16}

12. 101110010_{BCD}

Self-Check 3

1. 961

2. 1101101b

3. 5Ah

4. 10000011b

5. 1001100b

6. 01001101b

7. 93

8. 65h

9. 11259375

10. AAh = 170

Self-Check 4

1. 139

2. 213_8

3. 10001011_2

4. $22D_{16}$

5. 1000101101_2

6. 1055_8

7. 4681

8. 11111_8

9. 1249_{16}

Questions and Problems

1. 100_2 101_2 110_2 111_2 1000_2

3.

66_8	67_8	70_8	71_8	72_8	73_8	74_8
75_8	76_8	77_8	100_8	101_8	102_8	103_8
104_8	105_8	106_8	107_8	110_8		

5.

DD_{16}	DE	DF	E0	E1	E2	E3	E4
E5	E6	E7	E8	E9	EA	EB	EC
ED	EF	F0	F1	F2	F3	F4	F5
F6	F7	F8	F9	FA	FB	FC	FD
FE	FF	100	101_{16}				

7.

10001001_{BCD}	10010000_{BCD}	10010001_{BCD}	10010010_{BCD}	10010011_{BCD}
10010100_{BCD}	10010101_{BCD}	10010110_{BCD}	10010111_{BCD}	10011000_{BCD}
10011001_{BCD}	100000000_{BCD}	100000001_{BCD}		

9. a. $2 \cdot 2 \cdot 2 \cdot 2 - 1 = 15$ **B. a.** $2^{16} - 1 = 65,535$
   b. $2 \cdot 2 \cdot 2 \cdot 2 = 16$ **b.** $2^{16} = 65,536$

D.

Octal	Hexadecimal	Binary	Decimal	BCD
36	1E	11110	30	110000
251	A9	10101001	169	101101001
22	12	10010	18	11000
143	63	1100011	99	10011001
103	43	1000011	67	1100111

F. a. 10110   **c.** 111011   **11. a.** 0011   **c.** -010111
   b. 10101   **d.** 110011   **b.** 00111   **d.** 01000010

13. Ones and zeros are easy to represent electronically.

CHAPTER 2

Self-Check 1

1. See Fig. 2-1 and 2-2.   **3.** See Fig. 2-15 and 2-16.   **5.** See Fig. 2-9.
2. See Fig. 2-7 and 2-8.   **4.** See Fig. 2-3.   **6.** See Fig. 2-17.

7.
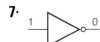 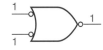 

 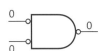

8. See Fig. 2-36.   **9.** All 1s in, 1 out.   **10.** All 0s in, 0 out.

Self-Check 2

1. See Fig. 2-23 and 2-24.   **3.** See Fig. 2-25.
2. See Fig. 2-30 and 2-31.   **4.** See Fig. 2-32.

5.
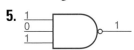 

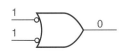 

6. See Fig. 2-36.   **7.** All 1s in, 0 out.   **8.** All 0s in, 1 out.

Self-Check 3

1. enables

2. inverted

3. 1, high

4. 1, unaltered

Self-Check 4

1. Connect the inputs together.

2. Connect the inputs together.

3. With an OR.

4. With an AND.

5. With an OR.

6. With an AND.

Questions and Problems

1. a.

$Y = \overline{A}$

b.

$Y = A + B$

c.

$Y = \overline{A + B}$

d.

$Y = A \cdot B$

e.

$Y = \overline{A \cdot B}$

3.

AND		OR		NAND		NOR	
AB	*X*	*AB*	*X*	*AB*	*X*	*AB*	*X*
00	0	00	0	00	1	00	1
01	0	01	1	01	1	01	0
10	0	10	1	10	1	10	0
11	1	11	1	11	0	11	0

5. a.

AB

b.

$\overline{A} + \overline{B}$

c.

AB	Y	
00	0	
01	0	
10	0	
11	1	Unique state

7. a.

$A + B$

b.

$\overline{\overline{A} \cdot \overline{B}}$

c.

AB	Y	
00	0	Unique state
01	1	
10	1	
11	1	

9. 1, 0, 0, 0

11. 0, 1, 1, 0

13. 0, 1, 0, 0

15. enables

17. Put a 1 on the control input.

19. 0

21. inverted

23. Put a 1 on the control input.

25. Locked up in the 1 state.

27. unaltered

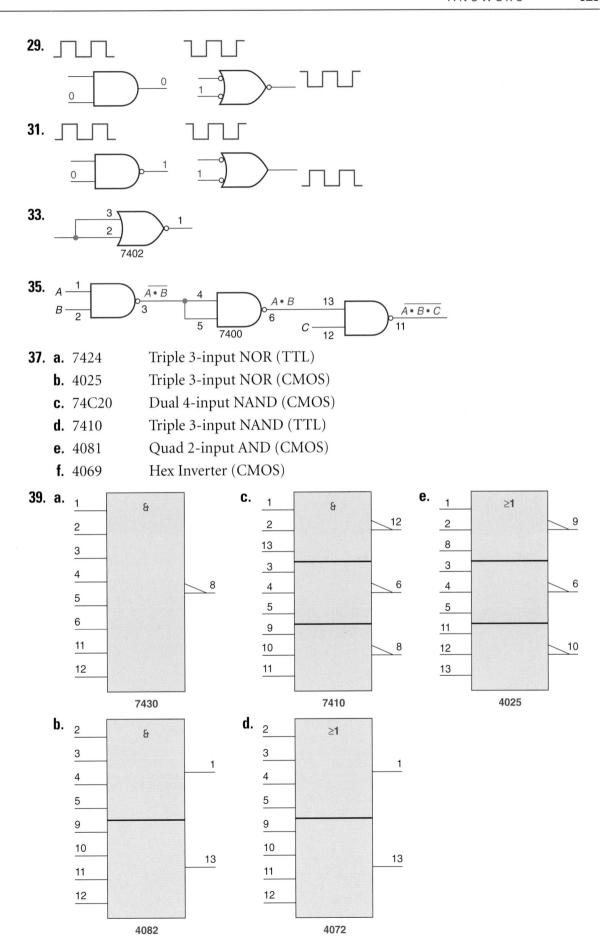

29.

31.

33. 7402

35.

37. **a.** 7424   Triple 3-input NOR (TTL)
     b. 4025   Triple 3-input NOR (CMOS)
     c. 74C20   Dual 4-input NAND (CMOS)
     d. 7410   Triple 3-input NAND (TTL)
     e. 4081   Quad 2-input AND (CMOS)
     f. 4069   Hex Inverter (CMOS)

39. **a.** 7430   **c.** 7410   **e.** 4025   **b.** 4082   **d.** 4072

41.

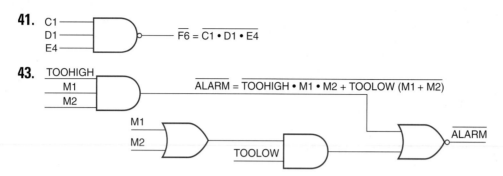

$$\overline{F6} = \overline{C1 \cdot D1 \cdot E4}$$

43.

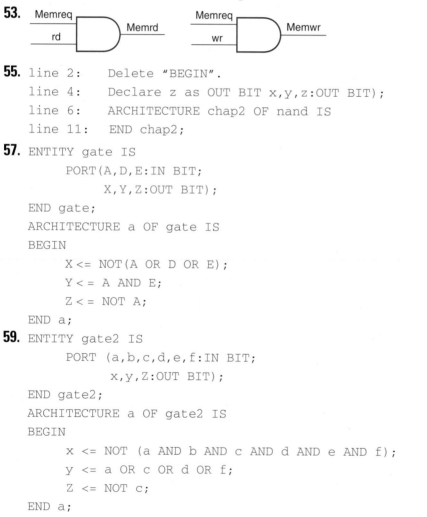

$$\text{ALARM} = \overline{\text{TOOHIGH} \cdot M1 \cdot M2 + \text{TOOLOW}(M1 + M2)}$$

45. Place a 1 on the control input to enable the AND and data passes unaltered. Place a 0 on the control input to inhibit the AND and the output is locked at 0.

47. Place a 1 on the control input to inhibit the OR and the output is locked at 1. Place a 0 on the control input to enable the OR and data passes unaltered.

49. If an input to a TTL NOR gate is left floating, the output will be LOW.

51. If an input to a TTL NAND gate is left floating, the gate will be enabled and the output will be the complement of the other input.

53.

55.
```
line 2:   Delete "BEGIN".
line 4:   Declare z as OUT BIT x,y,z:OUT BIT);
line 6:   ARCHITECTURE chap2 OF nand IS
line 11:  END chap2;
```

57.
```
ENTITY gate IS
      PORT(A,D,E:IN BIT;
           X,Y,Z:OUT BIT);
END gate;
ARCHITECTURE a OF gate IS
BEGIN
      X <= NOT(A OR D OR E);
      Y < = A AND E;
      Z < = NOT A;
END a;
```

59.
```
ENTITY gate2 IS
      PORT (a,b,c,d,e,f:IN BIT;
            x,y,Z:OUT BIT);
END gate2;
ARCHITECTURE a OF gate2 IS
BEGIN
      x <= NOT (a AND b AND c AND d AND e AND f);
      y <= a OR c OR d OR f;
      Z <= NOT c;
END a;
```

CHAPTER 3

Self-Check 1

1. 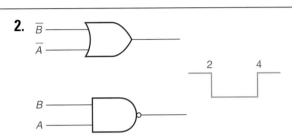   **2.**   **3.**

Self-Check 2

1. 1

2. 0

3. 1

4. $A + BC$

5. $\overline{A} \, B$

6. $\overline{C} \, (B + A)$

7. $\overline{\overline{A} \cdot C}$

8. $\overline{A \, B \, C}$

9. $\overline{A + B + C}$

Self-Check 3

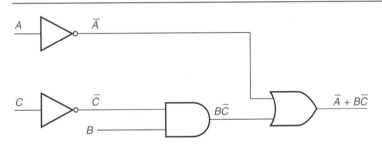

Self-Check 4

1.

	$\overline{C} \, \overline{B}$	$\overline{C} \, B$	$C \, B$	$C \, \overline{B}$
$\overline{A}$	0	0	1	0
A	0	1	1	1

$Y = C\,B + A\,B + A\,C$

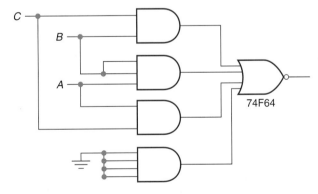

74F64

2. $Y = (\overline{A} + B + C)\,(A + \overline{B} + C)\,(A + B + \overline{C})$

$Y = (\overline{A}A + \overline{A}\,\overline{B} + \overline{A}C + BA + B\overline{B} + BC + CA + C\overline{B} + CC)\,(A + B + \overline{C})$

$\quad = (\overline{A}\,\overline{B} + BA + C\,(\overline{A} + B + A + \overline{B} + 1))\,(A + B + \overline{C})$

$\quad = (\overline{A}\,\overline{B} + BA + C)\,(A + B + \overline{C})$

$\quad = \overline{A}\,\overline{B}A + BAA + CA + \overline{A}\,\overline{B}B + BBA + BC + \overline{A}\,\overline{B}\,\overline{C} + BA\overline{C} + C\overline{C}$

$\quad = BA + CA + BA + BC + \overline{A}\,\overline{B}\,\overline{C} + BA\overline{C}$

$$= B(A + A + C + A\overline{C}) + CA + \overline{A}\,\overline{B}\,\overline{C}$$
$$= B(A + C + A\overline{C}) + CA + \overline{A}\,\overline{B}\,\overline{C}$$
$$= B(C + A(1 + \overline{C})) + CA + \overline{A}\,\overline{B}\,\overline{C}$$
$$= B(C + A) + CA + \overline{A}\,\overline{B}\,\overline{C}$$
$$= BC + BA + CA + \overline{A}\,\overline{B}\,\overline{C}$$

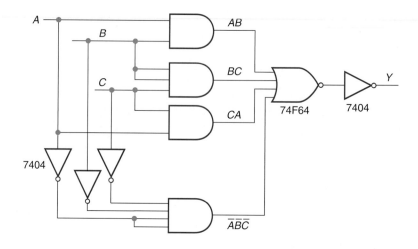

Self-Check 5

1. Each AND gate has 8 inputs.

2. The OR gate has 4 inputs.

3. Fuses, 0, 3, 5, 9, 10, 13, 16, 18, 20, 25, 27, 28

4. Each AND gate has 12 inputs.

5. Each OR gate has 4 inputs.

6. Fuses 1, 5, 9, 11, 13, 17, 20 22, 25, 28, 33, 34, 37, 40, 44, 47, 48, 52, 57, 59, 60, 65, 69, 70, 72, 77, 80, 83, 86

Questions and Problems

1.

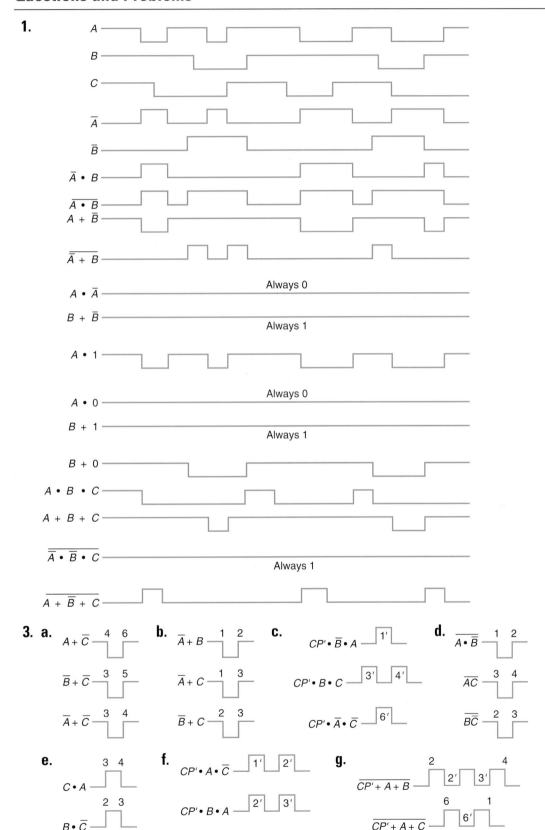

3. a.

$A + \overline{C}$ 4 6

$\overline{B} + \overline{C}$ 3 5

$\overline{A} + \overline{C}$ 3 4

b.

$\overline{A} + B$ 1 2

$\overline{A} + C$ 1 3

$\overline{B} + C$ 2 3

c.

$CP' \cdot \overline{B} \cdot A$ 1'

$CP' \cdot B \cdot C$ 3' 4'

$CP' \cdot \overline{A} \cdot \overline{C}$ 6'

d.

$\overline{A \cdot \overline{B}}$ 1 2

$\overline{AC}$ 3 4

$\overline{B\overline{C}}$ 2 3

e.

$C \cdot A$ 3 4

$B \cdot \overline{C}$ 2 3

$A \cdot \overline{C}$ 1 3

f.

$CP' \cdot A \cdot \overline{C}$ 1' 2'

$CP' \cdot B \cdot A$ 2' 3'

$CP' \cdot \overline{A} \cdot C$ 4' 5'

g.

$\overline{CP' + A + B}$ 2 2' 3' 4

$\overline{CP' + A + C}$ 6 6' 1

$\overline{CP' + B + \overline{C}}$ 5 5' 6

5.

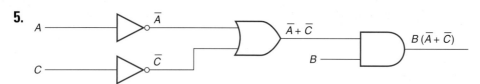

7.

9. a. $\overline{A}(B+C)$ **11. a.** $A+\overline{B}$ **13. a.** $\overline{B}+\overline{A}C$ **15.** $AC+\overline{A}\,\overline{C}$

   b. A **b.** $A+C$ **b.** $AB+\overline{A}\,\overline{C}$

   c. $A+D$ **c.** $A(\overline{C}+B)$

   d. 0 **d.** $A(\overline{C}+\overline{B})$

   e. $\overline{A}+B$

17. a.

	$\overline{C}\,\overline{B}$	$\overline{C}B$	CB	$C\overline{B}$
$\overline{A}$	1	0	0	1
A	1	1	1	1

$Y=A+\overline{B}$

b.

	$\overline{C}\,\overline{B}$	$\overline{C}B$	CB	$C\overline{B}$
$\overline{A}$	0	0	1	1
A	1	1	1	1

$Y=A+C$

c.

	$\overline{C}\,\overline{B}$	$\overline{C}B$	CB	$C\overline{B}$
$\overline{A}$	0	0	0	0
A	1	1	1	0

$Y=A\,\overline{C}+AB=A(\overline{C}+B)$

d.

	$\overline{C}\,\overline{B}$	$\overline{C}B$	CB	$C\overline{B}$
$\overline{A}$	0	0	0	0
A	1	1	0	1

$Y=A\,\overline{C}+A\,\overline{B}=A(\overline{C}+\overline{B})$

19. $\overline{Y}=\overline{C}\,BA+\overline{C}\,B\,A+C\,\overline{B}\,A+C\,B\,A$

   fuses   5 3 1   13 10 9   21 18 16   28 27 24

21. From Figure 3-54:

   $\overline{Y}_1=\overline{D}\,\overline{C}\,\overline{B}\,\overline{A}+\overline{D}\,\overline{C}\,B\,\overline{A}+\overline{D}\,C\,\overline{B}\,A+\overline{D}\,C\,B\,A$

   fuses   11 9 5 1   23 21 16 13   35 32 29 24   47 44 40 36

   $\overline{Y}_2=Y_1+D\,\overline{C}\,\overline{B}\,A+D\,\overline{C}\,B\,A+D\,C\,B\,A$

   fuses   50   70 69 65 60   82 81 76 72   94 92 88 82

23.

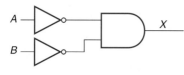

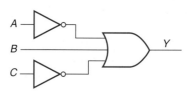

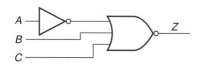

25. line 2: Delete "BEGIN"
   line 3: declare a,b,c, and d as IN BIT   (a,b,c,d:IN BIT;
   line 4: declare x,y,z as OUT BIT   x,y,z:OUT BIT);
   line 10: use parentheses to define function clearly, such as:

   z <= (NOT d) OR (a AND b) OR c;

27.
```
ENTITY p145_27 IS
    PORT(
        a,b,c,d,e:IN BIT;
        x,y,z:OUT BIT);
END p145_27;
ARCHITECTURE a OF p145_27 IS
BEGIN
    x<=(a AND NOT c AND e)OR(NOT b AND c AND d)OR(c AND NOT
    d AND e);
    y<=(NOT a AND NOT c AND e)OR(NOT b AND c AND NOT d)OR(c AND NOT
    d AND e);
    z<=(NOT a AND NOT c AND NOT e)OR(NOT b AND c AND NOT d)OR(c AND
    d AND e);
END a;
```

29.
```
ENTITY p145_29 IS
    PORT(
        A,B,C:IN BIT;
        Y:OUT BIT);
END p145_29;
ARCHITECTURE a OF p145_29 IS
BEGIN
    Y <= NOT ((NOT A AND B AND NOT C)OR(A AND NOT B AND C));
END a;
```

31.
```
ENTITY p145_31 IS
    PORT(
        A,B,C:IN BIT;
        Y:OUT BIT);
END p145_31;
ARCHITECTURE a OF p145_31 IS
BEGIN
    PROCESS (A,B,C)
    BEGIN
        IF (C = '0' AND B = '0' AND A = '1')THEN
                Y<='0';
        ELSIF (C = '0' AND B = '1' AND A = '1') THEN
                Y <= '0';
        ELSIF (C = '1' AND B = '1' AND A = '1') THEN
                Y <= '0';
        ELSE Y <= '1';
        END IF;
    END PROCESS;
END a;
```

33. Pin 5 of U2B is floating and the 7411 will process this input as a HIGH. When *A* is HIGH and *B* is LOW, the output of U2B is HIGH. The output of U3A is LOW and the LED does not light. Line 2 of the truth table produces a LOW instead of a HIGH.

35. The 7402 will process pin 2 as a HIGH. The output of the circuit will always be LOW and the LED will not light.

CHAPTER 4

Self-Check 1

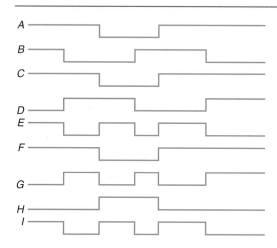

Self-Check 2

1. 0 **2.** 0

3.

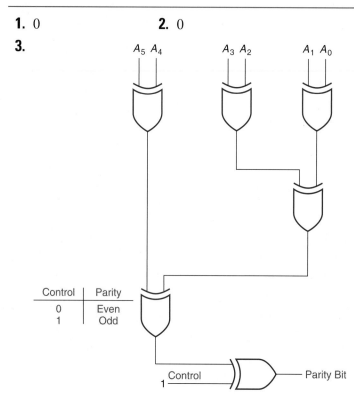

Control	Parity
0	Even
1	Odd

4.

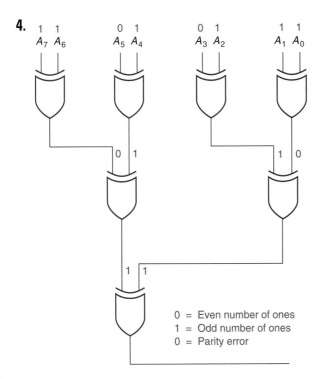

```
0 = Even number of ones
1 = Odd number of ones
0 = Parity error
```

Self-Check 3

1.

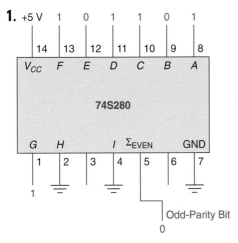

2.

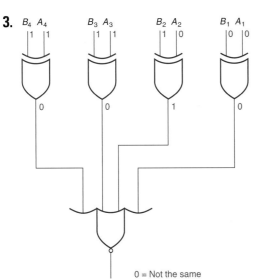

3.

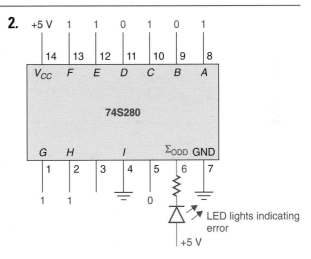

4.

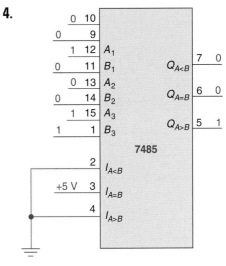

Self-Check 4

1.
```
ENTITY circuit IS
        PORT (
            A,B,C,D,E,F:IN BIT;
            W,X,Y,Z:OUT BIT);
END circuit;
ARCHITECTURE a OF circuit IS
BEGIN
    W <= NOT ((NOT B) AND C AND (NOT D));
    X <= NOT (B XOR (NOT A));
    Y <= NOT(((NOT A) AND B AND C AND (NOT E))OR(F AND (NOT B) AND
        C AND D));
    Z <= (NOT (A AND B AND C))OR(NOT (D AND E AND F));
END a;
```

Questions and Problems

1.

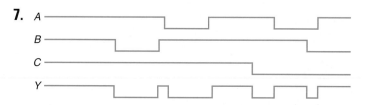

Inputs		Output
A	**B**	**Y**
0	0	0
0	1	1
1	0	1
1	1	0

5.

3. Inverted

7.

9.

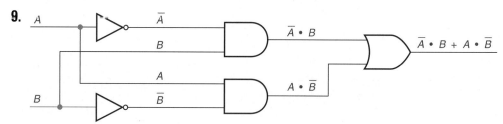

Alternate Solution

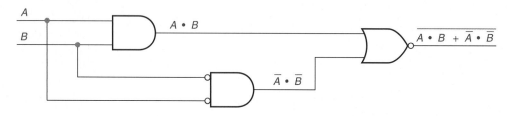

11. a. EVEN: <u>0</u> 101101
b. ODD: <u>1</u> 110000

c. EVEN: <u>0</u> 000011
d. ODD: <u>0</u> 110010

13.

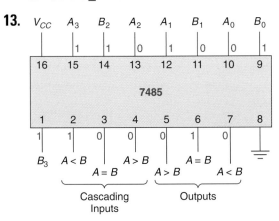

15.

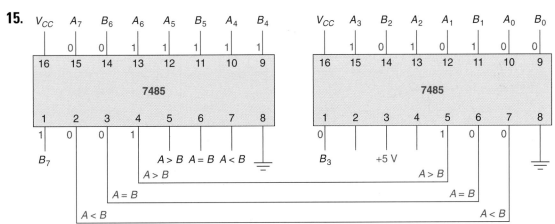

17.

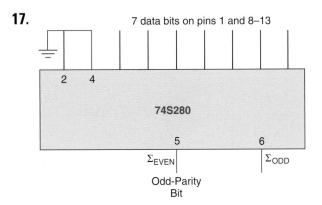

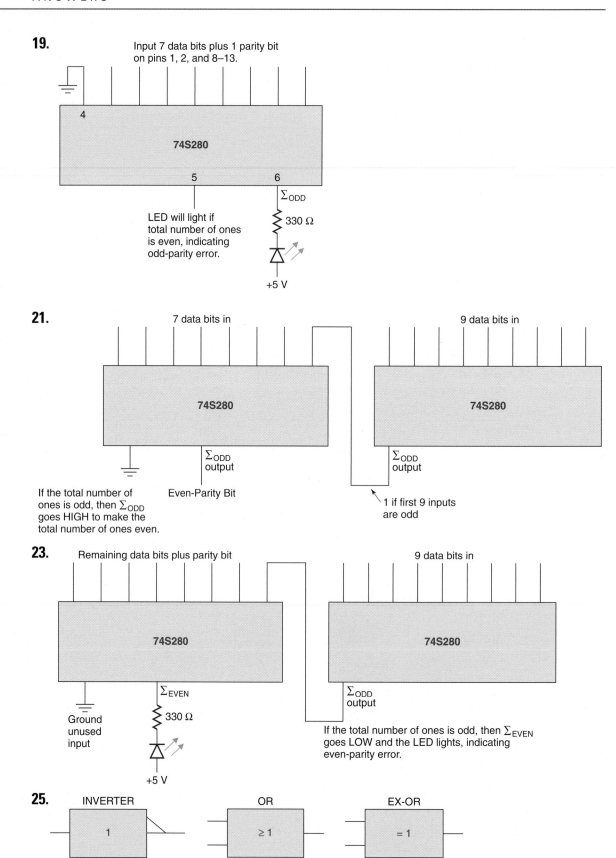

19.

Input 7 data bits plus 1 parity bit on pins 1, 2, and 8–13.

74S280

Σ_{ODD}

330 Ω

LED will light if total number of ones is even, indicating odd-parity error.

+5 V

21.

7 data bits in

9 data bits in

74S280

74S280

Σ_{ODD} output

Even-Parity Bit

Σ_{ODD} output

1 if first 9 inputs are odd

If the total number of ones is odd, then Σ_{ODD} goes HIGH to make the total number of ones even.

23.

Remaining data bits plus parity bit

9 data bits in

74S280

74S280

Σ_{EVEN}

330 Ω

Ground unused input

Σ_{ODD} output

+5 V

If the total number of ones is odd, then Σ_{EVEN} goes LOW and the LED lights, indicating even-parity error.

25.

INVERTER

1

OR

≥ 1

EX-OR

= 1

27.

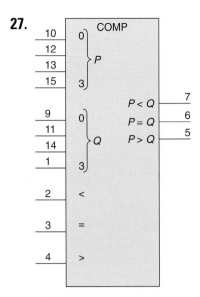

29. Create a folder to house the project. Use New Project Wizard to create a project with the same name as the folder. Create a new VHDL file. Write a VHDL program (.vhd). Compile the program. Create and configure a vector waveform file (.vwf). Create input waveforms for simulation. Simulate the program to check its operation. Download the program to the CPLD. Test the hardware.

31.
```
ENTITY prob31 IS
        PORT (A,B,C,D,E,F:IN BIT;
            M,N,O,P:OUT BIT);
END prob31;
ARCHITECTURE a OF prob31 Is
BEGIN
        M <= A AND B AND C AND D AND E AND F);
        N <= NOT(A AND B AND (NOT C) AND D AND (NOT E) AND F);
        O <= A OR B OR C OR D OR E OR F;
        P <= NOT ((NOT A) OR B OR C OR (NOT D) OR E OR F);
END a;
```

33.
```
ENTITY odd_parity_8 IS
PORT (
        A:IN BIT_VECTOR(6 DOWNTO 0);
        pb:OUT BIT);
END odd_parity_8;
ARCHITECTURE a OF odd_parity_8 IS
BEGIN
        pb <= NOT (A(6) XOR A(5) XOR A(4) XOR A(3) XOR A(2) XOR A(1)
        XOR A(0));
END a;
```

35.
```
ENTITY even_parity_checker_8 IS
        PORT (
            A:IN BIT_VECTOR(6 DOWNTO 0);
            pb:IN BIT;
            err:OUT BIT);
END even_parity_checker_8;
```

```
ARCHITECTURE a OF even_parity_checker_8 IS
BEGIN
      PROCESS (A,pb)
      BEGIN
          IF (A(6) XOR A(5) XOR A(4) XOR A(3) XOR A(2) XOR A(1)
          XOR A(0) XOR pb) = '0' THEN
                    err <= '0';
          ELSE err <= '1';
          END IF;
      END PROCESS;
   END a;
```

37.
```
ENTITY prob37 IS
      PORT (
            a,b,c: IN BIT;
            x,y,z: OUT BIT);
END prob 37;
ARCHITECTURE a OF prob37 IS
BEGIN
      x <= a XOR b XOR c;
      y <= (NOT a) XOR (NOT b) XOR (NOT c);
      z <= NOT (a XOR b XOR c);
END a;
```

39. line 1: ENTITY misspelled.

lines 2 and 3: Insert colons between signals and type:

  (a,b,c,d:IN BIT;   x,y,z:OUT BIT);

line 8: e used but not declared in line 2.

lines 7,8,9: wrong symbol for declaring relationships, change = to <=.

CHAPTER 5

Self-Check 1

1. A full adder has a third input, carry-in.

2. See Figure 5-1.

3. See Figure 5-5.

4. a.

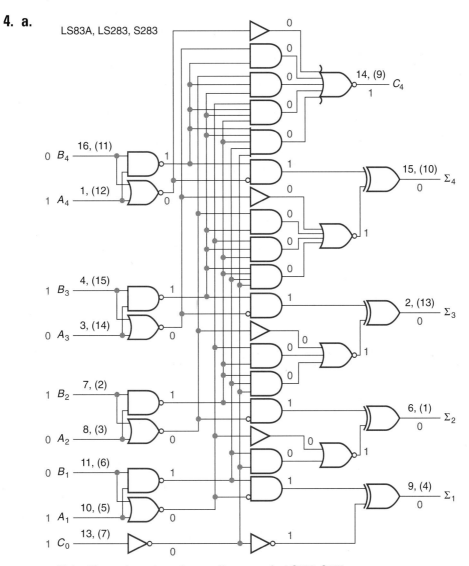

Note: Pin numbers shown in parentheses are for LS283, S283

4. b.

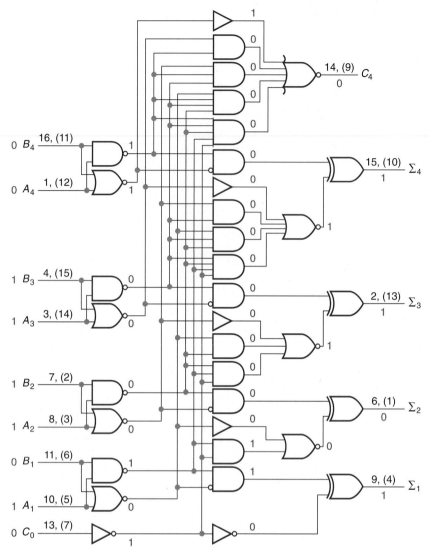

Self-Check 2

1. a.

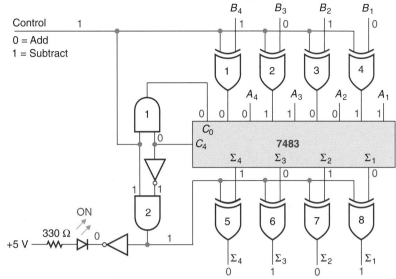

$$\begin{array}{r} 0101 \\ -1010 \\ \hline \end{array} \qquad \begin{array}{r} 0101 \\ +0101 \\ \hline 1010 \end{array}$$

No overflow; answer is -0101.

b.

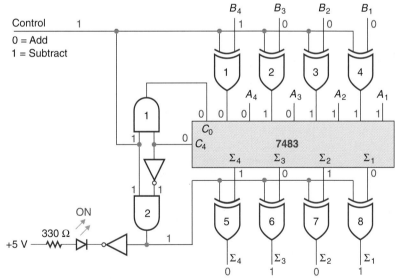

$$\begin{array}{r} 11 \\ -1000 \\ \hline \end{array} \qquad \begin{array}{r} 0011 \\ +0111 \\ \hline 1010 \end{array}$$

No overflow; answer is -101.

2. a.

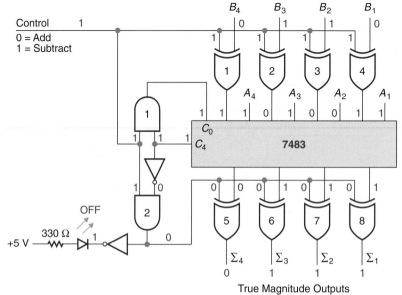

$$\begin{array}{r} 1101 \\ -0110 \\ \hline \end{array}$$

$$\begin{array}{r} 1101 \\ +1001 \\ \hline 10110 \\ +1 \\ \hline 0111 \end{array}$$ Answer is $+0111$.

b.

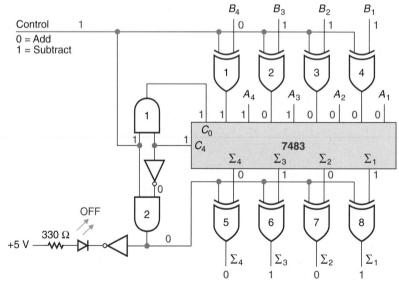

$$\begin{array}{r} 1100 \\ -111 \\ \hline \end{array}$$

$$\begin{array}{r} 1100 \\ +1000 \\ \hline 10100 \\ +1 \\ \hline 101 \end{array}$$ Answer is $+101$.

Self-Check 3

1. a.

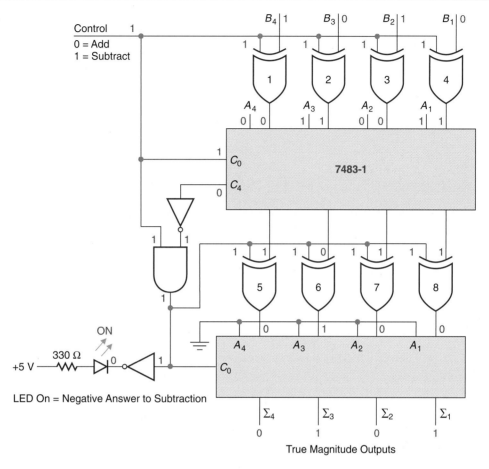

$$
\begin{array}{r}
0101 \\
-1010 \\
\hline
\end{array}
\qquad
\begin{array}{r}
0101 \\
+0110 \\
\hline
\underline{0}\ \ 1011
\end{array}
\qquad
\text{Answer is } -0101.
$$

1. b.

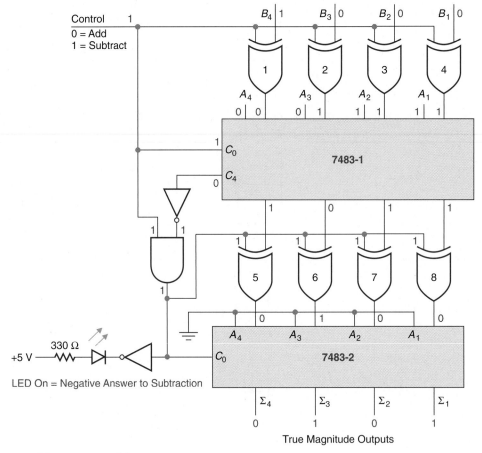

$$\begin{array}{r} 11 \\ -1000 \\ \hline \end{array}$$
$$\begin{array}{r} 11 \\ +1000 \\ \hline 0\quad 1011 \end{array}$$

Answer is -0101.

2. a.

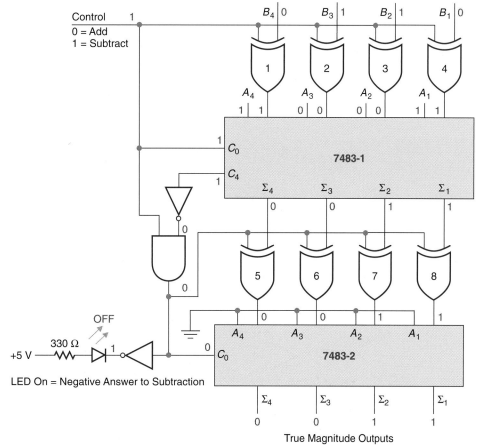

True Magnitude Outputs

$$\begin{array}{r} 1001 \\ -110 \\ \hline \end{array} \qquad \begin{array}{r} 1001 \\ +1010 \\ \hline 10011 \end{array} \qquad \text{Answer is } +11.$$

2. b.

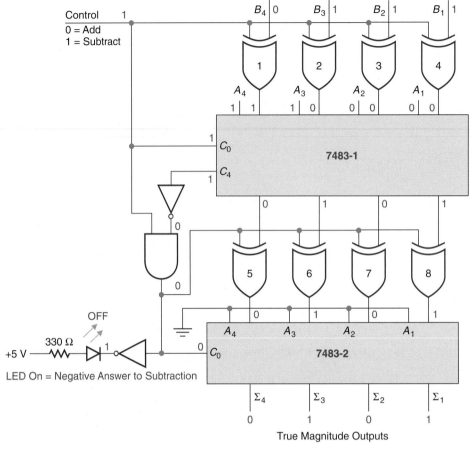

$$
\begin{array}{r}
1100 \\
-111 \\
\hline
\end{array}
\qquad
\begin{array}{r}
1100 \\
+1001 \\
\hline
1\ 0101
\end{array}
\qquad \text{Answer is } +101.
$$

Self-Check 4

1. $+11$ -74

2. 01100100   10011100

3. 1 01111101 (Incorrect)   11100110 (Correct)

4. 10001000 (Incorrect)   11100010 (Correct)

Self-Check 5

1. 1010   1011   1100   1101   1110   1111

2. If a carry-out is generated, or if the initial sum is not legitimate, 6 must be added.

3. a.

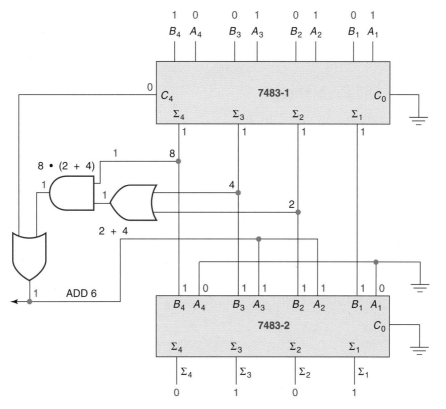

b.

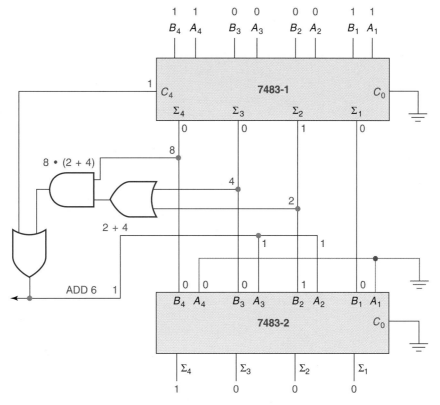

Self-Check 6

1. 1110 (-2 in 2's complement form)   **2.** $A \oplus B = 1100 \oplus 0100 = 1000$

Questions and Problems

1.

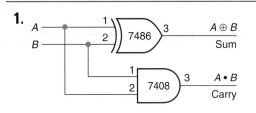

Inputs		Outputs	
A	**B**	**Carry**	**Sum**
0	0	0	0
0	1	0	1
1	0	0	1
1	1	1	0

3. a.
$$
\begin{array}{r}
1010 \\
-1000 \\
\end{array}
\qquad
\begin{array}{r}
1010 \\
+0111 \\
\hline
1\ \ 0001 \\
\text{EAC}\ \ 1 \\
\hline
+0010 \\
\end{array}
$$

b.
$$
\begin{array}{r}
10001 \\
-11101 \\
\end{array}
\qquad
\begin{array}{r}
10001 \\
+00010 \\
\hline
10011 \\
\end{array}
$$
$$
\begin{array}{r}
\text{1's complement} \\
-01100 \\
\end{array}
$$

5. a. 1111

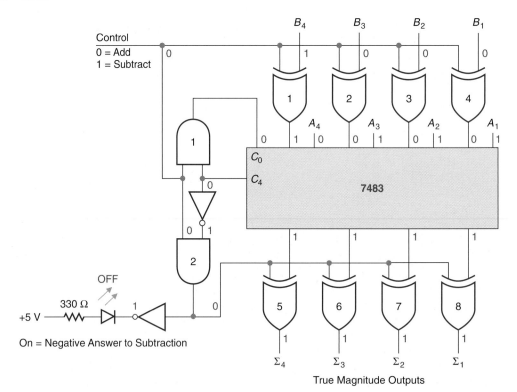

5. b. 0011

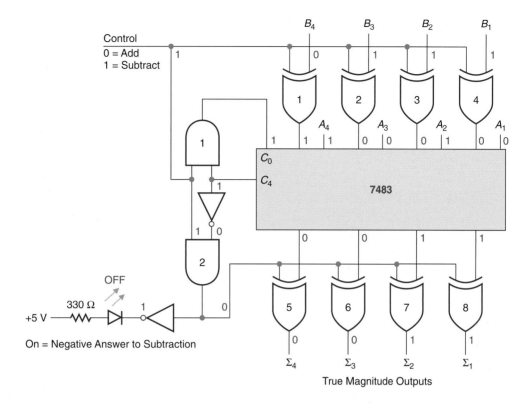

c. -0101

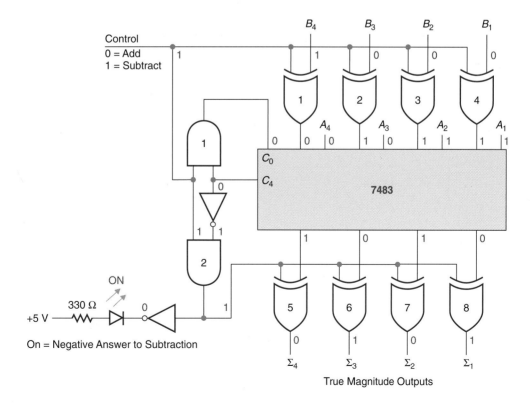

7. a. 100101   100101
    −1001   +110111
             1 011100

           +11100

b.   10101   10101
  −11000   +01000
            11101

    2's complement
            −00011

9. a.   0101
  +1000
   1101

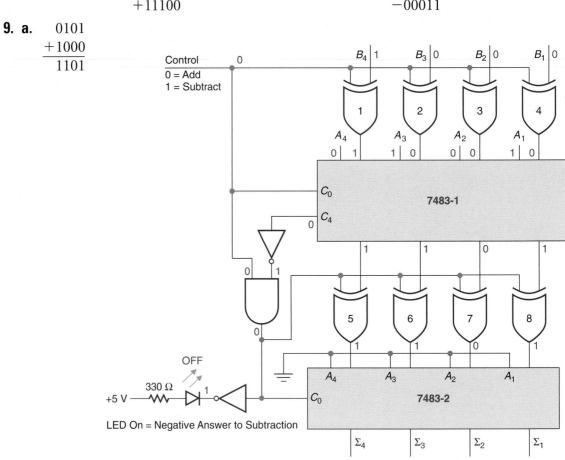

9. b.

$$
\begin{array}{r}
0101 \\
-0111 \\
\hline
\end{array}
$$

$$
\begin{array}{r}
1001 \\
+1001 \\
\hline
1 \quad 0010 \\
\end{array}
$$

$$+10$$

c.

$$
\begin{array}{r}
0011 \\
-1000 \\
\hline
\end{array}
$$

$$
\begin{array}{r}
0011 \\
+1000 \\
\hline
1011 \\
\end{array}
$$

$$-101$$

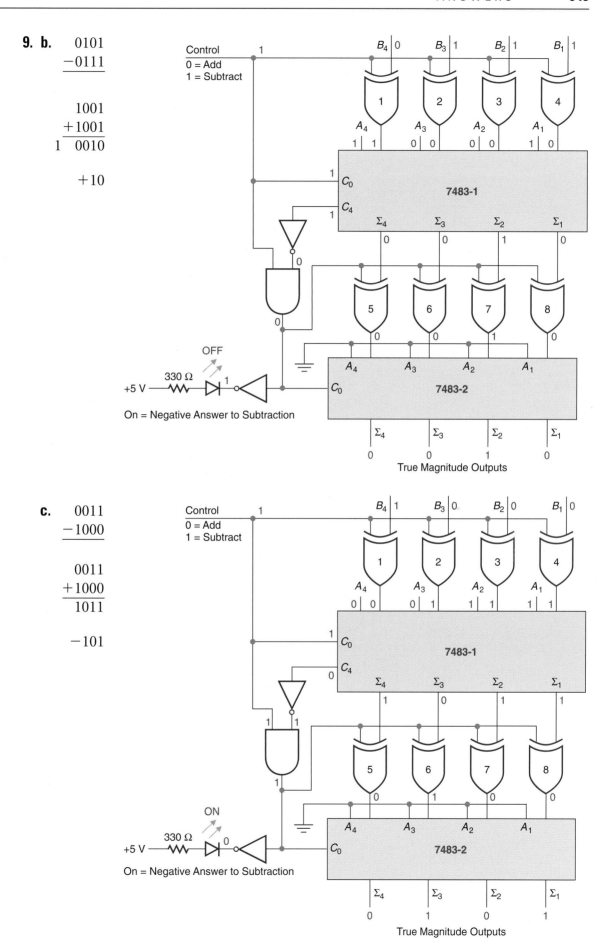

11. a.

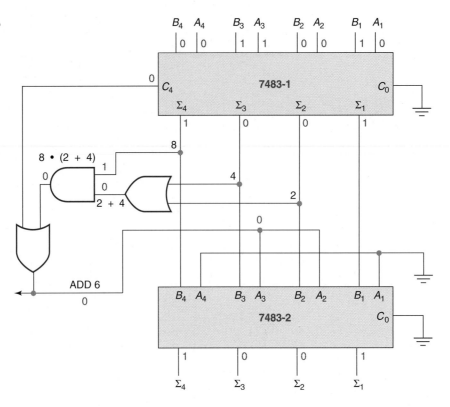

b. 1 0101

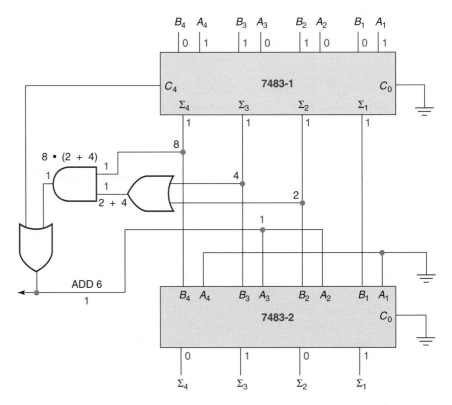

11. c. 1 0110

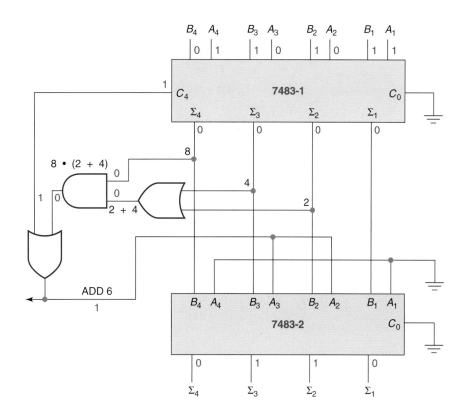

13.

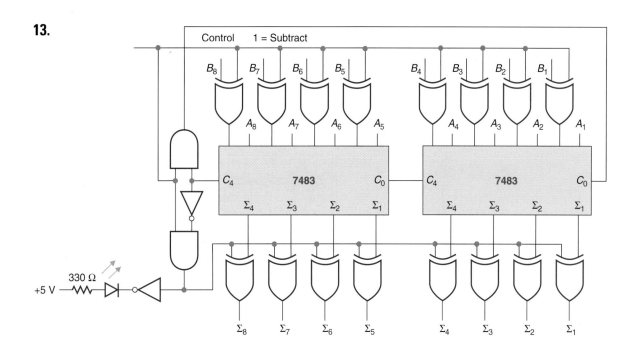

15.

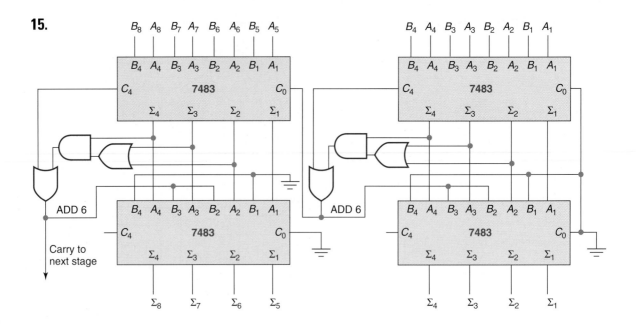

17. a.

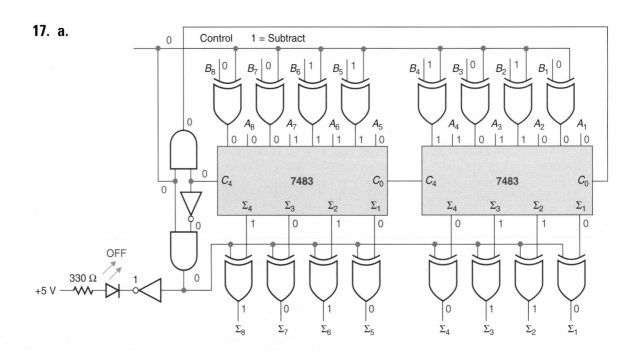

$$01101100$$
$$+00111010$$
$$\overline{10100110}$$

17. b.

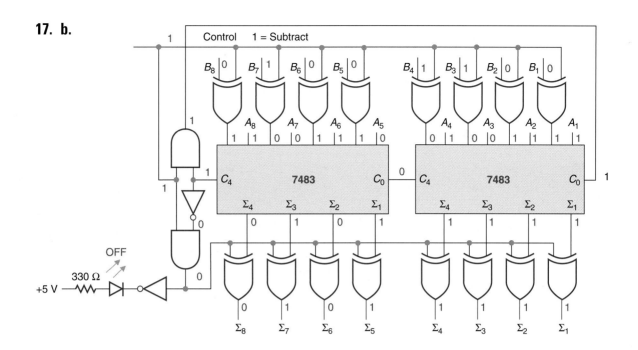

$$
\begin{array}{r}
10101011 \\
-01001100 \\
\hline
\end{array}
\qquad
\begin{array}{r}
10101011 \\
+10110011 \\
\hline
1\ 01011110 \\
\end{array}
$$

$$
\begin{array}{r}
\text{EAC} \qquad 1 \\
\hline
01011111
\end{array}
$$

c.

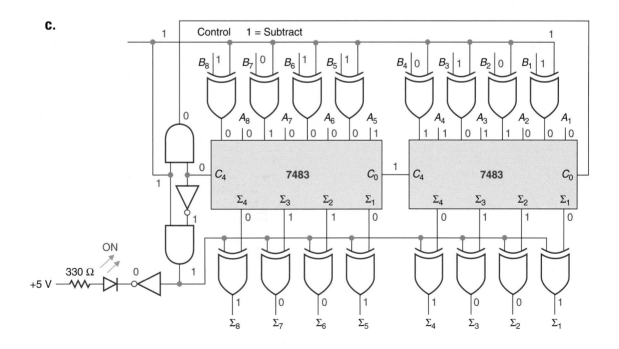

$$
\begin{array}{r}
00011100 \\
-10110101 \\
\hline
\end{array}
\qquad
\begin{array}{r}
00011100 \\
+01001010 \\
\hline
01100110 \\
\end{array}
$$

$$
\begin{array}{r}
-10011001
\end{array}
$$

19. a.

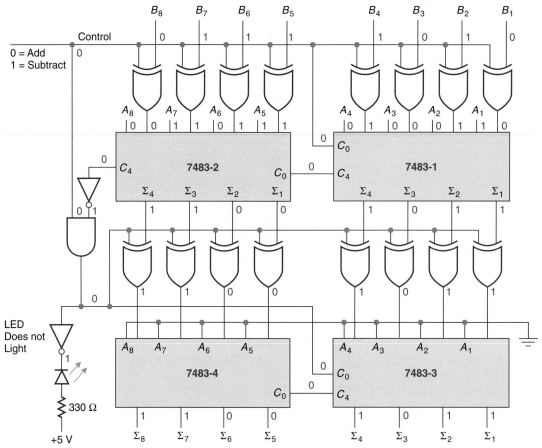

$$01010001$$
$$+01111010$$
$$\overline{+11001011}$$

19. b. 10100110

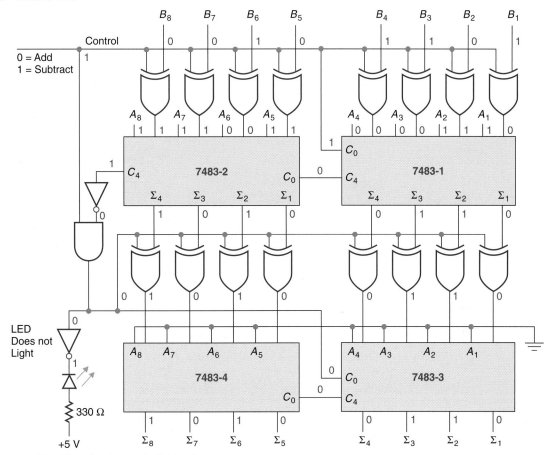

LED
Does not
Light

330 Ω

+5 V

On = Negative Answer to Subtraction

$$
\begin{array}{r}
11010011 \\
-00101101 \\
\hline
\end{array}
\qquad
\begin{array}{r}
11010011 \\
+11010011 \\
\hline
1\ \ 10100110 \\
\\
+10100110
\end{array}
$$

19. c. -00000001

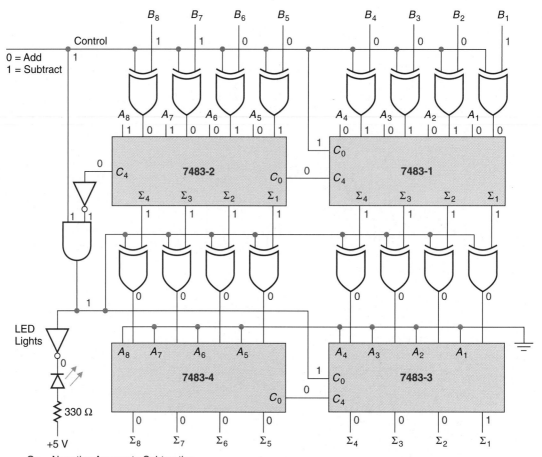

$$
\begin{array}{r}
11000000 \\
-11000001 \\
\hline
\end{array}
\qquad
\begin{array}{r}
11000000 \\
+00111111 \\
\hline
11111111 \\[4pt]
-00000001
\end{array}
$$

21. a.

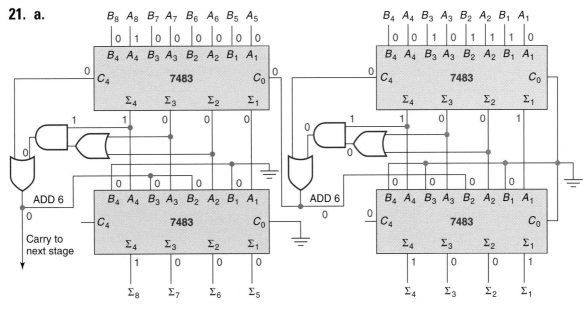

$$
\begin{array}{r}
10000010 \\
00000111 \\
\hline
10001001
\end{array}
\qquad
\begin{array}{r}
82 \\
+\ 7 \\
\hline
89
\end{array}
$$

b.

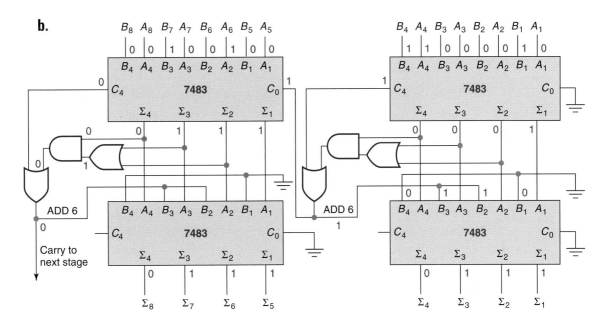

$$
\begin{array}{cc}
0010 & 1000 \\
0100 & 1001 \\
\hline
0111 & 0001 \\
& 0110 \\
\hline
& 0111
\end{array}
\qquad
\begin{array}{r}
28 \\
+49 \\
\hline
77
\end{array}
$$

The circuit is not designed to handle a subtraction problem. This is the solution to an addition problem.

21. c.

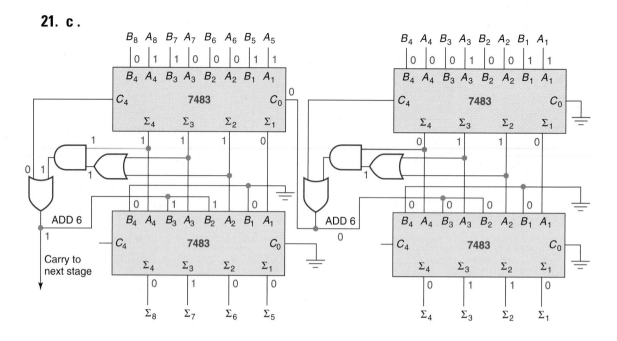

$$
\begin{array}{r}
1001\ 0101 \\
+0101\ 0001 \\
\hline
1110\ 0110 \\
+0110 \\
\hline
1\ 0100
\end{array}
\qquad
\begin{array}{r}
95 \\
+51 \\
\hline
146
\end{array}
$$

23. A 1 on the control line (subtract) enables AND gate 1. If $C_4 = 1$ (overflow), the output of AND gate 1 is 1. The 1 is fed into C_0 to accomplish the end-around carry.

25. A 1 on the control line (subtract) and $C = 0$ (no overflow) require that the sum be complemented to obtain the true magnitude of the answer. AND gate 2 provides a 1 in this case to cause Exclusive-OR gates 5, 6, 7, and 8 to invert the output of the 7483. The true magnitude appears at $\Sigma_4\Sigma_3\Sigma_2\Sigma_1$.

27. In a subtraction problem with no overflow, the 2's complement must be taken to obtain the true magnitude of the answer. In this case, the AND gate provides a 1, which causes Exclusive-OR gates 5, 6, 7, and 8 to invert the output of the 7483-1 to begin the 2's complement process.

29. The function of the 7483-2 is to add 1 during the 2's complement process. The 1 is added via the C_0, so inputs $A_4A_3A_2A_1$ are grounded.

31. The output of the AND gate is a 1 in a subtraction problem with no overflow. A 1 indicates that the 2's complement must be taken to obtain the true magnitude of the answer.

33. The sum of two BCD numbers can be a legitimate BCD number even though overflow has occurred. In that case, C_4 provides the ADD 6 signal.

35. A floating input on a TTL IC is often taken as a HIGH.

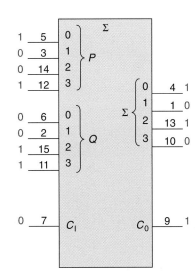

37. a. 50 = 00110010 binary   2's complement = 11001110

    b. 43 = 00101011 binary

    c. 2 = 00000010 binary   2's complement = 11111110

    d. 8 = 00001000 binary

    e. 120 = 01111000 binary   2's complement = 10001000

    f. 83 = 01010011 binary

39. a. 00010011   (answer correct)   61 − 42 = 19

    b. 00110000   (answer correct)   103 − 55 = 48

    c. 00110111   (answer incorrect; no carry from column 7, but overflow)
                   −100 − 101 ≠ 55

    d. 11111110   (answer incorrect; carry from column 7, but no overflow)
                   127 + 127 ≠ −2

41. F = 1001

43. F = 1111

45. S = 0110   Cn = 0   M = 0

47.
```
LIBRARY ieee;
USE ieee.std_logic_1164.all;
USE ieee.std_logic_unsigned.all;

ENTITY ADD_8 IS
    PORT (
        a:IN STD_LOGIC_VECTOR(7 DOWNTO 0);
        b:IN STD_LOGIC_VECTOR(7 DOWNTO 0);
        sum:OUT STD_LOGIC_VECTOR (7 DOWNTO 0));
END ADD_8;
ARCHITECTURE a OF ADD_8 IS
BEGIN
    sum <= a + b;
END a;
```

49. —negative answers are given in true magnitude form

```
LIBRARY ieee;
USE ieee.std_logic_1164.all;
USE ieee.std_logic_unsigned.all;
ENTITY SUB_4 IS
      PORT (
              a:IN STD_LOGIC_VECTOR(3 DOWNTO 0);
              b:IN STD_LOGIC_VECTOR(3 DOWNTO 0);
              diff:OUT STD_LOGIC_VECTOR (3 DOWNTO 0);
              neg:OUT STD_LOGIC);
END SUB_4;
ARCHITECTURE a OF SUB_4 IS
BEGIN
      PROCESS (a,b)
      BEGIN
              IF b > a THEN
                      neg <= '1';
                      diff <= b - a;
              ELSE    neg <= '0';
                      diff <= a - b;
              END IF;
      END PROCESS;
END a;
```

51.
```
LIBRARY ieee;
USE ieee.std_logic_1164.all;
USE ieee.std_logic_unsigned.all;
ENTITY twos_comp_4 IS
      PORT (
              A:IN STD_LOGIC_VECTOR(3 DOWNTO 0);
              Acomp:OUT STD_LOGIC_VECTOR (3 DOWNTO 0));
END twos_comp_4;
ARCHITECTURE a OF twos_comp_4 IS
BEGIN
      Acomp<=NOT(A)+ "0001";
END a;
```

53.
```
LIBRARY ieee;
USE ieee.std_logic_1164.all;
USE ieee.std_logic_unsigned.all;
ENTITY twos_comp_8 IS
      PORT (
              A:IN STD_LOGIC_VECTOR(7 DOWNTO 0);
              Acomp:OUT STD_LOGIC_VECTOR (7 DOWNTO 0));
END twos_comp_8;
ARCHITECTURE a OF twos_comp_8 IS
BEGIN
      Acomp<=NOT(A)+ "00000001";
END a;
```

55. —When control = 0, the output is a + b.

—When control = 1, the output is a - b.

—Negative output goes HIGH on a negative answer to a subtraction problem.

—A negative answer to a subtraction problem is in true magnitude form.

```
LIBRARY ieee;
USE ieee.std_logic_1164.all;
USE ieee.std_logic_unsigned.all;
ENTITY ADD_SUB_8 IS
     PORT (
           cntl:IN STD_LOGIC;
           a:IN STD_LOGIC_VECTOR(7 DOWNTO 0);
           b:IN STD_LOGIC_VECTOR(7 DOWNTO 0);
           sum_diff:OUT STD_LOGIC_VECTOR (7 DOWNTO 0);
           neg:OUT STD_LOGIC);
END ADD_SUB_8;
ARCHITECTURE a OF ADD_SUB_8 IS
BEGIN
     PROCESS (a,b,centl)
     BEGIN
          IF cntl = '1' THEN
                IF b > a THEN
                     neg <= '1';
                     sum_diff <= b - a;
                ELSE neg <= '0';
                     sum_diff <= a - b;
                END IF;
          ELSIF cntl = '0' THEN
                     sum_diff <= a + b;
                     neg <= '0';
          END IF;
     END PROCESS;
END a;
```

57. line 3: Insert "USE ieee.std_logic_unsigned_all;"

line 9: Insert "a". ARCHITECTURE a OF

line 10: Declare signal "intsum".

```
          SIGNAL intsum STD_LOGIC_VECTOR (4 DOWNTO 0);
```

line 13: sum is 4 bits wide. sum <= intsum (3 DOWNTO 0);

line 14: END a:

CHAPTER 6

Self-Check 1

1. **a.** V_{OH} = 2.4 VOLTS MIN
   b. V_{IH} = 2.0 VOLTS MIN
   c. V_{OL} = 0.4 VOLTS MAX
   d. V_{IL} = 0.8 VOLTS MAX
   e. I_{CC} = 79 mA MAX
   f. t_{PLH} = 47 ns MAX (C0 to 3)
   g. t_{PHL} = 47 ns MAX (C0 to 4)
   h. I_{OL} = 16 mA MAX (except C4)
   i. I_{IL} = −3.2 mA

2. **a.** Noise margin = 0.4 V   **c.** 16/3.2 ≈ 5
   b. fan-out = 10

3. A pull-up resistor is a resistor that is used to provide a path to the power supply voltage. Open-collector gates do not have an internal path to the power supply, so an external path through a pull-up resistor must be provided.

Self-Check 2

1. **a.** V_{OL} = 0.26 V   **c.** I_{IN} = −0.1 uA   **e.** I_{OUT} = 5.2 mA
   b. V_{OH} = 5.48 V   **d.** I_{OUT} = 5.2 mA

2. fan-out = 5200

3. **a.**

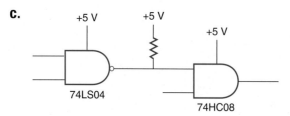

   b.

   c.

4. LVT is compatible with TTL.
5. LV and LVC can operate at 1.2 V.

Self-Check 3

1. V_{CC1} supplies current to the switching circuitry. V_{CC2} supplies current to the output stage. They are both grounded, which makes V_{EE} negative.
2. V_{EE} for a 10K IC is -5.2 V, and V_{EE} for a 100K IC is -4.5 V.
3. 10125
4. ECL is the fastest of the logic families. The 100K series is faster than the 10K series.
5. ECL consumes more power than the other logic families.
6. A gullwing lead extends downward and outward from an IC.
7. A J-lead extends downward from an IC and loops back under the IC in a J shape.
8. An LCC has no external lead protruding from the IC. Connection from printed circuit board to IC is through solder only. The SO and PLCC have external leads that solder to the surface of the printed circuit board.

Questions and Problems

1. 2.7 V
5. 18 ns
9. 4.6 mA

3. 5
7. 4.95 V

11.

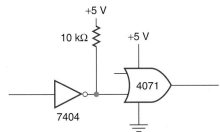

13.

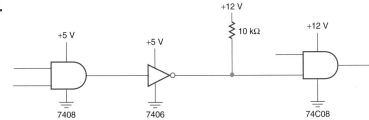

15.

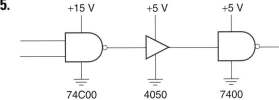

17.

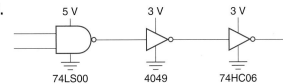

19. 4.5 to 5.5 V. **21.** Needs a pull-up resistor **23.** 10

25. ECL, AS, CMOS AC, FAST, S, ALS, LS, TTL, CMOS HC, CMOS

27. 0.001 Microamps **29.** 0.05 V

31.

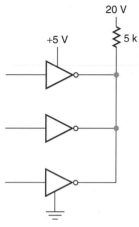

33.

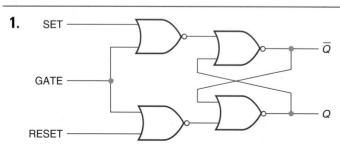

35. CE HIGH
DIR HIGH

37. Small Outline (SO) and Plastic Leaded Chip Carrier (PLCC)

39. A lower supply voltage reduces the power consumption of the circuit. Equipment can be smaller and lighter and run cooler.

41. LVT devices can sink 64 mA.

43. LVC and LVT devices can accept input signals from TTL devices.

45. LV, LVC, and LVT subfamilies offer basic gates.

47. Philips makes 74HC1Gxx and 74HCT1Gxx single-gate ICs, including AND, NAND, OR, and NOR 2-input gates and inverters. Pins 1 and 3 are inputs, 4 is output, 5 is V_{CC} and 2 is ground. Philips single-gate ICs have typical HC and HCT characteristics. Toshiba manufactures single-gate ICs that measure 1 mm × 1 mm, with a pin pitch of 0.35 mm.

CHAPTER 7

Self-Check 1

1. $Q = 0, \overline{Q} = 1$ **3.** $Q = 1, \overline{Q} = 0$
2. $Q = 0, \overline{Q} = 1$ **4.** $Q = 1, \overline{Q} = 1$, the unused state

Self-Check 2

1.

SET

GATE

RESET

$\overline{Q}$

Q

2.

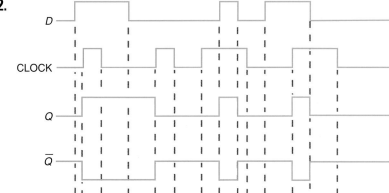

3.

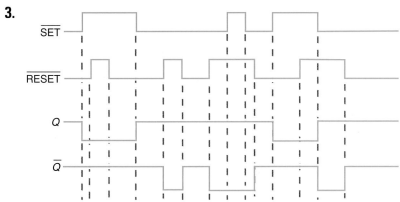

Questions and Problems

1.

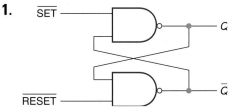

3.

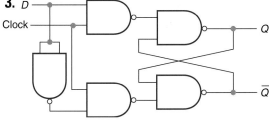 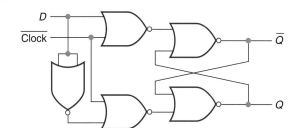

5.

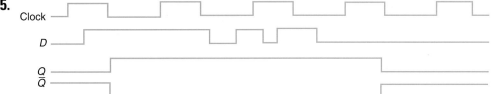

7.

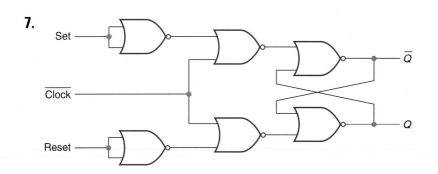

9.

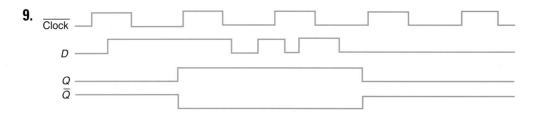

11.

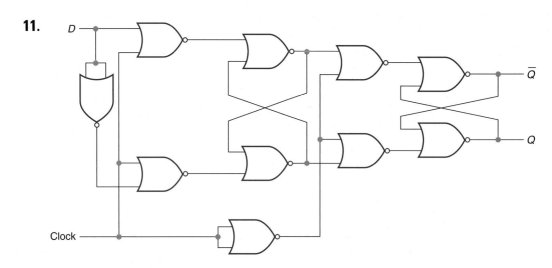

13.

15.

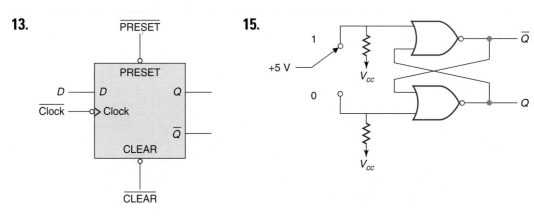

17.

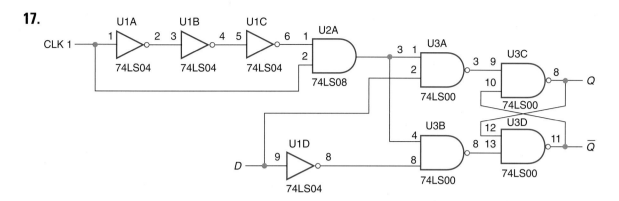

19. $Q = 1$

21. The oscilloscope will show the true waveform and not the average voltage at the pin.

23. $0 = .8$ volts or less and $1 = 2$ volts or greater on a TTL input.

25. They would both be high or logic 1.

27. No, not without external components, because the GAL does not have hysteresis as does a Schmitt trigger.

29. The slave part would not change states because the gating NAND gates would be disabled with a 0 on their inputs.

31.
```
ENTITY crossed_nor_ff IS
      PORT (
            reset,set:IN BIT;
            Q1,Q2:BUFFER BIT);
END crossed_nor_ff;
ARCHITECTURE a OF crossed_nor_ff IS
BEGIN
      PROCESS (set,reset)
      BEGIN
            IF set = '1' THEN
                  Q1 <= '1';
            ELSIF reset = '1' THEN
                  Q1 <= '0';
            END IF;
      END PROCESS;
      Q2 <= NOT Q1;
END a;
```

33.
```
line 1: need to involve the ieee library to use std_logic.Add:

      Library ieee;
      USE ieee.std_logic_1164.all;

line 1, 4, 5: Cannot use two underscores together in a name. Change
            set__reset to set_reset
line 7, 10, 13, 16: change "res" to "reset" to match input declaration
            in line 2.
```

CHAPTER 8

Self-Check 1

1. Connect the Q output to the D input.   **3.** $Q = 0$, $\overline{Q} = 1$

2. $J = 1$, $K = 1$

Self-Check 2

1.

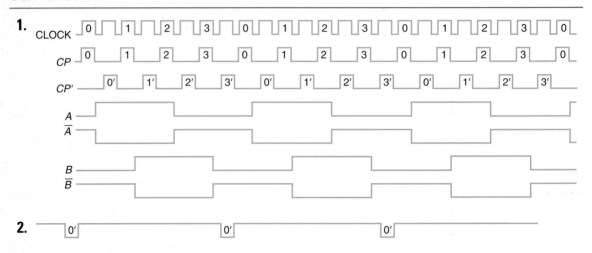

2.

Questions and Problems

1.

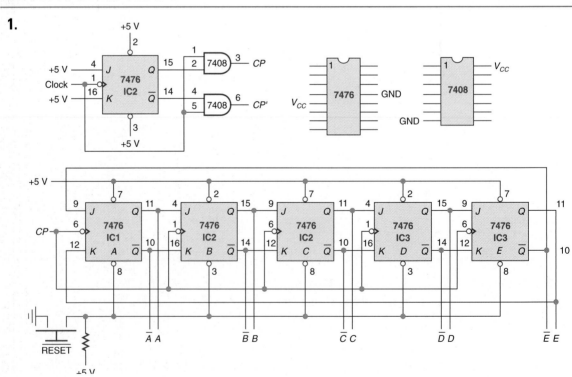

3.

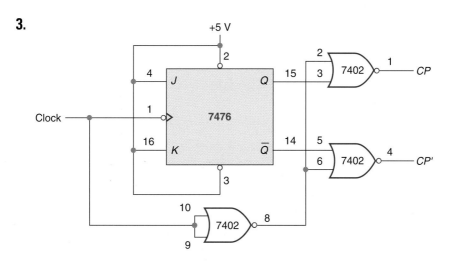

5.

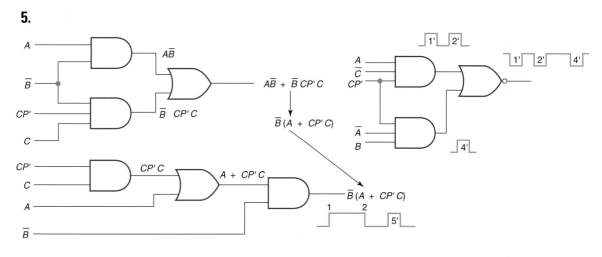

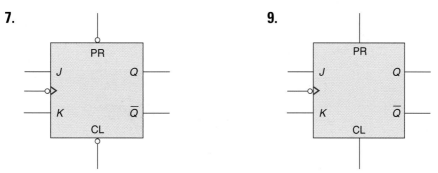

7.   **9.**

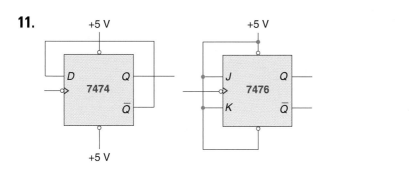

11.

13.

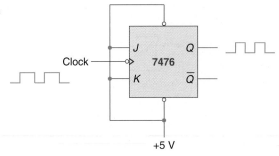

15. The same as the frequency of *CP*.

17.

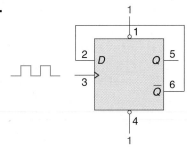

19.

21. The flip-flop would be positive edge triggered.

23. *J* and *K* must both be 0.

25. The film advance would be turned off and the film advance button would not change the state of the circuit.

27.

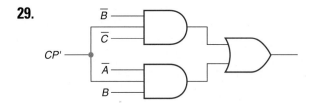

29.

31.
```
LIBRARY ieee;
USE ieee.std_logic_1164.all;
ENTITY TOGGLE_L_to_H IS
     PORT (
          clk:IN BIT;
          Q1,Q2:BUFFER BIT);
END TOGGLE_L_to_H;
ARCHITECTURE a OF TOGGLE_L_to_H IS
BEGIN
     PROCESS (clk,Q1,Q2)
     BEGIN
          IF clk'EVENT AND clk = '1' THEN
                  Q1 <= Q2;
          ELSE Q1 <= Q1;
          END IF;
     END PROCESS;
     Q2 <= NOT Q1;
END a;
```

33.
```
LIBRARY ieee;
USE ieee.std_logic_1164.all;
ENTITY delayed_clock IS
     PORT (
          clk:IN STD_LOGIC;
          cp1,cp2:OUT STD_LOGIC);
END delayed_clock;
ARCHITECTURE a OF delayed_clock IS
     SIGNAL Q1:STD_LOGIC;
     SIGNAL Q2:STD_LOGIC;
BEGIN
     PROCESS (clk,Q1,Q2)
     BEGIN
          IF clk'EVENT AND clk = '1' THEN
                  Q1 <= Q2;
          ELSE Q1 <= Q1;
          END IF;
     Q2 <= NOT Q1;
     cp1 <= Q1 AND clk;
     cp2 <= Q2 AND clk;
     END PROCESS;
END a;
```

35. line 5: Q1 and Q2 should be declared as BUFFERs.
```
          Q1,Q2:BUFFER STD_LOGIC);
```

line 7: insert "a". ARCHITECTURE a OF FF IS

line 12, 18; definitions reversed.

Line 20: insert "END IF;"

CHAPTER 9

Self-Check 1

1.

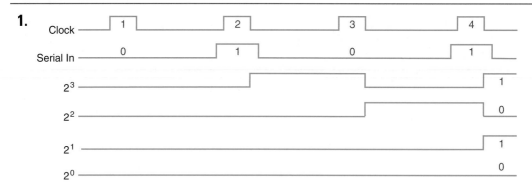

2.

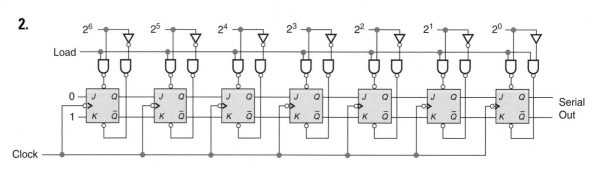

3.

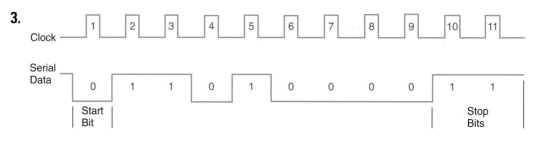

Self-Check 2

1. Recommended Standard

2.

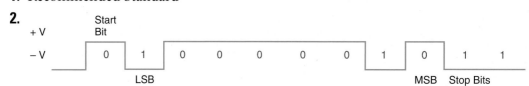

3.

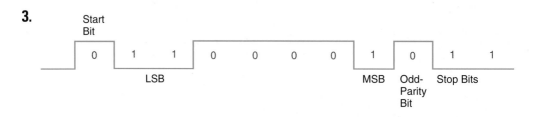

Questions and Problems

1.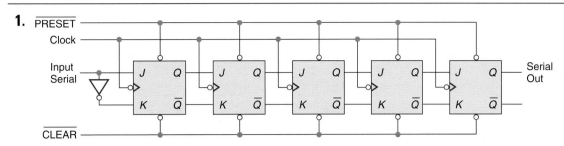

3.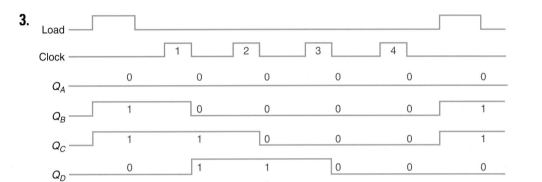

5.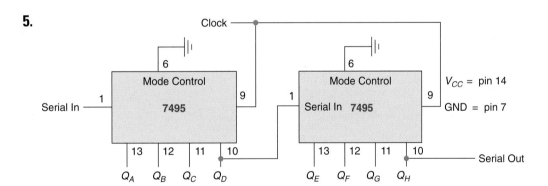

7.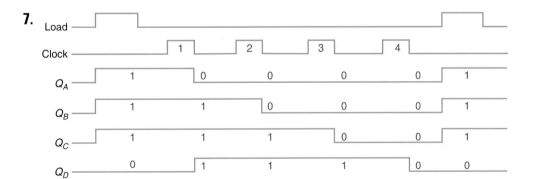

9.

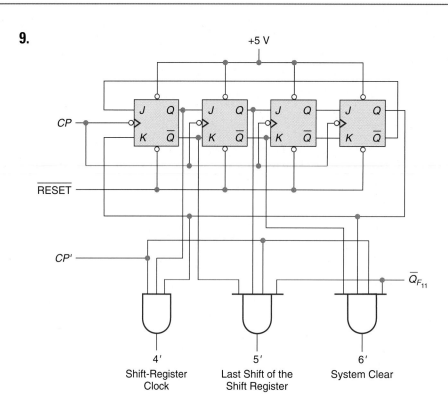

Latch Strobe

11.

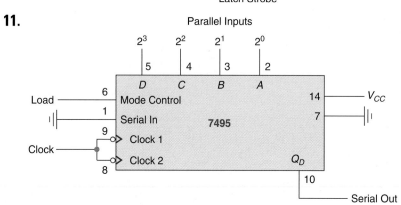

13.

D	I	G	I	T	A	L
44	49	47	49	54	41	4C

E	L	E	C	T	R	O	N	I	C	S
45	4C	45	43	54	52	4F	4E	49	43	53

15.

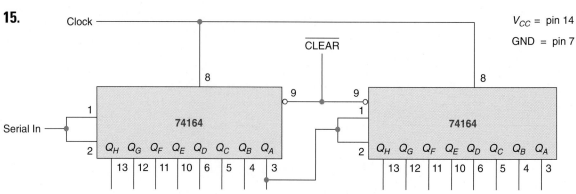

17. 128 or 2^7 **19.** NO!   **21.** 1.04 ms

23.

25. Pin 7   **27.** 00 hex   **29.** Yes

31.
```
LIBRARY ieee;
USE ieee.std_logic_1164.all;
ENTITY shift_counter IS
    PORT(
        clk,reset:IN STD_LOGIC;
        A1,A2,B1,B2,C1,C2:BUFFER STD_LOGIC);
END shift_counter;
ARCHITECTURE a OF shift_counter IS
BEGIN
    PROCESS (clk,reset,A1,B1,C1)
    BEGIN
        IF reset <= '0' THEN
                A1 <= '0';
                B1 <= '0';
                C1 <= '0';
            ELSIF reset <= '1' THEN
                IF clk'EVENT AND clk = '1' THEN
                        B1 <= A1;
                        C1 <= B1;
                        A1 <= NOT C1;
                END IF;
            END IF;
                    A2 <= NOT A1;
                    B2 <= NOT B1;
                    C2 <= NOT C1;
    END PROCESS;
END a;
```

33.
```
LIBRARY ieee;
USE ieee.std_logic_1164.all;
ENTITY kit IS
    PORT(
        clk,reset:IN STD_LOGIC;
        Q:BUFFER STD_LOGIC_VECTOR (7 DOWNTO0));
END kit;
ARCHITECTURE a OF kit IS
    SIGNAL direction:STD_LOGIC;
BEGIN
    PROCESS (clk,reset,Q,direction)
    BEGIN
        IF reset <= '0' THEN
                Q<= "10000000";
                direction <= '0';
```

```
                            ELSIF reset <= '1' THEN
                                IF clk'EVENT AND clk = '1' THEN
                                    IF direction = '0' THEN
                                        IF Q = "10000000" THEN
                                            Q <= "01000000";
                                        ELSIF Q = "01000000" THEN
                                            Q <= "00100000";
                                        ELSIF Q = "00100000" THEN
                                            Q <= "00010000";
                                        ELSIF Q = "00010000" THEN
                                            Q <= "00001000";
                                        ELSIF Q = "00001000" THEN
                                            Q <= "00000100";
                                        ELSIF Q = "00000100" THEN
                                            Q <= "000000100";
                                        ELSIF Q = "00000001" THEN
                                            direction <= '1''
                                        END IF;
                                    ELSIF direction = '1' THEN
                                        IF Q = "00000001" THEN
                                            Q <= "00000010";
                                        ELSIF Q = "00000010" THEN
                                            Q <= "00000100";
                                        ELSIF Q = "00000100" THEN
                                            Q <= "00001000";
                                        ELSIF Q = "00001000" THEN
                                            Q <= "00010000";
                                        ELSIF Q = "00010000" THEN
                                            Q <= "00100000";
                                        ELSIF Q = "00100000" THEN
                                            Q <= "01000000";
                                        ELSE  Q <= "10000000";
                                            direction <= '0';
                                        END IF;
                                    END IF;
                                END IF;
                            END IF;
                    END PROCESS;
                END a;

35. LIBRARY ieee;
    USE ieee.std_logic_1164.all;
    USE ieee.std_logic_unsigned.all;
    ENTITY half_8 IS
        PORT (
            load:IN STD_LOGIC;
            half:IN STD_LOGIC;
            I:IN STD_LOGIC_VECTOR (7 DOWNTO 0);
            Q:BUFFER STD_LOGIC_VECTOR (7 DOWNTO 0));
    END half_8;
```

```
ARCHITECTURE a OF half_8 IS
BEGIN
     PROCESS (half,Q,load,I)
     BEGIN
         IF load ='1' THEN
                  Q <= I;
         ELSIF load = '0' THEN
                  IF half'EVENT AND half ='1' THEN
                          Q(7) <= '0';
                          Q(6) <= Q(7);
                          Q(5) <= Q(6);
                          Q(4) <= Q(5);
                          Q(3) <= Q(4);
                          Q(2) <= Q(3);
                          Q(1) <= Q(2);
                          Q(0) <= Q(1);
                  END IF;
              END IF;
     END PROCESS;
END a;
```

37. line 4: Q is single-bit. Q:OUT BIT;
line 8: Declare number of bits. BIT_VICTOR (7 DOWNTO 0);
line 15: Need ' in event statement. IF (clk'event AND
line 27: Specify which bit. Q(0) <= internal(0);

CHAPTER 10

Self-Check 1

1.

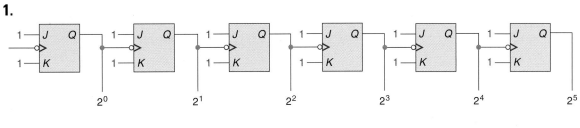

2.

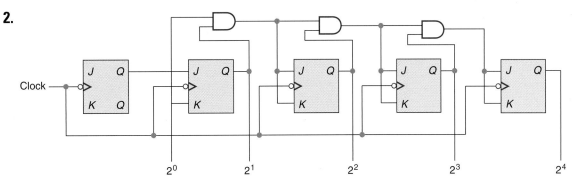

3.

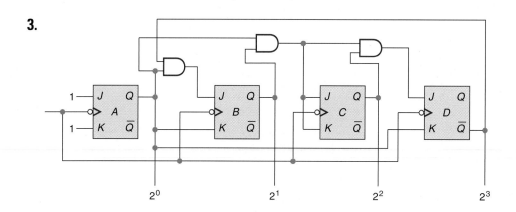

Q Before Clock				Q After Clock				D		C		B		A	
D	C	B	A	D	C	B	A	J	K	J	K	J	K	J	K
0	0	0	0	0	0	0	1	0	X	0	X	0	X	1	X
0	0	0	1	0	0	1	0	0	X	0	X	1	X	X	1
0	0	1	0	0	0	1	1	0	X	0	X	X	0	1	X
0	0	1	1	0	1	0	0	0	X	1	X	X	1	X	1
0	1	0	0	0	1	0	1	0	X	X	0	0	X	1	X
0	1	0	1	0	1	1	0	0	X	X	0	1	X	X	1
0	1	1	0	0	1	1	1	0	X	X	0	X	0	1	X
0	1	1	1	1	0	0	0	1	X	X	1	X	1	X	1
1	0	0	0	1	0	0	1	X	0	0	X	0	X	1	X
1	0	0	1	0	0	0	0	X	1	0	X	1	X	X	1

A: $J = 1$
$\quad K = 1$

B: $J = A\overline{D}$
$\quad K = A$

C: $J = AB$
$\quad K = AB$

D: $J = ABC$
$\quad K = A$

4.

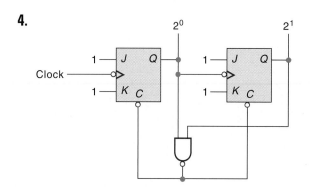

Self-Check 2

1.

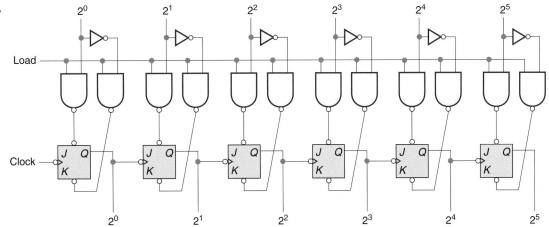

2.

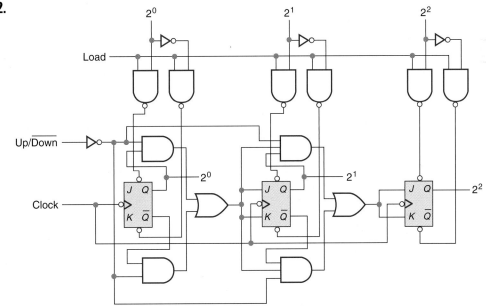

Questions and Problems

1.

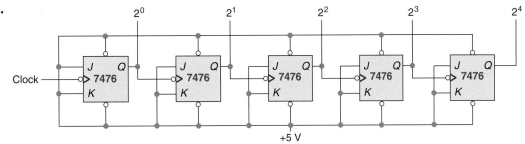

3.

Before Clock	After Clock	Before Clock	
Q	Q	J	K
0	0	0	X
0	1	1	X
1	0	X	1
1	1	X	0

Negative-Edge *JK* Flip-Flop
X = 1 or 0

Before Clock			After Clock			Before Clock					
Q			Q			C		B		A	
C	B	A	C	B	A	J_C	K_C	J_B	K_B	J_A	K_A
0	0	0	1	0	0	1	X	0	X	0	X
1	0	0	0	0	1	X	1	0	X	1	X
0	0	1	0	0	0	0	X	0	X	X	1

$J_C = \overline{A}$ $J_B = 0$ $J_A = C$

$K_C = 1$ $K_B = 1$ $K_A = 1$

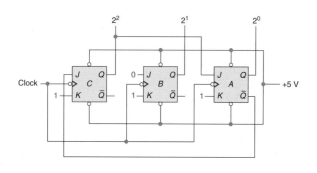

Clock Count			Outputs			
			2^2	2^1	2^0	
0	0	0	0	0	0	
0	0	1	1	0	0	Cycle 1
0	1	0	0	0	1	
0	1	1	0	0	0	
1	0	0	1	0	0	Cycle 2
1	0	1	0	0	1	
1	1	0	0	0	0	

Note: The *B* Flip-Flop could be eliminated

5.

CMOS	TTL
74C74	7474
74C174	74174
74C175	74175
74C374	74374
4013	
4027	
4042	
40174	
4723	

7. The answer is the same as for problem 3.

9.

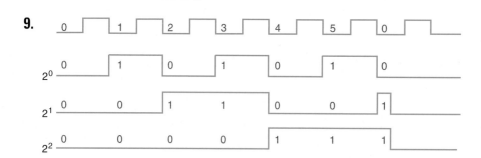

11.

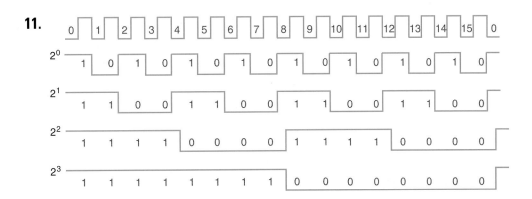

13.

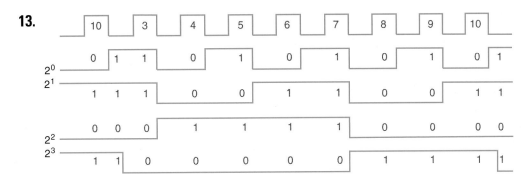

15.

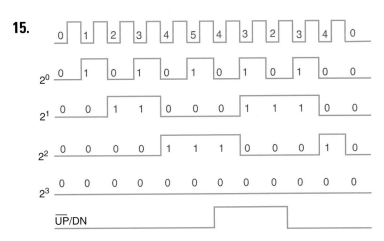

17.

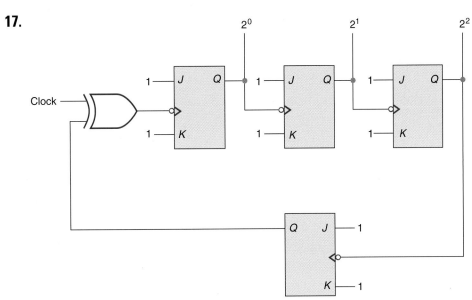

19. The speed of the counter is limited to the propagation delays of all the flip-flops. The decoder will produce a small spike in some outputs.

21. The counter would clear on the count of 8 and not on 9 because the disconnected NAND gate input would look like a one.

23. The count would not be correct for the *J* input of flip-flop *B*.

25.

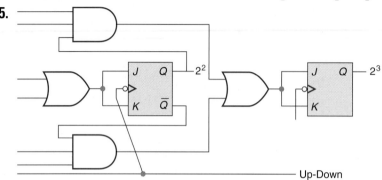

27. 60 Hz

29. No, they cannot toggle.

31.
```
LIBRARY ieee;
USE ieee.std_logic_1164.all;
USE ieee.std_logic_unsigned.all;
ENTITY divide_by_6 IS
     PORT(
          clk:IN STD_LOGIC;
          Q:OUT STD_LOGI);
END divide _by_6;
ARCHITECTURE a OF divide_by_6 IS
     SINGAL count:STD_LOGIC_VECTOR (2 DOWNTO 0);
BEGIN
     PROCESS(clk,count)
     BEGIN
          IF clk'EVENT AND clk = '1' THEN
               IF count = "101" THEN
                    count <= "000";
               ELSE count <= count + "001";
               END IF;
          END IF;
     END PROCESS;
     Q <= count(2);
END a;
```

33.
```
LIBRARY ieee;
USE ieee.std_logic_1164.all;
USE ieee.std_logic_unsigned.all;
ENTITY divide_by IS
     PORT(
          clk:IN STD_LOGIC;
          Q32,Q16,Q8,Q4,Q2:OUT STD_LOGIC);
```

```
END divide_by;
ARCHITECTURE a OF divide_by IS
     SIGNAL count:STD_LOGIC_VECTOR (4 DOWNTO 0);
BEGIN
     PROCESS (clk,count)
     BEGIN
         IF clk'EVENT AND clk = '1' THEN
                count <= Count + "0001";
         END IF;
     END PROCESS;
     Q32 <= count(4);
     Q16 <= count(3);
     Q8 <= count(2);
     Q4 <= count(1);
     Q2 <= count(0);
END a;
```

35.
```
LIBRARY ieee;
USE ieee.std_logic_1164.all;
USE ieee.std_logic_unsigned.all;
ENTITY down_count_8 IS
     PORT(
         clk:IN STD_LOGIC;
         Q:BUFFER STD_LOGIC_VECTOR (7 DOWNTO 0));
END down_count_8;
ARCHITECTURE a OF down_count_8 IS
BEGIN
     PROCESS (clk,Q)
     BEGIN
         IF clk'EVENT AND clk = '1' THEN
                Q <= Q - "00000001";
         END IF;
     END PROCESS;
END a;
```

37.
```
LIBRARY ieee;
USE ieee.std_logic_1164.all;
USE ieee.std_logic_unsigned.all;
ENTITY up_down_count_8 IS
     PORT(
         control:IN STD_LOGIC;
         clk:IN STD_LOGIC;
         Q:BUFFER STD_LOGIC_VECTOR (7 DOWNTO 0));
END up_down_count_8;
ARCHITECTURE a OF up_down_count_8 IS
BEGIN
     PROCESS (control,clk,Q)
```

```
        BEGIN
            IF clk'EVENT AND clk = '1' THEN
                    IF control <= '0' THEN
                            Q <= Q - "00000001";
                    ELSE
                            Q <= Q + "00000001";
                    END IF;
            END IF;
        END PROCESS;
    END a;
```

39. line 9: Q must be defined as signal or variable. SIGNAL Q:STD_LOGIC
line 14: Q <= "0000";
line 16: Parentheses omitted. IF Q = "1001" THEN
line 21: Q is single-bit. Specify which bit. clk2 <= Q(3);

CHAPTER 11

Self-Check 1

1. 4.527 KHz

2.

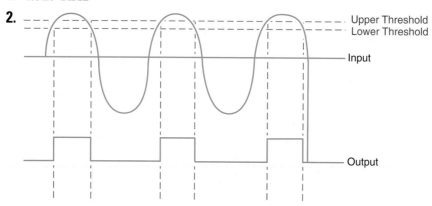

3. 1 volt

Self-Check 2

1. .32 UF

2.

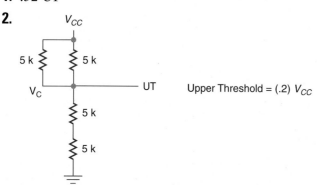

Upper Threshold = (.2) V_{CC}

Questions and Problems

1.

	Min.	Typ.	Max.
Upper Threshold	6.0 V	6.8 V	8.6 V
Lower Threshold	1.4 V	3.2 V	4.0 V

3.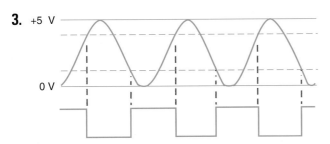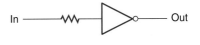

5. $F = \dfrac{0.66}{RC}$

7. When the capacitor charges, it charges through the resistor and the input to the Schmitt-trigger inverter. When the capacitor discharges, it discharges through only the resistor and therefore takes longer.

9.

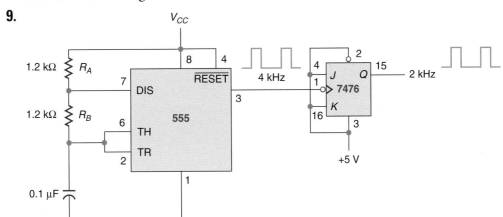

11. $F = \dfrac{3.45}{C(2R_B + R_A)}$

13.

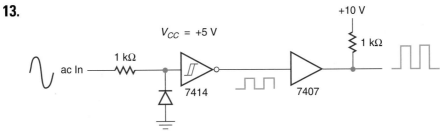

15. $F = \dfrac{0.529}{RC}$

17. If the external resistor is too large, the feedback voltage to the Schmitt trigger will not fall low enough to cross the lower threshold.

19. LT = .8 V   UT = 1.8 V

21. Lower threshold of $\frac{1}{3}$ V_{CC} and an upper threshold of $\frac{2}{3}$ V_{CC}

23. Open collector

25. The input impedance of the input to a CMOS device is very high and will not affect the operation of the *RC* timing used to create oscillation.

27.

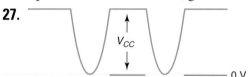

29. The *D* flip-flop would stop toggling and the 500-Hz signal would stop.

CHAPTER 12

Self-Check 1

1. .079 μF

2. A one-shot that will start a new timing cycle each time a trigger occurs.

Self-Check 2

1. R = 100 KΩ   C = 9.1 μF

2.

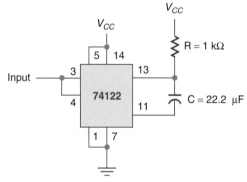

Questions and Problems

1. 34.184 ms   **3.** 78.74 kHz

5.

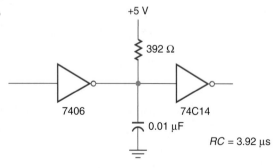

7.

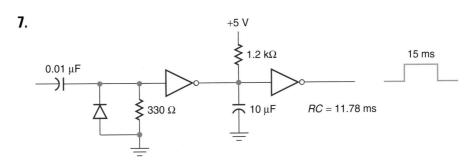

9.

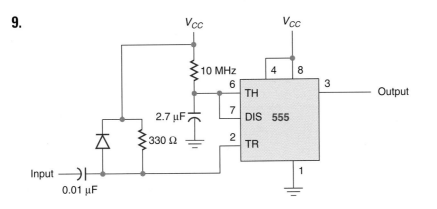

11. $22.\overline{2}$ kHz

13. 74C221, 4528, 4521, 4047

15. R = kΩ   C = 1.1 μF

17. 2.88 msec.

19. 5814 Ω

21. The pulse stretch would get longer.

23. The pulse width of the one-shot made from the *RC* timing and the *D* flip-flop would get shorter. This is because the TTL input would lower the resistance of the *RC* timing.

25. The 7406 could be damaged.

27. To prevent the input voltage to the LM339 from going below ground by a great amount.

29.

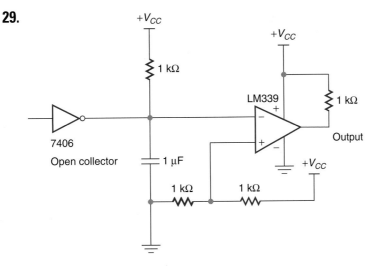

CHAPTER 13

Self-Check 1

1. Large binary numbers are not practical. Accurate resistor values are not easy to obtain.
2. .1875 volts
3. **a.** The voltage will not reach absolute ground on the $2R$ ladder.
   b. The pull-up resistors add error.

Self-Check 2

1. Successive approximation
2.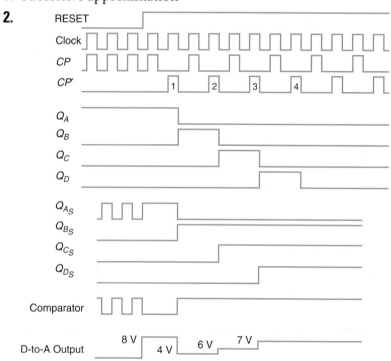

3.

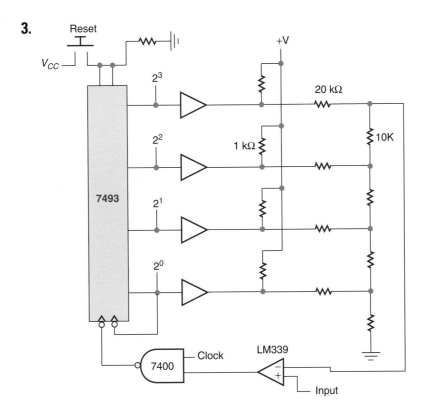

Questions and Problems

1.

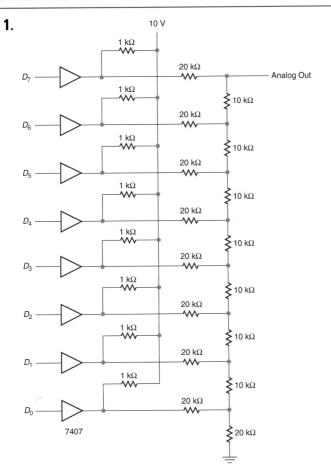

3. If $V_S = 5$ V, the voltage increment is 0.3125 V.

If $V_S = 10$ V, the voltage increment is 0.625 V.

If $V_S = 32$ V, the voltage increment is 2.0 V.

5.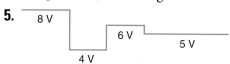

7. The purpose of the op amp is to prevent the $2R$ D-to-A converter from being loaded and distorting its output.

9.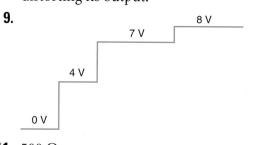

11. 500 Ω

13. $V_S = 22.5$ V

15.

17.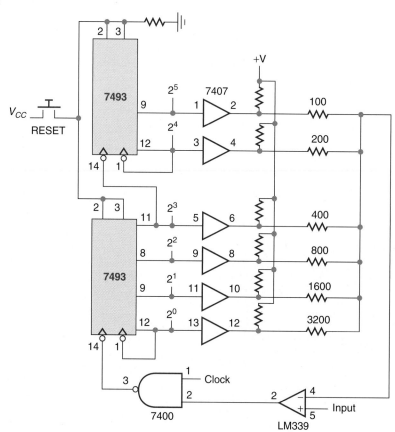

19. .059 V

21. There would be 8 steps in the output instead of 16.

23. The output would count from 0 to F and start over again.

25. The A-to-D converter would not start to convert the voltage because the flip-flops would not be set if they are left at a logic one.

27. This would be done by adding another flip-flop to the shift register and another digit to the least significant digit of the converter.

29. This cannot be determined accurately because the *RC* network has a pot which will affect the timing of the circuit.

CHAPTER 14

Self-Check 1

1.

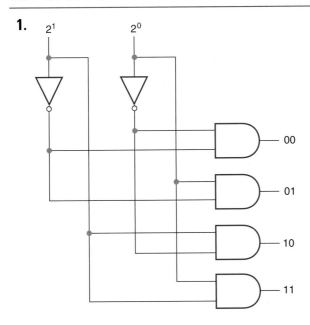

2.

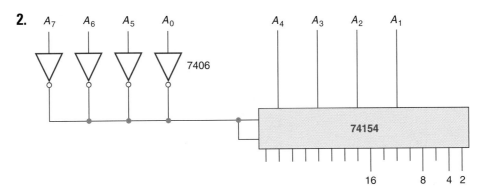

3.

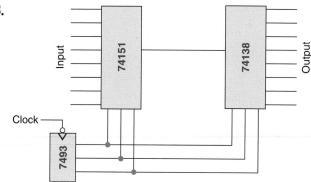

Self-Check 2

1.

2. Not enough current drive

3. 1.75 V

Questions and Problems

1.

2^2 2^1 2^0

000
001
010
011
100
101
110
111

3.

	E	D	C	B	A	Y	INPUT	VALUE
0	0	0	0	0	0	0	0	A
1	0	0	0	0	1	1		
2	0	0	0	1	0	1	1	1
3	0	0	0	1	1	1		
4	0	0	1	0	0	0	2	0
5	0	0	1	0	1	0		
6	0	0	1	1	0	1	3	$\overline{A}$
7	0	0	1	1	1	0		
8	0	1	0	0	0	1	4	$\overline{A}$
9	0	1	0	0	1	0		
10	0	1	0	1	0	0	5	A
11	0	1	0	1	1	1		
12	0	1	1	0	0	1	6	1
13	0	1	1	0	1	1		
14	0	1	1	1	0	0	7	0
15	0	1	1	1	1	0		

	E	D	C	B	A	Y	INPUT	VALUE
16	1	0	0	0	0	0	8	0
17	1	0	0	0	1	0		
18	1	0	0	1	0	0	9	0
19	1	0	0	1	1	0		
20	1	0	1	0	0	1	10	1
21	1	0	1	0	1	1		
22	1	0	1	1	0	0	11	A
23	1	0	1	1	1	1		
24	1	1	0	0	0	1	12	$\overline{A}$
25	1	1	0	0	1	0		
26	1	1	0	1	0	0	13	A
27	1	1	0	1	1	1		
28	1	1	1	0	0	1	14	1
29	1	1	1	0	1	1		
30	1	1	1	1	0	1	15	$\overline{A}$
31	1	1	1	1	1	0		

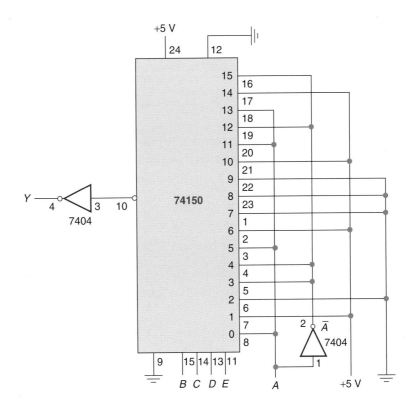

5.

7. Dynamic LCD and Field-Effect LCD

9. LED

11.

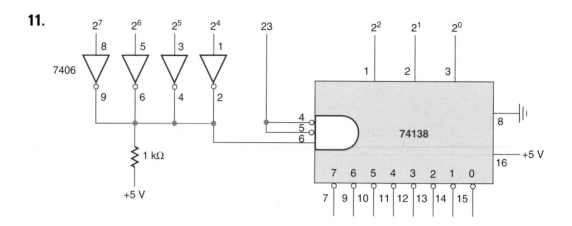

13.

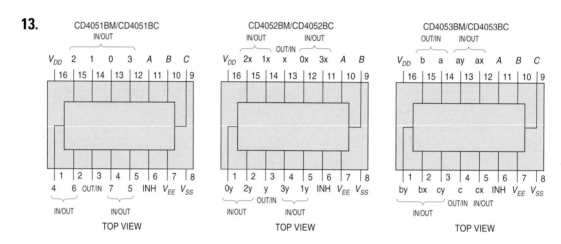

15. The 74C945 is a 4-digit counter for directly driving LCD displays. It contains a 4-decade up-down counter, output latches, counter/latch, select multiplexer and seven-segment decoders, backplane oscillator/driver, segment drivers, and display blanking circuitry.

17. 74151, 74S153, 74ALS157, 74ALS158

19. Glass which will only pass light of a single polarization.

21. The decoder would never have an active output.

23. The 74151 and the 74150

25. The numbers 2, 3, 6, and 7 for the most significant digit would not be displayed.

27. 32 bytes

29. Pins 1 and 2, shorted on JP1, JP2, and JP3. Pins 2 and 3 on JP4 and JP5.

31.
```
LIBRARY ieee;
USE ieee.std_logic_1164.all;
ENTITY mux_4_to_1 IS
     PORT(
          sel:IN STD_LOGIC_VECTOR (1 DOWNTO 0);
          input:IN STD_LOGIC_VECTOR (3 DOWNTO 0);
          Q:OUT STD_LOGIC);
END mux_4_to_1;
ARCHITECTURE a OF mux_4_to_1 IS
```

```
BEGIN
    PROCESS (sel,input)
    BEGIN
        IF sel = "00" THEN
                Q <= input (0);
        ELSIF sel = "01" THEN
                Q <= input(1);
        ELSIF sel = "10" THEN
                Q <= input(2);
        ELSE
                Q <= input(3);
        END IF;
    END PROCESS;
END a;
```

33.
```
LIBRARY ieee;
USE ieee.std_logic_1164.all;
ENTITY demux_1_to_4 IS
    PORT(
        sel:IN STD_LOGIC_VECTOR (1 DOWNTO 0);
        input:IN STD_LOGIC;
        Q:OUT STD_LOGIC_VECTOR (0 TO 3));
END demux_1_to_4;
ARCHITECTURE a OF demux_1_to_4 IS
BEGIN
    PROCESS (sel,input)
    BEGIN
        IF sel = "00" THEN
                Q(0)<= input;
                Q(1 TO 3) <= "000";
        ELSIF sel = "01" THEN
                Q(1) <= input;
                Q(2 TO 3) <= "00";
                Q(0) <= '0';
        ELSIF sel = "10" THEN
                Q(2) <= input;
                Q(0 TO 1) <= "00";
                Q(3) <= '0';
        ELSIF sel = "11" THEN
                Q(3) <= input;
                Q(0 TO 2) <= "000";
        END IF;
    END PROCESS;
END a;
```

35.
```
LIBRARY ieee;
USE ieee.std_logic_1164.all;
ENTITY decode_2_TO_4 IS
        PORT(
                sel:IN STD_LOGIC_VECTOR (1 DOWNTO 0);
                Q:OUT STD_LOGIC_VECTOR (0 TO 3));
END decode_2_to_4;
ARCHITECTURE a OF decode_2_to_4 IS
BEGIN
        PROCESS (sel)
        BEGIN
                IF sel = "00" THEN
                        Q(0) <= '0';
                        Q(1 TO 3) <= "111";
                ELSIF sel = "01" THEN
                        Q(1)<= '0';
                        Q(2 TO 3) <= "11";
                        Q(0) <= '1';
                ELSIF sel = "10" THEN
                        Q(2) <= '0';
                        Q(0 TO 1) <= "11";
                        Q(3) <= '1';
                ELSIF sel = "11" THEN
                        Q(3) <= '0';
                        Q(0 TO 2)<= "111";
                END IF;
        END PROCESS;
END a;
```

CHAPTER 15

Self-Check 1

1. Open-collector and tri-state.
2. From pins 2 to 18.
3. To allow use of a signal bus by many devices.

Self-Check 2

1.

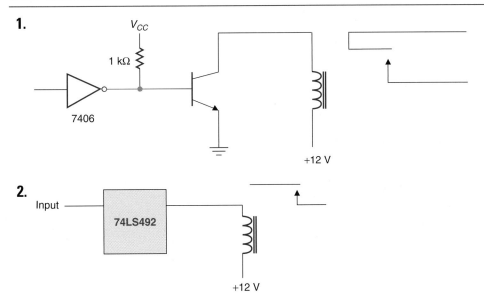

2.

3. Very good electrical isolation and high current. Poor speed.

Questions and Problems

1.

70C95/80C95 70C96/80C96 70C97/80C97 70C98/80C98	Tri-State Hex Buffers
4503	Hex Noninverting Tri-State Buffer
54C240/74C240 54C244/74C244 54C941/74C941	Octal Tri-State Buffers
4076	Tri-State Quad *D* Flip-Flop
74C374	Octal *D* Flip-Flop with Tri-State Outputs
74C373	Octal Latch with Tri-State Outputs
4043	Quad Tri-State NOR R/S Latches
4044	Quad Tri-State NAND R/S Latches
4048	Tri-State Expandable 8-Function, 8-Input Gate
4094	8-Bit Shift/Store Tri-State Register

3.

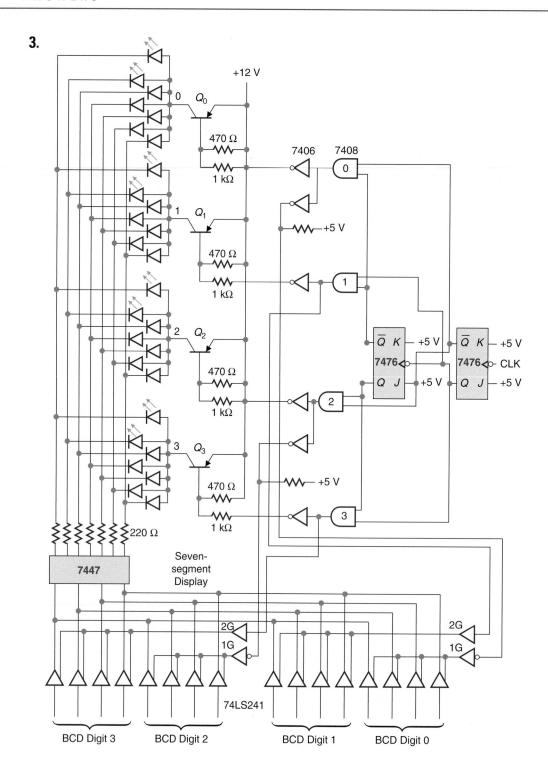

5.

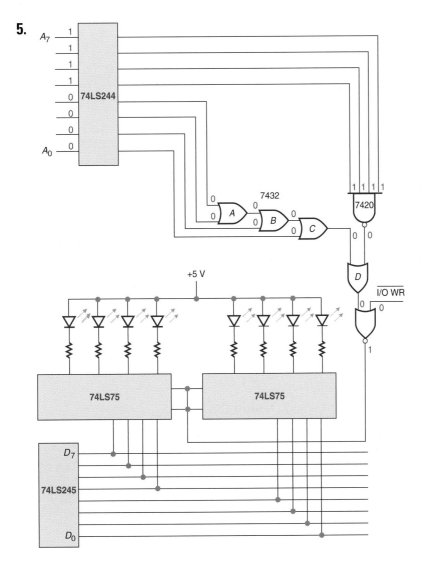

7. To select devices that the computer CPU wishes to use.

9. To prevent a large negative voltage spike.

11.

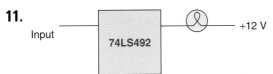

13.

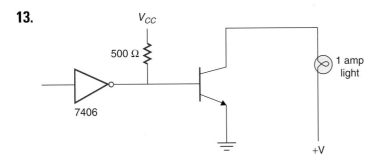

15.

17. 74C173   Tri-state quad *D* flip-flop
74C374   Tri-state octal *D* flip-flop

19. They are driven by active-low outputs of the 7447. Their common connection must be connected to a positive voltage. They are common anode LEDs.

21. There would be a dead short between the output of the top and bottom tri-state gates. This would put the input of the OR gate in no-man's-land as far as TTL input logic voltages go.

23. The two seven-segment LEDs are common anode and must be turned on and off at the +12 volts.

25. Open collector

27. The output of the Schmitt trigger would stop.

29. To allow the current to flow from collector to emitter. This is to supply a current path for reverse current produced by inductive loads.

CHAPTER 16

Self-Check 1

1. CPU, memory, I/O, program

2. Electrical, erasable, read-only memory

3. Static memory does not have to be refreshed.

4. 4

Questions and Problems

1. 2 MHz

3. The Z-80 CPU will finish the current instruction and then place the address bus, data bus, and the control signals in HiZ state.

5. 8000 H to 87FF H

7. There are four present and four more can be added.

9.

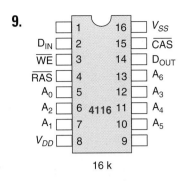

16 k

11. CPU, memory, I/O, program.

13. Read-only memory

15. Static RAM does not need to be refreshed.

17. 7

19. A short code for CPU instructions.

21. Only the upper half of the 64k of memory could be selected.

23. The CPU that was trying to write data to the seven-segment LEDs would not be able to write data to them.

25. A microcontroller is a small computer with ROM and RAM and I/O ports.

27. This chip will convert an RS-232C negative logic signal to a standard TTL signal.

29. EPROM (Erasable Programmable Read-Only Memory)